The Plovers,
Sandpipers, and Snipes
of the World

The Plovers, Sandpipers, and Snipes of the World

Paul A. Johnsgard

UNIVERSITY OF NEBRASKA PRESS
Lincoln and London

To Charles G. Sibley

Copyright © 1981 by the University of Nebraska Press
All rights reserved
Manufactured in the United States of America

Library of Congress Cataloging in Publication Data

Johnsgard, Paul A
   The plovers, sandpipers, and snipes of the world.

   Includes bibliographical references and index.
   1. Plovers,   2. Sandpipers.   3. Snipes.   I. Title.
QL696.C4J63   598′.33   80-22712
ISBN 0-8032-2553-9

Contents

Illustrations

Maps

Figures

Color Plates

❖ ❖ ❖ xiii

Black and White Plates

Preface

Throughout most of my life, the shorebirds occupied only a peripheral part of my world. As a child, I always watched with unrestrained eagerness for the first appearance of marbled godwits and pasque flowers on prairie remnants along the ancient shorelines of glacial Lake Agassiz, even though I was certain that only the geese could confirm that spring had really arrived in North Dakota. Later I watched in childish delight as killdeers performed "broken-wing acts" along railroad rights-of-way, and I thrilled at my first sightings of willets and upland sandpipers near prairie marshes. But it was probably not until my first trip to the Alaskan tundra at Hooper Bay that I confronted the shorebirds in their true heartland and finally realized that they are a group with every bit as much beauty and grace as my beloved waterfowl, and that one day I must devote proper attention to them.

In the fall of 1977, as I was finishing the draft of my *Ducks, Geese, and Swans of the World,* I began to think about an appropriate group to begin working on as a companion volume to that work. Repeatedly the shorebirds came to mind, and repeatedly I rejected taking on the group, believing it was much too large and complex an assemblage for me to challenge. Yet they persistently haunted me, and I finally decided that in spite of the difficulties they offered I must try to deal with them. No worldwide approach to the shorebirds had been attempted in nearly a century, since Henry Seebohm (1888) monographed them from a zoogeographic and taxonomic standpoint. Since that time an enormous technical literature had accumulated and several new species and dozens of subspecies had been described; yet nobody had monographed this fascinating group of birds.

Thus, from early 1978 to early 1980 I devoted most of my research time to assembling materials and writing the text for this book. It soon became apparent that the text would have to be slightly different from my earlier book on the Anatidae, owing to the absence of readily available reference sources for species identification and the generally lower incidence of information available on the biology of each species. I thus decided that rather more exten-

sive descriptions of plumage would be necessary, as well as comprehensive identification keys to species and higher taxonomic categories. Short taxonomic diagnoses of families, subfamilies, tribes, and genera also seemed desirable, inasmuch as no single modern source of these exists for the shorebirds. Further, owing to the unsettled state of shorebird taxonomy and the unusual diversity of mating systems in the group, introductory chapters on shorebird classification and reproductive biology seemed important components of the book. Space considerations precluded the inclusion of my envisioned comparative chapters on population dynamics, migrations, and molting patterns, but, because of the apparent taxonomic significance of the natal plumage patterns of shorebirds, I have illustrated the downy young of as many species as possible.

I had initially hoped it might be possible to follow one of the recent taxonomic revisions of most of the groups, and I have used the recommendations of Bock (1958) and Jehl (1968a) as my primary guidelines. But to varying degrees I have strayed from both of these authors, and I have deviated even more greatly from Peters's (1934) classification. For many groups the evidence is still not adequate to construct convincing taxonomies, but in all controversial cases I have tried to explain the basis for my actions.

Whenever they were available, I have attempted to provide photographic illustrations of representatives of major taxonomic categories (families, subfamilies, and tribes). When such were not available, I have provided drawings, and in almost all cases these have been based on photographs of live birds or, in a few instances, drawn from museum specimens. Only a very few species were drawn from museum specimens with little or no documentary evidence on the appearance of the bird in life.

Since there is no single authoritative source of English vernacular names, I have for some species or groups had to devise or choose suitable English names that did not conflict with my taxonomic decisions. While my choices may not please everyone, I have followed certain guidelines. Thus I have generally tried to avoid the use of "common" and the use of geographically inappropriate names for species. I have also avoided patronymic names when more descriptive names were available and have tried to keep the vernacular names of closely related species as similar as possible. I have kept species names fairly short so that, when necessary, subspecies can be designated by the simple addition of a descriptive modifier.

To a much greater degree than was true of my earlier books, I have had to rely on the assistance and advice of other people and of institutions. For early encouragement and advice I owe special thanks to Drs. Joseph Jehl and Gordon Maclean, and I have repeatedly had to turn to Dr. Jehl for help on difficult matters. Visits to the United States National Museum and the American Museum of Natural History were made possible by a grant from the F. M. Chapman Foundation, and the staffs of these museums were invariably helpful. In particular, Drs. Lester Short, Jr., John Farrand, Jr., and George Watson greatly aided my museum efforts. Although I was unable to visit the Museum of Vertebrate Zoology at Berkeley, important specimen data were made available to me through the assistance of Ms. Victoria Dziadosz and a National Science Foundation grant (BMS 7200102) to the University of California Department of Ornithology. Additional data were obtained from the museums of Louisiana State University and the University of Kansas. I was given free access to the collections and library of the British Museum (Natural History), and I especially thank Dr. P. J. K. Burton for arranging the photocopying of certain library materials. Likewise, Mr. Anthony Cheke at the Edward Gray Institute of Oxford very kindly allowed me extensive use of that library, including the photocopying of such rare items as Seebohm's monograph on the shorebirds. Dr. A. A. Kistchinski read certain portions of the manuscript dealing with Siberian species and made numerous helpful comments. Data on weights of Australian species were provided by D. Purchase of the CSIRO and by the staffs of the National Museum of Victoria and the Western Australian Museum. Dr. R. S. Phillips, Mrs. Sylvia Reed, and Dr. Rodney Hay provided unpublished data on various New Zealand species, and D. M. Skead and P. A. Clancey assisted on some African forms, as did J. C. Daniel for Indian species. A most valuable source of help was the lending of various Russian translations by Dr. Edward H. Miller and Leslie M. Tuck, and Wolfgang Grummt provided additional aid with German and Russian literature. Dr. and Mrs. Robert Mengel kindly provided me with photocopies of various materials in the University of Kansas libraries, and tape recordings of particularly interesting species were provided by Robert Stjernstedt (painted snipe) and J. P. Myers (pectoral sandpiper).

To a greater extent than was necessary for my earlier books, I have had to rely on the assistance of other photographers to provide a fairly comprehensive photographic coverage of the shorebirds. A very large number of persons volunteered or agreed to help me in this endeavor, but I owe a very special debt of gratitude to J. B. Bottomley and S. Bottomley, the British wader photographers par excellence, whose superb monochrome prints put my own efforts to shame. Additionally, Mr. Ed. Bry sent me a large selection of prints of North Dakota shorebirds, and both Kenneth Fink and Frank Todd generously offered me the use of transparencies from their large collections. Among the many other persons or institutions who provided slides or prints are the following: Bombay Natural History Society, Peter Alden, Vladimir Flint, E. O. Höhn, A. A. Kistchinski, Joseph Jehl, Frans Lanting, Tom Lowe, Kerry Muller, David Parmelee, Richard Phillips, Jean and Ed Schulenberg, Larry Stevens, Stuart Tingley, and Fred Zeillemaker. Mr. Zeillemaker also carefully read the final manuscript draft.

The fine painting of the Eskimo curlew was done on request by James McClelland, and two paintings of shorebird downy young were very kindly executed for me by Jon Fjeldså.

Last, I thank the University of Nebraska for allowing me time to complete this work, and the secretaries of the School of Life Sciences for typing the final manuscript copy.

Black-Tailed Godwit

Taxonomy and Evolutionary Relationships

SUBORDERS AND SUPERFAMILIES OF SHOREBIRDS

In his landmark classification of the Charadriiformes, Peters (1934) used the following organization for the waders or shorebirds:

Suborder Charadrii
  Superfamily Jacanoidea
    Family Jacanidae (Jacanas, 7 spp.)
  Superfamily Charadrioidea
    Family Rostratulidae (Painted Snipes, 2 spp.)
    Family Haematopodidae (Oystercatchers, 4 spp.)
    Family Charadriidae (Plovers and Lapwings, 61 spp.)
    Family Scolopacidae (Sandpipers and Snipes, 83 spp.)
    Family Recurvirostridae (Stilts and Avocets, 7 spp.)
    Family Presbyornithidae (known only from fossils; probably part of Recurvirostridae)
    Family Phalaropodidae (Phalaropes, 3 spp.)
  Superfamily Dromadoidea
    Family Dromadidae (Crab-plovers, 1 sp.)
  Superfamily Burhinoidea
    Family Burhinidae (Thick-knees, 9 spp.)
  Superfamily Glareoloidea
    Family Glareolidae (Pratincoles and Coursers, 17 spp.)
  Superfamily Thinocoroidea
    Family Thinocoridae (Seedsnipes, 4 spp.)
  Superfamily Chionidoidea
    Family Chionididae (Sheathbills, 2 spp.)

To varying degrees, most subsequent workers have followed this organization, as did Wetmore (1960), who used exactly the same sequence and followed the same family limits. However, this uniformity of procedure by no means indicates that all the major issues of higher-level charadriiform relationships are settled, and indeed most of the more recent studies have tended to disrupt this seemingly straightforward classification. In an excellent review of the literature, Sibley and Ahlquist (1972) summarized the history of classification of the waders over an entire century between the 1860s and the 1960s. They pointed out that, though the Charadriiformes are certainly a closely related assemblage of birds, adaptive radiation has been extremely great and has resulted in many cases of uncertain relationships. The primary questions posed by these authors concerning the typical or presumptive shorebirds (Charadrii) are the following:

1. Are the jacanas (Jacanidae) more closely related to the painted snipes (Rostratulidae) and thus to the other charadriiform groups, or are they instead more closely related to the rails of the order Gruiformes?
2. Are the seedsnipes (Thinocoridae) members of the Charadriiformes?
3. Are the pratincoles and coursers (Glareolidae) a monophyletic group, and are they related to the gull-like or plover-like charadriiform assemblages?
4. What are the relationships of the crab-plover (Dromadidae) and the sheathbills (Chionididae)?
5. Are the thick-knees (Burhinidae) related to the plovers or the bustards?

With respect to the jacanas, they concluded that these birds and the painted snipes are probably more closely related to one another than either is to any other group of Charadriiformes. Inasmuch as the painted snipes are now invariably included in the Charadriiformes, it thus seems reasonable that the jacanas should also be included in this book. However, their review of the available evidence did little to fix the evolutionary positions of the seedsnipes, glareolids, crab-plover, or sheathbills. The nearest relatives of the thick-knees were judged by Sibley and Ahlquist to be the oystercatchers and stilts, though the relationships are by no means clear, and Yudin (1965) has recently proposed that they be returned to the Gruiformes.

Although it appeared before Sibley and Ahlquist's

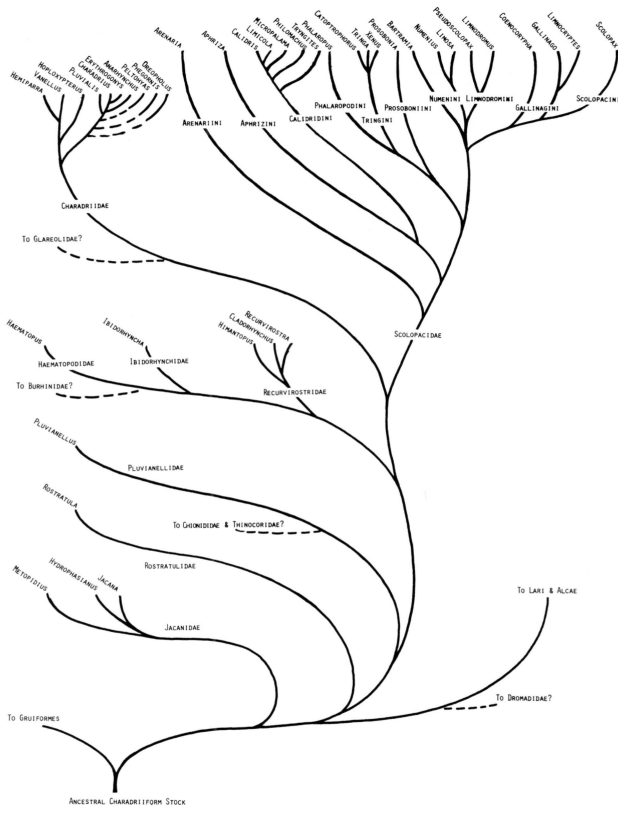

1. Diagram of evolutionary relationships of the tribes and genera of shorebirds.

review, the study by Jehl (1968a) deserves special mention. While focusing the comparative study of variations in downy plumage patterns and the taxonomic significance, Jehl also provided a general review of probable shorebird relationships and a presumptive phylogeny of the group. His proposed taxonomy differs in only a few respects from that of Peters (1934) and may be summarized as follows:

Suborder Charadrii
  Superfamily Jacanoidea
    Family Jacanidae (6 genera)
    Family Rostratulidae (2 genera)
  Superfamily Dromadoidea
    Family Dromadidae (1 genus)
  Superfamily Charadriidea
    Family Haematopodidae (1 genus)
    Family Ibidorhynchidae (1 genus)
    Family Burhinidae (2 genera)
    Family Glareolidae (6 genera, 2 subfamilies)
    Family Charadriidae
      Subfamily Vanellinae (1 genus)
      Subfamily Charadriinae (7 genera)
    Family Scolopacidae
      Subfamily Triginae (9 genera, 3 tribes)
      Subfamily Arenariinae (1 genus)
      Subfamily Phalaropodinae (1 genus)
      Subfamily Scolopacidinae (1 genus)
      Subfamily Gallinagoninae (4 genera)
      Subfamily Calidridinae (7 genera)
  Superfamily Chionidoidea
    Family Thinocoridae (2 genera)
    Family Chionididae (1 genus)

The most recent major approach to the classification of the Charadriiformes was that of Strauch (1976, 1978), who used cladistic methods and primary osteological characteristics to construct possible phylogenies using character compatability analyses. His conclusions differed markedly from those of earlier workers, and in particular he challenged the generally accepted view that the ploverlike birds are closely related to the sandpipers and plovers. Instead, he suggested that three phyletic lines exist in the Charadriiformes, which he treated as suborders. The first of these is the Scolopaci, including the forms usually included in the Jacanidae, Rostratulidae, Scolopacidae, Phalaropodidae, and Thinocoridae. The second group, the Charadrii, consists in Strauch's view of two lineages, one leading to the gulls and terns and the other leading to the remaining shorebirds. The third suborder, the Alcae, includes a single family, the Alcidae. An abbreviated

version of his proposed classification may be summarized as follows:

Suborder Scolapaci
  Superfamily Jaconidea
    Family Jacanidae (2 genera)
  Superfamily Scolopacoidea
    Family Thinocoridae (1 genus)
    Family Rostratulidae (1 genus)
    Family Scolopacidae (13 genera, 4 subfamilies)
Suborder Charadrii
  Superfamily Laroidea
    Family Laridae (5 genera, 4 subfamilies)
  Superfamily Charadroidea
    Family Dromadidae (1 genus)
    Family Burhinidae (1 genus)
    Family Pluvianidae (1 genus)
    Family Glareolidae (3 genera, 2 subfamilies)
    Family Chionididae (2 genera, 2 subfamilies)
    Family Charadriidae (6 genera, 5 subfamilies)
Suborder Alcae

Several other major taxonomic or morphological studies have recently been done on the shorebirds, including those of Yudin (1956), Burton (1974), Ahlquist (1974), and Stegmann (1978). Conclusions from these studies will be mentioned in the discussions of the individual families. On the basis of such studies, I have prepared a simplified phyletic dendrogram (fig. 1) that summarizes my personal interpretations of the currently available evidence as to the higher-level affinities of the groups included in this book. It differs in some respects from that proposed by Jehl (1968a) and also from that of Fjeldså (1977), though it comes considerably closer to the latter in most respects. In particular, this dendrogram and that of Fjeldså suggest that the plover assemblage is generally more primitive than and also directly ancestral to the scolopacid assemblage. However, Jehl (1968a) pointed out that very limited fossil evidence supports the view that these two groups were already separated by late Cretaceous times, well before the plover group became diversified. Supporting this position are the findings of Strauch (1976), who found no evidence that the charadriids and scolopacids are closely related. Yet McFarlane (1963) noted that the sperm of the Scolopacidae (at least of five genera) differ from those of the Recurvirostridae and Charadriidae, and he suggested that the Scolopacidae may be of more recent origin than these other groups. We may hope that future studies will help resolve these numerous taxonomic problems.

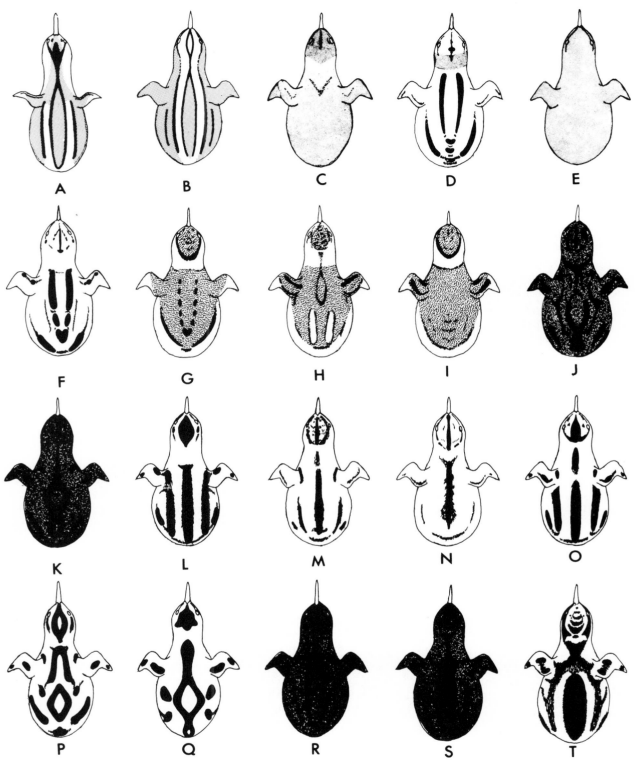

2. Diagram of downy patterns of *A*, African jacana; *B*, painted snipe; *C*, Magellanic plover; *D*, oystercatcher; *E*, ibisbill; *F*, stilt; *G*, spur-winged lapwing; *H*, lesser golden plover; *I*, semipalmated plover; *J*, ruddy turnstone; *K*, rock sandpiper; *L*, northern phalarope; *M*, Terek sandpiper; *N*, spotted sandpiper; *O*, lesser yellowlegs; *P*, whimbrel; *Q*, Hudsonian godwit; *R*, short-billed dowitcher; *S*, snipe; *T*, American woodcock. Heavy stippling indicates presence of powder-puff down. Adapted from Jehl 1968a.

FAMILIES OF TYPICAL SHOREBIRDS

The following families of relatively typical shorebirds have been selected for inclusion in this book. They include all the extant families of the superfamilies Jacanoidea and Charadroidea as defined by Peters (1934) and Wetmore (1960).

1. Family Jacanidae (Jacanas)

As I noted earlier, the jacanas are certainly the most controversial of the families discussed in this book, since they have been included in the Gruiformes almost as frequently as they have been placed in the Charadriiformes, and they certainly represent a highly peripheral group. The raillike features of jacanas have been stressed by Lowe (1925), and a more dubious argument was made more recently by Verheyen (1957), who placed the jacanas and some other raillike forms in a special order between his order "Ralliformes" and the Charadriiformes. Jehl (1968a) believed that the distinctive and consistent downy plumage pattern of the jacanas (fig. 2) indicated common ancestry with the painted snipes, and he suggested that the two groups should be placed in the superfamily Jacanoidea. Strauch (1976) believed they constitute a monophyletic charadriiform group and felt that no more than two genera should be recognized. Both Yudin (1965) and Ahlquist (1974) supported the view that the jacanas should be included in the Charadriiformes. My views on the generic and familial relationships of the jacanas are illustrated in figure 1.

2. Family Rostratulidae (Painted Snipe)

Although they have traditionally been included within the Charadriiformes, the painted snipes have also had a rather controversial taxonomic history. Lowe (1931a) concluded that they should be regarded as a family of the suborder Limicolae. Both Yudin (1965) and Ahlquist (1974) suggested that the painted snipes might be more closely related to the sandpipers and true snipes than to other shorebirds, and Strauch (1976) also included the painted snipes in his family Scolopacidae. Yet this was evidently based on a single shared character, and he admitted that the apparent alliance might be the result of an inadequate character set. Jehl (1968a) doubted whether the painted snipes are closely related to any of the other shorebird families except the Jacanidae. The egg-white proteins of painted snipes and jacanas are very similar, and they differ from those of all other cha-

radriiforms, thus supporting this general view of isolation from the typical snipes and other sandpipers. Burton (1974) found the anatomical features of the painted snipes to be "puzzling," and no obvious relationships were apparent to him.

3. Family Pluvianellidae (Magellanic Plover)

Until very recently, this little-studied species was simply considered to be one of the many plovers, but Jehl (1968a) assembled morphological and behavioral evidence that, in his view, warrants familial separation. He listed thirteen characteristics that differ from the usual plover condition, including several unique features such as a crop, a singular foraging behavior, courtship feeding, regurgitation of food for the young, and unusually blunt claws. He suggested that the species is an early offshoot of the forms that gave rise to the plovers, but that its closest relationships may lie with the Haematopodidae, the Chionididae, or possibly even the lapwing of the Charadriidae. Strauch's (1976) anatomical findings supported the idea of sheathbill relationships, and he actually placed this form in the Chionididae as a separate subfamily. Burton (1974) included the Magellanic plover in the Charadriidae but noted that it has a highly distinctive feature in the form of an extremely flexible (rhynchokinetic) upper mandible, whereas other plovers are only moderately rhynchokinetic.

4. Family Haematopodidae (Oystercatchers)

The highly distinctive oystercatcher group is notable for the extreme compression of the mandibles, associated with a specialized foraging adaptation. Their evolutionary affinities are distinctly obscure. Lowe (1931b) suggested on the basis of osteology that they are probably closely related to the Recurvirostridae, and there is a very limited degree of behavioral similarity, as well as some protein similarity, between these groups. Jehl (1968a) also suggested that downy patterns tend to link the oystercatchers with the Recurvirostridae and the Burhinidae.

5. Family Ibidorhynchidae (Ibis-bill)

This single species of shorebird, with a long and distinctively decurved bill, has traditionally been included with the avocets and stilts, though some authorities have placed it close to the oystercatchers (Seebohm 1888), and at times it has been regarded as constituting a separate subfamily of the Charadriidae (Lowe 1931b). Jehl (1968a) could find no evidence linking it any more closely with the stilts and avocets than with other groups such as those just mentioned, and he advocated retaining it in a separate family until it can be more thoroughly studied. Strauch (1976) found morphological evidence associating it with the oystercatchers, avocets, and stilts, and he suggested that these groups all be considered subfamilies of the Charadriidae. Burton (1974) included it in the Recurvirostridae and found no marked anatomical differences between it and *Himantopus* or *Recurvirosta*. It seems clear that the ibis-bill is most closely related to the stilt and avocet line, as is indicated in the phyletic dendrogram (fig. 1).

6. Family Recurvirostridae (Stilts and Avocets)

In spite of their distinctively long legs and bills, the avocets and stilts apparently are related not very distantly to other shorebird groups, and Sibley and Ahlquist (1972) reported egg-white protein similarities between these forms and oystercatchers, thick-knees, and curlews (*Numenius*). The stilts and avocets intergrade with one another, and Seebohm (1888) even placed them all in a single genus. There are some very limited structural and behavioral similarities with the oystercatchers (Jehl 1968a), and it seems most likely that these groups, together with the ibis-bill, must have come from a common source near the ancestral plover stock. Fjeldså (1977) has recently presented a phyletic dendrogram supporting this general view, and my own dendrogram (fig. 1) resembles his in several respects.

7. Family Charadriidae (Lapwings and Plovers)

The lapwings and plovers are an easily recognizable group of shorebirds, all of which have bills that are generally less specialized and less flexible than those of the Scolopacidae. Burton (1974) reports that the bills of this family are only moderately rhynchokinetic, as opposed to the highly rhynchokinetic bills in many scolopacid groups. At least in that regard, the Charadriidae can be regarded as more primitive than the scolopacid groups, and most authorities believe the lapwings are a more generalized assemblage (having more primitive characteristics) than are the typical plovers. Fjeldså (1977) suggested that plover evolution began in the late Cretaceous, when various lapwinglike forms existed that gave rise not only to the typical plovers but also to the coursers and the more primitive scolopacid waders such as

Aphriza and *Calidris*. This seems a reasonable hypothesis, and like most other modern taxonomists I believe it is prudent to accept two subfamilies (Vanellinae and Charadriinae) of lapwings and plovers. The two groups do approach one another, and such lapwings as *Hoploxypterus cayanus* are either convergent with *Charadrius* (Burton 1974) or possibly are transitional forms. Strauch (1976) actually includes this species in the genus *Charadrius*, while Fjeldså (1977) apparently considers it a typical lapwing, as did Bock (1958). Bock has strongly argued for the adoption of a single genus of lapwings, rather than the highly "split" taxonomy used by Peters (1934), Wolters (1975), and various other writers.

The relationships within the lapwings are greatly obscured by the repeated independent evolution of wing spurs, wattles, and other display structures that are of no taxonomic importance at the species level. The sequence of species that I have used is based on other criteria, including zoogeographic considerations. Bock's (1958) estimation of relationships was to a large degree based on similar criteria.

The more typical plovers are usually regarded as a distinct subfamily and include a very large number of species that are best regarded as constituting a single genus *Charadrius*. Additionally, there are several aberrant southern hemisphere forms usually included in the group as monotypic genera, and some relatively large arctic-breeding forms that seem to be closely related and that Bock has suggested placing in a single genus *Pluvialis*. Bock believed it was impossible to judge which of the *Charadrius* species might be most primitive, but he recognized seven distinct groups within the genus and clearly considered the "ringed plovers" a central assemblage. It is probable that such conditions as the presence or absence of a breastband have evolved independently several times in the group, although in contrast to the Scolopacidae the bill structures have remained remarkably uniform through the family (Burton 1974). My views on the probable evolutionary affinities of the genera of the Charadriidae are indicated in figure 1.

8. Family Scolopacidae (Sandpipers and Snipes)

This is by far the largest single assemblage of shorebirds and in many respects the most specialized. Burton (1974) noted that many of the species possess highly rhynchokinetic bills that have evolved independently in several lines. He envisioned two major pathways of evolution. The first leads toward such groups as the typical sandpipers (Calidridinae), fairly small species that take much of their prey from the surface of muddy or other soft substrates and that probe relatively little for food. The second major pathway leads toward the shallow-water foragers, and the earliest stages in this trend are shown in those moderately large forms that capture actively swimming prey, including the typical tattlers (*Tringa*). More highly flexible bills were evolved among species that began to exploit sedentary prey at the bottoms or edges of shallow water, such as snipes and woodcocks, having highly rhynchokinetic bills but strongly reinforced upper jaws.

In addition to the evolution of long, probing bills, some members of this family have evolved remarkably long and straight or decurved bills. These include the curlews (*Numenius*) and godwits (*Limosa*), which have only occasionally been regarded as close relatives. Jehl (1968a) supported this view strongly. Burton (1974) judged that the godwits were closely related to the dowitchers (*Limnodromus*), but that there was less evidence supporting the inclusion of the curlews in this group. Similarly, Ahlquist (1972) judged the dowitchers and godwits to be closely related, with the curlews not only not members of this group but actually perhaps related to the avocets and stilts. Strauch (1976) believed the curlews were part of the general scolopacid assemblage, most closely related to the upland sandpiper (*Bartramia*) and the Polynesian sandpipers (*Prosobonia*).

Evolutionary relationships in this large and varied group are often extremely difficult to separate from cases of convergent evolution, and such a question as whether the dowitchers are more closely related to the snipes than they are to the godwits is but one example of many such problems. These questions will be dealt with later in the text, within the discussions of each species. However, my own estimate of probable phyletic relationships of some of the major subdivisions (genera and tribes) of the Scolopacidae is presented here (fig. 1).

MINOR ABERRANT CHARADRIIFORM FAMILIES

Five small families of birds that at times have been included near or among the groups considered in this book have been excluded here, primarily owing to their extremely uncertain evolutionary relationships and because they are generally rather poorly studied

and do not contribute anything vital to an understanding of the general evolutionary trends of the typical shorebirds. The groups are briefly summarized here, but the interested reader will have to look elsewhere for details of their biology.*

1. Family Burhinidae (Thick-knees)

This family of medium to large birds occurs over temperate and tropical portions of Europe, Africa, Asia, Australia, and tropical America. Nine species, placed in 2 genera, are usually recognized. The birds are highly terrestrial and are usually associated with relatively dry habitats. They are also mainly crepuscular and nocturnal in their activities. The sexes are alike, and the birds are monogamous. The clutch is usually 2 eggs, rarely 3, and the young are downy and may either remain in the nest for some time or leave it after about a day. Jehl (1968a) believed they were part of an evolutionary group including the Recurvirostridae, Haematopodidae, Ibidorhynchidae, Glareolidae, and Charadriidae and placed them between the Recurvirostridae and Glareolidae in taxonomic sequence. Strauch (1976) included them in his superfamily Charadroidea, between the Dromadidae and the courser group (Pluvianidae and Glareolidae), finding no evidence linking them to the Gruiformes. Ahlquist (1974) believed they were closer to the Charadriiformes than to the Gruiformes and thought they might be the most primitive group of the former order. Morony, Bock, and Farrand (1975) placed the family between the Recurvirostridae and the Glareolidae. Sibley and Ahlquist (1972) placed them between the Dromadidae and the Glareolidae.

2. Family Dromadidae (Crab-plover)

This family consists of a single medium-sized shorebird species of the coastlines of the Indian Ocean from Natal to Ceylon. The species is entirely black-and-white, with a heavy, somewhat gull-like bill and long lapwing-like legs. It feeds entirely on crabs, other crustaceans, and mollusks, and it nests in underground burrows. Only a single large white

---

* In this section and in the taxonomic diagnoses, "very small" = species with wings under 100 mm, "small" = those with wings 100–200 mm, "medium" = those with wings 200–300 mm, "large" = those with wings 300–350 mm, and "very large" = those with wings over 350 mm.

egg is laid, and the hatchlings are downy but are fed by the parents and reared in the burrow. The downy young are countershaded but unpatterned, and Jehl (1968a) believed that burrow-nesting is a fairly recent adaptation for avoiding intense solar radiation. He placed the species in a separate superfamily Dromadoidea, though most other workers have placed it near the Burhinidae. Although Strauch (1976) noted similarities with both the Glareolidae and the gulls, he believed it possessed many primitive characteristics aligning it with the Charadriidae and the other aberrant minor groups summarized here.

3. Family Glareolidae (Pratincoles and Coursers)

This family of small, highly terrestrial birds occurs from the Mediterranean Basin eastward and southward through Africa, India, and Indochina to Australia. There are 17 species in two quite distinctive groups (usually considered subfamilies). The typical pratincoles have short bills, short legs, long, narrow wings, and forked tails. Coursers have long bills and legs and shorter and broader wings and tails. The sexes are similar but often differ in size. The birds forage largely on insects, sometimes in extremely dry environments, and are convergent with several of the arid-adapted species of plovers. One Australian species, *Peltohyas australis,* is so similar to the coursers that it has frequently been considered (Peters 1934; Jehl 1968a) part of this family. The eggs of pratincoles and coursers are usually laid on bare sand or rock, sometimes being buried in sand, and are usually incubated by the female, with the male assisting in some species. The young are downy and may be raised in the nest for some time or may leave it in a day or so. They strongly resemble the downy young of various plovers, and Jehl believed the family was closely related to the Charadriidae. Strauch (1976) was unable to make a decision as to the glareolids' closest relatives, but he considered the genus *Pluvianus* sufficiently distinctive to warrant family status, and he placed the pratincoles and coursers between the Burhinidae and the Chionididae.

4. Chionididae (Sheathbills)

This family of medium-sized antarctic to subantarctic birds includes only two species, which are quite gull-like but lack fully webbed toes, have crops, and have bare, fleshy areas on the face and at the base of the bill. They are omnivorous scavengers and predators, and the sexes are alike in appearance, the female often somewhat smaller than the male. They are monogamous, laying 2 or 3 white eggs in a hole, burrow, or rock crevice, and both sexes incubate. The downy young are quite different from those of other waders but slightly resemble those of gulls or plovers. Jehl (1968a) followed most previous workers in believing that the sheathbills and seedsnipes are related groups, and he included them as separate families in a superfamily Chionidoidea. Strauch (1976) questioned this relationship and included the sheathbills and the Magellanic plover as separate subfamilies in a single family Chionididae. Ahlquist (1974) suggested affinities with the gulls but not with the seedsnipes or the typical waders.

5. Family Thinocoridae (Seedsnipes)

This family of four small Andean birds occurs from Ecuador to the southern tip of South America. The birds have short, pointed bills, very short legs, long, pointed wings, and rather short tails. They possess crops and feed mainly on vegetable matter. Although short-legged, they run well and have a snipelike flight pattern. The sexes differ in appearance, and the male apparently plays no role in incubation or brooding. The nest is a simple scrape on the ground, and a 4-egg clutch is produced. The young are downy and nidifiguous; at least one of the two genera has a downy pattern that is quite ploverlike. However, Jehl (1968a) did not believe that the downy pattern indicated the relationships of the seedsnipes, and he agreed that they are probably most closely related to the sheathbills. On the other hand, Strauch (1976) judged them to be a "sister group" of the Scolopacidae, and Ahlquist (1974) suggested affinities with both the painted snipes and the sandpipers. Various workers have most often aligned them with the Glareolidae, but Sibley, Corbin, and Ahlquist (1968) were unable to define their relationships within the Charadriiformes by using biochemical techniques.

GENERA AND SPECIES OF SHOREBIRDS

A complete review of all of the problems of generic and species limits is impossible in an introductory overview, and in any event it is more appropriate later in the text. However, a synoptic summary of the complete taxonomy to the species level is appropriate here, as an easy means of comparing the classification used in this book with those proposed by the workers discussed earlier in the chapter.

SYNOPSIS OF THE SHOREBIRDS OF THE WORLD

Superfamily Jacanoidea
Family Jacanidae (Jacanas)

Metopidius capensis	Lesser Jacana
[*M. africana**	African Jacana
[*M. albinucha*	Madagascan Jacana
M. gallinacea	Comb-crested Jacana
M. indicus	Bronze-winged Jacana
Hydrophasianus chirurgus	Pheasant-tailed Jacana
Jacana spinosa	American Jacana

Superfamily Charadroidea
Family Rostratulidae (Painted Snipes)

Rostratula benghalensis	Painted Snipe
R. semicollaris	South American Painted Snipe

Family Pluvianellidae (Magellanic Plover)

Pluvianellus socialis	Magellanic Plover

Family Haematopodidae (Oystercatchers)

[*Haematopus ostralegus*	Oystercatcher
[*H. leucopodus*	Magellanic Oystercatcher
H. unicolor	Variable Oystercatcher
H. fuliginosus	Sooty Oystercatcher
H. ater	Black Oystercatcher

Family Ibidorhynchidae (Ibis-bill)

Ibidorhyncha struthersii	Ibis-bill

Family Recurvirostridae (Stilts and Avocets)

Himantopus himantopus	Stilt
Cladorhynchus leucocephala	Banded Stilt
Recurvirostra andina	Andean Avocet
R. avocetta	Eurasian Avocet
[*R. americana*	American Avocet
[*R. novaehollandiae*	Red-necked Avocet

Family Charadriidae (Plovers and Lapwings)
Subfamily Vanellinae (Lapwings)
Tribe Vanellini (Typical Lapwings)

Hemiparra crassirostris	White-faced Lapwing
Vanellus coronatus	Crowned Lapwing
V. gregarius	Sociable Lapwing
V. leucurus	White-tailed Lapwing
V. cinereus	Gray-headed Lapwing
V. malabaricus	Yellow-wattled Lapwing
V. superciliosus	Brown-chested Lapwing
[*V. lugubris*	Lesser Black-winged Lapwing
[*V. melanopterus*	Black-winged Lapwing
V. vanellus	Eurasian Lapwing
V. chilensis	Southern Lapwing
V. resplendens	Andean Lapwing
[*V. armatus*	Blacksmith Plover
[*V. spinosus*	Spur-winged Lapwing
V. melanocephalus	Spot-breasted Lapwing
V. tectus	Black-headed Lapwing
V. indicus	Red-wattled Lapwing
V. tricolor	Banded Lapwing
V. miles	Masked Lapwing
V. macropterus	Javanese Lapwing
V. senegallus	Wattled Lapwing
V. albiceps	White-crowned Lapwing

Tribe Hoploxypterini (Pied Lapwing)

Hoploxypterus cayanus	Pied Lapwing

Subfamily Charadriinae (Typical Plovers)
Tribe Charadriini (Typical Plovers)

[*Pluvialis apricaria*	Greater Golden Plover
[*P. dominica*	Lesser Golden Plover
P. squatarola	Gray Plover
Charadrius obscurus	Red-breasted Dotterel
C. montanus	Mountain Plover
[*C. leschenaultii*	Greater Sand Plover
[*C. mongolus*	Mongolian Plover
[*C. asiaticus*	Lesser Oriental Plover
[*C. veredus*	Greater Oriental Plover
C. modestus	Rufous-chested Dotterel
C. morinellus	Dotterel
C. bicinctus	Double-banded Dotterel
C. falklandicus	Two-banded Plover
C. tricollaris	Three-banded Plover
C. melanops	Black-fronted Dotterel
C. vociferus	Killdeer
C. wilsonius	Thick-billed Plover
C. placidus	Long-billed Plover
[*C. hiaticula*	Ringed Plover
[*C. semipalmata*	Semipalmated Plover
C. melodus	Piping Plover
C. dubius	Little Ringed Plover
C. peronii	Malaysian Ringed Plover
C. collaris	Collared Plover
C. thoracicus	Madagascan Sandplover
[*C. pecuarius*	Kittlitz Sandplover
[*C. sanctaehelenae*	St. Helena Sandplover

* Forms considered a superspecies are connected by braces.

C. pallidus	Chestnut-banded Sandplover
C. alexandrinus	Sandplover
C. marginatus	White-fronted Sandplover
C. ruficapillus	Red-capped Sandplover
C. cucullatus	Hooded Dotterel
C. novaeseelandiae	New Zealand Shore Plover
Erythrogonys cinctus	Red-kneed Dotterel
Anarhynchus frontalis	Wrybill
Peltohyas australis	Inland Dotterel
Phegornis mitchellii	Diademed Sandpiper-plover
Oreopholus ruficollis	Tawny-throated Dotterel

Family Scolopacidae (Sandpipers, Snipes, and Allies)
Subfamily Calidridinae (Sandpipers, Surfbirds, and Turnstones)
Tribe Arenariini (Turnstones)

Arenaria interpres	Ruddy Turnstone
A. melanocephala	Black Turnstone

Tribe Aphrizini (Surfbird)

Aphriza virgata	Surfbird

Tribe Calidridini (Typical Sandpipers)

Calidris tenuirostris	Eastern Knot
C. canutus	Red Knot
C. acuminata	Sharp-tailed Sandpiper
C. melanotos	Pectoral Sandpiper
C. fusicollis	White-rumped Sandpiper
C. bairdii	Baird Sandpiper
C. temminckii	Temminck Stint
C. pusilla	Semipalmated Sandpiper
C. mauri	Western Sandpiper
C. pygmeus	Spoon-billed Sandpiper
C. minuta	Little Stint
C. ruficollis	Rufous-necked Sandpiper
C. minutilla	Least Sandpiper
C. subminuta	Long-toed Stint
C. ferruginea	Curlew Sandpiper
C. alpina	Dunlin
C. maritima	Purple Sandpiper
C. ptilocnemus	Rock Sandpiper
C. alba	Sanderling
Limicola falcinellus	Broad-billed Sandpiper
Micropalama himantopus	Stilt Sandpiper
Philomachus pugnax	Ruff

Tryngites subruficollis	Buff-breasted Sandpiper

Subfamily Tringinae (Tattlers and Allies)
Tribe Phalaropini (Phalaropes)

Phalaropus tricolor	Wilson Phalarope
P. lobatus	Northern Phalarope
P. fulicarius	Red Phalarope

Tribe Trigini (Tattlers)

Catoptrophorus semipalmatus	Willet
Tringa erythropus	Spotted Redshank
T. totanus	Redshank
T. flavipes	Lesser Yellowlegs
T. melanoleuca	Greater Yellowlegs
T. nebularia	Greenshank
T. guttifer	Spotted Greenshank
T. stagnatilis	Marsh Sandpiper
T. ochropus	Green Sandpiper
T. solitaria	Solitary Sandpiper
T. glareola	Wood Sandpiper
T. brevipes	Siberian Tattler
T. incana	Wandering Tattler
T. hypoleucos	Eurasian Sandpiper
T. macularia	Spotted Sandpiper
Xenus cinereus	Terek Sandpiper

Tribe Prosoboniini (Polynesian Sandpipers)

Prosobonia cancellatus	Tuamotu Sandpiper
P. leucoptera	Tahitian Sandpiper

Tribe Numenini (Curlews and Godwits)

Bartramia longicauda	Upland Sandpiper
Numenius minutus	Little Curlew
N. borealis	Eskimo Curlew
N. phaeopus	Whimbrel
N. tahitiensis	Bristle-thighed Curlew
N. tenuirostris	Slender-billed Curlew
N. arquata	Curlew
N. americana	Long-billed Curlew
N. madagascariensis	Eastern Curlew
Limosa limosa	Black-tailed Godwit
L. haemastica	Hudsonian Godwit
L. lapponica	Bar-tailed Godwit
L. fedoa	Marbled Godwit

Tribe Limnodromini (Dowitchers)

Pseudoscolopax semipalmatus	Asian Dowitcher
Limnodromus scolopaceus	Long-billed Dowitcher
L. griseus	Short-billed Dowitcher

Subfamily Scolopacinae (Snipes and Woodcocks)
Tribe Gallinagini (Snipes and Semiwoodcocks)

Coenocorypha aucklandica	Subantarctic Snipe
Gallinago undulata	Giant Snipe
G. stricklandii	Cordillerian Snipe
G. imperialis	Imperial Snipe
G. nemoricola	Himalayan Snipe
G. media	Great Snipe
G. macrodactyla	Madagascan Snipe
G. nobilis	Noble Snipe
G. gallinago	Snipe
G. andina	Puna Snipe
G. nigripennis	African Snipe
G. hardwickii	Japanese Snipe
G. megala	Forest Snipe
G. stenura	Pintailed Snipe
G. solitaria	Solitary Snipe
Limocryptes minimus	Jack Snipe

Tribe Scolopacini (Woodcocks)

Scolopax saturata	Dusky Woodcock
S. minor	American Woodcock
S. rusticola	Woodcock
S. rochussenii	Indonesian Woodcock

GENERA RECOGNIZED BY PETERS BUT SYNONYMIZED HERE

Actitus = Tringa
Actophilornis = Metopidius
Aechmorhynchus = Prosobonia
Afribyx = Vanellus
Anomalophrys = Vanellus
Belonopterus = Vanellus
Chettusia = Vanellus
Chubbia = Gallinago
Crocethia = Calidris
Eiseyornis = Charadrius
Ereunetes = Calidris
Erolia = Calidris
Eudromias = Charadrius
Eupoda = Charadrius
Eurynorhynchus = Calidris
Heteroscelus = Tringa
Hoplopterus = Vanellus
Irediparra = Metopidius
Lobipes = Phalaropus
Lobipluvia = Vanellus
Lobibyx = Vanellus
Lobivanellus = Vanellus
Microparra = Metopidius

Microscarcops = Vanellus
Nycticryphes = Rostratula
Philohela = Scolopax
Pluviorhynchus = Charadrius
Pseudototanus = Tringa
Ptiloscelys = Vanellus
Rogibyx = Vanellus
Sarciophorus = Vanellus
Squatarola = Pluvialis
Steganopus = Phalaropus
Stephanibyx = Vanellus
Thinornis = Charadrius
Tylibyx = Vanellus
Xiphidiopterus = Vanellus
Zonibyx = Vanellus
Zonifer = Vanellus

GENERA NOT ACCEPTED BY PETERS BUT RECOGNIZED HERE

Pseudoscolopax = Limnodromus semipalmatus of Peters

SPECIES RECOGNIZED BY PETERS BUT HERE REDUCED TO SUBSPECIES OR SYNONYMIZED

Capella delicata = Gallinago gallinago delicata
Capella paraguaiae = Gallinago gallinago paraguaiae
Charadrius alticola = Charadrius falklandica alticola
Chubbia jamesoni = Gallinago stricklandii jamesoni
Hoplopterus duvaucelii = Vanellus spinosus duvaucelii
Lobibyx novae-hollandiae = Vanellus miles novaehollandiae
Prosobonia parvirostris = Prosobonia cancellatus
Scolopax celebensis = Scolopax rochussenii celebensis

FORMS CONSIDERED RACES BY PETERS BUT HERE GIVEN SPECIES STATUS

Capella paraguaiae andina = Gallinago andina
Charadrius alexandrinus marginatus = Charadrius marginatus
Charadrius alexandrinus ruficapillis = Charadrius ruficapillus
Charadrius hiaticula semipalmatus = Charadrius semipalmatus

Haematopus ostralegus unicolor = Haematopus unicolor

Limnodromus griseus scolopaceus = Limnodromus scolopaceus

OTHER MISCELLANEOUS DEVIATIONS FROM PETERS'S CLASSIFICATION

Charadrius venustus = Charadrius pallidus
Erolia testacea = Calidris ferruginea
Rogibyx tricolor = Vanellus macropterus
Peltohyas australis is included in the Charadriidae rather than the Glareolidae.

By way of brief summary, Peters (1934) divided the shorebirds included in this book into 7 families, 77 genera, and 167 species. I have recognized 8 families, 40 genera, and 165 species. Other approaches to the group are those of Morony, Bock, and Farrand (1975), who recognized 7 families, 44 genera, and 181 species, and Wolters (1975), who recognized 8 families, 70 genera, and 176 species. Although they did not extend their classifications to the species level, Jehl (1968a) recognized 7 families and 45 genera, and Strauch (1976) proposed a taxonomy of 4 families and 22 genera among the forms included in this book's coverage.

Ringed Plover
(distraction display)

BREEDING DISTRIBUTIONS AND PERIODS OF VULNERABILITY

Collectively speaking, the shorebirds exhibit a cosmopolitan pattern of breeding distribution and range from tropically breeding species to those that nest in high alpine or high arctic environments. Most of them, however, are open-country birds, associated with wetlands among grasslands, savannas, coastlines, or tundra communities. They are absent from the mainland of Antarctica and are more poorly represented in the southern hemisphere than in the northern hemisphere. Indeed, two major subfamilies (Calidridinae and Tringinae) are almost totally lacking as breeding species in the southern hemisphere. One family of shorebirds, the Jacanidae, is essentially pantropical, while the Calidridinae are circumpolar in the northern hemisphere. Two species, the stilt and oystercatcher, exhibit nearly cosmopolitan breeding distributions, although both these forms exhibit such great interpopulation variation in plumages that species limits remain rather controversial. On the other hand, the most highly restricted range of any major taxonomic group is the Pluvianellidae, consisting of a single species limited to a small portion of Isla Grande de Tierra del Fuego, and adjoining portions

of Santa Cruz Province, Argentina. Even more restricted in present distribution is the tribe Prosoboniini, which includes a single surviving species that is now found on a few of the most remote atolls of the Tuamotu Archipelago in French Polynesia. Almost equally remarkable is the occurrence in the Himalayas of a specialized river-dwelling species, the ibisbill, whose bill has been modified to allow for probing under rounded, water-worn boulders in swiftly flowing mountain streams.

If the breeding distributions of the shorebirds are grouped according to zoogeographic regions (table 1), some additional patterns become evident. The most obvious of these is that the two major northern hemisphere land masses, the Holarctic and the Palaearctic, support by far the largest number of species of breeding shorebirds, while the tropical and southern hemisphere areas are all appreciably poorer in species diversity. The groups that, at least in this book, are considered to be of earlier origins (Jacanidae, Rostratulidae, Pluvianellidae, and Vanellinae, in particular) mostly have tropical or southern hemisphere affinities, while the Calidridinae and Tringinae are distinctly northern in distribution.

There are no clear patterns of geographic affinities for the Haematopodidae, Recurvirostridae, Chara-

Table 1 Distribution of Breeding Shorebird Species by Zoogeographic Regions

Species	Palaearctic	Nearctic	Ethiopian	Neotropical	Australian	Oriental
Jacanidae (7 spp.)	—	—	3	1	1	3
Rostratulidae (2 spp.)	1	—	1	1	1	1
Pluvianellidae (1 sp.)	—	—	—	1	—	—
Haematopodidae (5 spp.)	1	1	1	3	3	—
Ibidorhynchidae (1 sp.)	—	—	—	—	—	1
Recurvirostridae (6 spp.)	2	2	2	2	2	1
Charadriidae						
Vanellinae (23 spp.)	6	—	12	3	3	3
Charadriinae (38 spp.)	12	8	7	8	9	12
Scolopacidae						
Calidridinae (26 spp.)	18	16	—	—	—	—
Tringinae (37 spp.)	21	18	—	—	2	—
Scolopacinae (20 spp.)	9	2	2	6	3	3
Totals	70	47	25	24	28	24

driinae, or Scolapacinae, all of which tend to have at least one species with extremely broad or nearly cosmopolitan breeding distributions as well as others that are highly restricted. Although the seemingly more generalized forms of the Scolopacinae are clearly southern in their distributions, Seebohm (1888) believed that this too is a northern group that extended its range southward and left various early isolated populations that have survived in the absence of competition from more advanced types of snipes and woodcocks. Seebohm likewise suggested a northern hemisphere (Asiatic) origin of his "subfamily" Charadriinae (which also included the Burhinidae and Glareolidae as now recognized, as well as the Vanellinae), and the concentration of contemporary Charadriinae species in the Palaearctic and Oriental regions also supports that position with respect to the subfamily as it is defined in this book. The family Haematopodidae is also a complex that is difficult to assess because of taxonomic problems, but I agree with Maclean (1972a) that it is probably a group that rose in the north temperate zone and only later underwent radiation in south temperate regions.

In conjunction with these general geographic patterns of distribution, it is of interest to compare the periods of egg and chick vulnerability relative to body size and time spans available for breeding. If the incubation and fledging periods have evolved in relation to available breeding periods, groups having temperate to arctic distribution patterns should exhibit shorter periods of vulnerability than those occurring in subtropical to tropical areas. Although information on such periods is not available for all species, enough information is at hand to test this idea for two major groups, the Charadriidae and the Scolopacidae, and one minor group, the Recurvirostridae. When the combined incubation and fledging periods for members of these groups are plotted against average adult weights, some interesting results emerge (fig. 3). First, there is indeed a marked separation between the Charadriidae and the Scolopacidae with respect to their relative periods of vulnerability, most of which is associated with differences in fledging periods rather than incubation periods and fits the expected pattern of a more prolonged period of development in the more tropically distributed Charadriidae. Indeed, the species of Charadriidae that fall within the area occupied by the Scolopacidae are all arctic-nesting forms, the greater golden plover, the gray plover, and the ringed plover. On the other hand, a few temperate-breeding species of Scolopacidae exhibit some surprisingly short periods of vulnerability, specifically the American and Eurasian woodcocks. The few species of Recurvirostridae for which data are available seem generally to fall between the Charadriidae and Scolopacidae in their vulnerability period characteristics. The single species of Haematopodidae that has

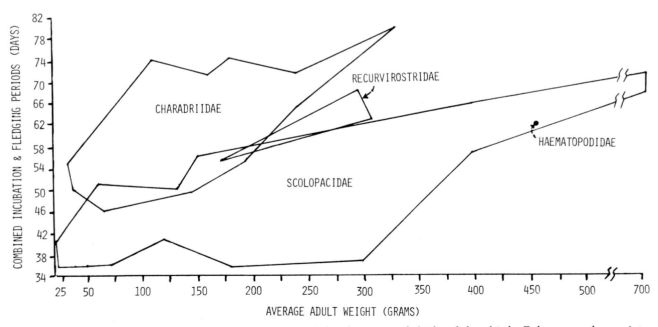

3. Relationship of adult weights to periods of vulnerability for eggs and chicks of shorebirds. Polygons enclose points representing species of indicated groups.

been plotted falls within the Scolopacidae's range of vulnerability, which is in agreement with its temperate-zone distribution.

VARIATIONS IN MATING SYSTEMS

The shorebirds show a greater diversity of mating systems, including monogamous, polygamous (polygyny and polyandry), and promiscuous mating patterns, than any other order of birds.

Prolonged Monogamy

Although indefinite monogamy is doubtless a rather rare type of mating among shorebirds, it nonetheless does occur in several species. For example, Harris (1967) reported that in a British population of European oystercatchers, the percentage of pairs in which both members survived until the following year and remated at that time ranged from 76 to 97 percent. In this population 18 color-marked pairs remained intact for at least three summers, while 22 additional pairs were mated for two consecutive years, within a population in which 40–43 pairs had been color-banded. In a German study of oystercatchers, 24 pairs remained unchanged for more than six years (Jungfer 1954). Possible indefinite pair-bonding was also reported by Barlow et al. (1972) for masked lapwings in New Zealand. In a study of Kentish sandplovers, Rittinghaus (1961) reported that 14 pairs of birds remated the following year, while 26 birds that had lost their mates paired with new ones and 42 individuals mated with new birds, even when their former mates were still available. Remating in this species is promoted by strong nest-site fidelity, and partner retention for as many as six years has been observed.

Seasonal Monogamy

Seasonal monogamy is probably the most common pattern of mating in shorebirds, especially those that have relatively short life-spans and long migration routes and that defend "typical" (type A) territories within which the nest is built and the young are reared, often with the help of the male. Of the species of Palaearctic and Nearctic shorebirds analyzed by Glutz et al. (1975, 1977), at least 24 were classified as seasonally monogamous for the most part, while 7 others exhibit at least brood monogamy and 4 have seasonal pair bonds of uncertain type. By comparison, one species (the oystercatcher) was considered to show a lifelong pair bond, while 7 species were classified as polygynous or promiscuous, and successive polyandry was noted as occasional or regular for several species but is probably not universal in any, as noted below.

Polygyny and Promiscuity

Within the rather indefinite category of "polygamy" one may include polygyny (the maintainence of two or more female mates by a male simultaneously or successively) and polyandry (the maintenance of two or more male mates by a female simultaneously or successively). The distinction between polygyny and promiscuity is a rather difficult one, but, as used in this discussion, promiscuity implies no individualized social bonding between the sexes other than that directly associated with copulation.

According to Glutz et al. (1975, 1977), promiscuous mating occurs in several western Palaearctic or vagrant Nearctic species, including the great snipe, European woodcock, white-rumped sandpiper, buff-breasted sandpiper, and ruff. These authors additionally believed that the sharp-tailed sandpiper exhibits either polygyny or promiscuity, and that the pectoral sandpiper has a mating system ranging somewhere between monogamy, polygyny, and promiscuity. I would prefer to consider both these species essentially promiscuous, based on my interpretation of available descriptions of breeding behavior. Additionally, it is possible that promiscuous mating occurs in the pintailed snipe, judging from its communal aerial courtship display (Tuck 1972), and it is also obvious that the North American woodcock is promiscuous (Sheldon 1967). The mating systems of the other woodcocks are completely unknown, as are those of other snipes, but promiscuity or polygyny in one or more species of each of these groups is quite probable. Finally, it has been suggested (Pitelka et al. 1974) that the curlew sandpiper may be polygynous, but Glutz et al. (1975) reviewed its biology and considered this a monogamous species.

Occasional cases of polygyny abound among the shorebirds, though some species seem to be more prone toward this behavior than others. Simultaneous bigamous matings by males seem to occur fairly often in the European lapwing and greenshank and have been reported at least twice in greater golden plovers (Bannerman 1961). Similar occasional or rare instances of male bigamy have been observed in several species including wrybills (Rodney Hay, unpublished MS), blacksmith plovers (Hall 1959b), and sandplovers (*Point Reyes Bird*

Observatory 45:4–5). Pitelka et al. (1974) distinguished those species of arctic sandpipers that they believed showed simultaneous polygyny (white-rumped, curlew, and sharp-tailed sandpipers) from those species that exhibit serial polygamy (either polyandry or polygyny), which they thought included two or three species (sanderling, Temminck stint, and perhaps little stint). Likewise, Emlen and Oring (1977) distinguished the latter type of mating system ("rapid multiple-clutch polygamy") from typical polygyny, including promiscuity. In their view, avian polygyny is brought about by males' controlling breeding resources through territoriality, controlling females directly through harem accumulation and defense, or through differential male dominance and attraction displays in breeding season aggregations such as leks. At least three species of shorebirds (great snipe, buff-breasted sandpiper, ruff, and possibly pintailed snipe) exhibit true lek behavior, while the other examples of polygyny among shorebirds seem to be associated with the differential abilities of dispersed males to attract females, perhaps on the basis of territorial quality (Pitelka et al. 1974).

Multiple-clutch Breeding Systems

As Emlen and Oring pointed out, the rapid multiple-clutch breeding system (in which females lay two or more clutches in rapid succession, with the female and her mate each incubating separate clutches of eggs) results in both sexes' having relatively equal opportunities for increasing their fitness. Although in its simplest two-clutch form, such as in the Temminck stint, this is a simple monogamous system, the system readily allows for the evolution of typical polyandry by the simple device of the female turning the second clutch over to a male, thus freeing her to produce yet additional clutches. The multiclutch breeding system is probably more common among the shorebirds than is realized at present; in addition to the three species mentioned by Pitelka et al. (1974), it probably also occurs in the rufous-necked sandpiper (Portenko 1972) and the mountain plover (Graul 1973). In the latter species the female sometimes remains with her original mate for the second clutch, while others apparently switch mates before laying the second clutch, thus showing the ready transition toward typical polyandry.

Sequential and Simultaneous Polyandry

Polyandry is the rarest form of avian mating system and so far has been reported to occur among only thirty-one bird species, mostly in the order Charadriiformes (Jenni 1974). According to Jenni, 14 species of Charadriiformes have been reported as probable examples of polyandrous mating, including 4 species of jacanas, 3 species of phalaropes, the Old World painted snipe, 2 species of plovers, and 4 species of scolopacids. In the case of the plovers and the scolopacids other than the spotted sandpiper, these instances of polyandry are all apparently rather infrequent and the probable result of double- or multiple-clutching associated with mate switching. Thus, though of great interest, they are probably not an intrinsic part of the species' regular breeding strategies. The mating system of the painted snipe is still inadequately studied, but like the phalaropes and jacanas it exhibits the expected reversal of sexual dimorphism in weight, and, in common with the phararopes (but not the jacanas), the females are appreciably more colorful than the males.

Jennie suggested that, among the phalaropes, sequential polyandry had been proved only for the northern phalarope, but it has since also been established for the red phalarope (Schamel and Tracy 1977), and there is circumstantial evidence for its occurring in the Wilson phalarope (Howe 1975b; Kagarise 1979). Sequential polyandry has been well verified for at least one species of jacana (the pheasant-tailed), and there is indirect evidence for it in at least two additional species. At the time of Jenni's review, the American jacana was the only shorebird so far proved to exhibit simultaneous polyandry, in which the female maintains simultaneous and persistent pair bonds with two or more males. Recently, Osborne and Bourne (1977) found some instances of probable simultaneous polyandry among a population of wattled jacanas (here considered conspecific with the American jacana), while Oring and Maxon (1978) have reported one example of simultaneous polyandry in a population of spotted sandpipers.

Effects of Polyandry and Polygyny on Egg and Body Weights

A recent survey by Ross (1979) has summarized data on relative egg weights versus adult body weights for a variety of shorebirds, supporting the idea that multiple-clutch breeding systems in these birds are associated with reduced egg size. Ross established that the reduction of egg weight relative to body weight in females of these species is achieved by a reduction of egg size as well as by an increase in female body size. However, there is no evidence that

Table 2 Relationship of Adult Male and Egg Weights to Female Weights

Breeding System and Species	Season or Age	Average Adult ♀ Weight (g)	Average Adult ♂ Weight (g)	♂ as % of ♀ Weight	Average Egg Weight (g)	Egg as % of ♀ Weight	Adult Weight Reference
Promiscuous or polygynous							
Philomachus pugnax	Fall	108 (40)[a]	185 (17)[a]	171.2	21.0	19.4	Glutz et al. 1975
Calidris melanotos	Summer	60 (10)	86 (25)	143.3	13.1	21.8	Irving 1960
Tryngites subruficollis	Summer	53 (6)	71 (4)	133.9	13.0	24.5	Irving 1960
Calidris fuscicollis	Summer	39.7 (7)	45.8 (6)	115.3	10.8[b]	27.2	Glutz et al. 1975
Calidris acuminatus	Fall	72 (8)	80 (8)	111.1	13.7	19.0	Shaw 1936
Calidris ferruginea	Summer	63.3 (16)	63.2 (11)	100.0	12.0	18.9	Glutz et al. 1975
Monogamous, single-brooded							
Calidris bairdii	Summer	39 (12)	39 (28)	100.0	9.6	24.6	Irving 1960
Calidris pusilla	Summer	26 (10)	24 (32)	92.3	6.9	26.5	Irving 1960
Calidris minutilla	Summer	22 (14)	20 (16)	90.9	6.4	29.1	Irving 1960
Calidris mauri	Summer	30.1(61)	26.3 (92)	87.3	7.5	24.9	Holmes 1972
Calidris canutus	Summer	147.9 (9)	125.5 (13)	84.8	19.0	12.8	Parmelee and MacDonald 1960
Calidris maritima	Summer	76.6 (6)	63.3 (6)	82.6	13.3	17.4	Glutz et al. 1975
Monogamous, double-clutching							
Calidris alba	Summer	54.4 (5)	52.4 (7)	96.3	11.2	20.5	Glutz et al. 1975
Charadrius montanus	Adults	100.2 (7)	93.8 (4)	93.6	16.5	16.5	Museum of Vertebrate Zoology
Calidris temminckii	Summer	26.4 (10)	23.6 (18)	89.3	5.8	21.9	Glutz et al. 1975
Calidris ruficollis	Fall	39 (29)	32 (62)	88.8	8.0	22.2	Shaw 1936
Calidris minuta	Summer	30.6 (4)	25.1 (11)	82.0	6.3	20.5	Glutz et al. 1975
Reportedly polyandrous							
Tringa erythropus	Aug.–Oct.	156.8 (14)	147.6 (13)	94.1	24.5	15.6	Glutz et al 1977
Calidris alpina	May–Sept.	55 (21)	50.6 (26)	92.0	11.2	20.4	Glutz et al. 1975
Phalaropus lobatus	Summer	35 (7)	32 (14)	91	6.3	18.0	Irving 1960
Rostratula benghalensis	Adults	130 (9)	117.6 (25)	90.5	12.5	9.6	This study
Eudromias morinellus	Summer	117.3 (5)	100.1 (10)	85.3	17.0	14.5	Glutz et al. 1975
Phalaropus fulicarius	Summer	61.0 (51)	50.8 (69)	83	7.5	12.3	Kistchinski 1975
Jacana jacana	Adults	142.8 (15)	108.3 (16)	75.8	8.3	5.8	Osborne and Bourne 1977
Actitis macularia	Summer	47.2 (3)	33.2 (3)	70.3	9.0	19.1	Glutz et al. 1977
Jacana spinosa	Adults	145.4 (12)	86.9 (16)	59.7	8.3	5.7	Jennie and Collier 1972
Hydrophasianus chirurgus	Adults	231 (3)	125 (5)	54.5	14.1	6.1	U.S. National Museum
Actophilornis africana	Adults	261 (5)	136.6 (8)	52.3	8.6	3.3	U.S. National Museum
Possibly polyandrous							
Phalaropus tricolor	Summer	68.1 (53)	50.2 (100)	73.7	9.4	13.8	Höhn 1967
Metopidius gallinacea	Adults	130 (7)	75.1 (6)	57.8	7.1	5.5	U.S. National Museum
Microparra capensis	Adult	41 (1)	—	—	4.5	11.0	Britton and Dowsett 1969
Actophilornis albinucha	Adult	239 (1)	—	—	12.7*	5.3	Benson et al. 1976
Nycticryphes semicollaris	Adult	76.6 (5)[c]	—	—	11.3	14.7	Sick 1962

[a] Number in parentheses = number in sample.
[b] Calculated from reported egg dimensions; other egg weights from Schönwetter 1963.
[c] Based on a mixed sample of both sexes.

polyandry or other multiple-clutch breeding systems are associated with reduced clutch size in shorebirds; at least 6 of 7 jacanas, all 3 phalaropes, and at least 1 of the 2 species of painted snipes consistently show the usual shorebird clutch of 4 eggs (Maclean 1972a). The strong tendency for shorebirds to consistently lay 4 similar-sized pyriform eggs is understandable in that the strategy minimizes the rate of heat loss when the eggs are uncovered. Thus shorebirds tend to invest available energy in producing larger and richer eggs rather than in increasing their clutch size (Miller 1979b).

Ross compared egg weight with adult male versus adult female weight but did not directly deal with the differences in adult male and female weights in conjunction with varied mating systems. Differing degrees of dimorphism in adult weights of the sexes relative to varying kinds and degrees of sexual selection in birds are quite evident and clearly can be detected among shorebirds. The most conspicuous example is that the ruff, but other regularly or occasionally polygynous species (buff-breasted sandpiper, pectoral sandpiper, sharp-tailed sandpiper, white-rumped sandpiper) all show a definite trend toward male dominance in average adult weights (table 2). This male weight dominance is lacking in the curlew sandpiper, which is probably normally monogamous. Interestingly, dominance in weights does not extend to the Scolopacinae, since polygynous mating occurs in at least two woodcocks and in the great snipe. Yet in both woodcocks the females average heavier than males, and in the great snipe the sexes are not obviously dimorphic in either plumage or weight.

Most of the other calidridine shorebirds are monogamous and single-brooded or at most occasionally double-clutching. The male/female and egg/female weight ratios of the apparently single-brooded versus the known double-clutching species of *Calidris* show no observable differences, and they collectively form a coherent series of egg/female weight ratios intermediate between the regular polygynous forms and the regularly or occasionally polyandrous ones (fig. 4). The line drawn in to fit these points is virtually identical to that established by Ross (1979) for nearly sixty species of plovers, sandpipers, and phalaropes.

Serial or sequential polyandry is obviously an adaptive extension of double-clutching and has been reported as variably common in the spotted sandpiper. There is a single case of sequential polyandry known in the spotted redshank (Raner 1972), and three cases of serial polyandry after the hatching of

the first clutch have been documented for the dunlin (Soikkeli 1967). Likewise, in the dotterel the usual monogamous mating system is occasionally varied with serial polyandry (Nethersole-Thompson 1973), and the same has been observed once for the mountain plover (Graul 1973). In this group of apparently facultatively polyandrous species (excluding the questionably typical cases of the spotted redshank and the mountain plover) the cluster of egg/female weight ratios lies below the line established for the monogamous and single-brooded or double-clutching species (fig. 4). Yet it closely approaches this line and suggests that normally monogamous calidridines could occasionally exhibit facultative polyandry, especially those that are regularly double-clutching species.

Serial polyandry, long assumed to occur in all three species of phalaropes, has only recently been documented for the red and northern phalaropes, and it is likely but still unproved for the Wilson phalarope. The egg/female weight data suggest that all three species may be facultatively polyandrous, and the male/female ratio of the Wilson phalarope actually approaches that of the American jacana. Thus this species is plotted (as an "X") among the polyandrous species in figure 4.

Beyond this group of apparently facultatively polyandrous species, there remain only the painted snipes and jacanas. Almost certainly the Old World painted snipe is serially polyandrous, and its egg/female weight ratio indicates an unusually small relative egg size. Although the egg/adult weight ratio (female weights are not available) of the South American painted snipe falls within the cluster of facultatively polyandrous forms, there is no evidence that polyandry exists in this species, and Sick (1962) indicated that the sexes appear not to differ in adult weight, which is also consistent with a monogamous mating system.

Among the jacanas, selection has obviously favored the evolution of eggs that are remarkably small compared with those of all other shorebirds (fig. 4), which promotes the evolution of polyandry. Thus, polyandry may well be universal in the group, although it is still unproved for several species. It is noteworthy that the jacanas should have carried egg-size reduction so far while still rigidly retaining a 4-egg clutch size. It seems that the larger species of jacanas could physiologically "afford" a larger clutch size or larger eggs, as a full clutch for these larger species represents only some 13–21 percent of the adult female's body weight, or less than the relative weight of a single egg of some of the smaller *Calidris*

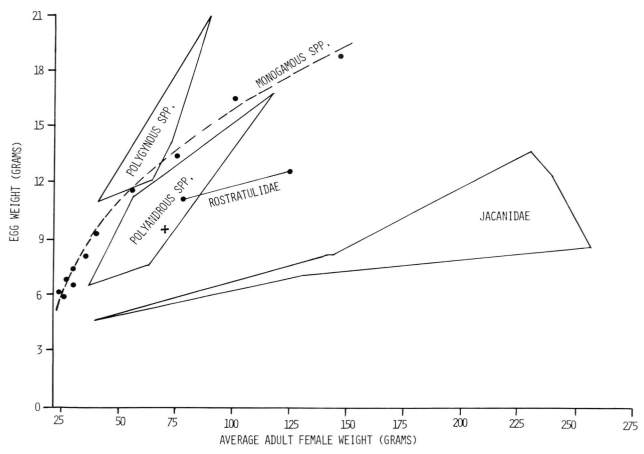

4. Relationship of adult female weights to egg weights of shorebirds. Polygons enclose points representing species of various groups, and curved line connects points representing monogamous species. *X* indicates *Phalaropus tricolor,* a presumed polyandrous species.

species. In that the females are considerably smaller than the males and also lay relatively small eggs, the jacanas resemble the Rallidae. Yet, clutch sizes and relative egg sizes in the rails average larger than in jacanas, even though rails feed their own young and jacanas do not. A comprehensive study of these aspects of jacana biology is certainly needed.

BREEDING BIOLOGY AND STRUCTURAL DIVERSITY

As two examples of the influence of breeding behavior on morphological diversity in shorebirds, the lapwings and snipes provide instructive cases of the taxonomic interpretation of "hard" structural characters in the light of breeding behavior variations. Although other examples could be chosen to illustrate these points, the lapwings and snipes have the advantage of being large and distinctive groups in which morphological traits have been used taxonom-

ically with little or no consideration of their possible behavioral or ecological significance. In the case of the lapwings, the species are highly distinctive and morphologically diverse, resulting in as many as twenty or more generic names proposed for about as many species. On the other hand, the snipes are a morphologically conservative group with only a few genera, but in which the species-level taxonomy has often been dependent upon seemingly nonadaptive variation in rectrix number or condition.

Wattle, Spur, and Toe Variations in Lapwings

One of the first significant lapwing classifications was that of Seebohm (1888), who divided the group into 2 genera. These consisted of 14 species of "wattled lapwings" in the genus *Lobivanellus*, and all the nonwattled forms, which he included in the genus *vanellus*, also considered to include 14 species. Seebohm admitted that this was to some degree a

❖ ❖ ❖ 19

classification of convenience, but he nonetheless regarded the presence of a lobe or wattle as an ancient trait. Peters (1934) was evidently more impressed by wattles, spurs, and similar variations than Seebohm had been, and he recognized no fewer than 19 genera and 25 species, with 15 of the genera being monotypic. Boetticher (1954), using the presence or absence of wattles, wing spurs, and a hallux as primary characters, proposed that only 4 genera should be recognized. Finally, although Wolters (1974, 1975) questioned the value of using the hallux and wing spur condition, he did use wattles and variations in plumage to produce a taxonomy consisting of 14 genera and 25 species. The first person to draw attention to the impracticality of such classifications was Bock (1958). He argued that, since no clearly separable groups of species could be found, a single large genus was at least as realistic as Peters's approach of recognizing 19 genera, most of which are monotypic. Bock's grouping of the species was based to a considerable degree on zoogeographic grounds, and among his few conclusions on relationships was the suggestion that *miles* and *novaehollandiae* should be regarded as conspecific (a view now generally accepted) and that *spinosus, armatus,* and *duvaucelii* probably constitute a superspecies. Between the classifications of Bock, Peters, and a recent revision by Wolters (1975), there is virtually no agreement as to lapwing affinities other than that (1) *leucurus* and *gregarius* are likely close relatives; (2) *lugubris* and *melanopterus* are likewise affiliated; and (3) *spinosus* and *duvaucelii* (and perhaps *armatus*) constitute a cluster of species.

Most recently, Strauch (1976) did a cladistic analysis of the lapwings. One major conclusion of his work was that *Hoploxypterus cayanus* is not a lapwing at all, but rather belongs with the typical plovers. His analysis did not result in a suggested linear sequence, but he did conclude that the presence of a wingspur constitutes "a true cladistic character." Although his findings are not yet published, Jon Fjeldså (in litt.) has examined the natal plumages of many charadriiform species, and with respect to the Vanellinae has concluded that 3 lapwing species come closest to the common ancestral wader-flamingo stem type and additionally believes that many of the living plovers included represent other "very ancient types."

A review of the ecological and behavioral literature on the lapwings makes it at once apparent that many aspects of their behavior remain unresolved, but at least some generalities and trends are evident. For example, most lapwings are birds of the open countryside, especially plains and savannas, where the values of contrasting plumage features and other visual signals are enhanced. Second, some but by no means all lapwing species are highly territorial, and aggressive intraspecific behaviors including fighting are common facets of early breeding behavior of the territorial forms. In such species, threat-enhancing signals such as wattles and wing spurs are clearly valuable attributes. So far as is known, lapwings are normally monogamous, with pair bonds that often persist from year to year and sometimes last as long as both members of a pair are available. Pair-forming behavior is evidently much like that of typical *Charadrius* plovers, and at least in some species it involves scraping behavior much like nest-building behavior. Wing-raising displays, often exposing contrasting underwing coloration or wing spurs, occur in

Table 3 Summary of Morphological and Ecological Traits in the Lapwings

	crassirostris	*coronatus*	*gregarius*	*leucurus*	*cinereus*	*malabaricus*	*superciliosus*	*lugubris*	*melanopterus*	*vanellus*	*chilensis*	*resplendens*	*armatus*	*spinosus*	*melanocephalus*	*tectus*	*indicus*	*tricolor*	*miles*	*macroptera*	*senegallus*	*albiceps*	*cayanus*
Wattles[a]	—	—	—	—	x	X	x	—	—	—	—	—	—	—	x	x	x	X	X	X	X	X	—
Wingspur[a]	x	—	—	—	—	—	—	—	—	—	X	x	X	X	—	—	x	—	X	X	X	X	X
Hallux[b]	X	—	X	X	X	—	—	—	—	X	X	—	—	—	X	—	X	—	X	X	X	—	—
Body size[c]	M	M	M	S	L	S	S	S	M	L	L	M	M	S	L	S	L	S	L	L	L	M	S
Arid/mesic	M	A	A	M	M	A	A	A	M	M	A	M	M	M	A	M	A	M	A	M	M	M	M
Territorial[a]	X	x	x	x	?	X	x	?	x	x	X	?	X	X	?	?	?	x	X	?	X	X	?

[a] X indicates that the characteristic is well developed, x poorly developed, and — absent

[b] X indicates that the hallux is present, — indicates its absence

[c] S indicates small (wing under 200 mm average weight up to 150 g); M indicates medium (wing 200–220 mm, average weight 150–225 g); and L indicates large (wing over 220 mm, average weight over 225 g).

at least two quite different contexts. Thus, wing-raising occurs as a postcopulatory display in some species such as *coronatus* and *chilensis* (Maclean 1972*b*). Wing spurs may thus serve as "courtship" structures as well as inter- or intraspecific weapons of attack (Bock 1958).

I believe facial wattles and spurs are primarily visual signal enhancement devices among those species that for ecological reasons have evolved rather strong territorial tendencies, while the absence of wing spurs and wattles simply reflects a lack of these ecological pressures. If that is so, the presence of such structures would have no taxonomic significance and indeed might often prove misleading. Likewise, the presence or absence of a hind toe seems readily subject to convergent or parallel evolution according to the importance of physical support for the foot on various substrates. Therefore, similarity in this feature cannot support a conclusion of homology and thus common ancestry. Lapwings also vary quite considerably in body size, and, though this characteristic has not been used as a taxonomic trait, it almost certainly has ecological consequences and is worth considering in terms of a general analysis. Although body weight is perhaps the best single statistic for measuring overall body size, such weight data are not available for all species, and so I have of necessity resorted to using available weight data in conjunction with wing measurements for assessing size variation. Thus I have estimated weights for three species by correlating wing measurements and known weights of the remaining forms.

A summary of these major traits occurs in table 3, together with an indication of the relative degree of territoriality each species displays wherever it has been possible to assess this from the literature or other sources. Additionally, each species is assessed for whether it is arid-adapted or mesic-adapted, again on the basis of the literature. Sometimes this is a judgment evaluation; thus Bock's (1958) summary might suggest that both *gregarius and resplendens* should be regarded as mesic, whereas I have interpreted both as arid-adapted. Further, he did not list spurs as characteristic of *crassirostris*, *resplendens*, or *indicus*, but small spurs are usually present on these forms, according to Seebohm (1888).

An examination of this table provides some clues to the ecological and taxonomic implications of these variables. Of the species that are neither spurred nor wattled, at least 5 are weakly territorial, and none has been reported as strongly territorial. Of those that are spurred and wattled, at least 3 are territorial, and none is known to be weakly territorial. Of those

with either spurs or wattles, 5 have been reported as territorial and 2 as weakly territorial. Spurs and wattles seem to be associated with species of mesic distributions (5 mesic, 0 arid), while those lacking wattles and spurs are generally more arid-adapted (4 arid, 2 mesic).

There seem to be no behavioral relationships with the presence or absence of a hallux, but 8 of 12 species that lack a hallux are arid-adapted, while at least 10 of 11 species (or all, if *gregarius* is considered mesic) having a hallux are mesic in distribution.

Finally, body size may be somewhat more closely related to habitat than to aggressiveness. Of 10 small to medium species, 6 are known to be weakly territorial, compared with 5 of 8 medium to large species reported being territorial. More convincingly, 9 of 15 small to medium species are arid as opposed to mesic in distribution, while 11 out of 15 medium to large species are mesic-adapted as opposed to arid-adapted.

It is thus fairly clear that wattles and spurs in lapwings should be considered as characteristics resulting from ecological sources of selection associated with territoriality or general aggressiveness rather than as reflecting relationships, and that the presence or absence of a hallux is likewise of no taxonomic significance.

Outer Rectrix Variations in Snipes

It has long been recognized that the variations in the number and condition of rectrices in the snipes have taxonomic utility; indeed, several species can be readily identified only by recourse to these traits. Further, though the genus *Gallinago* is remarkably conservative in most of its plumage and other morphological features, the variations exhibited in the numbers of rectrices (from twelve to twenty-eight) are among the greatest such variations found in any genus outside the influence of artificial selection. Seebohm (1888) commented on the fact that the snipes of Europe, Africa, and America have fourteen to sixteen rectrices, while those of eastern Siberia normally have eighteen to twenty-six. He suggested that it was "perhaps impossible to discover a rational explanation for these curious facts," discounting both natural selection and sexual selection as reasonable mechanisms.

In spite of this, there can be no doubt that these variations are indeed adaptive and are probably the result of intense natural selection associated with demands for effective acoustic advertisement under crepuscular conditions where visual signals are essen-

Table 4 Variations in Weight, Wing Length, and Rectrices among *Gallinago* Snipes

Species[a]	Approximate Adult Weight (gm)	Average Wing Length (mm)	Total Rectrix Number[b]	Width of Outer Rectrix (mm)[b]
G. imperialis	M	161	12[c]	ca. 8
G. undulata	300	160	14	5–7
G. s. stricklandii	—	155	14	6–7
G. s. jamesoni	230	160	14	5–7
G. g. gallinago	115	128	14	12.8
G. g. delicata	110	129	16	7.2 (6–9)
G. g. paraguaiae	110	119	16	4.4 (4–6)
G. g. magellanica	130	130	16	5.0 (4.5–5.5)
G. g. andina	100	114	16	5.0 (4.5–5.5)
G. g. nigripennis	115	127	16	5.0
G. media	200	140	16	ca. 7
G. nobilis	190	140	16	6.5
G. macrodactyla	215	144	16	ca. 8
G. hardwickii	150	162	18	4
G. nemoricola	170	146	18	ca. 4
G. solitaria	150	160	20 (18–22)	2–4
G. megala	140	142	20 (22)	3–4
G. stenura	125	128	26 (24–28)	0.6–1.0

[a] Arranged according to increasing rectrix number; data from various sources.
[b] Parentheses indicate range; otherwise, normal or mean is indicated.
[c] Based on a single specimen (John Farrand, Jr., pers. comm.)

tially worthless. It is of interest that in the woodcocks, which also display under similar seminocturnal conditions, the tail has remained unmodified for sound production, but in one species (*Philohela minor*) the three outer primaries have become modified for this purpose. In the woodcocks and rectrices are invariably twelve in total, whereas all snipes except *Limnocryptes* and *G. imperialis* (each with twelve) have at least fourteen. This includes the seemingly primitive sub-antarctic snipe, which seems intermediate in many ways between the snipes and woodcocks and represents a convenient postulated ancestral type. It thus appears probable that the ancestral snipes had twelve to fourteen rectrices, and that the larger numbers of most present-day snipes represent derived adaptations. That presumption being accepted, it is obvious that the pintailed snipe with twenty-four to twenty-eight rectrices represents the specialized extreme, while several large South American snipes (*G. undulata*, *G. stricklandii*, *G. s. jamesoni*, and *G. imperialis*) represent the most primitive type. Indeed, Seebohm regarded these species (and *Coenocorypha*) as "semi-woodcocks" and clearly considered them a link with the true woodcocks. Not only do these species have the "primitive" number of tail feathers, but their outermost rectrices are not obviously modified in shape or pigmentation from the conditions of the adjoining ones. The widths of the outermost rectrix in these forms (table 4) ranges from 5 to 9 mm, or about average for *Gallinago* as a whole. However, species with eighteen or more rectrices tend to have the outermost rectrix distinctly attenuated, and a reasonably good correlation can be drawn between the total number of rectrices and the average width of the outermost rectrix (fig. 5).

However, the recognition that outer rectrix width is related to the number of rectrices does not provide a complete explanation. Did selection favor the increase in rectrix number as such, secondarily resulting in reduction of the width of the outermost one, or did selection favor the evolution of a specific outer rectrix width and adjust the nubmer of rectrices accordingly? Obviously, a major function of the rectrices lies in steering and braking, and the faster and more agile snipes might well need to retain a maximum tail surface area, while at the same time gaining the advantage of increasing feather vibration potential associated with increasing speed. Excepting the forms constituting the superspecies *gallinago*, there does indeed seem to be a clear inverse correlation between each species' average body weight (as an indirect measurement of speed potential) and its number of rectrices (fig. 6). Thus, in general, small, speedy species have a large number of rectrices and generally also exhibit considerable narrowing of the

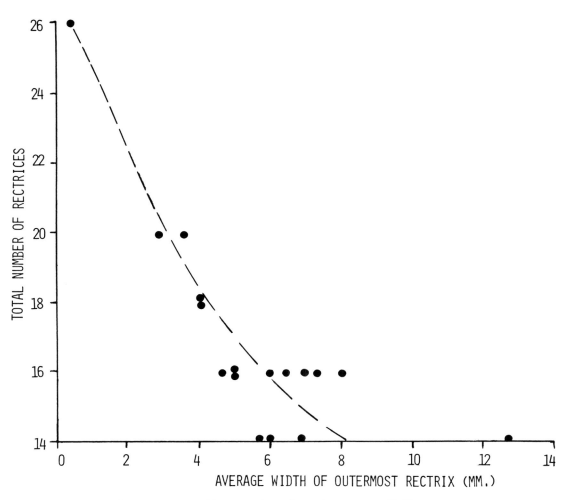

5.   Relationship of outer rectrix widths to total numbers of rectrices in *Gallinago*.

outermost ones, while larger, slow-flying species have a relatively small number of rectrices that exhibit little if any attentuation laterally (fig. 7). As noted, *gallinago* and its South American (*andina*) and African (*nigripennis*) counterpart forms that are here afforded specific rank do not follow this trend, and instead show a combination of small body size and rather low rectrix number (fourteen to sixteen). However, the outermost rectrices of these populations tend to be distinctly wider than is true of the species having eighteen or more rectrices, and perhaps the braking and maneuvering capabilities are thus retained. In this context it is interesting that the Eurasian snipe (*Gallinago g. gallinago*) has only fourteen tail feathers, and the width of the outermost one averages almost 13 mm, whereas in four New World subspecies (*delicata, paraguaiae, magellanica,* and *andina*) the normal number is sixteen and the width of the outermost rectrix varies from 3 to 9 mm (Tuck

1972). These differences in the widths of the outermost rectrices have apparently produced a distinct subspecific difference in the "bleating" sounds produced by the birds, with *gallinago* generating a low-pitched humming, while *delicata* produces a louder and more melodious sound (Thönen 1969). Recent studies by Reddig (1978) have proved beyond doubt that at least in this species these sounds are entirely the result of vibration of the outermost pair of rectrices, modulated by variable airflow past them as controlled by wing movements.

Although detailed descriptions of aerial display are not available for all species of *Gallinago*, information is available on at least some of the more specialized forms. "Bleating" is best known in *G. gallinago*; Tuck (1972) has provided an excellent description and review of the literature. It is evident that the sounds produced during aerial display of *gallinago* are almost entirely instrumental rather than vocal;

only under "intense excitement," such as during pair-formation, does the common snipe alternate its bleating with vocal "yakking." However, in the Japanese snipe calling regularly occurs during the display flight and is interrupted only by frequent "power dives" and an associated sound that is almost certainly instrumental. A very similar alternation of calling and plunging evidently occurs in the forest snipe, which approaches the pintailed rectrix condition of *solitaria* and *stenura*. In *solitaria,* there is also apparently a loud, jerky call uttered before a rapid descent with spread tail feathers and a buzzing sound, while in *stenura* the males apparently often perform in a communal aerial display flight, the males uttering rapid metallic notes and plunging progressively more vertically, until the vocal notes finally merge with the sounds of the tail feathers (Tuck 1972).

By comparison, aerial display in the larger and apparently more primitive snipes of South America consists of a rather extended period of calling (30–60 seconds in *jamesoni*), followed by a rather shallow dive that produces a muffled and low-pitched sound something like a cow's bellow (Vuillemier 1969). Ap-

parently this instrumental sound is of such a low pitch that it can barely be heard, which is not surprising given the bird's rather large size and the unspecialized condition of its outermost rectrices. Similarly, the display of the imperial snipe consists of repeated calling while flying in nearly level circles, followed by a shallow dive toward the end of each song bout resulting in a rush of air that is audible at close range (Terborgh and Weske 1972). There is no information available on instrumental bleating in the giant snipe, though it is said to have a three-syllable display call (Haverschmidt 1974).

It is thus tempting to suggest that as the snipes evolved they became progressively more dependent on aerial displays incorporating instrumental sounds that initially supplemented vocalizations and that eventually in some species almost replaced them. Further, the width of the outermost rectrix evidently became adaptively adjusted in width and shape to generate a sound appropriate to the species' velocity during its display flight, resulting in higher-pitched and louder sounds in those species that attain considerable speed during swooping or plunging performances.

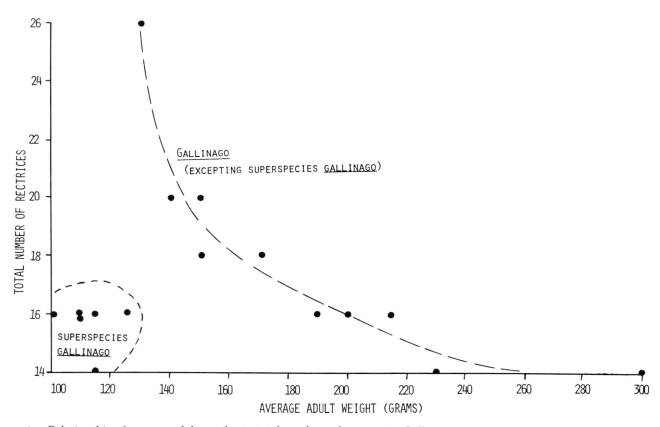

6.   Relationship of average adult weights to total numbers of rectrices in *Gallinago.*

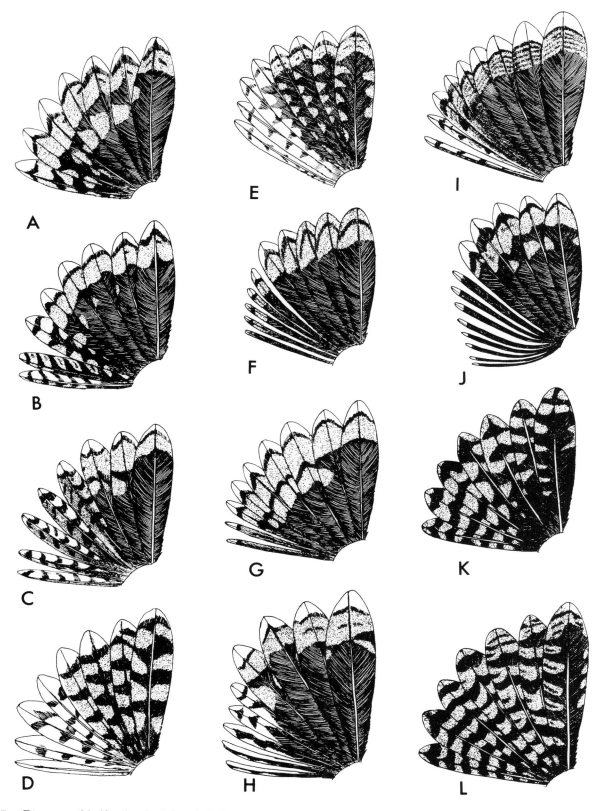

7. Diagram of half-tails of adults of *A*, Eurasian snipe; *B*, North American snipe; *C*, Paraguaian snipe; *D*, African snipe; *E*, great snipe; *F*, forest snipe; *G*, Himalayan snipe; *H*, solitary snipe; *I*, Japanese snipe; *J*, pintailed snipe; *K*, cordillerian snipe; *L*, giant snipe. In part after Seebohm (1888).

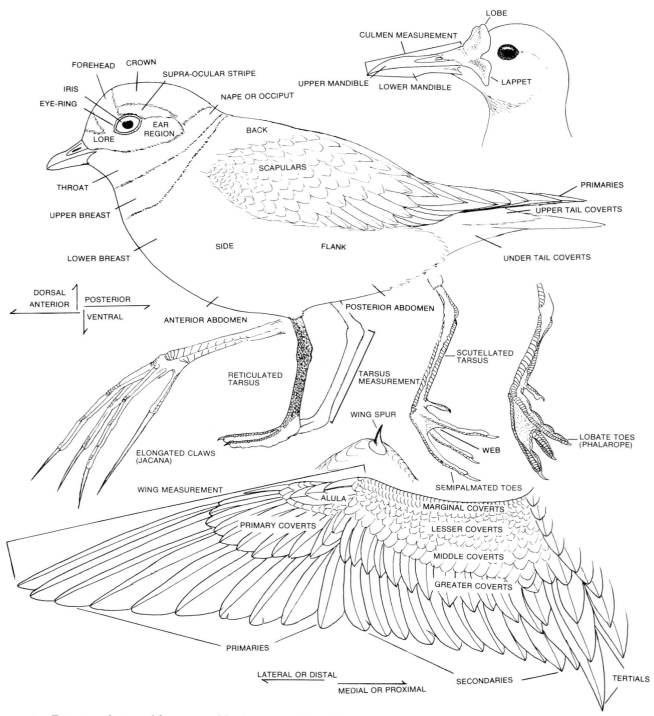

8. Diagram of external features and feather areas of shorebirds.

Key to Families, Subfamilies, and Tribes

A Posterior toes and claws greatly elongated, 10 rectrices (tail feathers) . . . . family Jacanidae (jacanas)

A′ Posterior toes and claws not elongated, 12–28 rectrices

  B Bill slightly to moderately decurved toward tip and fairly long (40–50 mm), with nostrils in deep grooves, all toes relatively long (middle toe at least 20 mm), 14–16 short rectrices, and with buffy "shoulder straps" extending from sides of breast to the posterior scapulars . . . . family Rostratulidae (painted snipes)

  B′ Bill length and curvature variable, anterior toes short to moderately long, hind toe often small or absent, number of rectrices variable, no "shoulder strap" pattern present.

   C Bill variable in length but not swollen distally or contracted in middle or near base and not compressed laterally, nasal openings narrow and in a long groove.

    D Tarsus with reticulated (networklike) scale pattern, hind toe small or absent

     E Bill strongly decurved, tarsus under 55 mm . . . . family Ibidorhynchidae (ibis-bill)

     E′ Bill straight to recurved, tarsus usually over 90 mm . . family Recurvirostridae (stilts and avocets)

    D′ Tarsus with scutellated scale pattern in front and usually also behind, hind toe well developed
    . . . . family Scolopacidae

     E Toes with lateral membrance, tarsus compressed . . . . tribe Phalaropodini (phalaropes)

     E′ Toes lacking a distinct membrane, tarsus not usually compressed laterally

      F Ear orifice anterior to the middle of the eye socket, which is completely surrounded by bone, bill flattened and pitted toward tip, with medial groove present, head strongly striped or barred
      . . . . subfamily Scolopacinae (snipes and woodcocks)

       G Crown barred or banded, tail feathers tipped with silvery white on undersides. . . . .
       . . . . tribe Scolopacini (woodcocks)

       G′ Crown longitudinally striped, tail feathers lacking silvery tips . . . . tribe Gallinagini (snipes)

      F′ Ear orifice posterior to the middle of the eye socket, which is bounded by bone; head sometimes striped but never transversely barred, bill gradually tapering toward tip and usually lacking medial groove

       G Nasal opening relatively broad posteriorly, extending into a narrow groove anteriorly that nearly reaches the tip of the bill. Mostly relatively small birds (bill + tarsus usually under 70 mm) with short legs and fairly short bills with depressed tips . . tribe Calidridini (sandpipers)

       G′ Nasal opening relatively narrow posteriorly, nasal groove usually not extending more than two-thirds the length of the bill, which is often long (bill + tarsus usually over 70 mm) and relatively pointed or elastic-tipped . . . . subfamily Tringinae (tattlers and allies)

        H Wings short (under 115 mm) and rounded (first primary no longer than fourth), bill short and slender, toes all cleft to base . . . . tribe Prosobonini (Polynesian sandpipers)

        H′ Wings relatively long and pointed (first primary much longer than fourth), bill variable in shape, toes often partly webbed at base

         I Bill relatively long and straight, narrowest in the middle and slightly flattened toward tip, with the flattened part somewhat pitted and having a medial groove. Lower back and rump whiter than mantle . . . . tribe Limnodromini (dowitchers)

         I′ Bill straight or curved, lacking pitting and medial groove near tip, lower back and rump not distinctly set off from mantle in most species

          J Bill essentially straight, never decurved or recurved, and usually under 50 mm long; if over 50 mm, then the wings with extensive white on flight feathers . . . .
          . . . . tribe Tringini (tattlers)

J′ Bill usually elongated (over 50 mm) and often decurved or slightly recurved; if straight, then either at least 70 mm long (at least a third as long as the wing), or fairly short (under 40 mm), with an elongated tail (over 75 mm, or nearly half as long as the wing) . . . . tribe Numenini (godwits and curlews)

C′ Bill either much longer than tarsus and strongly compressed laterally toward tip, or shorter than tarsus and not compressed, often swollen distally and contracted in the middle; nasal openings rounded and relatively broad toward anterior end

  D Bill longer than tarsus and compressed laterally; bill tip broad and rounded or blunt-tipped in lateral profile . . . . family Haematopodidae (oystercatchers)

  D′ Bill shorter than tarsus and not compressed laterally; often swollen toward tip

    E Hind toe absent or relatively small, tarsus reticulated or irregularly scutellated in front

      F Hind toe well developed, claws relatively blunt, maxilla gradually tapering toward tip. . . . .
        . . . . family Pluvianellidae (Magellanic plover)

      F′ Hind toe absent or minute, claws not blunt, maxilla swollen toward tip . . . .
        . . . . family Charadriidae (plovers and lapwings)

        G Wings and tail banded with black and white, often with wing spur and small hind toe, body plumage sometimes iridescent . . . . subfamily Vanellinae (lapwings)

          H Hind toe absent, wing no more than 150 mm, no crest or bare skin on face, upperparts gray, white, and black . . . . tribe Hoploxypterini (pied plover)

          H′ Small hind toe often present, crest, facial wattles, or lobes often present, upperparts varied but usually not black, gray, and white, wing at least 150 mm . . . tribe Vanellini (lapwings)

        G′ Wings and tail not banded with black and white, wing spurs, crests, lobes, or wattles never present, plumage never iridescent, hind toe usually absent, rarely minute . . . .
          . . . . subfamily Charadriinae, tribe Charadriini (typical plovers)

    E′ Hind toe well developed (about a third the length of inner toe), tarsus scutellated in front . . . .
      . . . . aberrant Scolopacidae

      F Maxilla ploverlike (swollen toward tip), legs with reticulated posterior tarsal surface, toes with lateral membranes . . . . tribe Aphrizini (surfbird)

      F′ Maxilla tapering toward an acute tip, tarsus with scutellated posterior surface, no lateral membranes on toes . . . . tribe Arenariini (turnstones)

Family Jacanidae (Jacanas)

African Jacana

Jacanas are very small to medium-sized tropical or subtropical shorebirds with a straight, compressed bill, a blunt or sharp wing spur, a long tarsus with large transverse plates in front and behind, and greatly elongated toes and claws. In at least some species the wing molt is simultaneous, as in some Gruiformes. The tail is usually short and is composed of ten rectrices. The downy young are strongly marked dorsally with black-edged whitish stripes. The sexes are alike as adults, but the female is larger than the male, and polyandry occurs in at least some species. The eggs are unusually shiny, with blackish streaks and scrawls. Three genera, including 7 species, are recognized here.

KEY TO GENERA OF JACANIDAE

A Wings knobbed but not spurred, no light wing stripe . . . . *Metopidius* (5 spp.)
A′ Wing spurs present, light wing stripe on flight feathers
  B Wing stripe white, no wattles . . . . *Hydrophasianus* (1 sp.)
  B′ Wing stripe yellow, wattles present . . . . *Jacana* (1 sp.)

Genus *Metopidius* Wagler 1832 (Typical Jacanas)

Typical jacanas consist of 5 Old World species of very small to medium-sized jacanas, having a relatively long and robust bill, often with a shield, comb, or lappet at its base. The wing is unspurred or has only a blunt spur at the wrist, and the flight feathers lack white or yellow markings.

KEYS TO SPECIES OF *Metopidius*

A Head lacking frontal shield or lobe, wing under 100 mm . . . . *capensis*
A′ Head with frontal shield, wing over 100 mm.
  B Frontal shield lacking free margins
    C Back of head white . . . . *albinucha*
    C′ Back of head black . . . . *africana*
  B′ Frontal shield with free margins
    C Shield large and crested, bill slender and shorter than head . . . . *gallinacea*
    C′ Shield small, bill stout and longer than head . . . . *indicus*

Lesser Jacana

Metopidius capensis (Smith) 1839
(*Microparra capensis* of Peters, 1934)

Other vernacular names. Pygmy jacana; smaller jacana; jacana nain (French); Zwergblatthühnchen (German).

Subspecies and range. No subspecies recognized. Resident in eastern Africa from Mali and the upper White Nile south to Cape Province. See map 1.

Measurements and weights. Wing: both sexes 81–98 mm. Culmen: both sexes 14–19 mm. Weights: 1 female 41.3 g (Britton and Dowsett 1969). Eggs ca. 24 × 18 mm, estimated weight 4.5 g (Schönwetter 1962).

DESCRIPTION

Adult males have the forehead and lores golden with a chestnut tinge; the cheeks and a broad band above

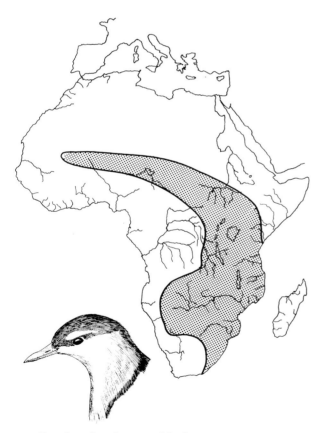

1. Breeding distribution of the lesser jacana.

the eye are white, separated by a chestnut line, and the crown and nape are bright chestnut. A broad black patch on the hindneck shades into deep purple on the mantle and is flanked by chestnut on each side. The sides of the neck are golden yellow, the back is brown with purplish gloss along the median line, the scapulars are brown, and the rump and upper tail coverts are chestnut, with purplish brown tips on some of the feathers. The wings are mostly brown, with the coverts having whitish margins. The primaries are blackish brown, and their coverts are dark brown with white tips. The outer secondaries are brown but are broadly tipped with white, and the inner secondaries are paler with narrower white tips. The under wing surface and axillaries are black. The throat, breast, and belly are white, with a yellow patch on each side of the breast, and the sides of the body are deep chestnut. The iris is brown, the bill is pale brownish olive, and the legs and feet are greenish olive. *Females* resemble males, but the chestnut on the crown is not so bright (Bannerman 1951). *Juveniles* have a blackish crown, and the nape is golden chestnut.

In the hand, the small size (less than 6 inches from bill to tail) serves to identify this species once it is recognized as a jacana (middle toe over 2 inches long). Further, its feathered forehead and lack of black on the crown separate it from the larger African jacana.

In the field (6 inches), the small size of this species is readily apparent, although young African jacanas (which are brown above and white on the underparts) might be confused with it. In flight, the white stripe on the trailing edge of the secondaries is conspicuous and a good field mark. Its call note is a sharp *knuk*.

NATURAL HISTORY

Habitats and foods. This species inhabits grassy swamps, quiet lagoons, lakes, and streams or similar aquatic habitats where there is an abundance of water lilies or other floating plants that provide support and shelter. It is said to require less emergent vegetation than does the larger African jacana and to colonize new bodies of water before the arrival of that species, but it is far less conspicuous and more likely to be overlooked. It is said to feed on insects and on water plants, primarily their seeds.

Social behavior. There seems to be no specific information on the social behavior of this elusive species,

which is unfortunate, considering its uniquely small size relative to the other jacanas. There is almost no sexual dimorphism in size or coloration, suggesting that the mating system is monogamous rather than nonmonogamous.

Reproductive biology. Like the other jacanas, the nest of this species is a small accumulation of water plants floating in the water, similar to but smaller than that of the African jacana. Records from Zambia suggest breeding from February through August, with the largest number of records (5 of 8) for February and March. Records from South Africa indicate breeding in March and April, but it has also been suggested that two broods may be raised in some areas. The clutch size seems to be from 2 to 4 eggs. The incubation period is reportedly 22 to 24 days, and incubation apparently is by both sexes. Both sexes are also said to tend the young (Campbell 1974).

Status and relationships. Although this species has a broad range, its distribution seems to be local, and because of its elusiveness its true status is impossible to judge. It is usually placed in the monotypic genus

Microparra, but the species' strong similarity to the young of *Metopidius africana* suggests that the two species are fairly closely related.

Suggested reading. Bannerman 1951.

African Jacana

Metopidius africanus (Gmelin) 1789
Actophilornis africana of Peters (1934)

Other vernacular names. Lily trotter; jacana à poitrine donee (French); Afrikanisches Blatthühnchen (German).

Subspecies and range. No subspecies recognized. Resident over most of Africa from the Senegal River and the Sudan south to the Cape. See map 2.

Measurements and weights. Wing: both sexes 133–82 mm. Culmen: both sexes 44–60 mm. Weights: males 115–56 g, average of 8 was 136.6 g; females 251–74 g, average of 5 was 261 g (Benson et al.

1976); Vernon (1973) indicates a similar range for each sex. Eggs ca. 32 x 23 mm, estimated weight 8.6 g (Schönwetter 1962).

DESCRIPTION

Adults of both sexes have a black stripe extending from the lores through the eyes, joining a black crown and hindneck. The sides of the face and neck are white, as is the throat, gradually becoming golden straw-colored on the sides of the neck and lower foreneck, the latter barred with black. The rest of the undersurface of the body is deep maroon chestnut, and under wing coverts are like the breast, becoming blackish on the under primary coverts. The general upperpart coloration is cinnamon rufous to chestnut, becoming maroon chestnut on the lower back. The primary coverts are blackish with rufous bases, and the primaries are black. The rufous color

increases toward the innermost secondaries, which are chestnut, and the tail is maroon chestnut (Sharpe 1896). The iris is brownish, the bill is bluish gray with a light blue shield, and the legs and toes are lead-colored. *Immatures* (until about 20 months old) have white underparts, and the breast is tinged with yellowish like the sides of the lower neck. The back is rich liver brown with a few olive-colored juvenile feathers, while the crown, nape, and hindneck are blackish with some more brownish juvenile feathers present.

In the hand, the combination of the long toes and claws, a bright blue frontal shield, and a bicolored black and white head serves to identify this species.

In the field (9–12 inches), the "lily-trotting" behavior and rather large size (at least 9 inches long) are unique in Africa. When seen close up, the bluish frontal shield should be evident. Immatures are brownish above and white below but never have the chestnut crown found in the lesser jacana. The birds are very vocal, and their calls include a series of short staccato notes, *ka-ka-ka,* uttered without change of tone.

NATURAL HISTORY

Habitat and foods. This species occurs almost anywhere there is well-developed surface vegetation on standing or slow-moving water, especially where water lilies occur. It occasionally feeds on open sandbanks of larger rivers and sometimes perches on the backs of hippos, catching insects and small invertebrates disturbed by these animals. In general its food is believed to consist of insects, aquatic larvae, snails, and plant seeds. Like the other jacanas, it usually forages by walking lightly over the leaves of floating plants, searching industriously for edible materials.

Social behavior. Outside the breeding season these birds are sometimes fairly gregarious, especially when drying waters force them into restricted locations. "Thousands" of individuals were once reported along the Niger River, presumably during the nonbreeding season; however, concentrations of breeding birds seem to be rather low. One estimate was of about 20 pairs on a lake about a mile long and a quarter of a mile broad, or roughly a pair per 8 acres. The pairs nested at well-spaced intervals, with no indication of colonial nesting (Bannerman 1951).

Reproductive biology. The age of initial breeding is still uncertain, but plumage changes suggest that it is

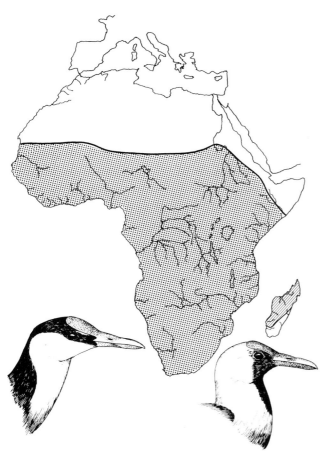

2. Breeding distributions of the African (*shaded*) and Madagascan (*hatched*) jacanas.

no earlier than 2 years. Nesting in southern Africa occurs over a very broad span, mainly from November to March (but also in June and July) in South Africa and from November to August in Zambia, with most records from February to June. The total span of the breeding activity in tropical Africa is from April to September, with most eggs found from June to August. In Kenya the birds normally breed in May, toward the end of the rainy season. The nest is built on floating vegetation, often the leaf of a water lily, but when no floating materials are available the birds may build a substantial foundation. The reported range of clutch sizes is from 2 to 5 eggs, but normally the clutch is 4 eggs. According to Cunningham-van Someran and Robinson (1962), only the female incubates and cares for the young, but the male may visit the nest regularly; however, Vernon (1973) reported a case of two males mated with a single female, the males apparently doing all the incubation. Likewise, Wilson (1974) believed that only one sex, probably the male, incubated. In spite of some early reports, the eggs are apparently not normally held between the wings but rather are incubated below the breast; however, the northern and pheasant-tailed jacanas have been observed incubating eggs by holding them against the wing, and so this behavior may also occur in the African species in unusually wet circumstances. The probable incubation period is 21 to 24 days. The young are not highly precocial, but by 3 days of age they can run agilely behind the parent. Hopcraft (1968) reports that parent jacanas carry as many as 3 chicks under their wings when escaping from danger. Chicks as old as 5 days have been seen being carried as far as 60 feet in this species, and similar behavior has been reported in the comb-crested jacana (Hopcraft 1968), while lifting but not actual carrying of the young has been seen in the New World form. The young of jacanas are similar to those of the painted snipe in that, unlike the other shorebirds, they are striped with whitish and black on the head and back. It has been reported that in this species as well as in the Madagascan jacana the birds undergo a flightless period during their postbreeding molt, much like ducks and rails (Gooders 1969).

Status and relationships. This species is apparently quite common over much of its extensive range. Its closest relative is almost certainly the Madagascan jacana, and it is probably somewhat less closely related to the lesser jacana.

Suggested reading. Miller 1951; Maclean 1972c.

Madagascan Jacana

Metopidius albinucha (I. Geoffroy Saint Hilaire) 1832
(*Actophilornis albinucha* of Peters, 1934)

Other vernacular names. Malagasy jacana; jacana à nuque blanche (French); Madagaskar-Blatthühnchen (German).

Subspecies and range. No subspecies recognized. Resident in Madagascar. See map 2.

Measurements and weights. Wing: both sexes 148–78 mm. Culmen (to end of shield): 46–54 mm. Weights: 1 female 239 g (Benson et al. 1976). Eggs ca. 37 x 25 mm (Rand 1936), estimated weight 12.7 g.

DESCRIPTION

Adults of both sexes have the hindcrown and occipital area black, while the nape, hindneck, and sides of the neck are pure white, becoming straw-colored on the neck. The side of the face, ear coverts, throat, and foreneck are black, and the remainder of the underparts are deep cinnamon rufous to chestnut, with a large patch of white on the lower flanks. The under wing coverts and axillaries are like the breast, the latter having ashy brown bases. The underparts are generally cinnamon rufous, including the upper wing coverts. The primaries are black, with rufous coverts, the secondaries are mostly like the back, and the rump is white. The upper tail coverts and tail are chestnut, with blackish feather tips. The iris is brown, and the bill, frontal shield, and feet are pearly gray. *Immatures* are paler in color, with black and white on the top of the head and nape; the neck and most of the face are white, and the back and tail coverts are rufous. The frontal shield is smaller than in adults (Sharpe 1896).

In the hand, the elongated toes and claws identify this as a jacana, and the white nape and hindneck readily distinguish it from its nearest African relative.

In the field (10–12 inches), this jacana is unlikely to be mistaken for any other species, since no other "lily-trotter" occurs in Madagascar.

NATURAL HISTORY

Habitat and foods. The few observations on this species suggest that it occurs from sea level to 750 meters, in small and large marshes, in marshy bays of lakes, and along river margins. Like its African counterpart, it is a noisy bird associated with open, floating vegetation (Rand 1936). Its foods and foraging behavior are probably similar but have not been studied.

Social behavior. Nothing specifically has been written on this.

Reproductive biology. Rand (1936) suggested that the breeding season probably includes at least the months of February, March, June, and December and noted that downy young were commonly seen during March and as late as April 4. A nest with a full clutch of 4 eggs was found in late June, suggesting that the breeding season is prolonged.

Status and relationships. No information is available on this species' status, but it is unlikely to be in any danger. The species is clearly a very close relative of the African jacana and may reasonably be regarded as having been derived from it.

Suggested reading. Milon et al. 1973.

Comb-crested Jacana

Metopidius gallinacea (Temminck) 1828
(*Irediparra gallinacea* of Peters, 1934)

Other vernacular names. Lotusbird, lily-trotter (Australia); jacana à crête (French); Australisches Blatthühnchen (German).

Subspecies and ranges. See map 3.
  I. g. gallinacea: East Indian comb-crested jacana. Resident in southeastern Borneo, Celebes, Mindanao, and adjoining islands.
  I. g. novaeguinae: New Guinea comb-crested jacana. Resident in New Guinea, Misool, and the Aru Islands.
  I. g. novaehollandiae: Australian comb-crested jacana. Resident in northern and eastern Australia.

Measurements and weights. Wing: both sexes 121–40 mm. Culmen (including comb): both sexes 37–44 mm. Weights: 6 males 68–84 g, average 75.1 g; 7 females 120–49 g, average 130 g (U.S.N.M.

3. Breeding distributions of the Australian (A), East Indian (E), and New Guinea (N) comb-crested jacanas.

specimens). Eggs ca. 30 x 21 mm, estimated weight 7.1 g (Schönwetter 1962).

DESCRIPTION

Adults of both sexes have a black stripe extending from the frontal shield to the upper back and downward to join a black breast, flanks, and underwing. The mantle and scapulars are drab brown, and the tail and flight feathers are black, while the rest of the upper wing surface is brown. There is a black streak from the eye to the base of the lower mandible; otherwise the face, throat, and foreneck are white, grading to bright orange ventrally and laterally. The belly and under tail coverts are white. The comb on the forehead and the base of the upper mandible vary from crimson to yellow; the base of the lower mandible is yellow. The rest of the bill is black, the iris is yellow, and the legs and feet are bluish black. *Immatures* have the top of the head rufous, the back feathers edged with rufous, and the underparts white.

In the hand, the long legs and toes readily identify this as a jacana, and the crimson, comblike bill is unique.

In the field (8–10 inches), the "lily-trotting" behavior identifies this as a jacana, and only on Mindanao might another jacana (the pheasant-tailed) potentially occur in the same region. There the bright comb and the absence of long tail feathers serve to

identify birds in breeding condition. In flight, the absence of white on the wings should separate these species. Calls include soft squeaks and loud and shrill repeated rattling or piping notes.

NATURAL HISTORY

Habitat and foods. Like the other jacanas, this species is associated with aquatic habitats rich in aquatic weeds—thus the common Australian names "lotus-bird" and "lily-trotter." They are generally thought to eat a variety of seeds, green vegetation, and insects, but one specimen examined from Australia had only small seeds in its gizzard.

Social behavior. Although some flocking might occur outside the breeding season, the birds are known to be very aggressive during nesting, and they vigorously defend territories. The age of initial breeding and pair-forming behavior has not been studied.

Reproductive biology. In eastern Australia this bird usually nests between September and January; it nests later, between January and May, in the northern part of Australia. This is apparently related to the timing of the monsoon rains in the northern areas. A late April nest with eggs has been reported in Borneo; probably in all tropical areas nesting is associated with the rainy season. The nest is built among water plants, often almost half-submerged, or sometimes is on a built-up platform near shore. The clutch is 3 or 4 eggs, with 4 typical, and like the eggs of other jacanas they are heavily marked with black and brown wavy lines. Their general appearance is rather shiny and wet-looking, which makes them extremely inconspicuous. It is reported that both parents tend the young; polyandrous mating has not yet been established for this species. The incubation period is not known but is probably close to the 21 to 24 days reported for the African species. Adults have been carrying both eggs and chicks by holding them under the wings and against the body. They sometimes also build extra nests for roosting (Frith 1976).

Status and relationships. Apparently this species is relatively common over much of its range, wherever local habitats are suitable. Although the species is unique in its vertically expanded crown, in most respects it is similar to the African jacana and is probably a close relative (Strauch 1976).

Suggested reading. Potter 1934; Hindwood 1940.

Bronze-winged Jacana

Metopidius indicus (Latham) 1790

Other vernacular names. None in general English use; jacana bronzé (French); Kupferspiegelblatthühnchen (German).

Subspecies and range. No subspecies recognized. Resident in India, Burma, and southern Annam, south to Cambodia and the Malay Peninsula; also on Java and Sumatra. See map 4.

Measurements and weights. Wing: males 153–77 mm; females 170–84 mm. Culmen (from base): males 37–41 mm; females 40–45 mm. Weights: 10 birds of both sexes 94–210 g, average 154.6 g (J. C. Daniel, pers. comm.). Eggs ca. 36 x 25 mm, estimated weight 11.9 g (Schönwetter 1962).

DESCRIPTION

Adults of both sexes have the head, neck, and lower parts of the abdomen black with a dark greenish gloss, the lower hindneck glossed with purplish, the back and wings olive brown, and the lower abdomen and thighs dull blackish brown. There is a white line above the eye and a white spot below it. The flight feathers are black with a dark green gloss, the lower back to the tail and the tail coverts are chestnut, and the tail itself is darker chestnut. The iris is brown,

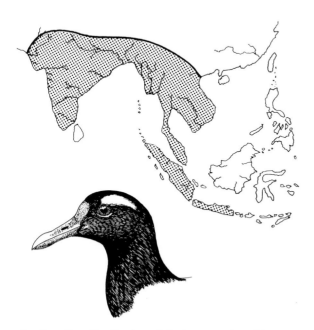

4. Breeding distribution of the bronze-winged jacana.

and the bill is greenish yellow, becoming reddish at the base and distally yellow brown with a black terminal band. The legs and feet are dull green at the tip; the frontal lappet is livid red. *Immatures* are mostly rufous white below, with a rufous brown crown and nape, and dull superciliary stripe. The back is like that of adults but paler, the upper tail coverts are barred brown and white, and the tail is rufous.

In the hand, the long toes and claws identify this as a jacana, and the white eye stripe and greenish brown wings with tubercular knobs are distinctive.

In the field (11–12 inches), the long toes and legs and "lily-trotting" behavior identify the bird as a jacana, and only the pheasant-tailed jacana is found in the same region. In flight, this species lacks white on the wings, and its feet trail behind the short tail. Calls include harsh cries or grunts and shrill, wheezy, or piping *seek-seek-seek* notes.

NATURAL HISTORY

Habitat and foods. Like other jacanas, this species is found on vegetation-covered waterways in the plains and plateau country of tropical Asia, generally in the same habitats as the pheasant-tailed jacana. This species, however, is more often associated with the larger and more vegetation-choked ponds and marshes than is the pheasant-tailed. The species is believed to be primarily vegetarian but also to eat insects and mollusks.

Social behavior. The only detailed account of this species' social behavior and evidence for polyandry comes from Mathew (1964). He observed a female and 2 males over a period of nearly 3 months. When his observations were made in late July, the female was associated with a male that was already incubating a full clutch of 4 eggs. Some 10 days before this clutch hatched, the female was seen with a second male, collecting nesting materials. He noted that each male defended an area of more than 2,000 square meters, from which all birds except the female were forcibly excluded. The female, however, helped defend the territories of both males. She would actively solicit copulation by standing before the male, becoming rigid, and exposing her vent. Both sexes helped build the nest, and the first egg of the second nest was laid on August 13, when the chicks of the first brood were hatching. Three eggs were laid, and incubation was begun on August 19, but this clutch was lost by August 21. A new nest was started about

6 feet from the first site, and 2 eggs were laid by late August. This nest was also lost, and a third one was begun on September 2. Incubation of this clutch proceeded for 2 weeks, but it too was eventually lost. The male continued to defend his territory until the end of September, but apparently no more nestings were attempted. The social system of this species is thus much like that of the pheasant-tailed jacanas, but it differs in that fewer clutches are apparently laid, and there is no indication that the female participates in any way in the care of the young (Mathew 1964).

Reproductive biology. Breeding in Burma and India occurs during the rainy season, starting in India in June, or shortly after the southwestern monsoons have begun, and lasting until September. The nest is a flimsy structure of grasses or weed stems, placed on a bed of matted vegetation or directly on the floating leaves of lotus or singara (*Trapa*). The nest evidently is constructed primarily by the male, with the female occasionally bringing in a few grasses, but like the other jacanas it appears that polyandry is character-istic of this species, and there is only a temporary liaison between the sexes. The clutch size is normally 4 eggs, patterned like those of the other jacanas, brownish with blackish sprawls and curls and with a shiny or wet appearance. The incubation is performed by the male alone (Mathew 1964), and its duration is still undetermined but is probably in the range of 21 to 24 days. The young birds are tended entirely by the male parent, and at least one instance is known of an adult carrying a newly hatched chick under its wing and scurrying off for some distance over the water before finally dropping it (Ali and Ripley 1969).

Status and relationships. This species is quite common over much of its range and is probably rarely, if ever, hunted or otherwise persecuted. It has usually been placed in the monotypic genus *Metopidius*. Like *Actophilornis*, it has wing knobs, and Strauch (1976) has recommended that *Metopidius* be merged with that genus, a view I share.

Suggested reading. Mathew 1964.

Pheasant-tailed Jacana
Genus *Hydrophasianus* Wagler 1832

The pheasant-tailed jacana is a medium-sized species of Asian jacana, having a relatively slender bill with no lappet, shield, or comb at its base. There is a sharp spur at the wrist joint, and the flight feathers have white markings. Additionally, the first and fourth primaries are highly modified, the first ending in a spatulate web and the fourth prolonged into an attenuated point. A distinct nuptial plumage is present, characterized by greatly elongated tail feathers.

Pheasant-tailed Jacana

Hydrophasianus chirurgus (Scopoli) 1786

Other vernacular names. None in general English use; jacana à longue queue (French); Wasserfasan (German).

Subspecies and range. No subspecies recognized. Resident in Southeast Asia from western India eastward to southern China, Taiwan, the Malay Peninsula, Sri Lanka, Java, Sumatra, and the Philippine Islands. See map 5.

Measurements and weights. Wing: both sexes 182–242 mm. Culmen: both sexes 25–29 mm. Weight: 5 males 113–35 g, average 126 g; 3 females 205–60 g, average 231 g (U.S.N.M. specimens). Eggs ca. 37 x 28 mm (Ali and Ripley 1969), estimated weight 14.1 g (Schönwetter 1962).

DESCRIPTION

Adults of both sexes in breeding plumage have white on the front and sides of the head, on the throat, and on the upper breast; this area is separated from a golden yellow nape and hindneck by a blackish line. The blackish line is continuous with blackish over most of the rest of the body and the elongated tail, with only the upper wing surface and the flight feathers a strongly contrasting white. The primaries are distinctive, the shaft of the first primary being ex-

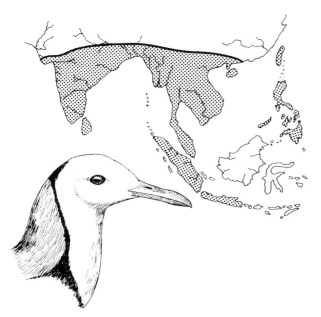

5. Breeding distribution of the pheasant-tailed jacana.

eye, down the side of the neck and across the breast, forming a gorget, and the golden neck patch becomes paler, being divided by a brown line down the back of the neck. The wings develop an olive brown patch on the median and greater secondary coverts, and the tail is much shorter, with white on the outer feathers and olive brown on the central ones. *Females* are not separable from males except by their slightly larger measurements. *Immatures* have no yellow on the side of the neck, the breastband or gorget is broken up with white, and the upperparts are more buffy.

In the hand, the elongated toes and claws identify this as a jacana, and no other jacana has white on the flight feathers.

In the field (12–22 inches), the long tail (about 10 inches) is unique among jacanas or other marsh-dwelling birds but is absent outside the breeding season. At that time the white on the wings is the best field mark. In flight the black-edged white wings contrast strongly with the dark body. Calls are numerous and include a nasal *tew, tew,* or *tewn, tewn* . . . , a catlike and high-pitched *miuu,* and a musical *hoo-hoo-hoo.* In regions where it occurs with the bronze-winged jacana, it is more likely to be seen in weed-free areas and is less shy than that species.

tended beyond the vane for about an inch and the shafts of the adjoining ones variably less so. The iris is pale yellow, and the bill is bluish in summer and dull greenish in winter. In *nonbreeding plumage* the underparts become white, the upperparts olive brown; a dark line passes from the beak through the

NATURAL HISTORY

Habitats and foods. This jacana is found in a variety of habitats ranging from small village ponds to larger marshes or lagoons, sometimes occurring in close proximity to humans and generally foraging in rather open environments. It is mostly found at lower elevations during winter, and during summer it moves as high as about 1,500 meters, with local movements depending on water conditions. It is especially prevalent where lotus, singara (*Trapa*), or water hyacinth (*Eichhornia*) plants are present. Foods are believed to be mostly vegetable, such as seeds and roots, but they also include mollusks, insects, and their aquatic larvae (Ali and Ripley 1969).

Social behavior. Outside the breeding season these birds are fairly gregarious and may be seen in flocks of from 50 to 100 individuals. In the breeding season, however, the birds are highly territorial, with each male establishing a territory, which the female also defends temporarily while she maintains her pair bond with that male. Each male establishes a *"Standplatz,"* a fixed mating place to which the male calls the female and where the displays that lead to copulation occur. During this period the same place is converted into the nest by the male, and to some extent the female, tearing loose and throwing up the nearby water plants in an unorganized manner. During the breeding period both sexes use two different calls, with that of the male somewhat higher in pitch. Additionally, the male has a unique note that serves as the call for attracting the female (Hoffmann 1950).

Reproductive biology. In India, the mating season is timed to occur during the southwestern monsoon, primarily from June to September, while in Kashmir it occurs from mid-May to July, and in Sri Lanka (Ceylon) it is mainly from March to July but also in January. In the vicinity of Peking, China, breeding begins in June and ends in August. In that area a single male can raise one brood, start incubating a second clutch, and, if it is destroyed, even begin incubating a third time. Females can produce 3 to 4 clutches a month and as many as 7 to 10 clutches during the entire breeding season. There seems to be, however, a high rate of egg destruction or mortality. Eggs are laid at 24-hour intervals, and incubation by the male begins with the deposit of the first egg. The eggs are held carefully between the breast and the inner surface of the wing, so that contact with water or wet vegetation is apparently avoided. Males sometimes push or roll their clutches as far as 15 meters from the original nest site if they are disturbed or in danger of flooding (Hoffmann 1950). This is done by pressing the egg between the throat and the breast or rolling it over the vegetation. Sometimes the eggs are even floated over areas of open water. At times the egg is also moved by holding the pointed end in the bill and dragging it backward (Ali and Ripley 1969). The incubation period is about 26 days. The chicks are tended by the male parent alone, and when danger threatens they may freeze while hidden under a leaf or even while completely submerged except for the bill. "Broken-wing" or other distraction displays may be performed by the male at this time. On rare occasions the female also has been observed protecting the young.

Status and relationships. There is no good information on this species' status, but it is apparently quite common in many areas. It has traditionally been placed on the monotypic genus *Hydrophasianus*, the only recent exception to this procedure being Strauch's (1976) recommendation that it be included in the genus *Jacana*. This is based on the fact that both these genera have sharp wing spurs, and both also have white or yellow wing markings. However, I am doubtful that the wing spurs represent an important criterion for judging relationships, and I am unsure of the significance of the wing markings. I believe the two genera should stand, but probably they are more closely related to each other than either is to *Metopidius*.

Suggested reading. Hoffmann 1950; Ali and Ripley 1969.

Genus *Jacana* Brisson 1760 (New World Jacanas)

New World jacanas include a single species (2 are frequently recognized) of Central and South American jacana, having a long, straight bill with a fleshy comb and lappet at its base. The wing has a sharp spur at the wrist joint, and the flight feathers have bright yellow markings.

American Jacana

Jacana spinosa (L.) 1758

Other vernacular names. Wattled jacana (forms other than *spinosa*); northern jacana (*spinosa* only); jacana d'Amérique (French); Jassana (German); gallita de laguna, gallita de agua frentirroja (Spanish).

Subspecies and ranges. See map 6.

J. s. spinosa: Northern jacana. Resident from southern Texas and northern Mexico south to

6. Breeding distributions of the Colombian (C), Ecuadorian (E), northern (N), Panamanian (Pa), Peruvian (Pu), southern (S) and Venezuelan (V) jacanas.

western Panama; also on Cuba, Isle of Pines, Jamaica, and Hispaniola. (Includes *gymnostoma* and *violacea*; this form is often considered a separate species.)

J. s. hypomelaena: Panamanian jacana. Resident in Panama and northern Colombia, possibly coexisting with *spinosa* in western Venezuela.

J. s. melanopygia: Colombian jacana. Resident in the Cauca Valley of Colombia, in Arauca, and in northwestern Venezuela.

J. s. intermedia: Venezuelan jacana. Resident in northern Venezuela, eastern Colombia, and eastern Ecuador; also on Trinidad.

J. s. scapularis: Ecuadorian jacana. Resident in western Ecuador and northwestern Peru.

J. s. peruviana: Peruvian jacana. Resident in northeastern Peru and adjacent Brazil.

J. s. jacana: Southern jacana. Resident in the Guianas and most of Brazil, southward to eastern Bolivia, Paraguay, Uruguay, and northern Argentina.

Measurements and weights. Wing (flattened): males 107–40 mm; females 111–39 mm. Culmen (from base of shield): males 27–37 mm; females 28–37 mm (Blake 1977). Weights (of *jacana*): males 89–118 g, females 140–51 g (Haverschmidt 1968); Osborne and Bourne (1977) reported that 16 adult males averaged 108.3 g and 15 females averaged 142.8 g (they also indicated that males and females have respective flattened wing measurements averaging 142.6 mm and 179.6 mm, or appreciably greater than the lengths reported by Blake). Breeding males and females of *spinosa* average 91 and 161 g, respectively (Jenni and Collier 1972). Eggs ca. 30 x 23 mm, estimated weight 8.3 g (Schönwetter 1962).

DESCRIPTION

Adults of both sexes have a frontal shield that is yellow and three-lobed (*spinosa*) or red and two-lobed (all other forms); in *spinosa* the rictal lappet is rudimentary or absent, but it is prominent in the

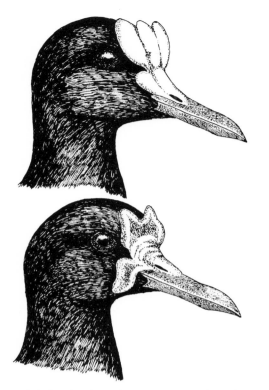

9. Adult heads of northern (*above*) and southern (*below*) jacanas. After Blake 1977.

others. The anterior parts and breast are black with a greenish gloss, while the body, wing coverts, inner flight feathers, and tail range from black to purplish maroon or chestnut. The belly is black or slightly chestnut. The primaries and secondaries are greenish yellow with dark tips, and the wing spur is bright yellow. The iris is brown, the bill is yellowish, and the legs and feet are gray to grayish green. *Immatures* have a small frontal shield or none, are mostly grayish brown to brownish above, and are white on the underparts and the sides of the head.

In the hand, the long toes and claws identify the bird as a jacana, and the yellowish green markings on the flight feathers are unique to this species.

In the field (8–9 inches), the "lily-trotting" behavior marks this as a jacana, the only one in the western hemisphere. On the Pacific slope of western Panama it is reported that both *spinosa* and *jacana* may occur together, which indicates that perhaps two species should be recognized. In this area the color and shape of the frontal lobe should be noted for critical determination. In flight, the yellowish areas of the flight feathers are extremely conspicuous. A variety of usually high-pitched squeaking or chattering notes

have been mentioned, said to be clacking rather than reedy as in gallinules.

NATURAL HISTORY

Habitats and foods. Throughout their range, the various forms of New World jacanas are associated with freshwater marshes, swamps, and lagoons, or with slow-moving rivers that are rich in floating aquatic plants. The birds also forage in shallow pools in pastureland and sometimes are seen around cattle. The best information on food and foraging is that of Osborne and Bourne (1977), who found that 20 percent of the food volume of *jacana* consisted of seeds, mostly of *Eleocharis*, while 80 percent of the volume was invertebrate materials, including weevils, grasshoppers, flies, and crickets. The birds forage in two ways, by pecking at food on the surface of vegetation and by pecking at food in the water.

Social behavior. It is known that both major types of New World jacanas (*spinosa* and *jacana*) are, at least at times, polyandrous; this was reported for the former by Jenni and Collier (1972) and for the latter by Osborne and Bourne (1977). The mating system is variable, however, and apparently is related to territory size and habitat diversity. Osborne and Bourne noted that, in *jacana*, birds breeding on lawns were "serially" monogamous, with the territories averaging about 1.1 hectares (2.6 acres), while those females breeding in the rice fields were polyandrous. Jenni and Collier (1972) report similar-sized territories for individuals of *jacana* breeding in Panama swamps but much smaller territories for polyandrous *spinosa* breeding in ponds in Mexico and Costa Rica. In Costa Rica the females defended areas of 0.88 acre and the males territories of only 0.37 acre. Aggressive territorial behavior is manifested mostly by aerial chases. Territorial defense is often indicated by "wing posturing," in which one or both sexes raise their wings, exposing the yellow patterning on the undersides (Osborne and Bourne 1977). This display occurs in both *spinosa* and *jacana*. In *spinosa*, adults have about six call types, of which one is apparently used exclusively between paired birds and males and their young (Jenni et al. 1974). At least in *spinosa*, copulation is often preceded by one of the birds' calling to its mate if it is off the territory or simply assuming the precopulatory posture if the mate is nearby. This soliciting posture is the same in both sexes, and when initiated by the male the female assumes the same posture at his side, after which the male mounts. After copulation is completed the female often throws bits of vegetation to one side or another. Although mating occurs at a variety of sites before nest construction begins, it later occurs on the nest itself. This female behavior thus may move materials either closer to or farther from the nest site (Jenni and Collier 1972). In *spinosa* the male does all the nest-building and perhaps selects the nest site, but in *jacana* it has been suggested that the female may select the site, that both sexes build the initial platform, and that the male plays the major role in building and adding to the nest itself (Osborne and Bourne 1977). It has been suggested that the polyandry of jacanas is related to patchy food distribution in marshes, with the larger females being better able to take advantage of spatial and temporal variations in food abundance. On the other hand, the serial monogamy found in *jacana* may be characteristic of species breeding in simple habitats, where strong mate fidelity exists, where territories may have abundant resources, and where heavy predation pressures exist (Osborne and Bourne 1977).

Reproductive biology. In studies of *jacana* in Guyana, it was found that 4-egg clutches are typical, and each egg is laid about 40 minutes later on successive days. These eggs are clustered into a tight group by the male, who then devotes more and more time each day to incubating them. The recycling time between successive nests by a pair ranged from 2 to 15 days, averaging 8 days (Osborne and Bourne 1977). In a population of polyandrous *spinosa* studied in Costa Rica, one study area had from 2 to 5 females present over a 10-year period and from 6 to 9 males, or an average of 3.0 females per 7.3 males. Female territories are large and superimposed on from one to four male territories. Males build their nests at one to several sites within their territories, and females often, but not always, select one of these sites in which to lay their 4-egg clutches. Males typically incubate the clutch by holding 2 eggs between each of their wings and their body. Females sometimes shade the clutch but apparently never incubate it (Jenni and

Betts 1978). The estimated incubation period is 22 to 24 days; the eggs may all hatch within an hour or two, or there may be as much as a day separating the hatching of the first and last eggs. Apparently the adults do not directly feed their young but do spend much time and effort in their brooding and care. Sometimes they crouch over the young chicks, holding them against the sides of their body, and occasionally they lift them above the substrate, in a manner similar to that described for the African jacana. Some brooding of the chicks may occur until they are about 10 weeks old. Although females do not incubate, they do sometimes help brood the young, especially during rainy periods, and they also perform antipredator behaviors at the same intensity as do males (Jenni and Betts 1978).

Status and relationships. Jacanas are relatively common over much of their range, though in Texas non-breeders are much more prevalent than breeders (Fleetwood 1973). The systematics of the jacanas is a most complex problem, and classifying the two major types either as species or as subspecies does not seem very satisfactory. Hybrid types have been reported from western Panama and from Costa Rica (Betts 1973); the differences between these two types in the area of overlap seem to be almost entirely associated with soft part development and coloration. Wetmore (1965) has argued that differences in the color of the frontal plate (bright yellow versus dark red) may be enough to maintain species distinctiveness, and additionally *spinosa* is more brown-backed than are most individuals of *jacana* from this area. Until the birds of this critical region of contact have been more thoroughly studied, I prefer to believe they should be regarded as subspecies. The relationships of the genus *Jacana* to the other jacanas are by no means clear, but I have accepted Strauch's (1976) opinion that *Hydrophasianus* is probably the nearest living relative.

Suggested reading. Jenni and Collier 1972; Osborne and Bourne 1977.

Family Rostratulidae (Painted Snipes)

Painted Snipe (threat display)

Painted snipes are small tropical to subtropical shorebirds whose bill is long and slender, with a tip that is hard, slightly swollen, and bent downward. Both mandibles are strongly grooved, and in at least one species the trachea is convoluted and the esophagus has a croplike enlargement. The tail is short and of fourteen to sixteen rectrices, and the wings are short and rounded, with extensive yellowish spotting. The female is larger and more brightly patterned than the male, but there is no seasonal plumage change. The downy young are strongly marked dorsally with black-edged tan stripes that extend from the forehead to the tail. One genus and 2 species are recognized here.

Genus *Rostratula* Vieillot 1816

This genus of 2 species has the characteristics of the family Rostratulidae.

Painted Snipe

Rostratula benghalensis (L.) 1758

Other vernacular names. None in general English use; rhynchée peinte (French); Buntschnepfe (German).

Subspecies and ranges. See map 7.
  R. b. benghalensis: Old World painted snipe. Resident in Africa, Madagascar, southern Asia from Asia Minor to eastern China and Japan, and south through India, Sri Lanka, the Malay Peninsula, and Cambodia; also the Philippines, Sumatra, and Java. Apparently a rare transient in Borneo.
  R. b. australis: Australian painted snipe. Resident in Australia and Tasmania.

Measurements and weights (of *benghalensis*). Wing: males 115–36 mm; females 130–46 mm. Culmen

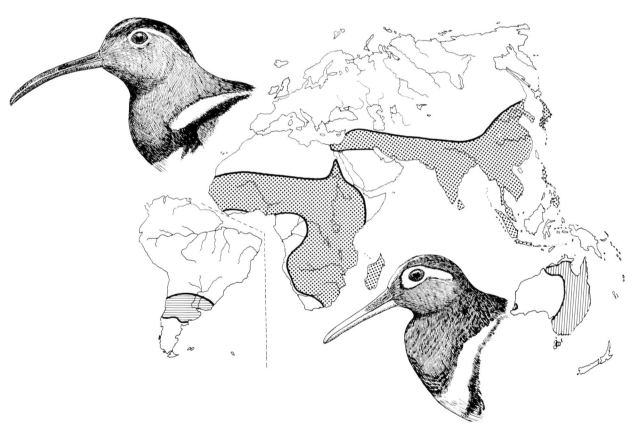

7. Breeding distributions of the South American (*horizontal hatching*), Australian (*vertical hatching*), and Old World (*shading*) painted snipes.

(from feathering): males 41–47 mm; females 45–50 mm. Weights: 25 males 90–164 g, average 117.6 g; 9 females 90–141 g, average 130.0 gm (various sources). Eggs ca. 35 x 26 mm, estimated weight 12.2–12.5 g (Schönwetter 1962).

DESCRIPTION

Adult males have a blackish brown crown finely barred with white and with a broad buffy central streak and a buffy eye ring extending to the side of the nape. The hindneck and upper breast are brownish gray, edged with buff "shoulder-strap" marks that extend downward to join the white breast. The scapulars are buffy orange; the lower back, wings, and tail are gray with white, buffy white, and cinnamon spotting; and the bases of the flight feathers are black. The lower face, throat, foreneck, and sides of the breast are brownish gray, speckled with white and edged with black, and the lower breast and belly are white. The iris is reddish brown, the legs and feet are greenish yellow, and the bill is grayish green basally, then brown, shading to an orange brown tip. *Breeding females* differ from males in having the face, neck, and upper breast reddish to blackish brown, the throat chestnut, and a patch of cinnamon rufous on the hindneck; the scapulars are dark greenish with fine black barring, and white feathers are concealed under the scapulars (Macdonald 1973). Outside the breeding season the sexes are more similar, but the female has a convoluted trachea, while that of the male is short and straight. Prater et al. (1977) reported that the median coverts of adult females are barred with black and glossy green, while those of adult males are barred with black and green buff to golden buff. *Juveniles* of both sexes are very similar to the adult male but have an entirely white throat and foreneck washed with brown, with some dusky streaks. Prater et al. (1977) reported that the only reliable character for separating adult males from young birds is the median coverts, which in young birds are smaller and have brown rather than black barring.

In the hand, the distinctive bill, which is long, slender, slightly swollen, and bent downward toward the tip, is characteristic, as are the short, rounded wings, with conspicuous yellow spotting on the flight feathers as well as on the tail.

In the field (10 inches), the birds appear raillike, with a slow and heavy flight. They resemble snipes as well, but the white markings around the eye and the distinctive "shoulder-strap" markings are conspic-

uous and unique. The typical call of the female is said to sound like the noise made by blowing into a large-mouthed bottle. It is usually given at night, at the rate of about four or five per second, in series of twenty to eighty and with a few minutes' break between runs. The male has a squeaky note, and in threat a hissing or growling sound is also produced.

NATURAL HISTORY

Habitats and foods. Favored habitats of this species consist of marshes and swamps having heavy vegetation, especially those with interspersed pools, muddy areas, and rather thick shrubby vegetation. It is mostly crepuscular in its activities, and this seminocturnal trait further decreases its conspicuousness. It feeds by probing in mud and ooze for worms, insects, mollusks, and crustaceans, and it also eats some vegetable matter such as grains and weed seeds. Rice paddies are a favored habitat in Japan and elsewhere, and there the birds probably eat some rice. Foraging is typically done by probing and by a lateral scything movement of the bill in shallow water, like that of an avocet. When foraging and also upon landing, the birds sometimes bob the hindquarters, like some *Tringa* species (Ali and Ripley 1969; Clancey 1967).

Social behavior. Although sometimes found in small groups outside the breeding season, these birds are usually encountered as pairs or individually. With the start of the breeding season, which is usually greatly prolonged and occurs virtually throughout the year in some areas, the female begins uttering a long series of *oook* calls at the rate of about one per second, according to published accounts. However, in a tape recording made in Zambia and provided me by Robert Stjernstedt, these calls were uttered at a rather constant rate of four to five per second. The fundamental frequency was about 1,000 cps, and there was no harmonic structuring. Females are said to sometimes utter single-note calls of the same type while in flight some 3 to 4 meters above the marsh. These are presumably territorial advertisement calls, and the female probably mates with any male attracted to her territory. Males are known to have a squeaking or whistling note, but its role in mating is still unknown. The lower-pitched voice of the female is apparently associated with her far more complex vocal resonating structures, which include a variably elongated and coiled trachea and an inflatable esophagus (Niethammer 1966), but the acoustic roles of these structures are still obscure. Females in threat situations (and reportedly, but not assuredly, in courtship situations) perform a remarkable frontal or "head-forward" display that fully exposes the complex wing and tail patterning (Muller 1975). A similar lateral display is performed by slowly retreating birds and apparently is a modification of the more intense frontal display. This display may be accompanied by a snakelike hissing; doglike growling sounds have also been heard.

Reproductive biology. Breeding in southern Australia occurs from October to December and in northern Australia from March to May. In southern Africa most breeding is from October to March, and in India most is from July to September, while in Sri Lanka it is from November to April. Breeding in China is from late June to early September and in Japan is from April to October. All these dates suggest a greatly prolonged breeding season. The nest is a simple structure, built on a raised area of mud surrounded by water and lined with vegetation to form a platform. It is often placed in a grass tussock and is very well concealed. Which sex constructs the nest seems to be unknown. The normal clutch size is 4 eggs, but extremes of 3 and 6 have been reported. The incubation period is 19 to 20 days. The clutch is incubated by the male alone, which presumably frees the female to mate with a new male and begin another clutch. Most information indicates that it is the male alone that cares for the young, although Lowe (1963) observed a female tending nestlings. The young are highly striped, in a manner reminiscent of jacanas, and are rather precocial; however, they are apparently tended for a prolonged period by the male (Beven 1913).

Status and relationships. This species is hunted in various areas such as Africa and India, but its relative abundance is impossible to judge with any certainty, even in areas where it is exploited as a game species. Banding data suggest that this species is successful at avoiding hunters and that about 40 percent of the birds banded survive the first year after banding (McClure 1974). As I noted earlier, the relationships of *Rostratula* are a fascinating problem in evolution, a problem that is made more difficult by the sex-reversal traits of the Old World species. Most of the current evidence summarized by Sibley and Ahlquist (1972) suggests that *Rostratula* is most closely related to the Jacanidae, although it has also recently been allied with the Scolopacidae (Ahlquist 1974; Strauch 1976).

Suggested reading. Lowe 1963; Ali and Ripley 1969.

South American Painted Snipe

Rostratula semicollaris (Vieillot) 1816
(*Nycticryphes semicollaris* of Peters, 1934)

Other vernacular names. None in general English use; rhynchée de Saint-Hilaire (French); Sudamerikanische Goldschnepfe (German); becasina pintada, aguatero (Spanish).

Subspecies and range. No subspecies recognized. Breeds in Paraguay, Uruguay, southeastern Brazil, northern Argentina, and central Chile. There are evidently some northward movements in winter. See map 7.

Measurements and weights. Wing (flattened): males 99–109 mm; females 105–23 mm. Culmen (from base): males 39–47 mm; females 40–47 mm. Weights: 5 adults of both sexes averaged 76.6 g (range 68 to 86 g), with no obvious sexual differences (Sick 1962). Eggs ca. 36 x 24 mm, estimated weight 11.3 g (Schönwetter 1962).

DESCRIPTION

Adults of both sexes have the head, neck, breast, and upper surface mostly deep chocolate brown. There is a narrow buff streak above the eyes and a broader one through the middle of the crown. The mantle and back are finely barred with black, and a prominent buffy "shoulder strap" extends from the rear scapulars downward to the side of the breast, where it becomes white and terminates without reaching the buffy white underparts. The wings are brown and black, with the primaries barred with white on the outer webs and the greater coverts white-tipped. The rump to the tail is brown, finely barred with black, and the under tail coverts are buffy white. The iris is brown, the legs and feet are green, and the bill varies from greenish to yellowish at the base and from yellow to black at the tip. *Females* average slightly longer in wing length but do not show significant differences in plumage from males (Höhn 1975). *Juveniles* are paler and more variegated than adults, their wing coverts are edged and tipped with buff, and the sides of the head, throat, and foreneck are blackish, streaked and mottled with white.

In the hand, the snipelike bill, distinctly downcurved at the tip, and the white-spotted wings, which are short and rounded, are distinctive. This species' bill is more decurved than that of the other form, and it is more expanded toward the tip; additionally,

10. Adult heads of Old World (*above*) and South American (*below*) painted snipes.

there is a slight web present between the middle and outer toes of this species that is lacking in the other. In both species the nostrils lie in a deep, narrow groove that extends more than half the length of the bill.

In the field (8 inches), the birds seem fluttery and "batlike" in flight; they flush silently, and after a few twists they generally rise no higher than about 10 meters before gliding down to a landing. Few calls have been heard; a hissing *wee-oo* is produced by captives when approached closely by an observer or another snipe, and a plaintive whistle has also been described.

NATURAL HISTORY

Habitats and foods. These birds are found in low-lying marshy habitats, including wet pastures, and in generally heavily grassy environments. Höhn (1975) found them in areas that were also utilized by pectoral sandpipers and in an area where coarse grasses grew only about 16 centimeters above the water, similar to areas used by true snipes. It is not known what foods the birds normally eat, but Höhn was able to keep one bird alive for a week by providing *Azolla* and various insects and insect larvae.

Social behavior. Although not obviously social, these birds are often found in pairs and may move about in small groups outside the breeding season. In Chile it has been noted that groups of 5 to 6 nests are sometimes found in an area of 2 to 3 acres, suggesting that territoriality may not be very strongly developed. Polyandrous mating has been suspected in this species but is not yet proved. Incubation patches were noted by Höhn (1975) as being present in males as well as females, which suggests that monogamy may occur or at least that the female regularly participates in incubation.

Reproductive biology. The breeding season seems to be relatively long in this species. In Chile, pairing takes place by midwinter, and incubation is under way by July or August. Data from Argentina indicate that nesting there perhaps occurs over a broad period between July and February, and one male collected in Mendoza province still had enlarged gonads in late March (Höhn 1975). The clutch size is almost invariably 2 eggs, rarely 3 (Johnson 1965). The nest is said to be less elaborate than those of the true snipes and placed in wetter locations. It usually consists of a small platform of reed stems and grass, resting at the edge of a stream or a patch of open water. Unlike the eggs of the true snipes, those of painted snipes are oblong in shape and are heavily blotched with black and dark browns. The incubation period is unknown, but in the Old World species it lasts about 20 days.

Status and relationships. This species is so inconspicuous that its status would be difficult to judge even in areas where it is probably common. It has usually been placed in the monotypic genus *Nycticryphes*, but Strauch (1976) could find no reasons for retaining two genera and recommended their merger. The bill of the South American species is more sharply curved and flattened near the tip, it has a wedge-shaped rather than rounded tail, and as noted earlier there is a small amount of webbing between the middle and outer toes, but these all seem rather trivial differences. Certainly *Rostratula benghalensis* has a large number of unique anatomical traits (Lowe 1931a), most of which have not yet been verified for *semicollaris*, but at least superficially the two painted snipes certainly appear to be congeneric.

Suggested reading. Johnson 1965.

Family Pluvianellidae (Magellanic Plover)

The Magellanic plover is a small South American shorebird having a well-developed hind toe, a short, thick and reticulated tarsus, blunt toes and claws, and a well-developed and inflatable crop. The sexes are identical in plumage as adults and do not vary seasonally. The downy young are an unpatterned greenish gray with a golden tinge. A single genus and species are recognized.

Genus *Pluvianellus* Gray 1846

This monotypic genus has the characteristics of the family Pluvianellidae.

Magellanic Plover

Pluvianellus socialis G. R. Gray 1846

Other vernacular names. None in general English use; pluvier Magellan (French); Magellanregenpfeifer (German); chorlito de Magallanes (Spanish).

Subspecies and range. No subspecies recognized. Breeds in extreme southern Chile and Argentina in the region of the Strait of Magellan, including adjacent Tierra del Fuego. Winters northward along the coast of Argentina. See map 8.

Measurements and weights. Wing (flattened): males 137–42 mm; females 132–42 mm. Culmen (from base): males 21–23 mm; females 20–22 mm (Blake 1977). Weights: males 79–102 g, average 89 g; females 69.5–87 g, average 79.5 g (Jehl 1975). Eggs ca. 35 x 25 mm, estimated weight 11.3 g (Schönwetter 1963).

DESCRIPTION

Adults of both sexes have the entire upper surface pale ashy gray, which extends downward across the breast as a more brownish tone and grades to white on the cheeks and throat. There is a dusky streak from the eyes to the bill. The lower breast, abdomen, flanks, axillaries, and under wing coverts are white, except for the middle pair of rectrices, which are blackish. The primaries are blackish with pale shafts and white toward the bases of the inner vanes, and the inner secondaries and the tips of the greater coverts are also white. The bill is black, with a small pink spot near the base, the iris is pink, grading to yellow around the pupil, and the legs and feet are coral pink. *Females* are almost identical to males in plumage but average slightly darker and heavier, with more blackish lore streaks and darker brown on the lower border of the chest. *Juveniles* are strongly tinted with golden dorsally, the breastband has dark brown streaking, the iris is creamy white with a lavender tinge, and the legs and feet are orangish. *First winter* birds ("first basic plumage") resemble adults but have mottled breastbands and extensive margining of grayish white on the wing coverts, chest feathers, tertials, and crown, and the legs and feet are orange yellow. There is a prominent yellow mark at the base of the black bill (Jehl 1975).

8. Breeding distribution of the Magellanic plover.

some dovelike notes and a *pip-wheet* alarm call (Jehl 1975).

NATURAL HISTORY

Habitats and foods. The breeding distribution of this unique species is limited to shallow ponds, lagoons, and lakes, primarily of glacial origin, in the steppe-like regions of Tierra del Fuego and southern Patagonia. It occurs in both freshwater and slightly brackish habitats, avoiding more saline areas near the coast and shallow lakes that dry up in summer or those with extensive surface vegetation. Lakes with stony beaches and with adjacent stretches of open shoreline are used for nesting. In winter the birds are found in coastal localities from the Strait of Magellan to at least the Váldez Peninsula, mostly in sheltered areas such as bays and river mouths. Evidently the larvae of chironomid flies are the primary winter foods, which the birds obtain by pecking at the sur-

In the hand, this species exhibits a distinctive, turn-stonelike bill, a hind toe, relatively short, stout legs with blunt claws, and unusually pale dove gray upper coloration that lacks black chest or belly markings.

In the field, this inconspicuous shorebird blends well with the background and frequently is found with *Charadrius* species of the same size, but its bright pink legs and lack of black breastbanding set it apart. The short legs result in a waddling gait, and the birds show a broad white wing stripe when in flight. They have a large array of calls, including

face and scratching with their strong claws. Sometimes they also feed like turnstones, by flipping shells and other debris at the base of small dunes of sand, or they may dig some distance into the sand. This digging behavior appears to be a unique foraging adaptation among the typical shorebirds (Jehl 1975).

Social behavior. Arrival on the breeding grounds occurs by early September, or possibly late August, but early stages of the breeding cycle are still unstudied. Thus, pair-forming displays are unknown, but territories are established along shorelines that contain suitable nesting habitat. Additionally, feeding territories well away from the nesting area are defended; at times these may even be on the shorelines of adjacent lakes. The territory is linear in shape and may range from about 300 to 500 meters. Territorial defense is performed by both members of a pair acting as a unit, resembling in this way the behavior of oystercatchers. Aggressive postures include several displays, such as an "upright" display with the bird holding the head high, inflating the chest, and pointing the bill toward the ground, sometimes interspersed with stiff "bowing" movements. In one observed instance, copulation was preceded by an elaborate "arc dance," with the male performing a series of sidestepping movements from one side of the crouching female, behind her tail, to the other side and back again, before finally mounting. Later the probable male was observed to perform a scraping display before the slightly crouching female, and still later both sexes were seen performing spinning and digging movements that produced numerous scrapes. The scrape that is actually used is later lined with small bits of gravel (Jehl 1975).

Reproductive biology. Nests are usually only a few feet from water, in fully exposed situations. The eggs are probably laid from early September to about the middle of November. There are normally only 2 eggs, with 1-egg clutches probably the result of egg loss and 4-egg clutches almost certainly the result of two females laying in a single nest. Incubation is performed by both sexes, but the length is unknown. Unlike typical plovers, these birds do not exhibit distraction displays at any stage in the breeding cycle. Typically the young hatch over an extended period, with the first-hatched young emerging from 8 to 14 hours before the second one. The young are surprisingly weak, and in contrast to typical shorebirds they are fed directly by the adults, usually by regurgitating food from the well-developed crop. The young grow slowly and for several weeks remain close to the nest, apparently being almost entirely dependent on their parents during this time. Starvation of the weaker young, normally the second-hatched, is frequent, and fledging does not occur until the young are 28–30 days old. They remain with their parents for some time thereafter and may be fed until they are at least 40 days old (Jehl 1975).

Status and relationships. Although it is difficult to judge the abundance of this inconspicuous and remote species, Jehl (1975) suggested that the total world population may not exceed 1,000 individuals. This would make the species one of the rarest of all shorebirds. The species' relationships are also quite uncertain. Although having a somewhat turnstone-like bill, it is clearly not closely related to that group, and anatomically seems nearest the true plovers (Burton 1974). However, Jehl (1975) suggested that it is probably an offshoot of the branch giving rise to the plovers and may actually be more closely related to the sheathbills (Chionididae) or oystercatchers than to the true plovers. Strauch (1976, 1978) has also suggested a relationship with the sheathbills.

Suggested reading. Jehl 1975.

Family *Haematopodidae* (Oystercatchers)

Atlantic Oystercatcher

Oystercatchers are cosmopolitan coastal-dwelling, medium-sized shorebirds that have a highly compressed and elongated reddish bill that is rounded or truncated at the tip, and that have mostly black or pied plumage. The tarsus is short and reticulated, there is no hind toe, and the anterior toes are only partially webbed. The sexes are identical in plumage as adults, with only minor seasonal variations. The downy young are grayish and are relatively unpatterned with black stipples and spots. There is a single genus, and five species are recognized here, but several melanistic populations exist that sometimes are considered distinct species.

Genus *Haematopus* Linné 1758

This genus of 5 species has the characteristics of the family Haematopodidae.

KEY TO SPECIES OF *Haematopus*

A Plumage entirely blackish or brownish, white lacking or restricted to base of primaries
  B Iris color red; Old World forms
    C Bill shorter (culmen under 78 mm), African . . . . *H. ostralegus*
    C′ Bill longer (culmen over 75 mm, usually over 80 mm)
      D Limited to Australia . . . . *H. fuliginosus*
      D′ Limited to New Zealand . . . . *H. unicolor*
  B′ Iris color yellow, New World forms
    C Breast dark sooty brown, bill relatively slender (maximum depth of maxilla under 15 mm), North American . . . . *H. ostralegus bachmani*
    C′ Breast black, bill heavier (maximum depth of maxilla at least 15 mm), South American. . . . . . . *H. ater*
A′ Plumage having white on wings and underparts of body
  B Entire rump and lower back white
    C White markings of breast and lower back variably smudged with black, averaging larger in size (wings over 260 mm) . . . . *H. unicolor reischeki*
    C′ White of lower breast well demarcated from black, averaging smaller in size (wings usually under 260 mm) . . . . *H. ostralegus*
  B′ Rump and lower back brown to blackish
    C Mantle and back brownish, lighter than head and neck; bill relatively stouter (maxilla over 10 mm deep) . . . . *H. ostralegus*
    C′ Mantle and back black, like head and neck; bill relatively slender (usually less than 10 mm at maximum depth) . . . . *H. leucopodus*

Oystercatcher

Haematopus ostralegus L. 1758

Other vernacular names. Black oystercatcher (*moquini* and *bachmani*); pied oystercatcher (*finschi*); redbill; variable oystercatcher; huîtrier pie (French); Austernfischer (German); ostrero común (Spanish).

Subspecies and range. See map 9.
  H. o. bachmani: American black oystercatcher. Breeds locally in western North America, north to Bristol Bay (*Condor* 78:115), and south to Baja California, winters from southern Alaska to Baja California. Often regarded as a full species.

  H. o. frazeri: Mexican oystercatcher. Resident on Baja California and the west coast of Mexico.

  H. o. palliatus: Atlantic oystercatcher. Breeds on the Atlantic and Gulf coasts of the United States south from New Jersey through eastern Mexico and perhaps Central America. Also in South America from Colombia to Brazil and Uruguay. A resident population in the West Indies is included here but sometimes is recognized as distinct (*pratti*).

  H. o. galapagensis: Galápagos oystercatcher. Resident on the Galápagos Islands.

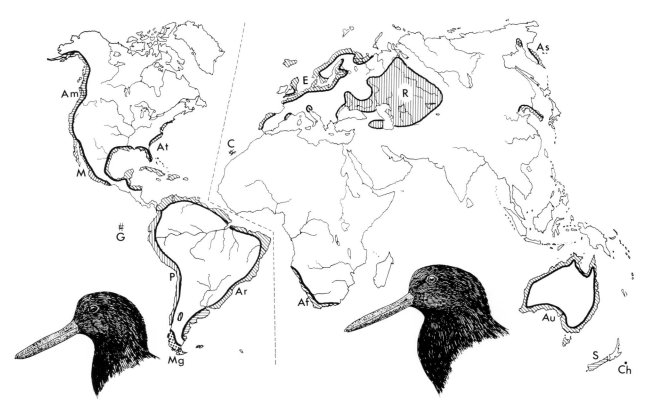

9. Breeding distributions of the African (Af), American (A), Argentine (Ar), Asian (As), Atlantic (At), Australian (Au), Canary Island (C), Chatham Island (Ch), European (E), Galapagos (G), Magellanic (Mg), Mexican (M), Peruvian (G), Russian (R), and South Island (S) oystercatchers.

H. o. pitany: Peruvian oystercatcher. Resident on the Pacific coast of Panama, Ecuador, Peru, and Chile. Casual or accidental in western Argentina.

H. o. durnfordi: Argentine oystercatcher. Breeds in coastal Argentina south to Santa Cruz.

H. o. meadewaldoi: Canary Island oystercatcher. Resident on the Canary Islands. Possibly ex-

tinct; reported only once since 1914. Sometimes considered a subspecies of *moquini.*

H. o. moquini: African black oystercatcher. Resident on the coast of Africa from Gabon to Natal. Often regarded as a full species, as for example Heppleston 1973.

H. o. ostralegus: European oystercatcher. Breeds on Iceland, Faeroes, and the coasts of Europe west to the Scandinavian Peninsula and south to Greece and Turkey. Winters south to Sierra Leone, Mozambique, the Arabian Sea, and sometimes to India. Populations breeding in Britain (*occidentalis*) and in Iceland and the Faeroes (*malacophaga*) are sometimes regarded as distinct races, as by Baker (1977).

H. o. longipes: Russian oystercatcher. Breeds from eastern and southern Russia east to western Siberia, and south to Russian Turkistan. Wintering areas are uncertain but probably include the Persian Gulf, Arabian Sea, and northeastern Africa.

H. o. osculans: Asian oystercatcher. Probably breeds on the east coast of Kamchatka, the west coast of Korea, and the Gulf of Penzhin at the north end of the Sea of Okhotsk. Winters on the coast of China.

H. o. longirostris: Australian oystercatcher. Resident on the coasts of Australia, Tasmania, and probably locally in southern New Guinea and adjoining islands.

H. o. finschi: South Island oystercatcher. Resident on South Island of New Zealand. Sometimes considered a full species. Regarded by Peters (1934) as a synonym of *H. ostralegus unicolor,* which in this book is considered a distinct species.

H. o. chathamensis: Chatham Islands oystercatcher. Restricted to the Chatham Islands. Regarded as a subspecies of *H. unicolor* by Baker (1977) and the New Zealand Checklist Committee (1970).

Measurements and weights. Wing: males 232–72 mm; females 235–300 mm. Culmen (from feathers): males 64–86 mm; females 73–99. Baker (1974c) reports significant differences in culmen length between the sexes of *finschi* for all age categories and for adults of *chathamensis.* Weights (of nesting *ostralegus*): males 425–560 g; females 445–90 g. The overall range of birds between September and May varying from first-year to third-year or older was 405–720 g (Glutz et al. 1975). Comparable data not available for the other races, but Baker (1975) provided some weight data for *finschi,* reporting that 120 males averaged 539 g, compared with 561 g for 107 females. Eight males of *chathamensis* averaged 540 g, compared with 640 g for 8 females. Some weights for *moquini* are provided by Summers and Cooper (1977). Eggs ca. 57 x 40 mm, estimated weight 43–54 g (Schönwetter 1962).

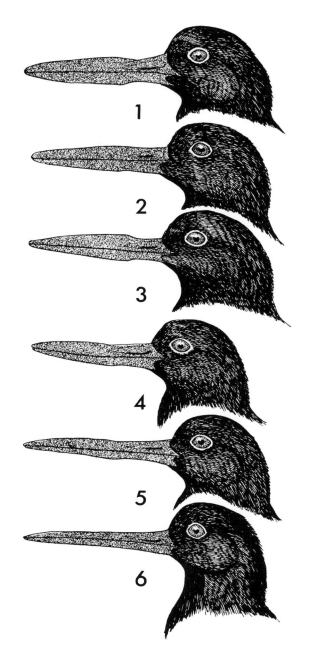

11. Adult heads of *1,* black; *2,* Atlantic; *3,* American black; *4,* African black; *5,* sooty; and *6,* variable oystercatchers.

DESCRIPTION

Adults of both sexes (of *ostralegus*) have the head, neck, chest, wings, and tip of the tail black, the wings and tail slightly brownish, and the lower back, rump, upper tail coverts, basal portion of tail, greater wing coverts, and underparts of body white. The bill is vermilion, tipped with yellow, the iris and eyelids are crimson, and the legs and feet are from flesh-colored to pinkish. Other races vary greatly in the extent of white; in *palliatus* the rump and lower back are grayish brown to black; in *frazeri* and *galapagensis* the same area is much darker, and the breast, sides, flanks, and under tail coverts may be blotched with blackish and the white of the wings more restricted, with no white on the outer webs of the innermost primaries. The other South American races *pitanay* and *durnfordi* similarly show either reduction or elimination of white markings on the inner primaries (*pitanay*) or have white-marked primaries but a mottled line of demarcation between the black and white of the breast (*durnfordi*). The iris color of all these American races is bright yellow. The darkest extremes occur in *bachmani*, which is entirely blackish gray to sooty brown with a yellow iris; in *meadewaldoi*, which is entirely glossy black except for the white basal portions of the inner primaries; and in *moquini*, which is entirely black, with the primaries lighter toward the base but not actually white. In these last two races the adult iris color is red and the eyelids are orange to orange vermilion. *Females* of all races are identical in plumage to males, but at least in *finschi* the sexes of adults, subadults, second-year birds, and juveniles can be determined by differences in absolute bill length and by ratios between bill length to bill depth and bill length to bill width (Baker 1974c). Likewise, these four age-classes can be recognized by color characters. *Subadults* (third or fourth year) differ from adults in having red rather than scarlet iris color, orange rather than bright orange bills, and pink rather than coral pink to purple leg color. *Second-year* birds have brownish black rather than black dorsal plumage, orange red rather than red iris color, and light pink leg color. *Juveniles* have very brown dorsal plumage, brown iris color, gray legs, and a bill that is pale orange with a dark tip (Baker 1974c).

Prater et al. (1977) have provided aging and sexing information for nominate *ostralegus* as well as for *palliatus*, *bachmani*, and *moquini*. In *palliatus* and *bachmani* females are reportedly larger than males, and in *ostralegus* a 5–10 mm difference is present in bill lengths of birds as early as their first winter. *Juveniles* of all forms may be recognized by pale buff tips on their upperparts, and probably by their brownish legs and dark-tipped bills. *First-winter* birds retain buff tips on some inner median wing coverts and sometimes the upper tail coverts. At least in *ostralegus*, birds have a white collar and a dull iris, bill, and legs. *Second-winter* and *third-winter* birds have black upperparts and a large white neck collar throughout the year. The eyelids gradually become more orange, the bill tip becomes more reddish, and the legs change from darkish gray to pinkish.

In the hand, identification is likely to be a problem only in those few areas where another species of oystercatcher also occurs. It is easily separated by its white underparts from the South American blackish oystercatcher, the Australian sooty oystercatcher, and the dark form of the New Zealand variable oystercatcher. Distinction from the pied variable oyster-

catcher is dealt with in that species' account, and in South America possible confusion exists with the Magellanic oystercatcher. South American races of *ostralegus* have heavier bills (maxilla at least 10 mm at maximum depth), slightly longer toes, and a brownish rather than black back and scapular coloration as is typical of *leucopodus*.

In the field (15–17 inches), the distinctive long, brightly colored bill and the predominantly black coloration quickly serve to identify these birds as oystercatchers. As noted above, confusion with another oystercatcher species is possible in some areas, but except in South America it should be feasible to separate these species using field marks. All oystercatchers have similar call notes, which include loud, piping *klee-eep* notes.

NATURAL HISTORY

Habitats and foods. The breeding habitats of oystercatchers as a species include a rather wide range of habitats, including rocky or pebbly beaches and sandy seacoasts, dry coastal meadows, the shores of wide estuaries with sandbanks and mud flats within the tidal zone, muddy coastal lagoons, coral reef walls, the sandy shorelines of salt lakes, sand dunes, and sand steppes, pebbly shorelines and meadows along riverbanks, and wet inland meadows or the grassy shorelines of eutrophic lakes (Voous 1960). Although oystercatchers are primarily coastal birds, some inland breeding populations do exist, and these birds preferentially breed on tilled land rather than natural grasslands (Heppleston 1972). Additionally, the "black" populations of this species (*moquini*, *bachmani*, and probably also *meadewaldoi*) tend to favor mixed or rocky shorelines over sandy areas (Summers and Cooper 1977; Hartwick 1974). The Chatham Island population is somewhat intermediate in this regard (Baker 1974a). In the case of *moquini*, the birds mainly feed in rocky sites but breed on sandy beaches, while in *bachmani* nesting occurs on open rock surfaces and shell beds, or even on the tops of isolated rock pinnacles (Webster 1941; Hartwick 1974). Nests of *meadewaldoi* were reportedly on sand or in "dents" in rocks with a sprinkling of sand (Bannerman 1963). On the other hand, nominate *ostralegus* tends to breed on flat open land, with habitats offering high visibility and loose substrates for nest sites being favored (Heppleston 1972). Sandy spits in riverbeds are favored for breeding by *finschi* in New Zealand, and sandy beaches are also used by the Australian races of this species (Oliver 1955; Frith 1976). Many populations of this species are nonmigratory, but in any case the birds are never far from their primary food sources, which are primarily mollusks and to a smaller extent include marine worms and echinoderms. The birds immobilize open mussels (*Mytilus*) by a stab through the posterior adductor muscles and spread them apart by inserting the closed bill and forcing the valves open. Closed mussels are struck with the bill until a hole is made and the muscles are cut. Cockles (*Cardium*) are located by sight or probing and are likewise opened by repeated blows. Limpets are also dislodged from rocks by sharp blows or pried away by using the bill as a lever. Crabs are killed by stabbing, and worms are captured by probing techniques (Baker 1974*a*; Burton 1974).

Social behavior. Oystercatchers are highly social birds, and the relatively long period to sexual maturity means that an extended period of yearlong flock behavior is typical before initial pair-formation occurs. The process of pair-formation is extremely subtle, and at least initially it is not related to copulation, since this behavior occurs promiscuously in winter flocks (Makkink 1942). It likewise is probably not directly related to the well-known "piping" behavior, which during the breeding season is associated with territorial defense, and during winter flocking may be related to maintenance of individual distance (Heppleston 1970). Nevertheless, monogamous pair bonds are eventually formed, and in at least two populations there is considerable mate fidelity from year to year. For example, in a banded British population, 76 percent of the pairs in which both birds had survived the intervening winter remated in one year, and in the next year this percentage was 97 percent (Harris 1967). A small sample of banded *moquini* also suggests mate fidelity in that population (Summers and Cooper 1977). Thus, however pair bonds are formed, they eventually become very stable. Copulation may be initiated by either sex and is similar or identical in the several populations for which descriptions are available. The male's precopulatory display consists of a lowered tail and withdrawn head, associated with a soft, repeated call and a "stealthy" approach toward the female. The female, if receptive, raises her tail, draws in her neck, and holds her bill horizontally. The male jumps up from behind, fluttering his wings to maintain balance, and utters a distinctive call. He also pecks at the female's nape (Makkink 1942; Webster 1941; Baker 1974*b*). No specific postcopulatory displays are present. Only rarely is the piping display associated with copulatory behavior. In this display two

or more birds point their bills downward, often while standing side by side and moving forward, and utter a distinctive piercing cry. In the three New Zealand oystercatchers (*finschi*, *unicolor*, and *chathamensis*), the piping displays and vocalizations are all very similar (Baker 1974*b*), suggesting that the behavior is not related to species recognition. Piping occurs in juveniles as well as adults and sometimes even occurs while the birds are in flight. Paired birds establish and hold breeding territories, within which nests are situated. These territories are flexible and often very small, but they tend to be occupied annually, with the males apparently more faithful to their territories than are females (Harris 1967). The organization of the territories appears to be very flexible in this species, depending on the density of the population and the characteristics of the habitat. Foraging often occurs well beyond the territorial limits (Webster 1941). Territories often contain a rock or other elevated site that is used as a lookout, and intruding oystercatchers are vigorously threatened (Heppleston 1972).

Reproductive biology. Nests are in scrapes that may be used year after year, but often the location of the nest shifts annually within the territory (Harris 1967). In an English study it was found that the birds preferentially select flat, open land for their nest sites and also prefer loose over firm substrates (Heppleston 1972). In a colony of *moquini* the average distance between nests was nearly 30 meters, but in extreme cases the nests were as close as 3 meters apart. Both sexes help in nest-building, which is sometimes a very lengthy process (Hartwick 1974). The clutch size is apparently quite variable between populations. In *bachmani*, a sample of 167 nests ranged from 1 to 3 eggs, with 98 of the nests having 2 eggs (Hartwick 1974). A 2-egg clutch was apparently typical of *meadewaldoi* (Bannerman 1963). In *moquini*, 139 nests also ranged from 1 to 3 eggs, with 111 of the nests having 2 eggs. However, in European populations of *ostralegus* a 3-egg clutch is the most common pattern, with clutches of 2 or 4 eggs exceptional and with rare clutches of 1 or 5 eggs (Glutz et al. 1975). A 3-egg clutch is also typical of *palliatus* in North America and of *finschi* in New Zealand, while in the Australian race *longirostris* the normal clutch is 2, rarely 3, eggs. On the Galápagos Islands the clutch is only 1–2 eggs (Le'vẽque 1964). In all populations it is believed that both sexes incubate, and the incubation period is 24–29 days, while fledging occurs in 28–35 days in European birds (Glutz et al. 1975), about 40 days in *bachmani* (Hartwick 1974), and about 35 days in *palliatus* (Stout 1967). In many populations there are high egg losses to avian predators, and repeat nesting after the loss of an initial clutch or even of small chicks is not uncommon. In some cases as many as four or five nesting attempts will be made in a single season. In spite of this early high mortality, annual survival of adult oystercatchers is very high, possibly as high as 90 percent (Gooders 1969; Harris 1967). There is a prolonged period of sexual immaturity, and initial breeding by females probably normally occurs in the fourth year of life.

Status and relationships. Apart from the questionably extinct race *meadewaldoi* of the Canary Islands (Bannerman 1963), the status of this species is probably fairly secure. The European race is certainly in healthy condition, with a recent wintering population of about half a million birds (Prater 1974*b*). The population of *moquini* in about 40 percent of its known breeding range was estimated to be about 3,000 birds in the early 1970s (Summers and Cooper 1977). The population of *finschi* in New Zealand numbered about 50,000 birds in the early 1970s, while the Chatham Island form totaled only about 50 individuals (Baker 1973). The Galápagos Islands race is also probably very small, possibly totaling under 100 pairs. The specific and subspecific relationships of *Haematopus* are extremely complex and have yet to be thoroughly worked out. The review by Heppleston (1973) provides a useful summary of the many problems. The studies by Baker (1975, 1977) on Australian and New Zealand populations support the view that the presence of melanism in a population is of little taxonomic significance and indicates that high morphological variation exists within and between these populations. Suggested relationships of *ostralegus* to other species of *Haematopus* are tentative at best, but I believe that *leucopodus* is certainly a very close relative although its displays are apparently quite distinctive (Miller and Baker 1980). Larson (1957) has postulated an evolutionary history of *Haematopus* that proposes that the dark forms of this genus were the ancestral types, which later gave rise to light-colored forms, rather than the reverse, as is usually hypothesized. He considered these differences to be related to climatic adaptations to cold and heat rather than ecological adaptations. Baker (1973) supported this view to a limited extent but suggested that habitat selection is also a potent factor influencing the distribution of plumage types.

Suggested reading. Harris 1967; Heppleston 1972.

Magellanic Oystercatcher

Haematopus leucopodus Garnot 1826

Other vernacular names. Black and white curlew (Falkland Islands); Fuegian oystercatcher; huîtrier de garnot (French); Magellanausternfischer (German); ostrero Austral (Spanish).

Subspecies and range. No subspecies recognized. Breeds from Chiloé Island, Chile, and Chubut, Argentina, south to Tierra del Fuego; also resident on the Falkland Islands. Heppleston (1973) has suggested that this form might best be considered a subspecies of *H. ostralegus*. (See map 9.)

Measurements and weights. Wing (flattened): males 237–59 mm; females 244–60 mm. Culmen (to feathering): males 68–76 mm; females 72–85 mm. Weights: 2 males 585 and 610 g; 3 females 585–700 g; average 632 g (J. Jehl, pers. comm.). Eggs ca. 59 x 40 mm, estimated weight 51.3 g (Schönwetter 1962).

DESCRIPTION

Adults of both sexes are similar to *H. o. palliatus*, but the head and neck to the breast and rump are uniform blue black, and the wings have immaculate white secondaries. The primaries are black, with no white on the shafts or webs. The under wing coverts are black and the upper tail coverts and lower breast to under tail coverts are white, as is the basal half of the tail, which is broadly tipped with black. The iris and eyelids are yellow, the bill is scarlet to orange red, and the toes and legs are grayish white to faintly rosy. *Immatures* are similar but more dusky, less bluish on the back, and the dorsal feathers are edged with rusty. The bill is also broadly tipped with dusky (Blake 1977). Chicks have a brown iris, which gradually changes to yellowish and finally to clear yellow.

In the hand, the slender bill (the maxilla usually less than 10 mm at maximum depth) and uniformly glossy black head, neck, mantle, lower back, and rump separate this form from the South American races of *ostralegus*. The feet are also unusually small (middle toe and claw 37–41 mm vs. 42–44 mm for *o. durnfordi*). In particular, the toes are unusually short and stubby, and the claws are especially broad.

In the field (17 inches), this oystercatcher might be most likely confused with *durnfordi*, since the ranges of these two forms probably come into contact or overlap on the Patagonian coast and also in the vicinity of Chiloé Island, Chile. It is unlikely that these could be distinguished in the field. Besides occurring

12. Adult heads of Magellanic (*above*) and black (*below*) oystercatchers.

on muddy to gravelly beaches, the birds also are found on inland lagoons and grassy meadows some distance from salt water. The flight call is a drawn-out and plaintive whistle, *peee*, often wavering, quite different from the loud and abruptly ending *keep* or *keeup* of the black oystercatcher. However, a loud and sharp *keep* note is uttered by the Magellanic oystercatcher during display on the ground (Woods 1975).

NATURAL HISTORY

Habitats and foods. This species is mostly associated with coastlines during winter, and with coastal habitats as well as freshwater lagoons and grassy flats during spring and summer, often ranging well inland on mainland South America. On the Falkland Islands, however, it is sometimes found on freshwater ponds in winter, but it nests close to the beaches, using both sandy and shallow rocky beaches having mussel beds. On Tierra del Fuego the birds have been found feeding on "large white larvae" for which they apparently probe, while on the Falkland Islands a major part of the diet consists of "sandworms," limpets, and mussels. They also are known to eat

"lice" and stranded jellyfish. Sometimes they obtain very large limpets by breaking part of the shell with the bill and then levering the animal from the rock surface (Woods 1975).

Social behavior. Highly social outside the nesting season, flocks of hundreds to several thousands of these birds have been observed along the coastal flats of the Strait of Magellan, while in the Falkland Islands flocks of up to a hundred are seen from January to August, sometimes in company with black oystercatchers. Woods (1975) reported that during display one or both members of a pair will walk or run with the bill lowered until it touches the ground, raising and fanning the tail and uttering a repeated sharp *keep* note. A second high, squeaky call is also uttered during display. Very recently published descriptions of displays (Miller and Baker 1980) suggest that this is an aberrant species of oystercatcher in its behavior and calls, with several unique displays and calls.

Reproductive biology. Nesting in the Falkland Islands occurs from mid-September to mid-December, with the peak of nesting from late September to the latter part of October. On Tierra del Fuego there are egg records from as early as late October, but probably most nesting is from November onward, and chicks have been reported as late as early February. Nests on the Falkland Islands are typically at the top of beaches above the high-water mark, often exposed or partly sheltered under vegetation. Some nests are also nearer the beach, in short grassy cover. There seems to be a uniform 2-egg clutch. Information on the incubation period and on other details of nesting is lacking, but they are presumably the same as in other species.

Status and relationships. This species is common and widespread on the Falklands, probably being more numerous than the black oystercatcher, and on the mainland it is generally more common than the black oystercatcher on Tierra del Fuego, or than the Peruvian oystercatcher in southern Chile. It is presumably most closely related to *ostralegus* on the basis of its plumage pattern. Baker's (1977) phenetic analysis for males indicates a close affinity with *ostralegus finschi*, and his data for females also suggest closest affinities with the taxa here included in the species *ostralegus*. Heppleston (1973) suggests that it should be considered conspecific with *ostralegus*, but this is not supported by behavioral evidence.

Suggested reading. Woods 1975; Miller and Baker 1980.

Variable Oystercatcher

Haematopus unicolor Forster 1844
(*Haematopus ostralegus unicolor* of Peters, 1934)

Other vernacular names. Pied oystercatcher (*reischeki*); huîtrier variable (French); Neuseeland Austerfischer (German).

Subspecies and ranges. See map 10.

H. u. unicolor: Black variable oystercatcher. Resident in coastal habitats on both islands of New Zealand, northward as far as Parengarenga, Great and Little Barrier Islands, Kapiti Island, and D'Urville Island.

H. u. reischeki: Pied variable oystercatcher. Largely limited to North Island of New Zealand, and most common on the sandy coastline of Northland, but occurring as far south as Kaikoura and Fiordland on South Island. Interbreeding between *unicolor* and *reischeki* occurs in some areas.

Measurements and weights. Wing: both sexes 260–77 mm. Culmen: both sexes 76–93 mm. Baker (1974*c*) reports an average male culmen of 81.7 mm, compared with 90.6 mm for females. Weights: males average from 678 g (black phase) to 717 g (pied phase), and females from 724 g (black phase) to 779 g (intermediate phase) (Baker 1975). Eggs ca. 62 x 43 mm, estimated weight 58.5 g (Schönwetter 1962).

DESCRIPTION

Adults of both sexes are alike; the females differ only in being slightly larger and having more pointed bills. Typical *unicolor* are entirely sooty black, with sooty brown primaries and secondaries and brown on the undersurface of the tail. Typical *reischeki* adults are similar to those of *H. o. finschi,* having the inner webs of the primaries with gray areas, the outer webs of the secondaries mainly white, and the rump and tail coverts white to mottled. The bill is orange to coral red, the iris scarlet, the eyelids are orange, and the legs and feet are coral pink. *Immatures* have a duller bill and leg color and pale edging on the body feathers, especially on the back. Four age-classes can be identified: juveniles have a brown dorsal plumage, a brown iris, and gray legs; second-year birds have an orange red iris and pale pink legs; subadults have a dull red iris and pink legs; and adults have a scarlet iris and bright coral pink legs. At least in adults,

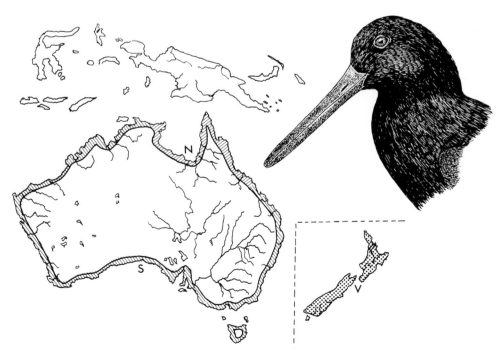

10. Breeding distributions of the northern (N) and southern (S) sooty oystercatchers, and the variable oystercatcher (V).

females can be identified by their longer bill measurements and various bill ratios (Baker 1974c).

In the hand, specimens of typical *unicolor* can be readily separated from other New Zealand oystercatchers. Specimens of *reischeki* tend to be larger than those of *finschi* (wing over 260 mm) and show a greater blurring between the white and black markings.

In the field (17 inches), distinction between *reischeki* and *finschi* is usually possible by the former's somewhat larger size, the more smudgy edges of the white and black markings, and the usual absence of the white recess on the "shoulder" that is typical of *finschi*.

NATURAL HISTORY

Habitats and foods. The black phase of this species is characteristically found on rocky shores, especially on South Island. On North Island, black, pied, and intermediate birds are usually found where sandy beaches alternate with rocky areas, and, wherever pied oystercatchers (*finschi*) are abundant in areas of species overlap, the variable oystercatchers tend to retreat to more rocky habitats. Whereas the pied oystercatchers feed largely on bivalve mollusks, the variable oystercatchers tend to take rather large prey items such as mussels and limpets from rocky shorelines. Where the two species coexist, the effects of competition between them are reduced by these differences in foraging niches (Baker 1974a,b).

Social behavior. The social behavior and displays of the variable oystercatcher are very similar to those of the other oystercatchers. Thus, the "social piping" display is similar in all three major types of New Zealand oystercatchers (*finschi, unicolor,* and *chatamensis*), including the details of their vocalizations. Vocalizations in this species average slightly lower in pitch than in the other New Zealand forms, but it is unlikely that this minor difference is important in maintaining species isolation. Social piping and other courtship begins about a month later in *unicolor* than in *finschi,* in association with a later breeding season. Pair-formation in *unicolor* occurs by September and precedes territorial establishment. Copulation behavior is the same as that of the other two New Zealand forms and is not preceded or followed by elaborate posturing. Evidently, pair bonds in this species, once formed, tend to persist through the year, and, as a result, winter copulations are not uncommon. Large winter flocks, such as occur in *finschi,* are not typical of *unicolor.* Probably the temporal separation in the timing of pair bonds and the fairly permanent pair-bond relationship that exists, at least in *unicolor,* are important in maintaining reproductive isolation between *unicolor* and *finschi* (Baker 1974b).

Reproductive biology. Variable oystercatchers breed from October to February, with a peak in December. This compares with a breeding spread of August to December in the pied oystercatcher, which certainly is important in preventing hybridization between these types. Further, though pied oystercatchers breed in inland locations, the variable oystercatchers are strictly coastal breeders, frequently nesting on offshore islands. Adult variable oystercatchers either remain on their territories throughout the year or congregate in relatively small flocks near feeding areas. Nests are simple scrapes in sand or shingle, often with little or no lining. There are usually 2 or 3 eggs, and presumably the incubation period is comparable to that of the pied oystercatcher. Young birds remain with their parents for several months, gradually becoming independent. They may wander considerably in their second or third years of life before forming pair bonds and becoming localized (Baker 1974b; Oliver 1955).

Status and relationships. Baker (1973) reported that variable oystercatchers have a scattered distribution, with flocks never exceeding 150 birds. He estimated that there may be about 1,300 black-phase birds of this species, 300 pied-phase, and 400 intermediate-phase individuals. These were considered conservative estimates, and in general oystercatchers have been increasing in numbers since their shooting was prohibited in 1940. Like other black oystercatchers, the relationships of *unicolor* are confounded by the fact that melanism has obviously evolved several times in this genus and is of ecological rather than taxonomic significance. Baker's (1977) phenetic analysis does not provide clear clues to the real affinities of this species, but it seems reasonable to imagine that it has derived from an *ostralegus*like ancestor.

Suggested reading. Baker 1974a,b.

Sooty Oystercatcher

Haematopus fuliginosus Gould 1845

Other vernacular names. Black oystercatcher; redbill; huîtrier fuligineux (French); Australischer Austernfischer (German).

Subspecies and ranges. See map 10.

> *H. f. fuliginosus:* Southern sooty oystercatcher. Resident throughout most of coastal Australia and Tasmania, mainly on rocky coastlines and more common toward the south. Sometimes considered conspecific with *H. unicolor.*

> *H. f. ophthalmicus:* Northern sooty oystercatcher. Resident in the Gulf of Carpentaria and the Cape York Peninsula.

Measurements and weights. Wing: both sexes 262–97 mm. Culmen (to feathering): both sexes 78–82 mm. Weights: 2 unsexed adults, 550 and 774 g. Eggs ca. 65 x 44 mm, estimated weight 69.5 g (Schönwetter, 1962).

DESCRIPTION

Adults of both sexes are entirely sooty black. The bill, eyes, and eyelids are red, and the legs and feet are pinkish. The race *opthalmicus* is said to differ from *fuliginosus* in that a wider area of bare skin is present around the eye, but this needs verification. *Females* probably have longer bills than males, but detailed data are lacking. *Immatures* have duller leg and bill coloring and probably also have edging on the dorsal feathers. Probably the same age criteria indicated for the variable oystercatcher can be applied to this species.

In the hand, the uniformly dark color of this species at once separates it from the pied form *longirostris.*

In the field (17–18 inches), this species is more likely to be found on rocky shorelines than on sandy beaches like *longirostris.* However, the two species sometimes associate, at which times the white markings of *longirostris* are readily apparent. The voices of the two species are reported to be similar.

NATURAL HISTORY

Habitats and foods. Like the other black oystercatchers of the world, these birds are associated with rocky coastlines, where they feed on bivalves and limpets. They are usually found along rocky promontories or stony beaches, and at times they occur on small offshore islands and rocky islets around the entire Australian shoreline, typically breeding on isolated promontories or offshore islets (Frith 1976).

Social behavior. This species is said to court and pair in the same manner as does *ostralegus,* and it has the same piping display. Apparently both species are highly territorial but will tolerate one another when they are nesting in close proximity. Evidently they rarely interbreed and produce hybrids (Frith 1976).

Reproductive biology. Breeding in this species occurs from August to January, during the same period as breeding in *ostralegus.* As noted, nesting is usually among boulders in isolated locations, and the nest is a simple scrape, sometimes with a lining of bivalve and limpet shells. There are usually 2 or 3 eggs, rarely as many as 4. The incubation period is not reported but probably is comparable to that of *ostralegus,* and like that species the adults have a well-developed injury-feigning display when disturbed by terrestrial predators. The chicks are unusually dark-colored and are very difficult to see against their usual background of dark boulders (Frith 1976).

Status and relationships. The ecological specializations of this species and its restriction to rocky shorelines place distinct limits on its population, but no estimates are available. The relationships of this form are in some doubt, but, as Baker (1977) has pointed out, the melanism it exhibits has doubtless evolved independently in many areas and is no evidence for close relationship with other black-bodied oystercatchers. Baker's phenetic analysis suggested that the affinities of *fuliginosus* may be with the New World rather than Old World oystercatchers, but the entire group seems to be so closely interrelated that I find it impossible to suggest any specific phyletic histories.

Suggested reading. Frith 1976.

South American Black Oystercatcher

Haematopus ater Vieillot and Oudart 1825

Other vernacular names. Black curlew (Falkland Islands); blackish oystercatcher; huîtrier noir (French); schwarzer Austernfischer (German); ostrero negro (Spanish).

Subspecies and range. No subspecies recognized. Breeds on rocky coastlines of South America from northern Peru and northern Argentina south to Tierra del Fuego; also resident on the Juan Fernández and Falkland Islands. See map 11.

Measurements and weights. Wing (flattened): males 253–75 mm; females 251–80 mm. Culmen (from

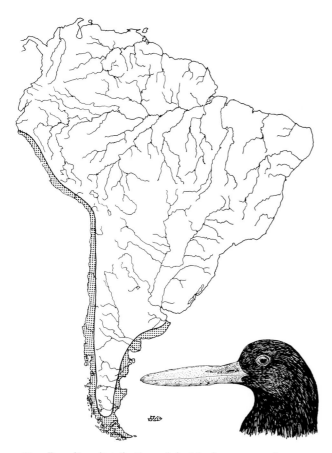

11. Breeding distribution of the black oystercatcher.

feathers): males 68–75 mm; females 72–84 mm (Blake 1977). Weights: 3 males 700–708 g, average 702 g; 2 females 585 and 700 g (J. Jehl, pers. comm.). Eggs ca. 62 x 41 mm, estimated weight 54.7 g (Schönwetter 1962).

DESCRIPTION

Adults of both sexes have the head, neck, and undersurface dull black, the latter tinged faintly with brown. The wings and back to the tail are a dull brown, which is darkest on the primaries and the tip of the tail. The iris and eyelids are yellow, the bill is red to orange red, and the legs and toes are white to pink. *Females* have longer bills (averaging 79.5 mm exposed culmen) than do males (71 mm). *Juveniles* are entirely brown, with a paler abdomen and with many of the feathers margined with buffy or ochraceous, especially on the wings and belly. The iris color of chicks is brown, which gradually changes to yellowish brown and finally to brilliant opaque yellow.

In the hand, the entirely black plumage separates this species from all other oystercatchers in South America. Further, the extreme compression and depth of the bill (maximum depth 14–18 mm) separate it from other black oystercatchers.

In the field (17–18 inches), the totally black coloration distinguishes this species from other South American oystercatchers. It is especially associated with rocky coastlines or shingle beaches rather than with sandy areas.

NATURAL HISTORY

Habitats and foods. In the Falkland Islands, this species lives almost exclusively on rocky beaches, but it sometimes also is seen on shallow tidal beaches where mussels are abundant, and during the winter small parties of mainly immature birds are found along sheltered creeks. On Tierra del Fuego it is also associated with rocky or shingle coasts and rarely occurs on inland habitats. Its foods have not been well studied, but on the Falkland Islands limpets and mussels are apparently the major part of the diet, with moderate-sized individuals (usually between 1 and 2 inches, or about 45 mm) being favored; only a few shells as large as about 75 mm in diameter are taken (Woods 1975). The same is evidently true on Tierra del Fuego and the mainland; additionally, a specimen collected in Peru contained the remains of barnacles, a small fish, a crab, and gastropod mollusks (Murphy 1936).

Social behavior. This species is reportedly less gregarious than are the other two South American oystercatchers, rarely being seen in groups of more than 4 or 5 at a time in Chile, while in the Falkland Islands the largest group reported by Woods (1975) was 8 individuals. The courting period in Peru evidently begins in September or early October. Displays and other behavior have not been well studied, but like other oystercatchers this species has the usual piping display when a third bird joins a pair, typically ending with erect tail and lowered head. This is associated with a chorus of short trilled notes that start fast, and then swell, and finally slacken and fall away. Apparently the same call may be heard from birds in flight, and Woods describes two birds that faced each other about 2 feet apart, bowed until their bills nearly touched the ground, then quickly lifted their heads. The performance lasted some 30 seconds

and was accompanied by shrill vibrating whistles. A pair in flight called similarly while one of them flew with slow wingbeats, then glided with its head bent down and the wings held up at about a 45° angle.

Reproductive biology. In the Falkland Islands, pairs tend to remain on or near their territories throughout the winter, with eggs being laid there as early as the end of October and as late as the end of January. In general, the nesting season is about a month later than that of *leucopodus*. On Tierra del Fuego eggs have been found from mid-November on, and adults with well-grown young have been observed on Chiloé Island, Chile, in April and May. The nests are typically on the beach just beyond the high-water line, and the nest may be a simple scrape in the sand or be hidden among beach rocks. Frequently the nests are near a colony of gulls such as kelp gulls (*Larus dominicus*) or South American terns (*Sterna hirundinacea*). In spite of the black plumage and bright bill, the birds are said to often be almost invisible among the dried stalks of giant kelp that sometimes occur around the nest. The incubation period is unreported, but both parents strongly defend their nest and young (Woods 1975; Murphy 1936).

Status and relationships. This species is confined to rocky and shingle coastlines and thus is much less widespread and less numerous than is the Magellanic oystercatcher, but it is in no obvious danger from a population standpoint. Baker's phenetic studies produced a cluster analysis in which both sexes of *ater* grouped most closely with *fuliginosus*. The taxonomic importance of melanism in this form, as in the other melanistic and rock-nesting oystercatchers, is virtually nil, and thus any assessment of its evolutionary affinities must exclude this feature from consideration.

Suggested reading. Murphy 1936; Woods 1975.

Family Ibidorhynchidae (Ibis-bill)

The ibis-bill is a medium-sized Asian shorebird that has a long, decurved reddish bill, with slitlike nostrils in a groove that extends more than half the length of the bill, a short, reticulated tarsus, no hind toe, and a very small web between the outer and middle toes. The sexes are identical in plumage as adults and do not vary seasonally. The downy young are whitish to fawn or rufous, minutely freckled with black. There is a single genus and species.

Genus *Ibidorhynchus* Vigors 1832

This monotypic genus has the characteristics of the family Ibidorhynchidae.

Ibis-bill

Ibidorhynchus struthersii Vigors 1832

Other vernacular names. Sicklebill; ibidorhynque de Struthers (French); Ibischnabel (German).

Subspecies and range. No subspecies recognized. Breeds from Chinese and Russian Turkistan eastward in the mountains to eastern China and probably southwestern Mongolia. Winters in the foothills of the Himalayas and mountains of Turkistan and China. See map 12.

Measurements and weights. Wing: males 224–43 mm; females 236–74 mm. Culmen: males 61–74 mm; females 70–78 mm. Weights; 4 males 270–320 g, average 300 g; 3 females 280–300 g, average 287 g (Shaw 1936). Eggs ca. 51 x 37 mm, 37 g (Dementiev and Gladkov 1969).

DESCRIPTION

Adults of both sexes have the anterior part of the face black, including the throat. The rest of the upperparts are mostly ashy gray brown, the rump feathers having basal black markings. The tail is ashy gray, with narrow wavy blackish crossbars that are broad near the tips of the rectrices. The sides of the head, neck, and upper breast are bluish gray. There is a broad black gorget, separated from the upper breast

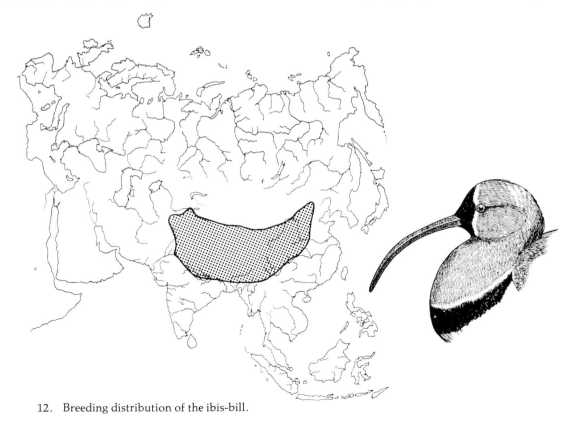

12. Breeding distribution of the ibis-bill.

by a narrow white band. The rest of the lower plumage is white. The wing is mostly grayish brown, but a white band is present at the base of the primaries. The iris is crimson, the bill is dull reddish brown to scarlet red, and the legs and feet are blood red (or livid greenish gray) in breeding adults. In nonbreeding adults the legs and feet are said to be pinkish gray (Ali and Ripley 1969). *Females* are apparently identical to males, but birds with culmens of at least 75 mm are probably females. *Juveniles* have brown on the head and throat, and the upperparts have faint pale buffy scales. The breastband is brown rather than black, and the legs and feet are pinkish gray. *First-winter* birds resemble adults but retain some buff-tipped wing coverts, especially inner medians, until spring (Prater et al. 1977).

In the hand, the strongly decurved bill and distinctive body coloration should serve to identify the bird in any plumage.

In the field (16 inches), the birds are usually found near rapidly flowing water, where feeding occurs near boulders. When alarmed, they "bob" the head and neck, and slowly wag the tail, in the manner of a greenshank. In flight this species somewhat resembles a stone curlew, but the curved bill and white wing bar are distinctive and conspicuous. In flight the birds utter a loud, ringing single note that is quickly repeated, described as *tee-ti-ti-ti-tee*, and a musical *klew-klew* is also produced.

NATURAL HISTORY

Habitats and foods. The breeding habitat consists of mountainous river valleys between 1,600 and 4,400 meters elevations. They usually use rather wide valleys having rapidly flowing streams, avoiding both very wide and sandy valleys and those with very narrow valleys or thicket-lined shorelines. Rather placid stretches of river with shingle banks seem to be ideal. The birds forage by wading breast-deep and probing with the bill under submerged pebbles, sometimes using the curved bill to slip under the front or side of a water-worn boulder. The foods are said to consist of insects, mollusks, crustaceans, and perhaps small fish, but the details remain unstudied (Ali and Ripley 1969; Dementiev and Gladkov 1969). In winter the birds sometimes occur at lower elevations at the foothills of the Himalayas, but in some areas they seem to be year-round residents.

Social behavior. These birds are reportedly usually to be found in pairs or in groups of no more than 6 to 8 birds, perhaps consisting of family groups. Little is known of their general social behavior, but they occupy their breeding areas quite early, from March (Kansu) to early April (Tibet), even before the ground is free of snow. The birds evidently space themselves out in pairs that each occupy about 1,000 meters of stream length (Dementiev and Gladkov 1969), suggesting a rather high degree of territoriality, as seems typical of torrent-dwelling birds.

Reproductive biology. The nest is a simple structure, consisting of a hollowed-out place in vegetation-free shingle cover. Sometimes it is between large stones, but often it is on the crest of a ridge of shingle that divides the main stream of a torrent, thus providing some safety from flooding. The eggs are nearly the identical color and shape of the pebbles that line the scrape, and it has been reported that when only a single egg is in the nest a stone of the same size may be placed next to it, better concealing the egg. When the incubating bird settles on the nest, its head is lowered onto its neck and its red bill is placed along the curved contour of its breast so that its bluish gray back blends with the rocks around it (Phillips 1945). There are usually 4 eggs, but extremes of 2 and 5 have been reported. Incubation is by both sexes, and Ali (1962) noted that as one of the pair members (the male?) approached the nest to take over incubation he uttered a trilling song, which was answered by the sitting bird with a series of loud *titee* notes. One nest observed by Ali (1962) had 3 birds apparently associated with it, suggesting that monogamous mating may not be invariable. The incubation period has not been specifically determined, but in the southern USSR complete clutches occur in late April and hatching begins in mid-May. Some information on the development of young has recently been provided by Abdusalyamov (1971).

Status and relationships. There is no information on the population status of this species, which is bound to be limited by habitat considerations. Its evolutionary relationships are distinctly obscure. Jehl (1968a) suggested placing it in a monotypic family until it had been better studied, a recommendation that Maclean (1972a) supported. Since then, Strauch (1976–78) has produced evidence supporting the idea that the ibis-bill, oystercatchers, avocets, and stilts are of common descent and are all apparently derived from a ploverlike ancestor. It thus seems realistic to consider this group as consisting of three closely related families, all rather closely allied to the Charadriidae.

Suggested reading. Bates and Lowther 1952; Ali 1962.

Family *Recurvirostridae*
(Stilts and Avocets)

Eurasian Avocet

Stilts and avocets are medium-sized cosmopolitan shorebirds having long, slender bills that are straight or recurved, extremely long legs with the tarsus reticulated, no hind toe or only a very small one, and variable webbing between the front toes. The sexes are identical in plumage as adults, but some species show seasonal plumage variations. The downy young are irregularly "pebbled" with black dorsally. Three genera and 6 species are recognized here.

KEY TO THE GENERA OF RECURVIROSTRIDAE

A Bill not recurved, no hind toes, front toes weakly webbed . . . . *Himantopus* (1 sp.)
A′ Bill recurved, front toes webbed
  B No hind toe . . . . *Cladorhynchus* (1 sp.)
  B′ Small hind toe present . . . . *Recurvirostra* (4 spp.)

Genus *Himantopus* (Brisson 1760) (Typical Stilts)

This genus consists of one highly variable species (up to 8 are recognized by some authorities) of temperate to tropical shorebirds with extremely long legs and no hind toe, the middle toe being joined to the outer toe with a broad web and to the inner toe with a narrower one. The bill is straight and slender, with slitlike nostrils. The downy young are grayish above, with a pair of dark lines dorsally. Adult plumages vary seasonally and sometimes also by sex. Melanism and albinism are frequent in different populations, as summarized by the following key to major variations in male breeding plumages.

KEY TO MALE BREEDING PLUMAGE VARIATIONS IN *Himantopus*

A Black absent from hindneck or, if present, separated from the back by a white area
  B Little or no black present on the back of the neck in breeding males . . . . *himantopus* and *meridionalis*
  B′ Extensive black markings present on back of the neck in breeding males
    C Black of hindneck not reaching any part of the head . . . . *leucocephalus*
    C′ Black of hindneck reaches the ear coverts . . . . *melanurus*
A′ Black of neck continuous with that of the back, and extending forward at least to the top of the crown
  B Underparts white, including the axillaries
    C Black extends from the top of the crown down to immediately in front of the eyes, tail pale grayish, sides of neck white . . . . *mexicanus*
    C′ Black extends from the front of the crown downward to include the posterior half of the lores, tail and upper tail coverts tipped with blackish, sides of the neck black . . . . *knudseni*
  B′ Underparts black, including the axillaries . . . *"novaezeelandiae"* (here considered a melanistic form of *leucocephalus*)

Stilt

Himantopus himantopus (L.) 1758

Other vernacular names. Black stilt (New Zealand); black-necked stilt (North America); black-tailed stilt (South America); black-winged stilt (Africa); Hawaiian stilt (Hawaii); pied stilt (New Zealand); white-headed stilt (Australia); échasse blanche (French); Stelzenläufer (German); cigüenela (Spanish).

Subspecies and ranges. (See map 13.)

H. h. leucocephalus: Pacific stilt. Resident in Australia, New Zealand, New Guinea, Java, Celebes, Borneo, the Philippines, and associated Pacific islands. Melanistic birds in the New Zealand population are sometimes considered a separate species or subspecies *novaezeelandiae* but are here regarded as a dark color phase.

H. h. himantopus: Eurasian stilt. Breeds from Spain and northwestern Africa west locally

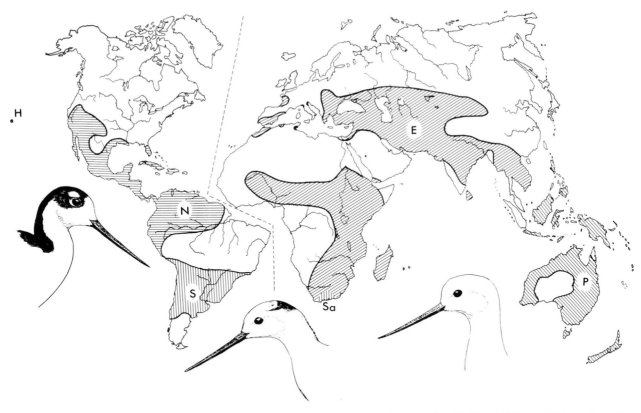

13. Breeding distributions of the Eurasian (E), Hawaiian (H), northern (N), Pacific (P), South African (Sa), and southern (S) stilts.

through southern Europe to the southern Ukraine and coasts of the Black Sea eastward through central Asia to China, India, and Indo-China. Resident in Egypt, Sudan, and much of Africa south of the Sahara excepting the Congo Basin and South Africa, and in Madagascar. Also resident on Sri Lanka (Ceylon) (a form sometimes separated as *ceylonensis*).

H. h. meridionalis: South African stilt. Restricted to South Africa.

H. h. mexicanus: Northern black-necked stilt. Breeds in western North America from Oregon to Louisiana and south through Mexico and Central America to Peru and Brazil. Also resident in Florida and the West Indies.

H. h. melanurus: Southern black-necked stilt. Resident in South America from eastern Peru, Bolivia, and southeastern Brazil south to southern Chile and southern Argentina.

H. h. knudseni: Hawaiian stilt. Mainly limited to Kauai, Maui, Niihau, and Oahu, but also on Hawaii and Molokai, of the Hawaiian Islands. Moves between islands.

Measurements and weights (for *himantopus*). Wing: males 242–51 mm; females 222–40 mm. Culmen (from feathers): males 60–68 mm; females 57–66 mm (Witherby et al. 1941). Weights (*mexicanus*): males 142–79 g; females 165–89 g (Haverschmidt 1968). Weights of 24 *leucocephalus* 136–204 g, average 173 g (D. Purchase, pers. comm.). A sample of 6 male *mexicanus* averaged 177 g, and 12 females averaged 160.3 g (Hamilton 1975). Two unsexed *melanurus* weighed 216 and 219 g (Sanft 1970). Eggs ca. 44 x 31 mm, estimated weight 21.8–23.8 g (Schönwetter 1963).

DESCRIPTION

Adults of both sexes (of *himantopus*) have an entirely white plumage except for a brownish or black back, black wings above and below, white axillaries, and a pale gray brown tail. The head and neck are white, except for blackish or dusky sometimes appearing on the head and neck. The bill is black, the legs and feet are carmine to rose pink, and the iris is carmine. Adults of *mexicanus* differ in having a black or nearly black crown, a black mark through the eyes, and a black hindneck. In *knudseni* the sides of the neck and the posterior portion of the lores are also black, and in *melanurus* the black of the hindneck is separated from the back by a band of white, and there is also a white forehead. In typical *leucocephalus* the black on the hindneck does not reach the head and is also separated from the back by a white band. The South

African race differs from *himantopus* only on the basis of a larger tail and shorter wings. The melanistic form of *leucocephalus* has the head, neck, and undersurface brownish black, shading to slaty gray on the face and at the base of the lower mandible, while the back, rump, and upper surface of wings and tail are glossy greenish black. Intermediates between this form and the typical one sometimes occur. *Females* have a brown back and scapular coloration, contrasting with more blackish wing coverts, but these are also duller than in males. At all ages males also average slightly larger than females, but measurements greatly overlap. *Juveniles* (of at least *himantopus* and *mexicanus*) have pale reddish buff fringes on the feathers of upperparts, and the secondaries and inner primaries have whitish tips. *First-winter* birds have brown upperparts like females but retain buff fringes to some wing coverts, especially the inner medians, until spring or summer (Prater et al. 1977).

In the hand this species is easily recognized by the combination of extremely long legs, a long, slender, straight, or slightly recurved beak, and unwebbed toes.

In the field (15 inches), stilts are unlikely to be confused with any other species except possibly avocets, which have considerably more recurved bills. In Australia the banded stilt and pied stilt might be confused, but the former have a chestnut breast and no black on the hindneck. Immature banded stilts lack the chestnut breast but are white rather than gray on the back of the head. The calls include various yelping notes, including a sharp and insistent *kik-kik-kik-kik* and sometimes a disyllabic *kiwik, kiwack, kyaak,* etc.

NATURAL HISTORY

Habitats and foods. This species has an extremely broad breeding range, extending from temperate to tropical climates, and from rain forest to desertlike environments. It typically breeds along the grassy shorelines of shallow freshwater or brackish pools having extensive areas of mud flats, but it also breeds on salt meadows, the brackish portions of coastal lagoons, swampy hayfields, and flooded rice or taro fields. It sometimes also occurs on the shorelines of salt lakes and salt pans where Eurasian avocets are absent, but usually these species overlap only in areas of brackish water (Voous 1960). Likewise, it breeds in the same areas as the North American avocet, but minor differences in foraging depths and

13. Breeding-plumage heads of *1,* black-phase Pacific stilt; *2,* northern black-necked stilt; *3,* Pacific stilt; and *4,* Eurasian stilt.

foraging methods may help reduce competition. In particular, stilts feed in slightly deeper water than do American avocets, while the avocets feed more deeply below the surface. Avocets use the "plunging" method of foraging (immersing the head and sometimes also the neck) more often than do stilts, and they use pecking methods less often. Likewise, scything movements are used rarely if at all, while that method is especially characteristic of foraging avocets. Both species rely heavily on brine flies during the breeding season, and likewise both species eat brine shrimp in quantity (Hamilton 1975). Stilts eat a rather wide range of foods, ranging in size up to tadpoles, small fish, and crustaceans. Insects and their larvae, including dragonflies and mayflies, earwigs, water beetles, and dipterans are eaten in large quantities, and young birds sometimes take considerable numbers of the larvae of beetles and flies.

Social behavior. Aggressive behavior patterns of this species have been described by Hamilton (1975), who reported that several hostile displays are shared with American avocets, including an aggressive upright posture and a crouched run. Unlike avocets, stilts assume an extremely tall and slim "giraffe" posture as a pecking threat display and also frequently perform an aggressive "head and legs down flight," involving hovering above an opponent with dangling legs. Stilts are more territorial than are American avocets, and unlike avocets they perform several aerial advertisement displays. They also defend more extensive territories around their nest and are more aggressive near their nests. Like avocets, stilts sometimes perform group interactions involving at least 3–4 birds that congregate in shallow water. They begin calling, often while facing in the same direction, with individuals sometimes taking off and flying above the group. This "mobbing" display is probably the counterpart of the "grouping" display of Eurasian avocets, and both are at least partially hostile in function. Sexual displays are evidently limited to those concerned with copulation; no nest-scraping ceremonies have been described for this species. Before copula-

tion one bird approaches the other, each pecking a few times at the water or at the ground if the birds are standing on land. This pecking behavior is frequently interrupted by very brief preening movements on the breast feathers. Eventually the female assumes a stationary invitational posture, with her head and neck parallel to the surface of the water or land, but the bill well above the surface. The male then becomes quite excited, walking quickly from side to side or around the female, often flicking water at her with his bill while doing so. He then jumps on her back and crouches down with bent legs, bill open, head slightly retracted, and the wings held upward for maintaining balance. Copulation is brief, and afterward the male stands beside the female with his neck feathers puffed out and momentarily crosses his bill over hers before they separate (Moon 1967). This description suggests that the copulatory behavior of stilts and avocets is nearly identical, except perhaps for a slightly different precopulatory posture in the female. Hamilton (1975) observed no differences in the copulatory behavior of stilts and the American avocet, but he observed only one completed copulation of the former species.

Reproductive biology. Stilts nest in loose colonies, and in a colony of 31 nests studied by Hamilton the average distance between nests was 21.9 meters, with a maximum distance of 42 meters. These nests were rather regularly distributed, suggesting definite spacing associated with territoriality. They are in rather open areas, close to foraging areas and in sites that usually allow for 360 degree visibility. Apparently both sexes help in nest-building, and although the nest-building movements have not yet been well described they appear to be comparable to those of avocets. Further, both stilts and avocets line their nests with whatever materials happen to be most readily available in the vicinity, including pebbles, stems of vegetation, shell fragments, and the like. Nesting material also accumulates near the nest as a result of nest-relief behavior, during which the departing bird tosses objects over its back while leaving the nest. Eggs normally are laid every day, but stilts sometimes skip a day between eggs (Hamilton 1975). There are usually 4 eggs; in a USSR colony 51 nests contained 4 eggs, 5 contained 3, and 2 contained 5. In this colony the nests were remarkably close together (from 30 centimeters to 2.3 meters apart), suggesting that territoriality was virtually lacking in this population (Mambetjumayev and Ametov 1973). Both sexes incubate, the partners typically changing places at the nest about every hour. The incubation is probably 22–26 days under natural conditions, and a 25-day incubation period was determined for some laboratory-incubated eggs. The chicks usually remain on the nest for no more than 24 hours, and often they begin to wander away even before they are completely dry. Like avocets, stilts perform strong distraction displays when the nest or young are approached, and in stilts the usual display in this circumstance is "wing-flagging" and crouching, which somewhat resembles the "broken-wing acts" of plovers. Both species also perform aerial distraction displays and ground displays resembling incubation behavior (Hamilton 1975). The fledging period is probably normally 28–32 days, rarely up to 37 days.

Status and relationships. This species exists as a large number of isolated populations and subspecies, of which only a few are perhaps rare enough to warrant special attention. The Hawaiian race *knudseni,* which is of somewhat questionable validity, is considered an endangered form, and in 1944 there may have only been about 200 birds in existence (Schwartz and Schwartz 1951). Protection soon allowed population increases, and censuses since 1968 have usually ranged from 1,000 to 1,500 birds. The only other extremely rare form is the New Zealand "black stilt," here considered a color phase of *leucocephalus.* In any case it is extremely rare, but it does occur in at least four breeding colonies in South Canterbury and North Otago. Apart from this species-level problem, the probable nearest relative of *Himantopus* is *Recurvirostra.* The occurrence of a captive-bred hybrid between *Himantopus* and *Recurvirostra americana* suggests that these are fairly closely related forms, as does their very similar copulatory behavior. The fertility of the hybrid is not known, but attempted mating behavior was observed (Princepe 1977).

Suggested reading. Hamilton 1975.

Genus *Cladorhynchus* (Gray 1840) (Banded Stilt)

The banded stilt is a medium-sized Australian shorebird with extremely long legs, no hind toe, and the anterior toes fully webbed. The bill is very long, very slightly recurved, and tapering, with slitlike nostrils. The downy young are whitish and lack obvious patterning. The sexes are alike as adults. There is a single species.

Banded Stilt

Cladorhynchus leucocephalus (Vieillot) 1816

Other vernacular names. Bishop snipe, Rottnest snipe; avocette à tête blanche (French); Schlammstelzer (German).

Subspecies and range. No subspecies recognized. Resident in southern Australia, except for Tasmania, mainly along the western coast and the southern interior. See map 14.

Measurements and weights. Wing: both sexes 190–203 mm. Culmen (to feathering): both sexes 68–73 mm. Weights: 7 males 213-50 g, average 232 g; 3 females 110–247 g, average 197 g (National Museum of Victoria specimens). Eggs ca. 55 x 40 mm, estimated weight 44 g (Schönwetter 1963).

DESCRIPTION

Adults of both sexes are mostly white, except for a chestnut band that extends from the bend of the wing downward across the breast and anterior abdomen, grading to a blackish abdominal patch. The upper wing surface is brownish black, except for a broad white edge on the inner flight feathers. The bill is black, the iris is brown, and legs and feet are pink. *Females* are identical in plumage to males. *Juveniles*

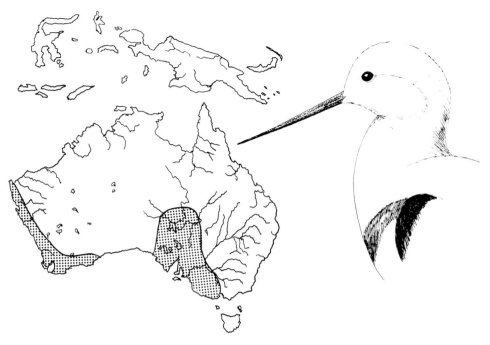

14. Breeding distribution of the banded stilt.

lack black or chestnut on the underparts and have more brownish upper wing coloration.

In the hand, the very slightly recurved bill, webbed feet, and absence of a hind toe provide identification for this species.

In the field (15 inches), this species is most likely to be confused with the Pacific race of the stilt, which never has chestnut in the plumage and has either a black hindneck or (in juveniles) pale dusky on the neck and posterior part of the head. In flight, the banded stilt shows a black upper wing surface and a white trailing edge on the inner flight feathers, and the birds frequently utter short yelping or barking notes as well as softer "creaking" calls.

NATURAL HISTORY

Habitat and foods. To a much greater extent than the other Australian stilt or avocet species, this form is limited to slaty environments, particularly salt lakes, estuaries and marshes, and to coastal areas such as Rottnest Island (where it is called the Rottnest snipe). Like typical avocets and in contrast to stilts, it has webbed toes, and so it swims effectively. Nonetheless, it usually feeds by wading in shallow water and picking up small organisms in a stiltlike manner. These items often consist of brine shrimps and similar aquatic invertebrates associated with salty or brackish habitats.

Social behavior. These birds are usually found in large flocks when they are away from their breeding areas, often numbering from a hundred to several thousand individuals. Some breeding colonies contain as many as 27,000 pairs, with nests as close as 2 meters (Howe and Ross 1931). Some of the largest or best-known colonies are near Menzies and at Lake Grace, Western Australia, and at Lake Callabonna, South Australia. Breeding in such colonies is mostly concentrated between May and December but may also occur at other times following extensive rains (Frith 1976).

Reproductive biology. Eggs are laid in depressions on sandy spits or islets, or in other soft-ground situations. Undoubtedly both sexes assist in incubation, but details of the breeding biology are still unreported. The early observations of McGilp and Morgan (1931), and the later ones of Jones (1945) still constitute most of what little is known of this interesting species. Incubation apparently requires 28 days, somewhat longer than the periods reported for other stilts or avocets, but otherwise there is little to suggest that this species deviates significantly from the others of its group.

Status and relationships. This species is locally common and is clearly dependent for survival on the continued presence of inland salt lakes. In its bill shape and general appearance the species is roughly midway between the typical stilts and the avocets; like the avocets it has fully webbed toes, but it lacks a hind toe, as do the typical stilts. Thus it seems to be a rather good transitional form between *Himantopus* and *Recurvirostra*.

Suggested reading. Jones 1945.

Genus *Recurvirostra* Linné 1758 (Typical Avocets)

Avocets are temperate to tropical medium-sized shorebirds with extremely long legs, a small hind toe, and the anterior toes deeply webbed with the webs notched in the middle. The bill is very long, tapering, and strongly recurved, with both mandibles flattened. The nostrils are slitlike, in ill-defined grooves. The downy young are light tan, with darker dorsal spots and short stripes. Four species are recognized.

KEY TO SPECIES OF *Recurvirostra*

A Mantle and wings black . . . . *andina*
A' Mantle and wings with white present
  B Innermost primaries white
    C Head and neck bright chestnut . . . . *novaehollandiae*
    C' No chestnut on head and neck . . . . *avocetta*
  B' Innermost primaries black . . . . *americana*

Andean Avocet

Recurvirostra andina Philippi and Landbeck 1861

Other vernacular names. None in general English use; avocette des Andes (French); Andensabelschnäbler (German); avoceta Andina (Spanish).

Subspecies and range. No subspecies recognized. Resident in the puna zone of the Andes of northern Chile, northwestern Argentina, western Bolivia, and southern Peru. See map 15.

Measurements and weights. Wing (flattened): males 239–58 mm; females 236–54 mm. Culmen: males

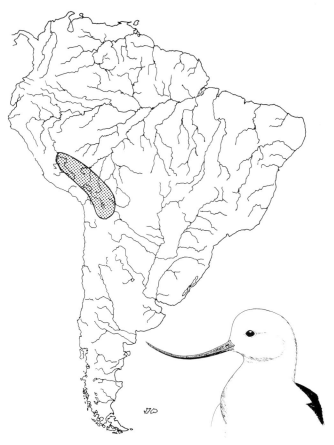

15. Breeding distribution of the Andean avocet.

combination of an upturned bill, webbed feet, and an entirely blackish wing pattern.

In the field (18 inches), this is the only avocet species found in the high Andean lakes and marshes and might perhaps only be mistaken for the common stilt, which has a straight bill and black on the head and occurs in much more tropical habitats. In flight, the Andean avocet frequently utters a barking *kieet-kieet* call, which is the basis for a Spanish vernacular name "caiti"; its uniformly dark wings contrast strongly with its otherwise white body.

NATURAL HISTORY

Habitats and foods. This species occupies the puna zone of southern Andes, generally at elevations over 12,000 feet (3,600 meters), primarily in saline marshes and lakes. Its feeding methods, and presumably also the foods taken, are very similar to those of its close North American relative.

Social behavior. Almost nothing is known of this. Johnson (1965) reports that the birds are found in small groups of up to 15 or 20 birds that remain resident in their breeding habitats throughout the year.

Reproductive biology. Although Johnson (1965) personally reported finding eggs of this species only in January, he was informed by a local guide that the birds nest twice a year, in January and again in September or October. This situation is typical of the stilts that nest in the same general region. Nests are simple scrapes in the ground, usually among clumps of coarse vegetation fringing salt lakes. They may be only a couple of yards from the water, or at times as far as 20 yards removed from it. There are normally 4 eggs, sometimes 3. The incubation period is unreported but almost certainly is the same as for the other species of *Recurvirostra*, about 22–24 days.

Status and relationships. No specific information is available on the status of this species, which is generally quite removed from human disturbance. Although this species has a typically recurved bill, it was pointed out by Seebohm (1888) that its black wings, mantle, and scapulars are very stiltlike, and its affinities are perhaps as much with that group as with the avocets, if not more. However, it exhibits the webbed feet and hind toe typical of *Recurvirostra* and so might be regarded as somewhat transitional between these two genera.

Suggested reading. None.

74–84 mm; females 68–79 mm (Blake 1977). Weights: 1 male 374 g (Niethammer 1953); 2 males 330 and 410 g; and 2 females 315 and 374 g (M.V.Z. specimens). Eggs ca. 50 x 37 mm, estimated weight 38.5 g.

DESCRIPTION

Adults of both sexes are entirely white except for brownish black on the upper back, the upper and lower wing surfaces, and the tail. The iris is orange, the eye ring is yellow, the bill is black, and the legs and toes are bluish gray. *Females* are identical to males but have slightly shorter and more distinctly recurved bills (culmen length averages 73 mm, vs. 79.5 mm in males). *Juvenile* plumages are undescribed but probably are similar to those of *avocetta*.

In the hand, this species may be identified by the

1. Downy young of (*left side, top to bottom*) wrybill, Andean lapwing, Madagascar jacana, (*right side, top to bottom*), ibis-bill, long-billed plover, American lesser golden plover, and Old World painted snipe. *Painting by Jon Fjeldså*

2. African jacana, adult. *Photo by author*

3. Northern jacana, adult. *Photo courtesy Kenneth Fink*

4. Magellanic oystercatcher, adult. *Photo courtesy Kenneth Fink*

5. South American black oystercatcher, adult. *Photo courtesy Frank Todd*

6. Galapagos oystercatcher, adult. *Photo courtesy Peter Alden*

7. Eurasian stilt, adult. *Photo by author*

8. Eurasian avocet, adult. *Photo by author*

9. American avocet, adult. *Photo courtesy Fred Zeillemaker*

10. Red-necked avocet, adult. *Photo courtesy Tom Lowe*

11. Crowned lapwing, adult. *Photo by author*

12. Southern lapwing, adult. *Photo courtesy Kenneth Fink*

13. Blacksmith plover, adult. *Photo by author.*

14. Red-wattled lapwing, adult. *Photo by author*

15. Banded lapwing, adult. *Photo by author*

16. Greater golden plover, winter plumage. *Photo by author*

17. Lesser golden plover, adult and young. *Photo by author*

18. Gray plover, nuptial plumage. *Photo by author*

19. Mountain plover, adult incubating. *Photo courtesy Ed Schulenberg*

20. Mongolian plover, adult incubating. *Photo courtesy Vladimir Flint*

21. Rufous-chested dotterel, adult. *Photo courtesy Kenneth Fink*

22. Double-banded dotterel, Falkland Islands. *Photo courtesy Kenneth Fink*

23. Black-fronted dotterel. *Photo courtesy Tom Lowe*

24. Ringed plover, male in nuptial plumage. *Photo by author*

25. Semipalmated plover, male in nuptial plumage. *Photo by author*

26. Red-capped sandplover, adult. *Photo courtesy Tom Lowe*

27. New Zealand shore plover, adult. *Photo courtesy R. S. Phillips*

28. Red-kneed dotterel, adult. *Photo courtesy Tom Lowe*

29. Wrybill, adults. *Photo courtesy Peter Alden*

30. Inland dotterel, adult. *Photo courtesy Tom Lowe*

Eurasian Avocet

Recurvirostra avocetta L. 1758

Other vernacular names. Avocet; black-crowned avocet (Africa); avocette élégante (French); Säbelschnäbler (German); avoceta comun (Spanish).

Subspecies and range. No subspecies recognized. Breeds locally from southern Sweden to southern Spain and eastward in northwestern and northern Africa, the southern Ukraine and Black Sea, across central Asia to outer Mongolia. Also breeds locally in east-central Africa and in South Africa. Winters in southern Europe, the Nile Valley, tropical and southern Africa, western India, and sometimes to eastern China and Burma. See map 16.

Measurements and weights. Wing: males 223–35 mm; females 219–32 mm. Culmen (from feathers): males 76–92 mm. Weights: males 219–435 g; females 228–320 g (Glutz et al. 1977). A sample of 15 postjuvenile birds averaged 318.7 g and ranged from 270 to 390 g (Summers and Waltner 1979).

Eggs ca. 50 x 35 mm, estimated weight 31.7 g (Schönwetter 1963).

DESCRIPTION

Adults of both sexes have the forehead, crown, lores, the region around the eye, a narrow line below the eye, and the nape black. The sides of the mantle and the inner scapulars are also black, the lower scapulars are brownish black, variably tipped with white, and the primaries are black with white bases except for the innermost ones, which are mostly white. The median and some lesser coverts are black, and the primary coverts are black with white bases. The rest of the plumage is white except for the central tail feathers, which are grayish, and the white mantle is sometimes suffused with pale gray. The bill is black, the iris is red to reddish brown (males) or hazel (females), and the legs and feet are bluish slate. *Females* are identical in plumage to males, but the iris color differences may help distinguish sexes. *Juveniles* have extensive brownish buff edging to both the

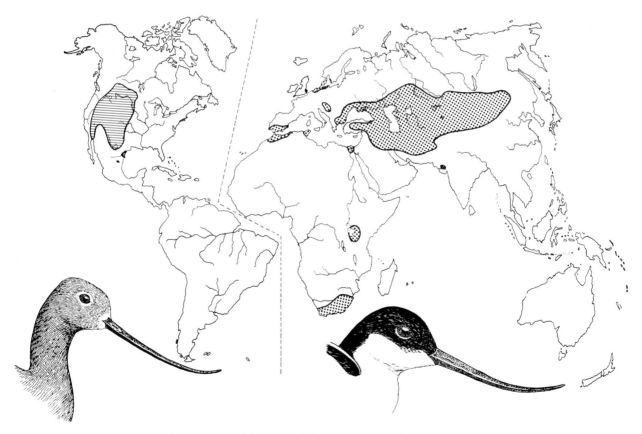

16.   Breeding distributions of the Eurasian (*shaded*) and American (*hatched*) avocets.

white and the dark feathers of the upperparts and a pale sepia cap, and there is a distinctive pattern of white on the primaries, with the fifth (counting outward) always having a whitish edge and the fourth a large white area. *First-winter* birds have paler caps than do adults and usually have some buff-tipped inner median wing coverts, as well as the distinctive juvenile primary pattern (Prater et al. 1977).

In the hand, the strongly recurved bill, the white on the intermost primaries, and the absence of chestnut coloration on the head serve to distinguish this species.

In the field (17 inches), this species is unlikely to be confused with any other, owing to its recurved bill and strong black-and-white pattern. In flight the black on the upper wing surface is discontinuous, in contrast to the pattern shown in the American avocet, and the black head markings are distinctive. The usual call is a loud, shrill *klee-eep*, or *kleep,kleep*, but there are softer notes and a special song associated with display flight.

NATURAL HISTORY

Habitats and foods. This extremely wide-ranging species occurs in a variety of climatic zones ranging from temperate to tropical, but it typically breeds in relatively arid habitats. The usual breeding habitats are the sandy or muddy shores of shallow salty or brackish waters in coastal lagoons, salt marshes, or inland salt or alkaline lakes. Two factors of probable importance in breeding habitat are water levels, which should gradually decline and expose additional feeding areas, and salt concentrations that are high enough to prevent excessive emergent and shoreline vegetation from developing. The species is distinctly colonial during the breeding season, and thus nesting densities are often very high. Colonies are usually of from 10 to 70 pairs, seldom as high as 200 or more pairs. The Minsmere Reserve in England has supported from 38 to 53 breeding pairs since attaining its present size of 20.5 hectares, a density of roughly 2 pairs per hectare. Some populations of avocets are nonmigratory, but many birds winter on the salt and alkaline lakes of the East African Rift Valley, where as many as 30,000 birds have been found. Throughout the year, foraging is done in much the same manner, although local foods probably differ. In deep water the birds perform "scooping" movements, modified pecking movements with the bill moved forward and upward through the water. In shallow water and on mud, "scything" is done, in which the head and bill are swung laterally, so that the recurved portion of the slightly open bill makes contact with the water or mud with each swing (Burton 1974). The foods thus taken include small crustaceans such as *Corophium*, which live in holes in the mud from which their long antennae extend. Many insects, mollusks, and worms are also eaten, but in general these are quite small species such as the larvae and adults of midges, brine flies, and other flies and beetles. The larvae, pupae, and adults of midges, brine flies, and other insects are also important foods for the chicks (Glutz et al. 1975). On Minsmere Reserve the primary foods are probably the larvae and adults of various flies and beetles and certain crustaceans (*Ibis* 109:474).

Social behavior. At least in southern England, about two-thirds of the avocets breed initially when 2 years old, with a small proportion starting when 3 or 4. There is evidently a rather weak attachment to the natal site, but once birds begin to breed on an area they tend to return to it (Cadbury and Olney 1978). The birds are seasonally monogamous, but the extent of mating with the mate of the previous year is still to be established. There is relatively little "courtship" behavior, inasmuch as no flight displays occur, but specific nest-scraping ceremonies and precopulatory displays do exist. Perhaps in contrast to most waders, the nest-scraping ceremony plays little or no role in pair-formation, but on the other hand the precopulatory ceremony is apparently important in establishing pair bonds. These are initially weak and easily broken, and copulation may occur with more than one partner in succession; but gradually the bond strengthens and a monogamous mating emerges. Copulation sometimes occurs on land, but typically it is done in shallow water. It begins with the two birds performing ritualized preening. This display differs from normal preening in that it often is interspersed with splashing and pecking movements in the water. The female rather suddenly lowers her head and neck to the water surface, then remains almost motionless or makes occasional lateral movements of the head as if foraging. The male continues his vigorous preening and water-splashing as he moves close to the female's side, then moves behind her and to her opposite side as he continues these activities. He continues to preen on alternate sides of the female for varying lengths of time, sometimes changing sides as many as twenty-five times before mounting. Finally he jumps sideways onto her back and sinks down on his tarsi. The female immediately begins to move her head in a wide arc. Copulation lasts only a few seconds, with the male waving his wings for balance, after which he jumps off and the two birds hold their wings slightly away from their bodies and move forward with their bills crossed. They then separate, either to resume foraging or to begin false preening once again. The complete sexual display cycle may occur again in only a few minutes (Olney 1970; Makkink 1936). Most sexual behavior occurs well outside the limited area around the nest that is defended as a territory.

Reproductive biology. The nest location is quite variable, and may be on bare sand, in dead vegetation, or in built-up mounds of debris. The size of the nest seems to be determined by the amount of available materials close at hand. Sometimes several nest scrapes are made before a final nest site is chosen, and there is little indication that territoriality plays any major role in nest spacing. Thus, nests are often as close as 3–4 feet apart in dense colonies. The nest-building activity is the same or a very similar behavior as that which occurs during "bowing," during which two or more avocets face each other and peck at objects or pick them up and throw them aside. Such ritualized nest-building also occurs during changeovers at the nest. The nest is thus built by the accumulating of materials in a rather desultory fashion. The normal clutch size is 4 eggs, but less often there are 2 or 3, and exceptionally as many as 5 may be present. In large colonies communal clutches having as many as 12 eggs have also been found. As many as 7 eggs have hatched from such double clutches. The incubation period is usually 23–25 days, and is performed by both sexes. Changeovers in incubation are rather frequent, often occurring at least once an hour. After hatching, the empty eggshells are carried to water and submerged. Both parents tend the young, which take to water shortly after hatching, sometimes swimming within an hour or two after emerging from the egg. The fledging period is quite variable but usually occurs between 32 and 42 days after hatching. A single brood is raised each year, though renesting after loss of the initial clutch sometimes occurs. Adult mortality rates are seemingly quite low, probably being between 10 and 22 percent annually in one population (Cadbury and Olney 1978).

Status and relationships. This species has a highly disjunctive breeding range, and, though its total population cannot be estimated, the complement in northwestern Europe was about 10,000 pairs in 1969 (Cadbury and Olney 1978). In some areas such as in Denmark, Great Britain, and Estonia there have been substantial population increases in recent decades, though in other areas reclamation projects have destroyed some local populations. This species is certainly fairly closely related to *americana*, as is suggested by the similarity in copulatory behavior, but I believe they probably are not closely enough related to be considered a superspecies.

Suggested reading. Makkink 1936; Olney 1970.

American Avocet

Recurvirostra americana Gmelin 1789

Other vernacular names. None in general English use; avocette Americaine (French); Amerikanischer Säbelschnäbler (German); avoceta Americana (Spanish).

Subspecies and range. No subspecies recognized. Breeds in North America from the prairie provinces of Canada southward to southern California, southern New Mexico, and northern Mexico. Winters from central California and southern Texas to southern Mexico and sometimes Honduras, rarely to Trinidad. See map 16.

Measurements and weights. Wing (chord): males 214–30 mm; females 212–26 mm. Culmen: males 75–97 mm; females 80–93 mm (Blake 1977). Weights: males 314–454 g; females 302–461 g (K.U. specimens); 16 males averaged 323 g; 17 females averaged 310 g (Hamilton 1975). Eggs ca. 50 x 34 mm, estimated weight 28.7 g (Schönwetter 1963).

DESCRIPTION

Adults in summer have most of the head, neck, and chest light cinnamon, grading to white on the front of the head. The upper wing surface and flight feathers (except for the anterior coverts, most secondaries, and the broad tips of the greater coverts), are black, and the tail is grayish white to pale gray. The rest of the plumage is white, including the wing areas mentioned above and the under wing coverts. The bill is black, the iris is brown, and the legs and feet are light grayish blue. *Adults in winter* are similar, but instead of being light cinnamon are white to bluish gray on

the head, neck, and chest. *Females* are smaller than males and have shorter but more strongly curved bills (Hamilton 1975). Birds having a culmen length of 90.5 mm are likely to be males, and those with shorter culmens are likely to be females. *Juveniles* have buffy brown fringing on their mantle and wing coverts, brown primaries, and pale pinkish chestnut tints on the head and neck. *First-winter* birds retain some buff-tipped inner median coverts until early spring and also retain brown, worn primaries (Prater et al. 1977).

In the hand, the combination of a strongly recurved bill, all the primaries black, and a black wing stripe extending without break from the primaries to the tertials serves to identify this species.

In the field (18 inches), no other North American species of shorebird has the distinctive black-and-white plumage pattern of this bird or exhibits the long and recurved bill condition. When disturbed, the birds utter sharp, repeated *wheek* or *kleek* notes.

NATURAL HISTORY

Habitats and foods. The favored breeding habitat of avocets consists of ponds or lakes with exposed, sparsely vegetated shorelines or mud flats that are closely adjacent to areas of shallow water. Such waters include alkaline ponds and lakes as well as subsaline semipermanent ponds and lakes (Stewart 1975). During migration and in the winter, the birds are sometimes also seen along marine shorelines. Animal foods constitute about two-thirds of the total diet and include phyllopod crustaceans, dragonfly nymphs, backswimmers, water boatmen, beetles, and flies and their larvae. The vegetable foods consist largely of seeds of plants associated with water. Feeding is done solitarily or in groups, with the bill immersed and swung scythelike from side to side, and at times the birds also pick up floating objects or items from the surface of mud flats (Wetmore 1925).

Social behavior. During the nonbreeding season these birds are very gregarious, often foraging in groups and nesting in loose colonies. Flocks of as many as 300 birds have been seen foraging as a group, and nesting colonies may contain as many as 15–20 pairs. In spite of this gregariousness, avocets establish and maintain discrete territories, which are defended by both pair members. Before nesting, territories center on feeding areas, which are sometimes well away from actual nest sites. During incubation the pair defends two territories, one that includes the nest and another that is used for foraging. Finally, after hatching occurs the territory is centered on the chicks and becomes a fluid area from 20 to 100 meters in diameter, around a foraging area. The birds are basically monogamous, but copulations with birds other than mates have been observed in both sexes. In at least one observed case, a pair bond apparently persisted several years, since a pair banded in 1967 was seen in 1968 and again in 1970 (Gibson 1971). Copulation is preceded by a rather simple ceremony involving breast-preening by the mate. Pairing probably occurs in late winter by the female associating with a particular male, even though she

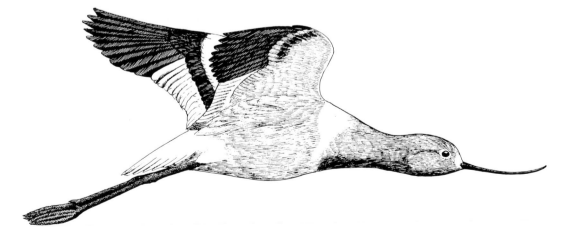

may initially be rebuffed by him. After the pair bond is formed the two birds forage closely together, copulate frequently between the time of arrival on the breeding grounds and the onset of incubation, and defend the territory together (Hamilton 1975; Gibson 1971).

Reproductive biology. Nesting in this species occurs in loose colonies, with the nests within the colony dispersed for maximum concealment. Apparently proximity to suitable feeding areas is important in the location of avocet nests, and a group defense rather than individualized nest defense seems to be typical of this species. Nests in Gibson's study area varied between different habitats in degree of dispersion, with the mean distance between nests ranging from 29.8 meters to 80.0 meters in four different study areas. The nests are simple scrapes; the amount of lining is variable, being greatest in sites subject to flooding. Both sexes participate in making the scrape and in adding lining material, usually gathering materials from no farther than 2–3 meters away. The normal clutch is 4 eggs (average of 111 nests was 3.7 eggs) although dump nests with 6 or more eggs are sometimes found, the result of two or more females' laying in the same nest. A full clutch of 4 eggs is usually laid in 5 days, and in cases of clutch destruction a new clutch may be begun only a day or two after egg loss, even late in incubation. Both sexes incubate, probably at about equal intensities, but the female is more often on the nest in late stages of incubation. The incubation period in Gibson's study averaged 24.2 days and ranged from 22 to 29 days, with all the eggs usually hatching within a day or two. The young may leave the nest within a few hours of hatching, but they often remain in it for at least a day if not disturbed. The adults tend their young closely and take turns brooding young chicks. Fledging occurs at 4–5 weeks of age (minimum 27 days), and parents continue to tend their young for some time thereafter. Postbreeding flocks begin to form soon after fledging, and the postnuptial molt of adults occurs at this time (Gibson 1971).

Status and relationships. Although there are no specific census data, the status of this species does not seem to be changing significantly. It utilizes saline to semisaline lakes that usually have little human disturbance or exploitation, and thus it is unlikely to suffer large-scale habitat destruction. Its relationships are clearly with *Recurvirostra novaehollandiae,* and the two constitute an obvious superspecies.

Suggested reading. Hamilton 1975; Gibson 1971.

Red-necked Avocet

Recurvirostra novaehollandiae Vieillot 1816

Other vernacular names. Australian avocet; avocette à cou marron (French); Rotkopfsäbelschnäbler (German).

Subspecies and range. No subspecies recognized. Resident throughout Australia except for the northern parts of the Northern Territory; occasional on Tasmania. Formerly bred on New Zealand, but not in the twentieth century. See map 17.

Measurements and weights. Wing: both sexes 224–30 mm. Culmen (to feathering): both sexes 82–95 mm. Weights: 5 males 276–360 g, average 308 g; 2 females 300 and 310 g (National Museum of Victoria specimens). Eggs ca. 50 x 36 mm, estimated weight 32.5 g (Schönwetter 1963).

DESCRIPTION

Adults of both sexes have a bright chestnut red head and neck, except for a whitish ring around the eye. The rest of the plumage is white, except for black on the scapulars, outer primaries, and upper wing coverts other than the posterior portions of the greater coverts, which are white. The bill is black, the iris is reddish brown, and the legs and feet are blue gray. *Females* are not separable from the males by their plumage, the bill is not noticeably more recurved, and the culmen averages only a few millimeters shorter in length. *Juveniles* are very pallid and are whiter than adults.

In the hand, the combination of an upturned bill, white on the innermost primaries, and chestnut head and neck pattern serves to identify this species.

In the field (18 inches), this species is likely to be confused only with stilts, which have straight bills and never have chestnut on the head and neck. In flight, the white wing stripe formed by the inner flight feathers and upper wing coverts is conspicuous and is lacking in both of the stilt species of Australia. The usual call is a sharp yelping sound; at times a melodious tooting call and single reedy notes are produced.

NATURAL HISTORY

Habitat and foods. Although it is widespread in Australia, this species' favorite habitats consist of rather salty to brackish marshes as well as coastal estuaries and inlets. The birds probably feed mostly on aquatic

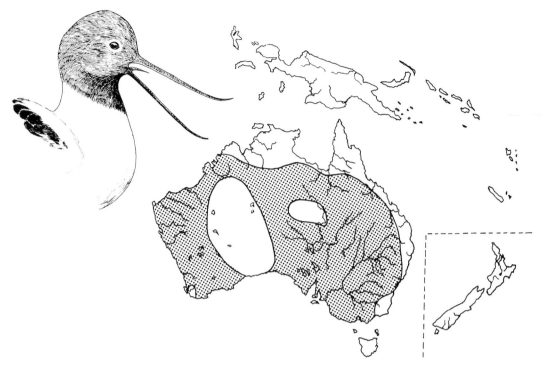

17. Breeding distribution of the red-necked avocet.

insects as well as other small aquatic invertebrates. Like typical avocets, they forage with side-to-side sweeping movements of the head, with the bill held slightly below the surface of the water. However, they sometimes also forage by swimming and upending.

Social behavior. These birds are usually to be found in flocks or small groups, and they remain fairly gregarious when breeding, with nesting done in colonies. Pair-forming and copulatory behavior is still undescribed but presumably is similar to that of the closely related North American species.

Reproductive biology. Avocet breeding occurs mainly between August and December, with the specific timing depending on the timing of the wet season. Breeding occurs in colonies, with the nests placed on the edges of swamps, on islands, or along the shorelines of shallow lakes or flooded ground. The nest is typically a simple scrape lined with a small amount of vegetation, or it may be built up of mud lined with grasses. Details of incubation length and behavior are still unknown, but probably incubation is performed by both sexes.

Status and relationships. This species is still fairly common over much of Australia, but it is rare in the interior and in the northeastern coastal region. At one time it occurred widely in New Zealand, but it has been seen there only a few times in this century. It is of interest that Australia is the only area to have three species of stilts and avocets, and this species often feeds in the company of banded stilts, presumably competing for the same foods. It is obviously a very close relative of the North American species *americana*, and the two clearly constitute a superspecies.

Suggested reading. Bryant 1947, 1948.

Family *Charadriidae* (Lapwings and Plovers)

Tribe *Vanellini* (Greater Lapwings)

Eurasian Lapwing

Lapwings and plovers are very small to medium-sized shorebirds of cosmopolitan distribution, having the bill shorter than the tarsus and not much deeper than wide, with the upper mandible usually swollen toward the tip. Foraging is often performed visually by picking objects off the surface or by shallow probing in sandy or muddy habitats. The hind toe is small or absent, and the tarsus is usually reticulated, but in a few species the scales approach a scutellate condition on the anterior surface. The downy young are usually marked with white napes and a "pebbly" dorsal color pattern. Many species nest in rather open and dry situations well away from water, and there is a fairly high proportion of species adapted to breeding in tropical or southern hemisphere environments. Less than 10 percent are associated with the arctic or alpine habitats of the northern hemisphere. The sexes are alike in adult plumage, sometimes varying seasonally. Two subfamilies and 3 tribes, containing 60 species, are recognized here.

SUBFAMILY VANELLINAE (LAPWINGS)

Lapwings are mostly medium-sized shorebirds of tropical to temperate distribution, with a relatively small and slender bill that is much shorter than the head, and the tarsus is usually considerably longer than the bill. The hind toe is small or absent, and the middle and outer toes are sometimes joined by webbing. The wing is usually rounded in shape and often has a carpal spur or knob, and the tail is fairly long and square-tipped, usually with black-and-white banding. Bare facial wattles, combs, or lappets are frequent. Two tribes are recognized here.

Tribe Vanellini (Greater Lapwings)

Greater lapwings consist of those larger species with a wing length greater than 150 mm, which are mostly of African or Asian distribution and which do not have scapular feathers that are black with white inner edges. Two genera are recognized here.

KEY TO GENERA OF VANELLINI

A The middle toe and claw greater than one-fourth of the wing, with a white face, a small hind toe, and a wing spur . . . *Hemiparra* (1 sp.)
A′ The middle toe and claw less than one-fourth the length of the wing, never with the combination of a white face, a small hind toe, and a wing spur . . . *Vanellus* (19 spp.)

Genus *Hemiparra* Salvadori 1865 (Long-toed Lapwing)

The long-toed lapwing is a medium-sized African shorebird with unusually long front toes, a small hind toe, and a small wing spur. There is extensive white on the head and wing coverts of adults, which lack bare facial lobes or wattles. The downy young are distinctive and somewhat stiltlike (Jon Fjeldså, pers. comm.). There is a single species.

White-faced Lapwing

Hemiparra crassirostris (Hartlaub) 1865

Other vernacular names. Long-toed lapwing; white-winged lapwing; vanneau a ailes blanches (French); Langzehenkiebitz (German).

Subspecies and ranges. See map 18.

V. c. crassirostris: Northern white-faced lapwing. Resident from Lake Chad and the southern Sudan to Tanzania and Malawi. Birds from the southern part of this region have sometimes been separated as *hybrida*.

V. c. leucoptera: Southern white-faced lapwing. Resident from Zaire and southern Tanzania to Mozambique.

Measurements and weights. Wing: both sexes 189–218 mm. Culmen: both sexes 31–35 mm. Weights: both sexes 187–225 g (Verheyen 1953). Eggs ca. 43 x 30 mm, estimated weight 19.1–20.0 g (Schönwetter 1962).

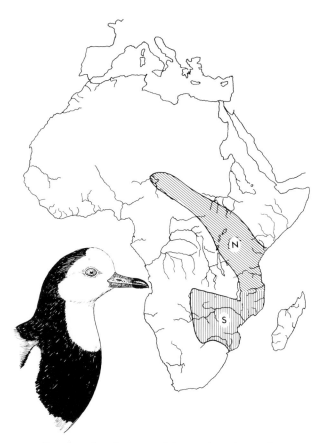

18. Breeding distributions of the northern (N) and southern (S) white-faced lapwings.

DESCRIPTION

Adults of both sexes have the entire face, throat, lores, ear coverts, and crown from behind the level of the back of the eye white. The hind part of the crown and back of the head are black, which extends around the neck as a broad collar and covers the breast. The rest of the upperparts are light earth brown, becoming darker on the rump, the middle feathers of which are tipped with white. The upper tail coverts are white, the tail is blackish with a blue gloss, and (in *crassirostris*) the primaries and secondaries are likewise black with blue gloss. The wing coverts, belly, under tail coverts, and axillaries are white. The iris is pale crimson, with a bright crimson (not yellow) outer ring, the bill is dull plum with a black tip, and the legs and feet are dull red and darkish brown (Bannerman 1951). The race *leucoptera* differs from the nominate form in that the secondaries and all but three of the outer primaries are white instead of black. *Juvenile* plumages are undescribed, but probably young birds have buffy feather edging as in other *Vanellus* species.

In the hand, the entirely white wing coverts and face serve to separate this species from its near relatives, as confirmed by the absence of wattles and the short wing spurs. The unusually long toes (more than one-fourth the length of the wing) are also distinctive.

In the field (12 inches), this species is likely to be encountered only in relatively moist habitats, and it is always seen near water. The birds often run on floating vegetation in the manner of jacanas, running rather than flying when frightened. In flight the large amount of white on the wings is very apparent. Their calls include a metallic *tip-tip* alarm note uttered on the wing and a loud, plaintive *wheet.*

NATURAL HISTORY

Habitats and foods. This species is almost entirely confined to marshes, rivers, and lakes where floating aquatic vegetation provides a jacanalike habitat. Thus it tends to be highly local in its distribution, but in favored localities it may be very abundant; a concentration of nearly 100 birds has been noted on the Kafue Flats of Zambia. Its foods are reported to include aquatic insects, maggotlike larvae, dragonfly nymphs and small snails in South Africa, and beetles, ants, and snail shells in Zambia (McLachlan and Liversidge 1957; Benson et al. 1971).

Social behavior. Little has been written on this, but judging from Saunders's (1970) account, the birds are highly territorial; they not only chase other species of lapwings such as blacksmith plovers, wattled lapwings, and white-crowned lapwings, but also chase predators such as fish eagles and gray-headed gulls. Walters (1979 and unpublished MS) reported rather small territories of 0.15–0.72 hectares in this species as compared with more typical lapwings.

Reproductive biology. There are few breeding records, but it is believed that in central Africa breeding probably occurs from December to March, while in southern Africa it may extend from June to November. Specific records include July to September in Zambia, late September in southern Rhodesia, and October in Zululand. Three nests were found by Saunders (1970) during late July in the swamps of the eastern Caprivi Strip of Botswana. All nests were along the main stretch of the Kasai Channel, which connects the Chobe and Zambezi rivers, either on flat expanses of mud (2 nests) or on a low, floating island of mud and aquatic plants. Each nest was in a shallow depression in wet mud and was lined with a few bits of dried vegetation. Two of the nests contained 4 eggs each, and the third had 4 newly hatched young. Earlier authors have suggested that the clutch size is only 2 eggs. Both sexes were in attendance at all 3 nests, with one bird usually standing nearby and uttering high-pitched *kick* alarm notes. A "broken-wing act" was performed by one bird when it was flushed from its nest. The incubation period is about 30 days, and the fledging period about 2 months (Jeffrey Walters, unpublished MS).

Status and relationships. This highly specialized species is more dependent on special ecological conditions than are the typical lapwings, and as such might be quite sensitive to habitat deterioration. It has often been placed in the monotypic genus *Hemiparra* on the basis of its jacanalike structural adaptations, and it is not an obvious close relative of any of the other *Vanellus* forms. In my view, it is perhaps the most specialized of any of the *Vanellus*-like species, and it might more reasonably be placed in a separate genus than any of the other forms I have included within *Vanellus*. On the other hand, Glutz et al. (1975) indicate that it possibly belongs with *gregaria*, *coronata*, *leucurus*, and *melanoptera* in a distinct genus (*Chettusia*).

Suggested reading: Saunders 1970; Walters 1979.

Genus *Vanellus* (Brisson 1760) (Typical Lapwings)

Typical lapwings are mostly medium-sized shorebirds with only moderately long front toes; the hind toe is small or absent, and a wing spur is present or absent. Facial wattles, combs, or lappets occur in many species, and crests or iridescent plumage are also frequent. The tarsus is relatively long (usually about one-fourth the length of the wing), and the toes are relatively short, with webbing limited to the middle and outer toes. The downy young are mottled dorsally with grayish brown and black, with a white nape band. Nineteen species are recognized here.

KEY TO SPECIES OF *Vanellus*

A Adults with wattles present in front of eyes
  B Wattles large, wing usually sharply spurred
    C Hind toe present
      D White forehead and crown . . . . *senegallus*
      D' Forehead and crown not white
        E Flanks blackish . . . . *macropterus*
        E' Flanks white . . . . *miles*
    C' Hind toe absent
      D Forehead and crown white, spur present on wing . . . . *albiceps*
      D' Forehead black, spurs lacking . . . . *malabaricus*
  B' Wattles small, spurs small or absent
    C Hind toe present
      D Wing coverts purple bronze . . . . *indicus*
      D' No purple on wing coverts . . . . *melanocephala*
    C' Hind toe absent
      D Purple bronze wing coverts, white throat . . . . *tricolor*
      D' No purple on wing coverts, throat not white
        E Outer tail feathers banded with black . . . . *tectus*
        E' Outer tail feathers lacking black band . . . . *superciliosis*
A' Adults lacking wattles
  B Hind toe present
    C Wing spur present . . . . *chilensis*
    C' Wing spur absent
      D Only central tail feathers banded, band absent on outermost feathers
        E Secondaries black, coverts iridescent . . . . *vanellus*
        E' Secondaries white, coverts not iridescent . . . . *gregarius*
      D' Entire tail white . . . . *leucurus*
  B' Hind toe absent
    C Spur lacking in adults
      D Crown longitudinally striped, primary coverts mostly white . . . . *coronatus*
      D' Crown lacking longitudinal stripe
        E Larger (wing over 190 mm), greater upper coverts white . . . . *melanopterus*
        E' Smaller (wing under 180 mm), greater upper coverts gray . . . . *lugubris*
    C' Spur present in adults
      D Wing coverts with purple iridescence . . . . *resplendens*
      D' Wing coverts lacking iridescence
        E' Cheeks white, crown black . . . . *spinosus*
        E' Cheeks black, crown white . . . . *armatus*

Crowned Lapwing

Vanellus coronatus (Boddaert) 1783
(*Stephanibyx coronatus* of Peters, 1934)

Other vernacular names. Crowned plover; pluvier couronné (French); Kronenkiebitz (German).

Subspecies and ranges. See map 19.

V. c. coronatus: Southern crowned lapwing. Resident in eastern and southern Africa from Ethiopia and Angola south to South Africa and east to Kenya. Includes *xerophilus,* described for South-West Africa by Clancey (1960).

V. c. demissus: Somali crowned lapwing. Limited to northern Somalia.

Measurements and weights. Wing: both sexes 185–218 mm. Culmen: both sexes 28–33 mm. Weights: 6 adults of both sexes 148–200 g, average 167 g (various sources). Eggs ca. 40 x 29 mm, estimated weight 17.0 g (Schönwetter 1962).

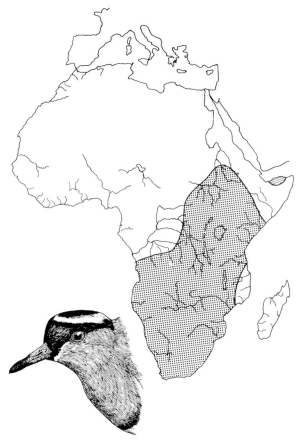

19. Breeding distribution of the southern (shaded) and Somali (hatched) crowned lapwing.

DESCRIPTION

Adults of both sexes are uniformly brown on the upperparts, chest, and breast, which are separated from the white belly by a narrow black line. The under tail coverts and under wing coverts are also white, and the tail is white with a broad subterminal band. The forehead is black and the crown is black, encircled by white and with a black median stripe. The iris is brownish orange (nonbreeding) to yellow (breeding), the bill is red with a brownish tip, and the legs and feet are red with the toes dark purple on the upper side. *Females* are apparently identical to males. *Juveniles* have the crown blackish with buffy edges, and the rest of the upperparts are also broadly edged with buff (Mackworth-Praed and Grant 1970). In addition, young birds have biscuit yellow iris color and greenish yellow bill and leg color.

In the hand, the distinctive crown coloration, which is evident even in nestling birds, is diagnostic.

In the field (12 inches), the inquisitive and noisy nature of this species, with its *kie-wieet* alarm call and its fearless threats, make it highly conspicuous. It is usually found in dry, open fields, and it alternates short runs, while holding the body horizontally, with a stationary upright stance.

NATURAL HISTORY

Habitat and foods. Crowned plovers are found widely over Africa, usually where the grass is short or has been burned; thus they occur in many areas of veld or semidesert. Not only are the eggs and downy young cryptically colored with sandy brown and blackish spots, but the plumage of the adults has also been described as a cryptic adaptation for desertlike environments. The birds no doubt forage opportunistically on a wide variety of invertebrate foods, but no extensive information is available. One specimen that was examined for food contents had consumed a large number of caterpillars and also had eaten some adult beetles and a bug. Termites have also been reported as a typical food, and the stomach contents of two other birds included great quantities of ants and adults of various beetles and a cricket (Skead 1955).

Social behavior. Crowned plovers are moderately gregarious outside the breeding season, sometimes occurring in flocks of as many as 36 birds, but most often occurring in small parties that seem to be loose aggregations that gain or lose members at random.

Even during the breeding season, territorial behavior is virtually absent, with pairs often nesting in rather close proximity. In one reported instance 3 nests were found in a triangular orientation with sides of 41, 37, and 9 meters, while in another instance 12 pairs of birds were found breeding in an area of about 10 hectares (*Ostrich* 44:262). In another instance, 25 nest scrapes were present in an area of about 1 square kilometer (Ade 1979). The birds also seem to be quite tolerant of black-winged lapwings, a species with similar ecological adaptations. Nothing specific has been written on pair-forming behavior, but copulation has been reported from July to September, as well as in January and March. Copulation seemingly is not preceded by elaborate displays, and it is followed by the male's raising one or both wings above his back. It is sometimes performed in the presence of other birds (Skead 1955).

Reproductive biology. In southern Africa breeding occurs in the spring months of July to October, with a secondary laying period in March. Most of the Zambian records are for August and September, with extremes of June and October. The nests are shallow scrapes in the soil, in grass cover only a few inches high, with a lining of vegetation and other debris that gradually accumulated during incubation. The nests are usually on flat ground, within 25 meters of a shade tree, and typically are close to mammal droppings that are similar in color to the eggs (Ade 1979). The same nest site may be used in subsequent years, or at least the location may be changed only slightly. There are normally 3 eggs, sometimes 2 or 4, and rarely 5. Incubation is by both sexes and requires 28 to 32 days for completion. During that time the birds tend to squat low in the nest to avoid detection, sometimes slipping quietly off the nest when a person approaches, or perhaps taking flight at the last moment and making a good deal of noise. The eggs are not covered constantly, and usually there are several changeovers each day. Nest-distraction behavior includes wing-spreading and tail-spreading, in either a standing or a nearly squatting position. The eggshells are apparently removed from the nest shortly after hatching, and as soon as hatching occurs the adults change from being quiet and undemonstrative to being extremely noisy and aggressive at the approach of a human. The chicks are quite precocious, and shortly after hatching they are led from the nest site. The parents typically divide the brood between them when danger threatens (Ade 1979). By the time they are about a month old, the young are as large as the adults, though still unable to fly, and the birds begin to mix with other adults and young (Skead 1955).

Status and relationships. This species is rather common over much of its range and certainly is in no special need of conservation attention. It has often been placed (Peters 1934) in the monotypic genus *Stephanibyx*, but there is little evidence to justify this approach (Bock 1958). My own guess is that the species is one of the more generalized of the African lapwings and is perhaps most closely related to the Asian sociable lapwing. However, Seebohm (1888) suggested that its nearest relative might be the black-winged lapwing.

Suggested reading. Skead 1955.

Sociable Lapwing

Vanellus gregaris (Pallas) 1771
(*Chettusia gregaria* of Peters, 1934)

Other vernacular names. Sociable plover; vanneau sociable (French); Herdenkiebitz (German); chorlito social (Spanish).

Suspecies and range. No subspecies recognized. Breeds on the steppes of central Russia and Kazakhstan, south nearly to the Caspian and Aral seas and Lake Balkash. Winters in Iraq and northeast Africa, sometimes to western India and rarely to Sri Lanka and central or western Europe. See map 20.

Measurements and weights. Wing: males 201–10 mm; females 196–212 mm. Culmen (from feathering): males 28–30 mm; females 26–30 mm. Weights (in breeding season): 2 males 245 and 260 g; 7 females 180–252 g, average 200 g (Glutz et al. 1975). Eggs ca. 46 x 34 mm, estimated weight 26.5 g (Schönwetter 1962).

DESCRIPTION

Adults in breeding plumage have the top of the head, lores, and a stripe behind the eyes black, while the forehead and superciliary area are white to creamy, and the lower part of the face has a more brownish tinge. The rest of the upperparts are light brownish gray, with this color extending down to the sides and front of the neck to the breast, which grades to a broad blackish lower breastband, edged posteriorly

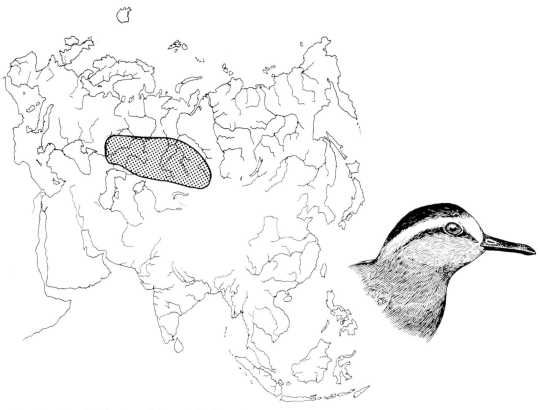

20. Breeding distribution of the sociable lapwing.

with rufous. The posterior abdomen and under tail coverts are white, as are the upper tail coverts, axillaries, and under wing coverts. The upper wing coverts are brown except for the greaters, which are tipped with white. The primaries are black, the inner ones sometimes with white edging, and the secondaries are white. The tail is white, with a broad black subterminal band that does not extend to the extreme outer pair of feathers. The iris is dark cinnamon, and the bill, legs, and toes are black, with a reddish tint on the legs and toes. *Females* can sometimes be identified by their brownish black belly, with some whitish feathers present, and they also have less chestnut around the vent. *Adults in winter* have the black areas of the head less well defined, with the black and rufous breast area replaced by smoky gray with brown spotting. *Juveniles* can be recognized by extensive buffy fringing and darker subterminal areas on the upper parts, and V-shaped flecks on the breast. *First-winter* birds retain buff fringing on their inner median coverts (Prater et al. 1977).

In the hand, the relatively small size, hind toe, absence of a crest or wing spurs, and the black eye

stripe and black lower abdominal area (in breeding plumage) provide a distinctive combination.

In the field (12 inches), this species shows the typical lapwing feature of a black-banded white tail, but it has a more pointed wing than does the Eurasian lapwing, and unlike the spur-winged plover its secondaries as well as the under wing coverts are white, contrasting strongly with the black primaries. The birds are relatively short-legged and thick-necked compared with the similar white-tailed plover and have blackish rather than yellow legs. The usual call is a short whistle and a harsh chatter, *krek-krek-krek.* . . . The Russian name "kretschetka" is onomatopoetic. There is usually a series of three rasping notes (variously described as *krek, etch, ketch,* and *reck*) that speed up when the birds are excited.

NATURAL HISTORY

Habitats and foods. The wintering habitats of this arid-adapted species consist of dry wastelands, plowed fields, and stubble in plains and plateau country. It is apparently not attracted to water at any time of the year. Its breeding range corresponds closely to the wormwood steppes, usually those with a scant to moderate cover of feather grass (*Stipa*) and fescue (*Festuca*), but where bare patches are also present. Steppes that are densely overgrown with grasses are avoided, as are highly saline soils that are nearly lacking in vegetation. Foods consist almost entirely of terrestrial insects such as beetles, weevils, locusts, and insect larvae. Foraging is done in typical plover fashion, by taking a series of quick forward steps, then stopping abruptly to pick up the food. Birds have also been observed trampling or pattering the ground with the feet while foraging (Riley and Rookse 1962). A very small amount of vegetation may also be consumed, but in general this species is considered highly beneficial to agriculture because of its consumption of pest insect species (Dementiev and Gladkov 1969; Glutz et al. 1975).

Social behavior. Spring flocks of these birds are usually rather small, often from 5 to 15 birds. The same is largely true in wintering areas, although flocks of 20–100 or more have sometimes been seen there. Fall aggregations are clearly the largest social groupings; during that season flocks of a thousand or more birds have been seen before migration in western Siberia and northern Kazakhstan. However, smaller groups are typical during fall migration itself; seldom are more than 20 birds seen together. A kind of group display is present, during which 5–10 individuals assemble on a flat area and males perform aggressive displays and flights with their bills, wings, and feet. These activities occupy considerable time immediately before the spring migration to the breeding areas (Glutz et al. 1975).

Reproductive biology. Frequently the breeding areas are still covered with snow when these birds arrive, but the birds are ready to breed immediately, and sometimes females arrive with an egg in the oviduct. At that time fights among males are frequent, and a peculiar type of flight occurs that presumably is a territorial flight. Rather loose colonies of from a few to as many as several dozen pairs are typical, but within such colonies each pair maintains a small but exclusive territory. The normal nesting cover consists of clumps of feather grasses on higher ground, under which the female makes a small scrape, often lining it slightly with grass. The usual clutch is 4 eggs, with 5 sometimes found. Incubation reportedly begins with the first egg and is typically performed by the female alone. The male remains in the general vicinity of the nest until a few days before hatching, when small groups of males begin to assemble. Incubation probably lasts about 25 days. The young are tended by both parents and feed largely on crickets and beetles, sometimes in the company of pratincoles (*Glareola*). The fledging period has not been precisely established, but in one case the first young were seen on May 24 with fledged young observed by July 5, so as in *V. vanellus* it probably requires about 35–40 days (Glutz et al. 1975).

Status and relationships. This species is apparently quite local, and even within its nesting range exhibits considerable yearly variations in populations. It is also an erratic vagrant in Europe during migration, sometimes appearing as far west as Great Britain. The species is generally regarded as a close relative of *leucurus*, but posture and proportional differences between these forms are considerable, and in this respect, as in head patterning, there are resemblances between *gregarius* and *coronatus*. Like other lapwings, the relationships are by no means self-evident.

Suggested reading. Bannerman 1961; Dementiev and Gladkov 1969.

White-tailed Lapwing

Vanellus leucurus (Lichtenstein) 1823
(*Chettusia leucura* of Peters, 1934)

Other vernacular names. White-tailed plover; vanneau à queue blanche (French); Weissschanzkiebitz (German); chorlito coliblanco (Spanish).

Subspecies and range. No subspecies recognized. Breeds from Transcaspia and Russian Turkistan southwest to Iran and locally to southern Iraq, and east across Kazakh to the vicinity of Lake Balkhash. Also breeds locally in central Turkey. Winters from northwestern India to Afghanistan, and in northern Egypt, sometimes to Sudan. See map 21.

Measurements and weights. Wing: males 172–84 mm; females 172–82 mm. Culmen (from feathering): males 27–32 mm; females 27–32 mm. Weights: both sexes 114–41 g (Glutz et al. 1975). Eggs ca. 39 x 28 mm, estimated weight 16.2 g (Schönwetter 1962).

21. Breeding distribution of the white-tailed lapwing.

DESCRIPTION

Adults of both sexes have a pinkish brown head and back, with a paler grayish white face and a whitish superciliary stripe. The chin, throat, and foreneck are ashy gray, and the breast is a darker gray, grading to a pinkish buff abdomen and rosy white under tail coverts. The upper wing surface is brownish, with gray median and greater coverts that have black subterminal bands and broad white tips on the greater and outer median coverts. The primaries are black and the secondaries white, but the outer secondaries are tipped with black, and some inner primaries are white-tipped. The upper tail coverts and tail are white. The iris is brown to blood red, the bill is black, and the legs and feet are pale yellow. *Females* cannot be externally differentiated from males. *Juveniles* have mantle patterns with black and yellow buff blotches, and the wing coverts are edged with buff. The breast and neck feathers also have paler edges, and the tail has a brownish subterminal band. *First-winter* birds lose their distinctive back pattern by early fall but retain buff-tipped inner median coverts until spring, and some outer tail feathers may have brownish marks until January (Prater et al. 1877).

In the hand, the entirely white tail, small hind toe, and absence of a wing spur serve to identify this species.

In the field (11 inches), these birds are usually found in grassy or marshy edges of large water areas, in some areas favoring slightly brackish lakes. The calls are said to include a lapwinglike *pi-wick,* a soft whistle, a *chee-viz,* and a loud *chetyre.* In flight, the entirely white tail is easily evident and should separate this species from the gray-headed lapwing and the sociable lapwing. The strong black-and-white wing pattern is also evident in flight. At a distance, the bird resembles the yellow-wattled lapwing, but it lacks a black cap and is always found near water.

NATURAL HISTORY

Ecology and foods. Although this species breeds in rather dry climates, it is rarely found far from water

94 ◈ ◈ ◈

and is said to occur along the swampy or grassy shores of brackish lakes and jheels, in areas overgrown with ephemeral plants and wormwood near salt or fresh water, and on herbaceous-covered islands in fairly deep lakes. Its food is said to include larval and adult insects, especially locusts, worms, small freshwater crustaceans, and mollusks (Ali and Ripley 1969; Dementiev and Gladkov 1969). Beetles (Hydrophilidae and Carabidae) have been found in the stomachs of spring migrants. The birds feed on dry ground or in water, taking prey from the water's surface and occasionally by submerging the head. When feeding on land they follow the typical lapwing method of taking a few forward steps and then stopping and pecking (Dean et al. 1977).

Social behavior. On their wintering ground, these birds are usually found in groups of from 6 to 20 birds, often in association with redshanks or other shorebirds. When such a flock is resting in a marsh, one of the members will often raise both wings upward back to back, apparently as a kind of contact signal (Ali and Ripley 1969) or perhaps as an agonistic display, as is common in other lapwing species. However, the birds are relatively gregarious and sometimes nest in mixed colonies of terns and pratincoles.

Reproductive biology. Rather few observations on nesting are available, but Evans (1920) located 2 nests. One was on an irrigated but uncultivated field, damp in some areas but with encrusted salt on the surface in others, and covered with grasses and suedas. The second was on a dry ridge near a marsh. Although as many as 21 birds were milling around Evans at a single time, he was unable to locate more nests, which are simple scrapes with little or no lining, with eggs that blend extremely well with the background. His nests were found in late May, and others have been reported as early as late April. The usual clutch size is 4 eggs, and nesting is usually colonial, in groups of from 4 to 10 (occasionally as many as 100) individuals. The incubation period has not been accurately determined but is probably about 22–24 days (Glutz et al. 1975).

Status and relationships. According to Dementiev and Gladkov (1969), these birds are rather abundant in spite of their restricted range. There is some indication that the birds may be extending their range westward. Besides breeding for the first time west of the Caspian Sea in the USSR in 1961, a small breeding population has recently been found in central Tur-

key. In 1971 one nest was found on the Goksu delta of the south coast, and two nests were located on the central plateau near Yarma. There was a major influx (at least 8 records) of birds in Europe in 1975, which may also be a reflection of this range expansion (Dean et al. 1977). The evolutionary relationships of this species seem to be with *gregarius*; Bock (1958) and other taxonomists have rather consistently aligned these two forms, which have complementary ranges and rather similar morphological features. They do, however, have rather different body proportions, and clearly *gregarius* is much more arid-adapted than is *leucurus*.

Suggested reading. None.

Gray-headed Lapwing

Vanellus cinereus (Blyth) 1842
(*Microsarcops cinereus* of Peters, 1934)

Other vernacular names. Gray-headed plover; vanneau à tête grise (French); Graukopfkiebitz (German).

Subspecies and range. No subspecies recognized. Breeds in Japan (Honshu), and possibly also central Manchuria and inner Mongolia. Winters in southern Japan and southeastern China south to India and the Indo-Chinese countries. See map 22.

Measurements and weights. Wing: both sexes 228–55 mm. Culmen (from feathering): both sexes 35–39 mm. Weights: 2 males 270 and 296 g; 5 females 236–95 g, average 265 g (Shaw 1936). Eggs ca. 46 x 33 mm, estimated weight 35.7 g (Schönwetter 1962).

DESCRIPTION

Adults of both sexes have a brown head and breast, suffused with ashy gray, and a chocolate and blackish band across the lower breast. The back is light brown, and the upper wing coverts are brown, with white tips on the greater coverts. The rump is white, as is the tail, except for a blackish brown subterminal band that is widest on the central part of the tail and absent on the outermost feathers. The primaries are black, and the secondaries are white. The underparts are white, including the under wing coverts. There is a small wattle behind the bill. The iris is red, the eye

❖ ❖ ❖   95

22. Breeding distribution of the gray-headed lapwing.

ring is yellow, the bill is yellow with a black tip, and the legs and feet are yellow. *Females* are not externally separable from males. *Winter adults* have a browner head and neck, the chin is whitish, streaked with brown, and the breastband is partly obscured. *Juveniles* are extensively fringed with buff on the upperparts, including the wing coverts, and have the breastband absent, poorly defined, or brownish. *First-winter* birds retain some juvenile inner median wing coverts until spring, sometimes to summer (Prater et al. 1977).

In the hand, the very small wing spur and the grayish to brownish gray head and neck, with a small yellow wattle, should identify the species.

In the field (13–14 inches), this species generally frequents fairly wet habitats, such as riverbanks or wet meadows. It is usually found in small groups or flocks up to 50 while on the wintering grounds and sometimes associates with other species of lapwings. When flushed, the birds utter a call sounding like "Did all eat?" and a plaintive *chee-it, chee-it* is also produced.

NATURAL HISTORY

Habitats and foods. Little specifically is known of the habitat needs of this species, but it is generally associated with rather wet environments such as marshy areas, rice stubble, and river flats. In India, the edges of jheels, riverbanks, wet grazing grounds, and plowed fields are used in winter. In northwestern Honshu, the birds most typically breed in swampy places and near rivers or other rather undisturbed areas. Yet they recently have begun to breed in the rice fields of central Honshu, apparently because of freedom from disturbance during the breeding season and also because the area provides a relatively vast and flat habitat, to which the birds are attracted. Foods of the species are still essentially unstudied, but Ali and Ripley (1969) indicate that insects, worms, and mollusks are probable foods.

Social behavior. In wintering areas this species is rather gregarious, usually being found in flocks ranging in size from about 5 to 50, and it also sometimes associates with red-wattled lapwings in India. Evidently, pair-forming behavior occurs on the wintering grounds or during spring migration, since the birds arrive on the breeding grounds as monogamous units. Evidently pair bonds are sometimes held more than a season, since one pair banded in 1968 returned the following year and built a nest in about the same location as the previous year. In central Honshu, territorial establishment occurs in early March, almost a month after their return from wintering areas. Even in a rather small study area, there were marked local differences in territorial density (Okugawa et al. 1973).

Reproductive biology. The nest of this species is a simple scrape, usually in a relatively exposed situation, often lined with twigs. Four eggs constitute the usual clutch size. Evidently both sexes share equally in incubation responsibilities, and each of the pair members feeds not only within its own territory but also sometimes well outside of it. When not incubating, the mate stands guard, and upon the approach of an aerial predator the sentinel bird utters an alarm call and together with its mate takes flight and attempts to mob the intruding bird from above. The incubation period, based on incubator studies, lasts 28–29 days, and the young chicks are very precocial. By about the middle of June the young birds fledge, and at that time several family groups merge and form a small flock, which later merges with other such flocks. At this time the birds gradually cease to attack airborne predators and instead ignore them (Okugawa et al. 1973).

Status and relationships. Although this species has a relatively restricted breeding range, it is not in any apparent danger at present. Increasing urbanization at the southern end of the species' breeding range may have an adverse effect on it, as does double-cropping rather than single-cropping in areas where it breeds in rice fields. Seebohm (1888) believed that this species provides an evolutionary connecting link between the wattled lapwings of his "Lobivanellus" group and the nonwattled typical Vanellus, with leucurus possibly being its nearest living relative. In plumage, the species more closely approximates V. resplendens, although from a zoogeographic standpoint it seems more likely that its affinities should lie with one of the Asian lapwings. I am thus placing it between malabaricus and resplendens in linear sequence, but I am unsure of its actual affinities.

Suggested reading. Okugawa et al. 1973; Sakane 1957–58.

Yellow-wattled Lapwing

Vanellus malabaricus (Boddaert) 1783
(*Lobipluvia malarbarica* of Peters, 1934)

Other vernacular names. Yellow-wattled plover; vanneau de Malabar (French); Malabarkiebitz (German).

Subspecies and range. No subspecies recognized. Resident from the lower Sind in Pakistan eastward to western Bengal and Bangladesh, and southward throughout the Indian Peninsula and on Sri Lanka. See map 23.

Measurements and weights. Wing: males 186–205 mm; females 181–205 mm. Culmen (from base): males 26–29 mm; females 28–30 mm. Weights: 3 adults 108, 109, and 203 g (J. C. Daniel, pers. comm.). Eggs ca. 36 x 27 mm, estimated weight 13.5 g (Schönwetter 1963).

DESCRIPTION

Adults of both sexes are black on the crown, which is edged narrowly with white behind the eyes. The chin is also black. The rest of the head, neck, upper breast, back, wing coverts, and inner flight feathers are light brown, except for a pale yellow wattle in front of the eyes. The outer primaries are black, with the bases of the feathers white on the inner webs. The white increases on the inner feathers and forms a white bar with the tips of the largest upper coverts. The rump is white, and the tail is also white to light brownish white, with a broad black tip near the tip on the central feathers, diminishing outwardly. The underparts are all white, and the brownish breast is separated from the white lower breast by a narrow black line. The bill is yellow or greenish yellow with a black tip, the iris is white to silver gray or pale lemon yellow, and the legs and toes are bright yellow. Females are virtually identical to males. *Juveniles* are pale sandy brown above, with narrow barring of darker brown on the upperparts and on the throat and upper breast and with a whitish chin.

In the hand, the yellow facial lappet and black crown stripe should easily identify this species in any plumage.

In the field (11 inches), this species is found in dry, open country and is often farther from water than the red-wattled lapwing. It is usually found in pairs or small groups and is less noisy and excitable than that species. The brown rather than black breast and the

23. Breeding distribution of the yellow-wattled lapwing.

yellow lappets should easily identify it. The usual call is a plaintive *dee-wit, dee-wit,* or *ti-ee, ti-ee,* and repeated *twit-twit-twit* notes.

NATURAL HISTORY

Habitats and foods. This species is much more arid-adapted than is the red-wattled lapwing, as a result it not only occurs in the same areas as that species, such as near jheels, but also is found on fallow fields, stubble fields, and relatively barren wastelands. Its foods are primarily insects, including grasshoppers, beetles, and the like. During the monsoon period these birds may migrate or disperse, and thus in some parts of India they occur only as dry-season visitors (Ali and Ripley 1969).

Social behavior. These birds typically are found only in pairs or at most in small family groups of 5 or 6 birds. The breeding season over most of the species' range extends from March or April to July, and in some areas such as Sri Lanka into August. The birds defend relatively large territories that in one study

averaged more than 2.5 hectares (over 6 acres). This study area of 15 hectares (37 acres) supported 4 or 5 pairs. Unlike the situation in *V. vanellus,* these territories are contiguous and nonoverlapping, with all foraging done within the territorial limits. As might be expected from the large wattles, no intraspecific intrusions are allowed in the territory; this expulsion included in one instance a fledged young of the first brood when the adults were incubating their second clutch (Jayakar and Spurway 1965*b*).

Reproductive biology. This species breeds during the dry, hot season, and its nests are simple scrapes on exposed, barren ground. This perhaps has the advantage of a less obstructed view for the incubating bird, but it exposes the nest to extremely warm temperatures. The usual clutch size is 4 eggs, and incubation duties are shared about equally by both sexes. During incubation the birds remain cool by gular fluttering, plumage fluffing, and other behavioral devices, and additionally they sometimes wet their eggs during the hottest part of the day by wetting their breast and belly feathers and then walking back to the nest. Such behavior has been observed only during the incubation of the second clutch of eggs. The incubation period has been determined to vary from 26 to 29 days, and the shortest observed fledging period was 32 days (Jayakar and Spurway 1965a, b).

Status and relationships. This is a rather common and widely distributed bird over much of its range and is apparently not in need of any special attention to its conservation. Seebohm (1888) regarded the species as an evolutionary connecting link between *V. cinereus* and the African form *V. senegallus.* I believe its nearest relatives are the Asian forms *gregarius, cinereus,* and *leucurus,* and although of course it has a more strongly developed wattle than any of these, the condition in *cinereus* is suggestive of this trend. Wolters (1975) included *malabaricus, miles,* and the Javenese wattled lapwing in the genus *Lobipluvia,* a quite different view of its possible affinities.

Suggested reading. Jayakar and Spurway 1965a, b, 1968.

white with black tips. The iris is yellowish or yellow brown, the bill is black, and the legs and feet are dusky red. *Females* cannot be distinguished externally from males. *Juvenile* birds have the feathers of the upperparts edged with pale chestnut, the chestnut superciliary stripe broader and more conspicuous, a dark brown crown, and less chestnut on the breast.

In the hand, the distinctive brown chest, small wattles, and absence of a hind toe or spurs should serve to identify this species.

In the field (12 inches), this species is found on open grasslands, dry ground, and recently burned fields. It is nomadic or migratory, and relatively wary. Its usual cry sounds like a rusty hinge squeaking and is uttered in flight. The brown chest band and dusky red legs together provide good field marks.

NATURAL HISTORY

Habitats and foods. This tropical species has a distribution that is bisected by the equator and is largely associated with "orchard bush" and savanna grasslands with scattered small trees. It is often found well

Brown-chested Lapwing

Vanellus superciliosus (Reichenow)
(*Anomalophrys superciliosus* of Peters, 1934)

Other vernacular names. Brown-chested wattled plover; vanneau caroncule (French); Rotbrustkiebitz (German).

Subspecies and range. No subspecies recognized. Resident in Africa, breeding from Togo to Zaire and Kenya. See map 24.

Measurements and weights. Wing: both sexes 181–99 mm. Culmen: both sexes 20–23 mm. Weights: no information, estimated ca. 150 g. Eggs ca. 36 x 26 mm, estimated weight 13.4 g.

DESCRIPTION

Adults of both sexes are olivaceous brown above, with a black crown and nape, a pale chestnut forehead and faint superciliary stripe, and a small yellow wattle in front of the eye. The lower part of the head, neck, and chest are grayish, the breast is bright chestnut, and the belly, under tail coverts, and under wing coverts are white. The tail is white, with a broad black band near the tip, and the secondaries are

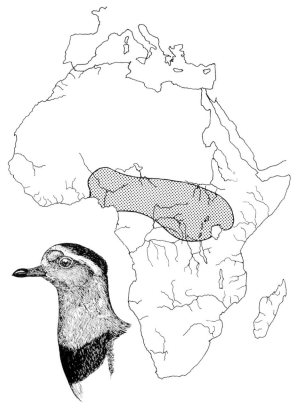

24. Breeding distribution of the brown-chested lapwing.

away from water, and other lapwings can frequently be found on newly burned ground. No information is available on its foods or foraging habits. Nesting in Nigeria has been noted in both grass savanna and orchard bush savanna habitats (Serle 1956).

Social behavior. Apparently this species is a transequatorial migrant, breeding in the northern savannas between about January and April and spending the remainder of the year in the southern savannas. At the time of their fall departure from Nigerian breeding areas, the birds form flocks of as many as 50 birds, but when they return as early as December they are already in pairs. Nothing is known of their pair-forming displays. They are evidently not very territorial, since as many as 3 pairs have been reported nesting within a radius of 200 yards. However, to some degree this may reflect not any social tendency but the necessity of placing the eggs on newly burned ground where they are safe from fire (Serle 1956).

Reproductive biology. The few records of nests with eggs for this species are all for January and February, with the birds selecting nest sites as soon as the savannas have been swept by fire. The nest is always a shallow scrape in rather stony red soil. The vegetation around it varies from charred grass tufts in grass savanna to scattered shrubs and trees either still black and leafless or just beginning to leaf out again following the fire. There are from 2 to 4 eggs in the clutch, most frequently 4, but apparently in at least one case the 2-egg clutch was a complete one. Most of the eggs found by Serle were decidedly erythristic or reddish, closely matching the lateritic soil on which they had been deposited. Erythrism in lapwing eggs has also been reported for the yellow-wattled lapwing in India, where it too locally nests in an area of lateritic soils. The incubation period lasts at least 24 days and probably begins with the last egg laid. Both sexes participate in incubation, and the nest relief is performed quickly and without obvious ceremony. Normally, human approach causes the brooding bird to leave the nest and move well away without making any effort to defend the site, but at other times, especially when small chicks were present, the adults perform wing-drooping displays or fly about while calling shrilly (Serle 1956).

Status and relationships. Although one of the least-studied African lapwings, this species is apparently not particularly rare and presumably is in no special danger. Ecologically, the species resembles the yel-

low-wattled lapwing of India, and furthermore in its plumage it also resembles *malabaricus* to some extent. Seebohm (1888) mentioned that *superciliosus* approaches *cinereus* and *gregarius* in some of its plumage characteristics, but I believe that *malabaricus* is more likely to be its nearest living relative.

Suggested reading. Serle 1956.

Lesser Black-winged Lapwing

Vanellus lugubris (Lesson) 1826
(*Stephanibyx lugubris* of Peters, 1934)

Other vernacular names. Lesser black-winged plover; Senegal plover; vanneau demi-deuil (French); Trauerkiebitz (German).

Subspecies and range. No subspecies recognized. Resident in Africa, breeding from Senegal to Kenya and south to the Congo River in the west and to Natal in the east. See map 25.

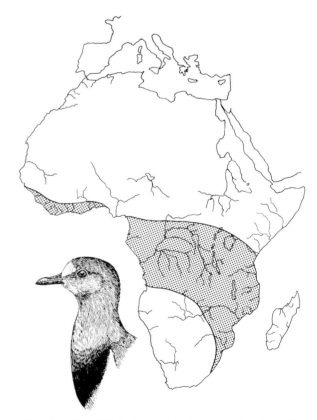

25. Breeding distribution of the lesser black-winged lapwing.

Measurements and weights. Wing: both sexes 165–85 mm. Culmen: both sexes 18–23 mm. Weights: both sexes 107–17 g. Eggs ca. 37 x 27, estimated weight 13.7 g (Schönwetter 1962).

DESCRIPTION

Adults of both sexes have a white forehead and an otherwise gray head, neck, and chest grading into a black breast. The mantle, wing coverts, and scapulars are brown, glossed with an oily green. The lower breast, belly, and under tail coverts are white, and the tail is white with a broad black band except for the outermost feathers. The primaries are black, the upper coverts are gray to brown, and the secondaries are entirely white, as are the axillaries and under wing coverts. The iris is orange yellow outwardly and brownish inwardly, the bill is purplish black, and the legs and feet are purplish black. *Females* are not externally separable from males. *Juveniles* have the feathers of the upperparts edged with pale buff (Mackworth-Praed and Grant 1970).

In the hand, the nearly uniformly gray head, neck, and breast and the absence of a spur, wattles, or hind toe should serve to identify this species. It closely resembles *melanopterus* but may be separated by its smaller measurements and by the gray rather than white primary under wing coverts and greater upper wing coverts.

In the field (9 inches), this small lapwing is most likely to be confused with the greater black-winged plover, except in eastern Africa, where their ranges do not overlap. This species is found in dry country and cultivated areas and is usually seen in pairs. In flight, the contrasting black and white primaries and secondaries are a good field mark, and the blackish band separating the gray breast from the white underparts is conspicuous. The calls include a clear, piping *tlü-wit* and a wailing alarm call.

NATURAL HISTORY

Habitats and foods. This rather widespread species is adapted to dry-land habitats such as dry plains and cultivated areas, particularly those that have recently been burned. It usually forages by making quick forward runs and pecking at the ground, much in the manner of *V. vanellus,* and it reportedly feeds on insects and their larvae as well as consuming some grass seeds.

Social behavior. Usually these birds are to be found in small flocks, sometimes numbering as many as 20

14.   Adult heads of black-winged (*above*) and lesser black-winged (*below*) lapwings.

birds. These are evidently quite mobile and often move from one place to another at night. The birds are locally migratory, or at least subject to periodic dispersal as weather conditions dictate. The breeding season in tropical portions of Africa is apparently irregular and prolonged, including records of April for Sierra Leone and Kenya, while farther south most of the records are from October to December. It is uncommon in Zambia, and breeding records there are only for September and October, while in northern Zululand (Natal) there are records for September and October. There is no information on breeding densities or territorial behavior.

Reproductive biology. The nest of this species is a simple scrape, sometimes placed on cultivated ground, often on open patchy soil with a grass cover. The clutch range is from 2 to 4 eggs, with 4 perhaps most typical. The incubation period is reported as 18–20 days, and the chicks are said to remain with their parents for about 2 months (McLachlan and Liversidge 1957). Surprisingly little specific information has been published on this species, in view of its wide distribution and relative tameness.

Status and relationships. This species is reported to be fairly common in at least some parts of its range. Bock (1958) considered the species part of an evolutionary complex that also included *melanopterus* and *coronatus,* and Seebohm (1888) considered it to be closely allied with and to be the West African representative of *melanopterus.* I believe these two forms constitute an obvious superspecies, a conclusion also reached by Snow (1978).

Suggested reading. Chapin 1939.

Black-winged Lapwing

Vanellus melanopterus (Cretzschmar) 1829
(*Stephanibyx melanopterus* of Peters, 1934)

Other vernacular names. Black-winged plover; vanneau à ailes noires (French); Schwarzflügelkiebitz (German).

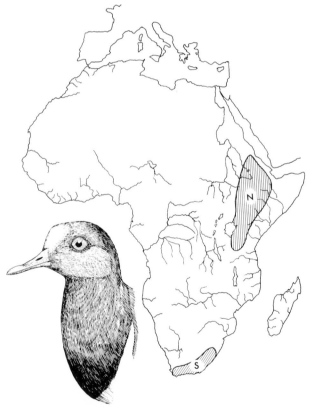

26. Breeding distributions of the southern (S) and northern (N) black-winged lapwings.

Subspecies and ranges. See map 26.

V. m. melanopterus: Northern black-winged lapwing. Resident in Africa, breeding from the eastern Sudan and Ethiopia to northern and central Kenya.

V. m. minor: Southern black-winged lapwing. Resident in Africa, breeding from Transvaal to the Cape.

Measurements and weights. Wing: both sexes 196–227 mm. Culmen: both sexes 25–28 mm. Weights: 3 adults of both sexes 163–70 g, average 167.7 g. Eggs ca. 41 x 29 mm, estimated weight 18.2 g (Schönwetter 1962).

DESCRIPTION

Adult males have the forehead and chin (sometimes also the lores) white and gray, with a darker crown, and the upperparts generally brown with a bronze to purplish cast. The primaries are black, and the secondaries are white with black tips except for the innermost, which are entirely white. The primary coverts are black, but the greater upper wing coverts are tipped with white and subterminally narrowly banded with black. There is a broad black band between the gray neck and the white belly. The rump, under tail coverts, and under wing coverts are also white, as is the tail except for a broad subterminal black band. The iris is yellow or brown, the eye ring is vinaceous red, the bill is black, and the legs and feet are brownish black tinged with red, especially on the toes. *Females* are similar to males but have a more ashy gray head and neck. *Juveniles* resemble females, but there is no black on the chest, which is entirely ashy brown, and the feathers of the upperparts are narrowly edged with buffy (Mackworth-Praed and Grant 1952).

In the hand, the combination of no hind toe, spur, or wattle and a grayish breast with a black band below will separate this species from all others except *lugubris.* The larger measurements (wing over 190 mm) and the white greater upper wing coverts separate it from that species. Additionally, in this species all the primaries and the outermost secondaries are black, while in *lugubris* the innermost primaries and the first six secondaries are broadly tipped with white.

In the field (10½ inches), this species is found in highland arid plains, especially burned areas and is usually found in small to large flocks. The black breastband and grayish head should separate it from all African species except *lugubris;* like that species, it is gray to brown above and white below, with a black breastband. The gray color of the under primary coverts of *lugubris* differs from the white ones in this species and should be visible in flight, and the white band on the upper wing formed by the greater coverts in this species may be visible in resting birds. The calls of the two species are similar but may be harsher in this species, which has a variety of loud screaming notes.

NATURAL HISTORY

Habitats and foods. Black-winged lapwings are largely limited to highland plains and grasslands, with the race *melanopterus* confined to areas above 7,000 feet, while *minor* often occurs at lower elevations. The birds seem to prefer recently burned grasslands, especially those areas used by big game for foraging. The species is ecologically similar to *coronatus* but occurs in moister areas. Wintering occurs at lower elevations, including coastal flats in some areas. Foods consist of insects and their larvae (Mackworth-Praed and Grant 1952).

Social behavior. These birds are locally migratory, and when found in the nonbreeding period they usually occur in flocks of from 6 to about 50 birds. There is disappointingly little information on the social behavior of this species during the breeding season, however, and nothing seems to have been written on pairing or territorial behavior. Maclean (1972*b*) illustrated an elaborate postcopulatory run display, with the pair members each raising one wing vertically. Given the absence of wing spurs or wattles, one might imagine that a low level of territorial behavior is present.

Reproductive biology. Breeding in Kenya is reported to occur from March to July, as well as in November and December. Breeding in southern Africa occurs from August to October. In the highlands of Ethiopia breeding of the race *melanopterus* is said to occur in April, while on the Ethiopian and Somaliland frontier it is reported for June. Breeding is said to occur in small colonies (Mackworth-Praed and Grant 1952), and the nests are poorly lined scrapes, usually in peaty soil or on a slight rise. Nests are often placed in recently burned areas, and there are usually 3 eggs, though clutches of 2 and 4 have also been reported. The incubation period remains unreported, as do the relative roles of the two sexes.

Status and relationships. Although the race *minor* is evidently rather common on the grasslands of East-ern Africa, the range of *melanopterus* is much smaller, and its status is unknown. This species is clearly a very close relative of *lugubris*, and although they overlap extensively they have been described as constituting a superspecies. Beyond that, *melanopterus* may be a part of the "*Chettusia*" group of lapwings (Glutz et al. 1975), although to some extent the similarities of this group of birds may be a reflection of their common structural and ecological adaptations to rather arid plains, their absence of wing spurs and wattles being apparently associated with weak territorial development.

Suggested reading. None.

Eurasian Lapwing

Vanellus vanellus (L.) 1758

Other vernacular names. Green plover; lapwing; peewit; vanneau huppé (French); Kiebitz (German); vanello (Spanish).

Subspecies and range. No subspecies recognized. Breeds in Europe and northern Asia from the British Isles (sporadically to the Faeroes, rarely to Iceland) to southern Ussuriland, south to northwestern Morocco, and east to northern Greece, Iran, and Mongolia. Winters from the breeding range south to the Mediterranean Basin, northwestern India, and eastern China. See map 27.

Measurements and weights. Wing: males 219–34 mm; females 216–33 mm. Culmen (from feathering): males 23–27 mm; females 22–27 mm (Witherby et al. 1941). Weights (in spring): males 174–259 g; females 220–80 g. Adult summer and fall weights from June to October average between 192 and 233 g (Glutz et al. 1975). Eggs ca. 46 x 33 mm, estimated weight 26.0 g (Schönwetter 1962).

DESCRIPTION

Adults in breeding plumage have an elongated crest, head, neck, and breast that are bluish black except for white on the sides of the head and neck that grades to gray on the hindneck. The back, scapulars, and upper wing surface are iridescent bluish to greenish, and the flight feathers are black except for the last inch or so of the outer three primaries, which are light gray with darker tips. The upper tail coverts are

27. Breeding distribution of the Eurasian lapwing.

cinnamon rufous, the tail is basally white and has a broad subterminal band decreasing outwardly. The underparts posterior to the chest are white, grading to light cinnamon on the under tail coverts, and the inner part of the under wing coverts is also white, contrasting with the black outer portion. The bill is black, the iris is dark brown, and the legs and toes are red. *Adults in winter* are more extensively white on the face. *Females* can sometimes be recognized by slightly shorter crests and by their greenish blue rather than blue lesser upper coverts. They also have broader and rounder wings than do males, which is sometimes evident when the birds are in flight. *Juveniles* may be recognized by their wing coverts, which have narrow buff fringes and a more brownish breastband. By the first winter these fringes are gone, and the birds are similar to adults but have entirely white outer tail feathers (Prater et al. 1977).

In the hand, this species' elongated crest, which appears in the juvenile plumage, a hind toe, and lack of wing spurs separate it from all other species in its range.

In the field (12 inches), the moderately large size, black-and-white pattern, and unusually rounded wings that flash black and white in flight are distinctive field marks. The wingbeat is rather slow, and the flight is relatively sluggish. The most common call is a *pee-wit* or *vee-veet,* and the birds are especially noisy during the breeding season.

NATURAL HISTORY

Habitats and foods. This species occupies a broad climatic range during the breeding season, from boreal to steppe and even desert climates. The favored breeding habitats are wide grasslands and arable fields, riverside meadows, grassy portions of moors, bogs, and swampy heathlands. Generally, the combination of low grasses and patches of bare or nearly bare soil where visual foraging is possible are the major habitat criteria for this broadly adapted species (Voous 1960). Breeding densities vary greatly, and although they are usually less than a pair per 10 hectares, in some areas "colonies" can develop that may reach a density of as many as 9 breeding pairs on less than a hectare (Glutz et al. 1975). In general, lapwings seem to seek out breeding areas that include one or more of the following features: absence of scattered trees or high wood plantings on the edges, low vegetation or bare ground, a grayish brown to grayish green color of the field or an uneven surface, and proximity to the place of hatching or the previous breeding ground (Klomp 1954). Foraging is done visually from the ground surface; the species primarily feeds on earthworms and insects, less often on mollusks and spiders. Beetles, including both larvae and adults, ants and other hymenopterans, orthopterans, and the larvae of dipterans are all important insect components of the diet (Glutz et al. 1975). This species often forages by alternating quick runs and pecking movements, and its skeletal structure has been modified for running by such features as its relatively wide pelvis, relatively long femurs, and relatively short lower legs (tibiotarsus plus tarsometatarsus), all of which aid in maintaining balance and speed (Klomp 1954).

Social behavior. Sexual maturity in this species is attained in the first year of life, though breeding is frequently deferred until the second or even third year. Seasonal monogamy is the usual breeding system, although variations (two or even three females associated with a single male) have at times been reported. In areas where the species is migratory, the first birds to return to the breeding grounds are often males. They soon take up territories and begin their aerial song flight display, which advertises their territory, warns off potential rivals, and challenges enemies (Bannerman 1961). These flights are made from the time territories are formed until about the time the eggs are hatched and the territories dissolve. The typical song flight consists of an initial "butterfly flight" phase, performed with deep and slow wingbeats, which is soon followed by an "alternating flight" slightly higher above the ground, with the bird revolving from right to left and following a zigzag course while producing nonvocal humming sounds. Dives toward the earth often occur during this phase. After the alternating flight the bird proceeds with slow wingbeats of very low amplitude just above the ground in a "low flight" phase. Toward the end of this phase the bird begins an ascent phase, rising 3–10 meters, then leveling off and beginning the "high flight" phase. At the end of this phase the bird suddenly turns on its back and dives more or less vertically toward the earth in a "vertical dive." After

this dive the bird levels out and remains close to the earth until it again begins the alternating flight phase. After this it may land or begin the sequence again. Besides the humming sounds of the alternating flight, the bird utters a three-motif song during the ascent, high flight, and vertical dive phases, with each motif associated with one of these phases. The entire song flight is relatively ritualized and is predominantly aggressively motivated as well as being ecologically adapted to effective transmission of signals over open terrain (Dabelsteen 1978). The nest-scraping ceremony is an important aspect of ground display, and there may be 20 or more scraping sites in a male's territory. When a female is attracted to a scrape and enters it, the male sometimes almost upends beside her, tilting his wing tips vertically and pulsating his raised tail. The female indicates her receptivity for copulation by bowing slightly forward, and the male may either fly directly onto her back or may sing and perform a corkscrew flight a few feet above ground before copulating. Or he may quickly run behind her and stand upright with crest thrust forward and gorget puffed out before mounting. Copulation may occur on the nest scrape, within a few feet of it, or well away from it. Copulation also is performed during incubation and may occur either within the territory or outside it (Bannerman 1961). Even after the chicks are hatched and well grown, copulation sometimes occurs, though single-brooding is apparently the rule throughout the species' range. Pair-bonding in this species is weak, and territorial boundaries are poorly defined, both of which tend to increase the likelihood of polygynous matings (Laven 1941).

Reproductive biology. Lapwings preferentially nest in low vegetation, often placing the nest on a site where the eggs closely match the color of the environment (Klomp 1954). The scrape is generally on a slightly raised site, and the female lines it with grass stems and other vegetation. Eggs are laid at intervals of 30–48 hours, with 4-egg clutches being completed in extremes of 5–10 days, but usually on the seventh or eighth day. The clutch size is normally 4 eggs (in about 90 percent of the clutches), with 3-egg clutches being the usual exception. Replacement clutches are also usually of 4 eggs and normally are begun within a week of clutch loss. Incubation is normally not begun before the last egg is laid and is performed by both sexes. Typically males do not incubate much during the first week, but they do so more steadily in the latter part of the incubation period. The female probably normally incubates at night. The incubation period ranges from 24 to 34 days, but it averaged 27.4 days in 17 instances. The pipping period varies considerably, ranging from about 36 to 96 hours in 31 cases. The chicks may remain in the nest for up to 40 hours if the weather is bad, but typically they leave within 12 hours of hatching. They are tended by both parents, with one of the parents, often the female, usually deserting them before they are fledged. Fledging occurs in 30–42 days, probably most often between 35 and 40 days (Glutz et al. 1975; Bannerman 1961).

Status and relationships. This species has adapted well to humans, and since the start of this century has significantly increased its European breeding range, primarily by northward expansion. This may be due to climatic amelioration during that period (Voous 1960), but is probably partly the result of improved protection and the increased area of cultivated or grazed lands. Probably the Netherlands supports 120,000–160,000 pairs, while Germany has at least 50,000–60,000, to mention only two major population centers (Glutz et al. 1975). The species' closest relationships appear to be with the South American lapwings *chilensis* and *resplendens* (Bock 1958) rather than with any African or Asian species, as might have been expected on zoogeographical grounds.

Suggested reading. Ennion 1949; Spencer 1953.

Southern Lapwing

Vanellus chilensis (Gmelin) 1789
(*Belonopterus chilensis* of Peters, 1934)

Other vernacular names. Chilean lapwing; vanneau tero (French); Cayennekiebitz (German); avefría Americana (Spanish).

Subspecies and ranges. See map 28.

V. c. cayannensis: Cayenne southern lapwing. Resident in Colombia, Venezuela, the Guianas, and northern Brazil.

V. v. lampronotus: Brazilian southern lapwing. Resident in Brazil south of the Amazon and southward to Bolivia, Paraguay, Uruguay, and northern Argentina.

V. c. chilensis: Chilean southern lapwing. Resident in Chile and Argentina south to Chloé and Chubut.

V. c. fretensis: Lesser southern lapwing. Breeds in southern Chile and southern Argentina from

28. Breeding distributions of the Brazilian (B), Cayenne (C), Chilean (Ch), and lesser (L) southern lapwings.

Tierra del Fuego, north to Aysén and Chubut. Migratory at southern end of range.

Measurements and weights. Wing (flattened): males 223–63 mm; females 222–61 mm. Culmen (to feathering): males 26–32 mm; females 26–34 mm (Blake 1977). Weights: 1 male of *cayannensis* 280 g; 1 unsexed adult of *fretensis* ca. 425 g *(Bull. Brit. Orn. Club* 93:117; Humphrey et al. 1970). Eggs ca. 45 x 32 mm (*cayannensis*) to 49 x 35 mm (*chilensis*); estimated weights 24.0–31.5 g (Schönwetter 1962).

DESCRIPTION

Adults of both sexes have a long, thin black or gray occipital crest and a black area extending from the forehead downward to the lower throat, where it comes to a pointed tip or merges with the black breast. Behind this is a white border, and the rest of the head, neck, and upper breast is ash gray to pale bluish gray. There is a broad black breastband, and the rest of the underparts including the under wing coverts are white. The gray color of the neck extends across the back and upperparts except for an area of coppery bronze on the scapulars. The lesser wing coverts are glossy green, the middle and greater coverts are extensively white, and the secondary coverts and inner secondaries are ash gray. The primaries and outer secondaries are bluish black, and the upper tail coverts and tail are white except for a broad black subterminal tail band. The iris and eye ring are red, the bill is red at the base and black toward the tip, and the legs and feet are dull reddish, or grayish brown with reddish joints. *Females* are identical in plumage to males but probably have shorter spurs. *Juveniles* have a more brownish breastband, are barred with brown and buff dorsally, lack black on the forehead, and lack crest and wing spurs.

In the hand, the combination of wing spurs, a narrow crest, and a small hind toe serve to distinguish this species from other *Vanellus* forms.

In the field (13 inches), this large and conspicuous shorebird can be seen almost throughout South America, where its loud calls and contrasting plumage pattern make it hard to overlook or to confuse with other species. Only the smaller pied lapwing is remotely similar to it, and that species lacks a crest, while the Andean lapwing lacks black on the face and breast. The call is loud and frequently consists of a repeated harsh two-noted whistle. The Spanish name "tero-tero" provides a good idea of its sound.

NATURAL HISTORY

Habitats and foods. Pasturelands and rather humid pampas are the preferred habitats of this species, though the birds have at times also been seen in open glades of wooded areas. The species is most common in the tropical zone but extends to temperate regions as a summer visitor. In Argentina the birds breed in shortgrass uplands and around marshy borders, while outside the breeding season they occur in grasslands, inland wetlands, sloughs, and tidal flats (Myers and Myers 1979). In Chile the species is largely associated with damp meadows and cultivated fields and at times is maintained in semicaptivity in farmsteads or even urban gardens. It is reportedly entirely insectivorous, often eating grasshoppers, but its food also includes earthworms and perhaps other terrestrial invertebrates (Humphrey et al. 1970; Johnson 1965). It is known that earthworms are a preferred food of young birds (Greer and Greer 1967).

Social behavior. These birds are fairly gregarious outside the breeding season; flocks of about 200 have been observed in autumn before leaving Isla Grande, Tierra del Fuego, while for much of the year in Chile they may be seen in loose flocks of a dozen or fewer individuals. The timing of pair-formation is not known, but in November the birds have been seen in Tierra del Fuego performing an aerial display about 50 feet in the air, flying with slow butterflylike wing movements. A similar display flight was noted by Wetmore (1926) in late October near Buenos Aires. It seems likely that this corresponds to the butterfly flight of *V. vanellus,* its probable nearest living relative. Maclean (1972*b*) has described three wing displays in this species. These include a conspecific defensive threat involving slight lateral wing spreading, a mutual advertising threat with both wings raised vertically by two birds running in parallel, and a defensive threat in which only one wing is raised while the other is held slightly away from the body. He found (in litt.) that the birds are loosely colonial, with individual pairs holding exclusive territories about 50 meters in diameter. Myers and Myers (1979) found that the birds are territorial but densely distributed, with nests sometimes less than 125 meters apart. In Chile breeding is said to begin in July, in the middle of the winter, with the smaller and more southerly race *fretensis* beginning two or three months later than the nominate form. On Isla Grande, eggs have been reported in October and young birds have been seen in late November and December (Johnson 1965; Humphrey et al. 1970).

Reproductive biology. The typical clutch size for this species is 4 eggs, with clutches of 3 also occurring occasionally. The nest is a simple scrape, sometimes lined with grasses, usually placed in a dry spot such as a slight rise within generally wet surroundings (Johnson 1965). The incubation period is probably about 27 days (Haedo Rossi 1969), and newly hatched chicks weigh about 20 g (Greer and Greer 1967). Both sexes incubate and care for the young, which are highly precocial. In one case a chick was found to increase in body weight from 148 to 250 g over a 23-day period, and on some days it consumed more than its own body weight in earthworms. The fledging period is still unreported but must be considerably more than a month. Recent observations by Walters (1980) in Venezuela have provided surprising evidence that cooperative breeding sometimes occurs in this species. Although most birds bred as monogamous, intensely territorial pairs that excluded conspecifics from their territories, 4 of 20 breeding groups contained one or two extra adults or well-grown young that behaved as a unit, collectively defending the territory or attacking predators. Walters suggested that habitat saturation and competition for suitable territories may have led to this behavior.

Status and relationships. Certainly this is one of the most conspicuous and numerous of shorebirds in the warmer and nonwooded portions of South America, and in many areas it might be regarded as abundant. It thus is in no need of attention from conservationists. Seebohm (1888) regarded this species as a direct descendent of the ancestral *V. vanellus,* which later also probably gave rise to *V. resplendens.* I agree that *chilensis* and *vanellus* are very close relatives, with *resplendens* slightly more removed from this pair and presumably the result of an early derivative of *chilensis* stock.

Suggested reading. Wetmore 1926; Hudson 1920.

Andean Lapwing

Vanellus resplendens (Tschudi) 1843

Other vernacular names. None in general English use; vanneau des Andes (French); Andenkiebitz (German); avefría Serrana (Spanish).

Subspecies and range. No subspecies recognized. Breeds in the high Andes from southwestern Colombia to extreme northern Chile and northwest-

ern Argentina. Winters at lower elevations, sometimes to sea level. See map 29.

Measurements and weights. Wing (flattened): males 210–25 mm; females 195–228 mm. Culmen (to feathering): males 35–41 mm; females 32–39 mm (Blake 1977). Weights: 6 males 201–30 g, average 210.8 g; 6 females 193–230 g, average 216.1 g (various sources). Eggs ca. 45 x 33 mm, estimated weight 25 g (Schönwetter 1963).

DESCRIPTION

Adults of both sexes have the head, neck, and breast ashy gray, grading to white on the throat. The back and scapulars are bronzy green, the lesser wing coverts are purplish violet, and the inner secondaries and tertials are ashy gray with greenish iridescence. The primaries and outer secondaries are black, and the middle and greater wing coverts are extensively white, as are the upper tail coverts and the basal half of the tail, which has a broad black subterminal

29. Breeding distribution of the Andean lapwing.

band. The abdomen, flanks, under tail coverts, and under wing coverts are entirely white. The iris is pink, the bill is red basally and brownish toward the tip, and the legs and feet are pink. *Females* are apparently identical to males. *Juveniles* exhibit little dorsal iridescence; the feathers are brownish with buffy tips, and the crown feathers are also edged with buffy.

In the hand, the combination of a wing spur, no hind toes, and iridescent upper wing coverts serve to identify this species from all other *Vanellus* forms.

In the field (13–14 inches), this species breeds in grassy habitats and near the shorelines of lagoons and rivers at relatively high altitudes, but nonbreeding birds rarely occur as low as sea level. It is the only South American shorebird with a strongly banded tail and almost uniformly gray head and breast coloration. It has loud characteristic calls, from which it apparently gets its Spanish name "Lique-lique."

NATURAL HISTORY

Habitats and foods. This Andean species is associated with grassy fields and open country, primarily in the temperate and puna zones between 3,000 and 4,000 meters, wintering at lower altitudes and sometimes to sea level. No information is available on foods.

Social behavior. Nothing specific is known of this.

Reproductive biology. The only information on the nesting of this species comes from Johnson (1965), who reported that in January and February, at an elevation of about 14,000 feet, he observed territorial birds in the provinces of Arica and Tarapacá of northernmost Chile. After unsuccessfully searching for nests in the grassy cover surrounding a lake and a river, he located one on a barren, waterless slope. In each of 2 nests found, there were 4 eggs present, and a greater amount of vegetational lining than usually occurs in the lowland species *chilensis*. No other breeding information seems to be available.

Status and relationships. There is no information on the population status of this species. It is most probably an offshoot of early *chilensis* stock, as has already been suggested by Seebohm (1888). The wing spur of *resplendens* is much smaller and blunter than that of *chilensis*, and the bird is also appreciably smaller overall. In this respect it seems convergent with some of the arid-adapted lapwing species of Asia, such as *leucurus* and *gregarius*.

Suggested reading. Johnson 1965.

Blacksmith Plover

Vanellus armatus (Burchell) 1822
(*Hoplopterus armatus* of Peters, 1934)

Other vernacular names. None in general English use; vanneau armé (French); Schmiedekiebitz (German).

Subspecies and range. No subspecies recognized. Resident in Africa, breeding from Kenya and Angola to South Africa. Subject to considerable seasonal movements. See map 30.

Measurements and weights. Wing: both sexes 194–223 mm. Culmen: both sexes 26–30 mm. Weights: 264 adults of both sexes 114–211 g, average 156 g (Skead 1977). Britton (1970) provided a range of 138–97 g for 47 birds, average 163.7 g. Eggs ca. 40 x 29 mm, estimated weight 16.5 g (Schönwetter 1962).

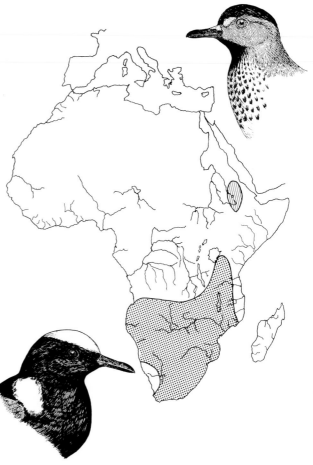

30. Breeding distributions of the blacksmith plover (*shaded*) and spot-breasted lapwing (*hatched*).

DESCRIPTION

Adults of both sexes have the forehead, crown, back of the neck, upper tail coverts, belly, under wing coverts, and rump white, with a terminal black band that is widest at the middle and narrows toward the edges. The primaries and secondaries are black, the legs and bill are dull black, and the iris is crimson. *Females* are identical to males. *Immatures* have a buffy to brownish crown, an off-white chin, and buff speckling on the areas that are black in adults.

In the hand, the large wing spurs, the absence of a hind toe, and the bluish gray wing coverts should serve to identify this species. The most similar species is the spur-winged plover, which has brown wing coverts and white cheeks with a black forehead, rather than black cheeks and a white forehead.

In the field (11–12 inches), this species is usually found on open but damp grasslands and tends to be very shy. When disturbed, it will often fly at an intruder and utter *klink* calls that sound like the striking of a hammer on an anvil.

NATURAL HISTORY

Habitats and foods. The most characteristic habitat of the blacksmith plover consists of relatively moist situations such as around the edges of vleis or on marshy grasslands. Nonbreeders frequently concentrate around sewage farms, where they may forage in sludge fields or other rather damp situations. They are said to feed on insects, worms, and small mollusks, and they often forage by making short runs terminated by rapid downward pecking movements. They have on occasion been observed to forage while wading in shallow water and have also been noted foraging in wet sludge fields by making tremoring movements with one leg extended forward, in a manner similar to that noted for the Eurasian lapwing (Hall 1959b).

Social behavior. Flocking during the nonbreeding season is well developed, and migratory or dispersal movements are common during that period. The apparent breeding season is quite varied geographically, apparently being from July to November in the Cape region, from March to June in the Transvaal, from April to June in Southern Rhodesia, and from May to October in Zambia. Farther north in Kenya it probably extends from April to August. In the area of the Cape, pair-forming behavior becomes evident in May and June, with copulation especially frequent during June and July. Observations by Hall (1964) suggest that pair-forming behavior may in part in-

volve pivoting and running, with the lead bird (perhaps the male) often carrying materials in the bill. This may promote nest site choice or stimulate copulation. Treading is typically preceded by the female's walking in front of the male, which takes a high and upright stance. She then crouches and is mounted, but unlike crowned plovers the birds usually perform no flapping or extension of the wings during or after copulation. A scraping ceremony, in which one bird or both presses the breast down onto the ground while making pecking movements, is also present and probably serves an important function in pair-bonding and nest site selection. Although nesting territories are well spaced and apparently do not overlap, there seems to be no clear-cut expulsion behavior toward conspecifics, and intruding birds have been seen foraging as close as 10 to 15 yards from a nest without stimulating any aggressiveness. However, other species of shorebirds often elicit attacks from birds having eggs or young, as do dogs and occasionally humans (Hall 1964).

Reproductive biology. Nests are almost invariably within a few yards of water and usually also are situated so that the incubating bird will have an unobstructed panoramic view. When placed on sandy soil, the nest is often very simple and virtually unlined, while in grassy cover it may show substantial construction. Most clutches (32 of 49) consist of 4 eggs, with extremes of 2 and 6, but clutches of more than 4 eggs are probably laid by two females. The average nesting territory size in Hall's study was found to be about 9 acres, and nests are usually several hundred yards apart. Incubation is performed by both sexes, with the average periods between nest relief about 50 minutes. Three records of incubation averaged 30 days, while in a fourth case incubation probably was between 25 and 27 days. Within a few days after hatching the young may be moved considerable distances from the nest, and one estimate of the fledging period is 41 days. At least in the Cape Town region there is good evidence that second nestings may be frequently attempted, with the pair remaining on the same territory (Hall 1959b, 1964).

Status and relationships. There is no reason to believe that this widespread and locally common species needs any special attention from the standpoint of its conservation, though it has a decidedly patchy distribution. Its nearest relative is almost certainly *V. spinosus*, and the two forms clearly constitute a superspecies.

Suggested reading. Hall 1959b, 1964.

Spur-winged Lapwing

Vanellus spinosus (L.) 1758
(*Hoplopterus spinosus* and *duvaucelii* of Peters, 1934)

Other vernacular names. Spur-winged plover; river lapwing *(duvaucelii)*; vanneau éperonné (French); Spornkiebitz (German); avefría espolada (Spanish).

Subspecies and ranges. See map 31.

V. s. *spinosus:* African spur-winged lapwing. Resident in Egypt and south of the Sahara from Ethiopia to Senegal and south to Nigeria and Kenya. Also breeds occasionally, but increasingly, in Greece and in coastal Turkey. It occurs on some of the Mediterranean islands as a migrant or wanderer, and it has nested rarely on Cyprus.

V. s. *duvaucelii:* Asian spur-winged lapwing. Resident from northern and eastern India east through Assam to the Malay Peninsula, Cambodia, and Hainan. Often considered a separate species.

Measurements and weights (for *duvaucelii*). Wing: both sexes, 185–205 mm. Culmen (from feathering): both sexes 26–28 mm. Weights: 6 adults of *spinosus* 127–59 g, average 148 g (Britton 1970). Eggs ca. 41 x 29 mm (Ali and Ripley 1969); estimated weights from 16.4 g (*spinosus*) to 18.0 g (*duvaucelii*) (Schönwetter 1962).

DESCRIPTION

Adults of both sexes have the forehead, crown, and a short nape crest black, with the black continuing down from the base of the lower mandible across the throat, either terminating there (*duvaucelii*) or extending to the belly, where it is continuous with a blackish abdomen and flanks (*spinosus*). The cheeks and the rest of the head are white to grayish, becoming (in *duvaucelii*) brownish gray on the lower breast and finally white on the flanks and abdomen, except for a black abdominal patch. The rest of the upperparts are gray to sandy brown, except for white upper tail coverts. The tail is white basally, with a

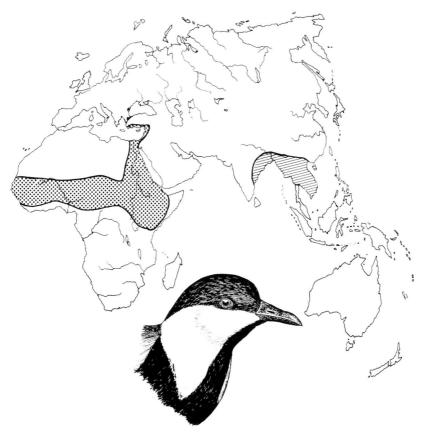

31. Breeding distributions of the African (*shaded*) and Asian (*hatched*) spur-winged lapwings.

15. Adult heads of African (*above*) and Asian (*below*) spur-winged lapwings.

In the hand, the combination of wing spurs, a short black crest, and no hind toe serves to distinguish this species.

In the field (12 inches), this species might be confused with the African black-headed lapwing, but that species has a longer crest and white flanks. In India it may perhaps be mistaken for the red-wattled lapwing, which, however, lacks a crest and has red wattles in front of the eyes. The species is found in dry river sand and shingle habitat and has a sharp, loud call sounding like a repeated *did,* or *did-did-do-weet,* resembling, but still distinct from, that of the red-wattled lapwing.

NATURAL HISTORY

Habitats and foods. The Asian race *duvaucelii* is largely associated with sandbars or shingle beds of flowing rivers above tidal limits and avoids stagnant waters or jheels. Its foods are said to include insects, worms, crustaceans, and even frogs or tadpoles (Ali and Ripley 1969). However, the foods of *spinosus* are believed to consist almost entirely of insects (Glutz et al. 1975). Further, the race *spinosus* is certainly not a river-dependent form, and it may instead be associated with marshy areas, mud flats along lakeshores, or seacoasts and similar habitats. Yet, in Africa as in Asia, the bird is never far from water, often being found along the larger rivers where they flow through arid country. The recent nestings in Greece, where the birds have been found nesting near salt water and in habitat dominated by salt-tolerant plant species such as *Salicornia, Juncus,* and the like, are probably not typical.

Social behavior. Outside the breeding season these birds are usually found in pairs or family groups, and at least in India they rarely occur in flocks of more than about 6 birds. However, where they are migratory in the Mediterranean Basin the situation may be somewhat different, and premigratory assemblages of a hundred or more birds have been observed. At least in some areas the birds nest in loose colonies, although intensive territorial activity seems to be the rule in these circumstances (Helversen 1963). A threat display, involving a highly erect posture and apparent exhibition of the black breast and undersides, is commonly performed at this time, as well as wing-lifting and actual fighting. Apparently, however, there is no aerial display flight in this species.

broad black terminal band. The axillaries and under wing coverts are white, the primaries and their coverts are black, the central secondaries are white, and the upper wing coverts are mostly ashy, becoming whitish on the greater coverts. The iris is carmine red, and the bill, legs, and feet are black. *Females* cannot be readily separated from the males by their plumage, but the spurs of females are shorter, usually under 7.5 mm compared with 11.5 mm or more on males. *Juveniles* can be recognized by their generally blackish brown color, and they retain the extensive buffy fringing on their inner coverts until spring (Prater et al. 1977).

Before copulation there is a ceremony of the male circling around the female that bears some similarity to the "parade-marching" of the small sandpipers. Copulation seems to be limited to the period of breeding and sometimes occurs at considerable distances from the nest (Glutz et al. 1975).

Reproductive biology. The nest is a simple scrape in the ground, often with little or no vegetational cover around it, and the contrasting plumage pattern of the adults makes incubating birds quite conspicuous. However, the birds brood with the head low, remaining as inconspicuous as possible, though when flushed off the nest they defend it with great vigor. There are typically 4 eggs, but clutches ranging from 2 to 5 have been reported. There is a rather long nesting period (from April to August for *spinosus* and from at least March to June for *duvaucelii*), and at least in some areas it is believed that the birds may have two broods per season. Both sexes incubate, but neither the incubation period nor the fledging period seems to have been established. The male probably performs the majority of the incubation, and the birds sometimes wet their belly feathers before starting to incubate, as a temperature or humidity control device. The young are brooded for roughly 3 weeks.

Status and relationships. Although the status of *duvaucelii* is not clear, it is at least clear that in *spinosus* there has been a recent range expansion. The birds began to become more common in Greece in the 1950s, and by 1959 they were definitely nesting. After initially colonizing the northeastern coast, the birds gradually extended their range to the Gulf of Salonika, and they now seem to be well established in Greece and have been seen as far north as Bulgaria and Romania (Hudson 1974). The two forms *spinosus* and *duvaucelii* have at times been considered full species (Bock 1958; Peters 1934) and at other times only subspecies (Ali and Ripley 1969). I believe the differences between them probably are not great enough to warrant species distinction, given the considerable plumage variation occurring in other subspecies of *Vanellus* (such as *miles*). However, it would be useful to have more specific information on behavior and ecology relative to this question. Beyond this, relationships within the genus *Vanellus* seem most probably to be with *armatus*, a species having considerable similarity in adult plumage pattern to these forms.

Suggested reading. Glutz et al. 1975; Hudson 1974.

Spot-breasted Lapwing

Vanellus melanocephalus (Rüppell) 1845
(*Tylibyx melanocephalus* of Peters, 1934)

Other vernacular names. Spot-breasted plover; vanneau d'Abyssinie (French); Strichelkiebitz (German).

Subspecies and range. No subspecies recognized. Resident in the highlands of northern and central Ethiopia. See map 30.

Measurements and weights. Wing: both sexes 230–45 mm. Culmen (to feathering): both sexes 23–27 mm. Weights: unsexed adults 199 and 228 g (USNM specimens). Eggs: no measurements available, described as brownish blue to smoke gray, with heavy black markings.

DESCRIPTION

Both sexes of adults are generally ashy brown above, with a faintly iridescent green tint. The forehead, crown, and nape are black, and the nape is elongated to form a slight crest. There is a white superciliary stripe, and a small primrose-colored wattle in front of the eye. The throat and front of the neck are black, with black extending as a series of short streaks downward into the white chest. The rest of the underparts are white, and the tail is white with a narrow black subterminal band. The upper wing surface has a white band formed by the greater upper coverts. The iris is yellowish gray with a dark yellow eye ring, the bill is black, becoming yellow toward the base, and the legs and feet are pale yellow. *Females* are apparently outwardly identical to males. *Juvenile* plumages are undescribed.

In the hand, the combination of a hind toe, no spur, a small reddish wattle, and a black throat surrounded by a gray neck should identify this species.

In the field (13–14 inches), this species is found only on the upland grassy plains of north central Ethiopia. The only other crested lapwing likely to be found in the same area is the black-headed lapwing, which has a white throat and long black breast plumes. These birds are said to be quite tame, flying and running like the Eurasian lapwing and having a *pewit* call similar to that species'. In addition, they are said to utter a *kree-kree-kre-krep-kreep-kreep*, and a *kueeep-kueep*.

Habitats and foods. This species has the smallest range of any African lapwing, usually being found above 3,050 meters (10,000 feet) in marshy grasslands and moorlands having giant heath, giant lobelia, alchemilla, and tussock grass in the western and southeastern highlands of Ethiopia. Nothing seems to have been reported of its foods, but it forages in a manner similar to that of *V. vanellus*, by making short runs and stops (Urban 1978).

Social behavior. In the nonbreeding season these birds are often seen in small flocks of up to 30 or 40 birds, but they otherwise occur in pairs or small parties, presumably family groups.

Reproductive biology. Very little is known of this species' reproduction, but it is known to breed in April in the Bale Mountains and during August in the Shoa region. Only a single nest has been described. It was a shallow scrape in a patch of grass and moss in the giant lobelia moorlands, and it contained 4 eggs (Urban et al. 1972; Urban 1978).

Status and relationships. This species is still regarded as locally common (Urban 1978), but its very restricted range and undoubtedly limited population qualify it for special attention from conservationists. Seebohm (1888) believed it might be quite closely related to both *V. cinereus* and *V. indicus* on the basis of its small wattle and lack of a wing spur, and in having a hind toe. I suspect it may be more closely related to the African form *V. tectus* and, to a lesser extent, *V. armatus*.

Suggested reading. None.

Black-headed Lapwing

Vanellus tectus (Boddaert) 1793
(*Sarciophorus tectus* of Peters, 1934)

Other vernacular names. Black-headed plover; vanneau coiffe (French); Schwarzkopfkiebitz (German).

Subspecies and ranges. See map 32.
> *V. t. tectus:* Northern black-headed lapwing, resident in Africa, breeding from Senegal to Ethiopia, Uganda, and Kenya, with stragglers reaching Egypt and sometimes farther north.

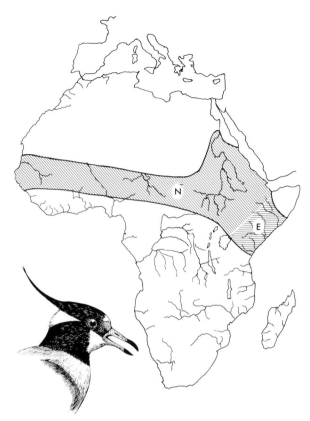

32. Breeding distributions of the eastern (E) and northern (N) black-headed lapwings.

> *V. t. latifrons:* Eastern black-headed lapwing. Resident in Africa, breeding from southern Somalia to Kenya.

Measurements and weights. Wing: both sexes 160–200 mm. Culmen: both sexes 23–27 mm. Weight: 1 female 100 g (Britton 1970). Eggs ca. 35 x 25 mm, estimated weight 12.3 g (Schönwetter 1963).

DESCRIPTION

Adults of both sexes are generally pale brown above, with a black crown that is extended into a short crest. There is a white streak behind the eyes, extending backward and joining on the nape under the crest, and a white forehead and throat. The neck all around the throat and between the base of the bill and the eye is black, and there is a small red wattle in front of the eye. The black of the neck extends down the center of the chest and breast and terminates in long plumes anterior to the white underparts. The tail is white with a broad subterminal black band. The inner wing coverts and inner secondaries are pale brown, grading to white on the outer wing coverts and secondaries. The innermost primaries and outermost secondaries are white on the basal half and black terminally, the other primaries being entirely black. The iris is yellow, the bill dull red with a black tip, and the legs and toes are carmine. *Females* are apparently identical to males. *Juveniles* resemble adults but have the feathers of the upperparts edged with buffy.

In the hand, the combination of a short black crest, small red wattles, no hind toe, and no spur should serve to identify this species. It is perhaps most likely to be confused with *melanocephalus,* but that species has a small hind toe.

In the field (10 inches), these birds are associated with dry, open country but not desert. The broad black streak extending down the middle of the breast and the white line around the crested crown provide good field marks. In flight, the white on the primary coverts and terminal half of the outermost secondaries should be visible. The flight is slow, and the birds are relatively tame. They are noisy in flight, and utter rasping cries said to be hollow-sounding and "peevish."

NATURAL HISTORY

Habitats and foods. This is a distinctly arid-adapted species, most often associated with dry sandy country that is covered by low brush with tufts of scattered grasses. The foods are little studied but are believed to include insects and mollusks. The birds are evidently not attracted to water for foraging, but instead tend to forage in dry and sandy areas. Probably much of the feeding is done at night.

Social behavior. Evidently this species is a local migrant, and it sometimes is found in small flocks. Breeding of the race *latifrons* in Kenya occurs in May, after the April rains. Additionally, breeding has been noted as from March to May in Nigeria, and it apparently extends from March to July in Ethiopia. Thus it is probably a rainy-season breeder in general.

Reproductive biology. Relatively little is known of this species' nesting biology. Nests are simple scrapes, typical of lapwings, but are well lined with grass or sometimes almost any kind of debris. Some times the eggs seem to be half-hidden in the lining, which may serve to reduce their visibility and may also help protect them from the sun's rays. The average clutch of *latifrons* is said to be only 2 eggs (North 1937), with as few as 1, but in the race *tectus* the

clutch is said to be 2 or 3 eggs. During the heat of the day, the brooding bird tends to fluff its feathers and squat over the eggs, thus shielding them rather than actually incubating them. Egg-wetting has not been reported, probably because the birds are only rarely close to a supply of surface water. The incubation period is unknown, and there is an unsubstantiated report of an adult's flying off with a chick between its legs (North 1937).

Status and relationships. Nothing specific can be said of this species' status, but, except for the restricted distribution of the race *latifrons,* the large overall range of this species seems to make it fairly secure. Its distinct plumage similarities to *melanocephalus* certainly suggest that these two are closely related; Wolters (1975) includes these two forms as well as *indicus* and *tricolor* in a single genus *Lobivanellus.* However, I suspect that *spinosus* and *armatus* may be more closely related to *tectus* and perhaps *melanocephalus,* though neither of these has wattles in front of the eye.

Suggested reading. North 1937.

Red-wattled Lapwing

Vanellus indicus (Boddaert) 1783
(*Lobivanellus indicus* of Peters, 1934)

Other vernacular names. Indian decorated lapwing; red-wattled plover; vanneau Indien (French); Indischer Lappenkiebitz (German).

Suspecies and ranges. See map 33.

V. i. aigneri: Western red-wattled lapwing. Resident from Iraq and eastern Iran to Sind and near the Indus River, and also in eastern Saudi Arabia and north to Turkmenistan. Migratory in the northern parts of its range.

V. i. indicus: Indian red-wattled lapwing. Resident in India from the Indus to Bengal and Assam, and north to the Himalayas.

V. i. lankae: Ceylonese red-wattled lapwing. Resident in Sri Lanka (Ceylon).

V. i. atronuchalis: Eastern red-wattled lapwing. Resident from the Brahmaputra River eastward through the Malay Peninsula and the Indo-Chinese countries.

Measurements and weights. Wing: males 211–38 mm; females 208–37 mm. Culmen (from base):

33. Breeding distributions of the Ceylonese (C), eastern (E), Indian (I), and western (W) red-wattled lapwings.

males 34–40 mm; females 33–38 mm. Measurements of *atronuchalis* average slightly less (wing 200–21 mm). Weights: 21 unsexed adults 110–230 g, averaging 181 g (J. C. Daniel, pers. comm.). Eggs ca. 42 x 30 mm (Ali and Ripley 1969), estimated weight 18.7–20.5 g (Schönwetter 1963).

DESCRIPTION

Adults of both sexes have a black breast, head, and neck with a broad black band extending from each eye and passing down the sides of the neck to join (except in *atronuchalis*) the white lower parts. The upper plumage and the upper wing coverts are brown glossed with greenish bronze and slightly with red. There is a white wing bar, but otherwise the flight feathers are black. The sides of the lower back, rump, and upper tail coverts are white, and the tail is white with a broad black subterminal band. The central tail feathers have this band bordered with brown on both sides, and the other feathers have white tips.

The iris is red brown, the eye ring and the wattles in front of the eyes are bright red, the bill is red with a black tip, and the legs are bright yellow. In *atronuchalis* the white on the head and neck is confined to the ear coverts and a narrow line between the nape and the mantle. *Females* are not externally separable from males. *Juveniles* resemble adults, but the dark area of the head and neck is brown, and the chin is white. The upper wing coverts have buffy edges. *First-winter birds* have buffy edges on some upper wing coverts, especially the inner medians, which may be held through the winter at least in some areas (Prater et al. 1977).

In the hand, the combination of the conspicuous red wattle, white ear coverts, and otherwise black head should identify this species.

In the field (13 inches), the birds are usually found in open, well-watered country, often in pairs or small groups. The birds are extremely noisy and inquisitive, and their loud calling when disturbed sounds like "Did you do it?" or "Pity to do it," accompanied by repeated *did-did-did* notes and diving attacks. In flight the bird closely resembles the spur-winged lapwing, but that species lacks a white-tipped tail.

NATURAL HISTORY

Habitats and foods. This species is associated with damp meadows, irrigated lands, and generally open and well-watered habitats, sometimes extending into grassy glades of open forest. Although it forages primarily on dry land, it is never found far from water. It is also frequently associated with cultivated lands, ditches, and temporary puddles, and at times it may be seen along the grassy shoulders of highways. Foods are diverse and evidently include ants, beetles, caterpillars, and other insects as well as mollusks and some vegetable materials (Ali and Ripley 1969). It is said to often feed on sandbars and dry meadows, where it consumes orthopterans (Dementiev and Gladov 1969).

Social behavior. These birds are essentially resident but undergo some altitudinal migration during the nonbreeding season in India, and at the northern edge of their range in the USSR they are migratory. Evidently even during the nonbreeding season the birds do not occur in large flocks, but rather in groups of from 2 to about a dozen, with most of their foraging activity occurring near dawn and dusk or even night in moonlight. In India the breeding season is primarily from March to August or September, while in the southern USSR breeding begins in mid-April. In Sri Lanka breeding occurs during the southwest monsoon, primarily during June. Nothing specific has been written on pair-forming behavior or territoriality.

Reproductive biology. The nest is usually near water on dry land, including fallow fields, stony wasteland, shingle riverbanks or islands, and the like. It is a simple scrape, sometimes with a lining of pebbles, goat droppings, or bits of cattle dung (Ali and Ripley 1969). There are typically 4 eggs, but as few as 2 have been reported. Incubation is performed by both sexes. During very hot weather the incubating bird will often wet the breast and belly feathers by dipping them nearby water and settling back down over the eggs, thus keeping them moist (Naik et al. 1961). The incubation period is still unreported. Nest-distraction display, in the form of "broken-wing" behavior, is sometimes performed by the adult, and the well-camouflaged young freeze on a vocal signal from the parents.

Status and relationships. This species is common over much of its range and occupies a broad and widely available ecological niche, so that it can exist even in densely populated areas. In its plumage and other characteristics it has some distinct similarities to *V. tricolor* of Australia, which I believe is probably its nearest living relative, and perhaps the two forms constitute a superspecies.

Suggested reading. Naik et al. 1961.

Banded Lapwing

Vanellus tricolor (Vieillot) 1818
(*Zonifer tricolor* of Peters, 1934)

Other vernacular names. Banded plover; black-breasted plover; vanneau tricoloré (French); Schildkiebitz (German).

Subspecies and range. No subspecies recognized. Resident throughout Australia including Tasmania, except for northern portions. See map 34.

Measurements and weights. Wing: both sexes 183–94 mm. Culmen (to base): both sexes 24–29 mm. Weights: 2 females 172 and 180 g; 3 males 182–97 g, averaging 187.7 g (specimens in National Museum of Victoria). Eggs ca. 45 x 31 mm, estimated weight 22.8 g (Schönwetter 1962).

DESCRIPTION

Adults of both sexes have a head that is black above the eyes, and a second broad black band extends from the base of the bill across the cheeks and down the sides of the neck to the breast, where it forms a broad band. Between these two black bands there is a white stripe across the back of the head from eye to eye, and there is a bulbous red wattle in front of the eye. The chin, throat, and upper breast are white, as are the underparts behind the black breastband, as well as the under wing coverts. The upper wing coverts are brown with a purple sheen, the primaries are black, the secondaries are white with black tips, and the inner tertials are entirely brown. The tail is white with a broad black subterminal band. The iris is yellow, with a yellow eye ring, the bill is yellow with a black tip on the upper mandible, and the legs and toes are dark red. *Females* are identical in plumage to the males. *Juveniles* have the wattle only poor-

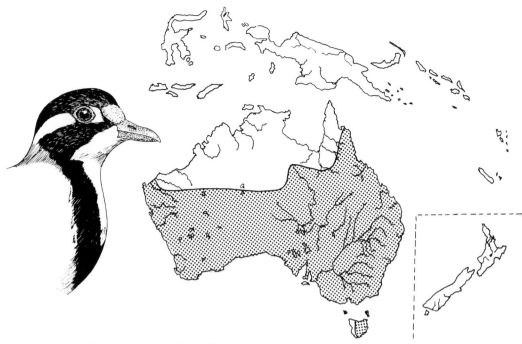

34. Breeding distribution of the banded lapwing.

ly developed, and there is brown and black mottling on the head and breast.

In the hand, the combination of no wing spurs, no hind toe, and a white throat serves to identify this species from other *Vanellus* forms.

In the field (12 inches), this species is likely to be confused only with the masked lapwing, which also has a black-and-white head, but which has a yellow wattle and an incomplete black breastband. The banded plover is found in dry, rather open areas with short vegetation. Its voice is thin and somewhat metallic, with a common call sounding like *a-chee-chee-chee* or *er-chill-char.* In flight, the white wing bar formed by the secondaries is conspicuous, which is lacking in the masked lapwing.

NATURAL HISTORY

Habitats and foods. Unlike the larger masked plover, this species is a bird of Australia's dry interior, generally found on dried mud flats, vegetation-free lakeshores, and in pastures where the grass is relatively short. Evidently, tall grasses impede movements of this short-legged lapwing, and when the grasses are damp they tend to make its breast feathers wet and matted. The species' foods consist mostly of insects and other small surface invertebrates, and it sometimes also feeds on small seeds, especially during cold weather (Frith 1969). When foraging, the birds tend

to run a short distance, then stop and pick up a food item. They have also been observed standing on one foot and quickly trembling the other, apparently to flush small prey.

Social behavior. During the nonbreeding period these birds range rather far and tend to congregate for feeding and roosting. During this period, flocks of 200 or more birds are sometimes seen. The birds also breed in localized colonies, and in this regard they evidently differ considerably from *miles,* which appears to be much less tolerant of nesting near others of its own species. The breeding season is not fixed but tends to follow rains, regardless of the time of year. However, much of the breeding occurs during the winter and spring period of July to December when the interior tends to receive most of its rainfall. However, autumn breeding has also been noted following unusual amounts of summer and autumn rainfall.

Reproductive biology. Nests are usually simple scrapes, or the eggs may be laid in a depression such as the mark made by a horse's hoof. The nest is often lined with small stones, bits of dung, or sometimes with dried grasses. The clutch is normally 4 eggs, but sometimes is only 3. Incubation is by both adults and lasts about 28 days. The young are cryptically patterned with sandy brown, darker brown, and black, and when disturbed they quickly freeze while both parents attempt to lure the intruder from the vicinity of the brood. Fledging required about 46 days in one zoo-raised bird (Gerrits 1956).

Status and relationships. The opportunistic breeding adaptations of this species, and its tolerance of arid habitats, assure it of continued abundance over most of the interior of Australia. It obviously is fairly closely related to the Australian masked plover *V. miles,* although clearly related to the red-wattled lapwing, *V. indicus,* which is perhaps its nearest living relative.

Suggested reading. Frith 1969.

Masked Lapwing

Vanellus miles (Boddaert) 1783
(*Lobobyx miles* and *L. novae-hollandiae* of Peters, 1934)

Other vernacular names. Masked plover; spur-winged plover (*novaehollandiae*); vanneau soldat (French); Soldatenkiebitz (German).

Subspecies and ranges. See map 35.
V. m. *miles:* North Australian masked lapwing. Resident in northern Australia, intergrading in central Queensland with *novaehollandiae.* Recently has extended its range to southern New Guinea. This form and intergrades with *novaehollandiae* also have recently colonized extreme western Australia. Nonbreeders occur widely over Australia, New Guinea, and adjacent islands.
V. m. *novaehollandiae:* South Australian masked lapwing. Resident in eastern and southern Australia including Tasmania. Also has recently colonized extreme southern New Zealand.

Measurements and weights. Wing: both sexes average from 210 (*miles*) to 247 mm (*novaehollandiae*), and culmen length averages 35–38 mm. Weights: males of *miles* 234–56 g, females 191–296 g (Hall 1974). Weights of both subspecies 216–411 g, average of 27 was 328 g (D. Purchase, pers. comm.). Eggs ca. 43 x 32 mm (*miles*) to 49 x 36 mm (*novaehollandiae*), estimated weights 21.4 g (*miles*) to 32.0 g (*novaehollandiae*) (Schönwetter 1962).

DESCRIPTION

Adults of both sexes have a head that is white from the level of the eyes downward, and a black crown above the eyes. This crown either terminates at the back of the head (*miles*) or extends down the back of the neck as a black line, expanding on the shoulders

35. Breeding distributions of the south Australian (S) and north Australian (N) masked lapwings. Areas of overlapping hatching show locations of intermediate populations.

16. Adult heads of south Australian (*above*) and north Australian (*below*) masked lapwings.

and sides of the breast to form an incomplete breast-band (*novaehollandiae*). There is a large yellow facial wattle in front of the eyes that hangs downward below the bill and forward as a rounded lobe in front of the eyes, and (in *miles*) extends backward to terminate above and behind the eyes. The upper wing coverts and tertials are olive brown, the primaries and secondaries are black, and there is a yellow wing spur with a brown tip. The under wing coverts are white, the tail is white with a broad black subterminal band, and the entire underparts are also white. The iris and eye ring are yellow, the bill is yellow with a brown tip, and the legs and feet are purplish red. *Females* are identical to male in plumage but sometimes have shorter wing spurs (Barlow et al. 1972). *Juveniles* have only partially developed wing spurs and wattles and have irregular dark and light brown spotting above (Frith 1969).

In the hand, the combination of long wing spurs, a

hind toe, and large yellow wattles serves to identify this species.

In the field (13–14 inches), this species is likely to be confused only with the banded lapwing, from which it is easily distinguished by the absence of a complete black breastband, its much larger yellow facial wattles, and its considerably longer legs. In flight it shows no white on the upper wing surface, and both sexes frequently utter a loud disturbance call, sounding like *keer-kick-ki-ki-ki*, or may produce other calls sounding like *krick-krick*, or *kitta-kitta*, and so forth.

NATURAL HISTORY

Habitats and foods. Widespread in Australia, this species is found on cleared land, along the margins of rivers, lakes and swamps, and on tidal flats. It is most numerous on grazing lands, especially those fairly near water. Outside the breeding season it occurs mostly along the margins of swamps and rivers, tidal flats, and farm ponds. During the breeding season, however, the birds are much more associated with the short grassy habitats of grazed pastures (Thomas 1969). Their diet is a wide array of worms, insects, and other invertebrates, and in cold weather they occasionally resort to eating seeds.

Social behavior. During the nonbreeding period, these birds are sometimes seen in flocks of as many as 500 birds, which evidently do not perform a true migration but nonetheless may move up to about 150 kilometers. Pairing occurs in the middle of the winter, and at least in Tasmania territorial behavior begins in May. However, in more tropical areas breeding occurs during the summer and autumn rainy period, rather than during spring as in the south. As territorial behavior develops the flocks decline, and only a few small flocks of presumed nonbreeders persist through the breeding season. Territories are contiguous and nonoverlapping; and, although pairs tend to return to the same area in successive years, relatively few territories were reoccupied in one study (Thomas 1969), probably because the vegetation had become too high. Females can breed when a year old and males in their second year, though some males may also breed as yearlings. It is probable that pairs do not break up until one member dies, although in one observed case a female remated within 2 weeks of the death of a mate of at least 3 years. There is one known case of a bird's breeding at almost 10 years of age, and another at 8 years (Barlow et al. 1972).

Reproductive biology. Most nests are in short grassy areas; nearly 60 percent of 255 nests in Tasmania were on grazed pastures. Drainage ridges, slopes, or other raised surface irregularities are used for nest sites in preference to flat or low-lying areas. Additionally, nests are often placed near some distinctive feature, such as a large stone or piece of wood. Nests are simple scrapes and are usually lined with vegetation or other materials available near the nest (Thomas 1969). Nest-building may begin as long as 12 days before the first egg is laid, and copulation has been seen from as early as 14 days before the first egg to 2 days afterward. The modal clutch size is 4 eggs in Tasmania (mean 3.53) and the same (mean 3.74) in New Zealand. Both sexes share incubation, and renesting is apparently frequent after nest or brood loss. Although Thomas (1969) found no evidence of second broods in Tasmania, there was at least one case of double brooding in a New Zealand study (Barlow et al. 1972). The incubation period has been determined as 28–32 days in Tasmania and 30–31 days in New Zealand. The fledging period was likewise estimated at 6–7 weeks in Tasmania and 7–8 weeks in New Zealand. Young birds remain with their parents until they are 7–8 months old.

Status and relationships. Although once regarded as two species, the two well-marked forms of masked lapwings are now generally believed to constitute races (Van Tets et al. 1967). Both seem to be flourishing, since the northern form has recently colonized southern New Guinea and the southern one has become well established in extreme southern New Zealand. A possible intergrade type has become locally established in extreme western Australia. The species' relationships with other *Vanellus* are not at all clear, though in plumage characteristics it is fairly close to *tricolor*, and it seems likely that the two are fairly closely related. Additionally, it may be a close relative of *macroptera*.

Suggested reading. Thomas 1969; Barlow et al. 1972.

Javanese Wattled Lapwing

Vanellus macropterus (Wagler) 1827
(*Rogibyx tricolor* of Peters, 1934)

Other vernacular names. Black-thighed wattled lapwing; Javanese wattled plover; vanneau hirondelle (French); Schwarzbauchkiebitz (German).

Subspecies and range. No subspecies recognized. Onetime resident on Java, and possibly on Sumatra and Timor. Current status unknown, possibly extinct. See map 36.

Measurements and weights. Wing: both sexes 237–49 mm. Culmen: both sexes 36–43 mm. Weights: no information, estimated ca. 325 g. Eggs (c.f. Hellebrekers and Hoogerwerf 1967) ca. 45 x 32 mm, estimated weight 22.8 g (Schönwetter 1962).

DESCRIPTION

Adults of both sexes have a glossy black crown and nape, followed by a grayish collar around the hindneck, which extends over the sides of the neck and lower throat. The sides of the face and throat are black, and the foreneck, breast, and abdomen are ashy brown, with the center of the abdomen and lower breast black. The vent, under tail coverts, under wing coverts, and axillaries are white. Above, the coloration is mostly light brown, including the wing coverts, while the greater coverts have slightly paler tips. The primaries and their coverts are black. The secondaries have paler bases and white on the inner webs, and the upper tail coverts and base of the tail are white. The rest of the tail is black with a white terminal band. The iris is brownish black, the lower unfeathered part of the tibia is orange, becoming yellow on the tarsus, and the scales of the tarsus and underparts of the toes are dark sepia brown. The bill is black, becoming pinkish at the base. The wattle is said to be white (yellow on museum specimens) with a darker tip. *Juvenile* plumages are undescribed.

In the hand, the unusual blackish plumage, as well as the presence of wing spurs and a hind toe, serve to identify this species.

In the field (11 inches), this extremely rare species, if not extinct, is associated with open areas near freshwater ponds. Its lapwinglike shape and blackish coloration would immediately separate it from all other shorebirds in the area.

NATURAL HISTORY

Habitats and foods. This species is said to inhabit open areas near freshwater ponds, and it presumably also occurs on agricultural lands. However, nothing more specific on its ecology or possible foods seems to have been written (Temple 1979).

Social behavior. No information is available.

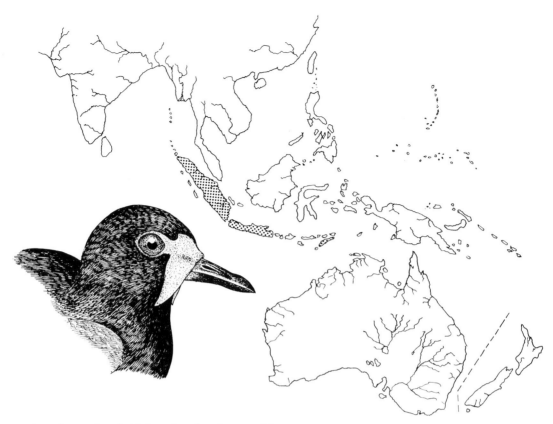

36.   Original probable breeding distribution of the Javanese wattled lapwing.

Reproductive biology. Breeding in West Java evidently occurred in May and June, based on information assembled by Hellebrekers and Hoogerwerf (1967). The eggs are said to resemble those of European lapwings and, except for size, Pacific stilts.

Status and relationships. According to Temple (1979), this species has only questionably been reported from Sumatra and Timor, and a reputed clutch of eggs from Sumatra is of doubtful validity. On Java, the species was collected near Djakarta in 1872 and observed off the north coast of that island shortly before 1885. It was also collected in the 1920s from the Tjitaroem delta and at Tangeran, west of Djakarta. The most recent report is from Meleman, on the south coast of East Java, in 1939; it has not been reported by any observers in recent years in spite of recent fieldwork in the Tjitaroem delta. Thus there is no way of knowing if the species still survives or is extinct. Its relationships are rather uncertain; Seebohm (1888) suggested that it was a derivative of early *malabaricus* stock, but I very tentatively suggest that *miles* may be its nearest relative. So little is

known of the bird that any judgment of its affinities must be extremely uncertain.

Suggested reading. Temple 1979.

Wattled Lapwing

Vanellus senegallus (L.) 1766
(*Afribyx senegallus* of Peters, 1934)

Other vernacular names. Wattled plover; Vanneau du Senegal (French); Senegalkiebitz (German).

Subspecies and ranges. See map 37.
   V. s. senegallus: Senegal wattled lapwing. Resident in Africa, breeding from Senegal and Gabon to the Central African Republic and the southern Sudan.
   V. s. lateralis: Southern wattled lapwing. Resident in southern Africa, breeding from the lower Congo and Angola east to Uganda and Natal.

Clancey (1979b) has recently separated part of this form in a new race *solitaneus*, with a range extending from Angola and South West Africa to Botswana, Zambia, and eastern Zaire.

V. s. major: Northern wattled lapwing. Resident in Ethiopia and northern Somalia.

Measurements and weights. Wing: both sexes 216–47 mm. Culmen: both sexes 32–37 mm. Weights: both sexes 197–277 g (Verheyen 1953). Eggs ca. 46 x 31 mm, estimated weight 21.2–25.4 g (Schönwetter 1962).

DESCRIPTION

Adult males are mostly olivaceous brown above and white below. The top of the head is blackish with a white center, and the forehead is also white. There are small dark red (above) and large yellow (below) wattles in front of the eyes, the neck is heavily streaked with black, and the chin and throat are black. The outer wing coverts are white, as are the lower belly, the under tail coverts, and the under wing coverts, while the tail is white with a broad

37. Breeding distributions of the northern (N), Senegal (Se) and southern (So) wattled lapwings.

subterminal black band. The iris is gray with a central ring of tawny brown and an outer ring of black, the eye ring is yellow, the bill is yellow with a black tip, and the legs and feet are yellow. The race *lateralis* is distinguished by having black between the brown belly and white underparts. *Females* are similar but have much less black on the throat. *Juveniles* have the chin and throat white with black streaks, only a little white on the head, and small face wattles (Mackworth-Praed and Grant 1952).

In the hand, the very large wattles, the small but sharp wing spur, and the distinctive facial patterns should easily identify this species.

In the field (13–14 inches), this species is usually found in savanna, fields, or cleared lands, usually quite near water. It is tame and usually is found in pairs or small groups. It somewhat resembles another wattled species, the white-crowned lapwing, but the black throat, streaked neck, and pale brown, rather than black, wings serve to separate the two species. It is a noisy bird, easily excited, and utters a screaming or yelping *peep-peep* and other shrill notes.

NATURAL HISTORY

Habitat and foods. This is a species of semiarid habitats. The birds are not usually found breeding in the vicinity of flats and pans, but rather are associated with locations having running water and adjacent higher and more rocky ground. However, it also feeds frequently in wetter areas, such as in short, damp grass and generally swampy grounds, as well as in recently burned areas that are sometimes rather far from water. Its foods evidently consist of insects, including large beetles and aquatic forms, and sometimes also grass seeds. Grasshoppers, ants, and other insects are reported as probable insect prey, and the bird captures them by walking slowly and deliberately, rather than by making short runs as do *armatus* and *vanellus* (Little 1968).

Social behavior. During the nonbreeding season these birds are fairly mobile, but they rarely seem to gather in very large numbers. In Zambia, groups of from 20–60 have been seen at certain times, usually between March and June, but there seems to be no definite migration pattern there. However, in the Transvaal the species is definitely migratory, though it rarely occurs in groups larger than family-sized units. Studies there by Little (1968) indicate that the species is relatively solitary, and the birds arriving in September are already paired. The pairs apparently return to certain localities each year and soon establish exclusive territories that are essentially feeding territories. These territories vary in size, but in Little's study they averaged about 300 yards long by 100 to 211 yards wide, or 30,000–60,000 square yards, and generally did not include the location of the nest itself. Most of the territorial defense is performed by the male, and most of the copulations occur within the territorial boundaries. Fights usually are the result of territorial trespass, and an "upright threat" with extreme neck-stretching is a high-intensity threat. Wing-raising is another form of threat, and a display flight, involving a downward swooping with a rolling motion toward an intruding bird, was also observed once. Copulation is usually preceded by a high-pitched call initiated by either bird then joined in by the other, followed by the birds' approaching one another and the female's crouching. After treading, the male typically raises both his wings vertically for a second or two, while also slightly fanning the tail and stretching the neck. As in some other lapwings, copulation often follows the expulsion of another bird from the territory. Apparently the male initiates nest-building by performing the scraping ceremony while simultaneously uttering a clucking call. As he pushes his breast into the ground, he simultaneously kicks backward with his feet in alternating sequence. This may stimulate the female to perform the same behavior, either at that site or a few feet away. Sideways throwing movements, which later occur during nest-relief ceremonies, are usually also part of the display (Little 1968).

Reproductive biology. Breeding north of the equator is reported to occur from March to June, and the same period is reported for Kenya, whereas to the south most nestings occur from August to November. In her Transvaal study area, Little (1967) noted that the nest is often outside the feeding territory and is usually situated so as to give the bird an unobstructed view of the surrounding countryside. The eggs are laid at approximately daily intervals and usually number 4. Incubation begins with the third egg, before which the eggs are covered and shaded from the sun. The incubation period is probably 30–32 days (*Ostrich* 36:224) and is performed by both sexes about equally. Both sexes guard the chicks, initially alternating brooding duties. Later the male takes the initiative in mobbing or threatening intruders while the female remains with the brood. At least some birds fledge 6 weeks (or 40 days) after hatching, but even when the young are fully grown they remain with their parents.

Status and relationships. Wattled lapwings are evidently common over much of their range and are unlikely to cause any conservation problems, since they are highly tolerant of human disturbance and agricultural practices. The species has frequently been placed in the monotypic genus *Afribyx*, and Wolters (1975) includes it and *albiceps* in the genus *Xiphidiopterus*. I believe these two species are probably quite close relatives, although their sharing of wattles and wing spurs should not be regarded as important evidence favoring this view.

Suggested reading. Little 1968.

White-crowned Lapwing

Vanellus albiceps (Gould) 1834
(*Xiphidiopterus albiceps* of Peters, 1934)

Other vernacular names. Black-shouldered plover; white-headed plover; pluvier à tête blanche (French); Langspornkiebitz (German).

Subspecies and range. No subspecies recognized. Resident in Africa, breeding from Senegal to the Sudan, south to northern South Africa, and east to Kenya and Mozambique. Subject to seasonal migratory movements. See map 38.

Measurements and weights. Wing: both sexes 198–231 mm. Culmen: both sexes 30–36 mm. Weights: 1 adult 201 g (Britton and Dowsett 1969). Eggs ca. 42 x 31 mm, estimated weight 20.4 g (Schönwetter 1962).

DESCRIPTION

Adults of both sexes are gray on the upper mantle, head, and neck, with a broad white streak extending from the forehead down the center of the crown. There is also a white spot below the eye and long greenish yellow pendant wattles in front of the eyes. The gray of the upper mantle is separated from the nape by a narrow black-and-white stripe, and there is also a white stripe along the outer scapulars separating the gray mantle from the black upper wing co-

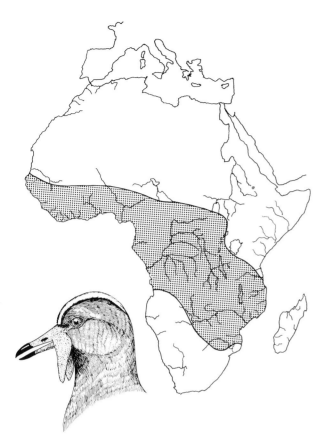

38. Breeding distribution of the white-crowned lapwing.

peated piping call, *peep-peep,* and the alarm note is a loud *whit-whit.*

NATURAL HISTORY

Habitats and foods. Throughout its range this species seems to be closely associated with water, specifically with rivers that have extensive sandbanks and sandy islands. The birds are essentially sedentary, but during times of flooding they leave the rivers and move to lagoons or temporarily flooded areas on drier ground. Their foods have not been well studied but are known to include insects such as mantids. A bone from a frog or fish has also been reported as a food item. The birds overlap ecologically with *V. senegallus,* but that species occurs not only along sandbanks but also more generally around swampy ground and adjacent grassy areas.

Social behavior. Although it has not been extensively studied, it is known that this species is highly territorial, and nests are typically no closer than a quarter-mile apart (Serle 1939). However, during the nonbreeding period the birds are more gregarious and have been reported migrating in "great flocks" (Bannerman 1951). In western Africa and equatorial Africa breeding usually begins near the end of April, with Nigerian records from March to May, Cameroon records for December and January, and Zaire records for February and August. In Zambia the records extend from July to October, with a peak in September, and in South Africa they are reported to breed during the periods of low water on the rivers, specifically during September and October on the Zambesi River.

Reproductive biology. Without exception, nests of this species are on sandbars or areas of shingle, usually only a short distance from the edge of the water. The nests are shallow scrapes, often lined with small sticks or pebbles that apparently gradually accumulate in the nest during incubation changeover ceremonies, as the departing bird picks up small stones and flicks them backward. There are from 2 to 5 eggs, usually 3, and typically the eggs are directly exposed to sunlight during most of the day. Thus the eggs must be kept damp by the incubating bird, through repeated trips to the water, where it dips its breast and belly into water and quickly returns to the nest to settle over the eggs (Reynolds 1968). Incubation is by both sexes, and copulation frequently oc-

verts. The secondaries'and inner primaries are white, and the wings are sharply spurred. The underparts from the chin posteriorly are white, and the tail is also white, with a broad black tip. The iris is yellow, the bill is yellow with a black tip, and the legs and feet are yellowish green. *Females* have a brown-bordered crown. *Immatures* resemble adults but have white tips on the wing coverts and have smaller spurs and facial wattles, and their brownish back is faintly barred with darker brown.

In the hand, the long spurs and long wattles, together with a white throat, should identify this species.

In the field (12 inches), this species is usually found on sandy riverbanks and less often in swamps, usually in small, noisy groups. It is relatively tame, and in flight it shows a great deal of white on the flight feathers, contrasting with the black upper wing coverts. The large wattles and white crown and underparts serve as field marks. It utters a loud and recurs even as late as the time of hatching. Copulation is not preceded by any obvious display, and it often occurs after an intruding bird has been driven off the territory. There is a high degree of aggression toward other shorebirds, including smaller *Charadrius* species such as three-banded plovers. The incubation and fledging periods are still unreported but probably incubation lasts more than 25 days (*Bokmakierie* 32:21). In one observed case, one parent took charge of 2 chicks that had already hatched while the other parent remained on the nest to tend a newly hatched chick and a pipping egg. Interestingly, during this period the adults continued to exchange nest-tending duties and also copulated several times (Reynolds 1968).

Status and relationships. The white-headed lapwing is considered common in much of its range but is ecologically restricted to certain rivers, primarily in the tropical zones. It has at times been placed in the monotypic genus *Xiphidiopterus* (Peters 1934); more recently Wolters (1975) expanded this genus to include *senegallus.* It seems likely to me that *senegallus* is the closest living relative to *albiceps,* and they also share some ecological similarities.

Suggested reading. Reynolds 1968.

Tribe Hoploxypterini (Pied Lapwing)

The pied lapwing is a small South American shorebird with no hind toe, no crest or bare skin on the face, and a relatively long tarsus (about one-third the length of the wing). The scapulars are black with white inner edges, forming a distinctive V-shaped dorsal plumage pattern. The downy young are *Charadrius*-like but have a broad black head band extending around the nape and indistinct dark stripes on the sides of the back (Helmut Sick, pers. comm.). One genus and species.

Genus *Hoploxypterus* Bonaparte 1856

This monotypic genus has the characteristics of the tribe *Hoploxypterini.*

Pied Lapwing

Hoploxypterus cayanus Latham 1790

Other vernacular names. Little white-winged lapwing; three-toed Cayenne lapwing; vanneau de Cayenne (French); Diademkiebitz (German); avefria de espolon (Spanish).

Subspecies and range. No subspecies recognized. Resident east of the Andes in South America from southeastern Colombia and Venezuela south and east to Paraguay and southeastern Brazil. See map 39.

Measurements and weights. Wing (chord): males 135–50 mm; females 132–49 mm. Culmen (to feathering): males 23–25 mm; females 22–24 mm (Blake 1977). Weights: 4 males 55–80 g, average 72.2 g; 2 females 82 and 84 g (various sources). Eggs: measurements evidently unreported.

DESCRIPTION

Adults of both sexes have the forehead, sides of the head, upper back, and a broad band across the breast black. This area is separated from a pale grayish brown crown and hindneck by a narrow white border, and the lower parts of the face, throat, and foreneck are also white. The scapulars are black, with a white medial border, forming a conspicuous V-mark, and the rest of the back, upper wing coverts, and tertials are pale grayish brown. The primaries are black, the secondaries are white, with the outer ones tipped with black, and the greater coverts are also tipped with white. The tail is white, broadly tipped with a black band, while the tail coverts, abdomen, and other underparts posterior to the breast are white. The iris is pink, with a pink eye ring (also reported gray with a scarlet eye ring in some sources), the bill is black, and the legs and toes are reddish pink to salmon red. *Females* are identical to males. *Juveniles* closely resemble adults but exhibit smaller breastbands, and the black scapular markings are also less well developed.

In the hand, the combination of a small but sharp wing spur, no hind toe, and a wing length of no more than 150 mm should identify this species.

In the field (9 inches), this species may resemble a *Charadrius* species more than the larger lapwings, but the strongly banded tail and the dark black V-

39. Breeding distribution of the pied lapwing.

mark formed by the scapulars should serve for identification. Its usual call is a clear, mellow whistle of two notes, the second lower in pitch. The call is more melodious than that of typical lapwings and is reminiscent of that of *Charadrius*, which is probably a reflection of the small body size.

NATURAL HISTORY

Habitat and foods. Other than the fact that this is an open country and savanna-dwelling species, nothing specific seems to have been written of its ecology.

Social behavior. The only good account of this species' behavior comes from Friedmann and Smith (1975), who reported that these plovers are usually found in groups of from 6 to 10 birds, but at times as many as 100 congregate around some larger ponds in northeastern Venezuela. They observed displays over a three-month period of approximately May through July. At this time birds could be seen flying over a particular spot, using a peculiar undulating flight and uttering a high *kee-kee-kee-kee-kee* call that rose and fell in pitch. The birds would then light and face one another, standing with their wings fully spread. The wings would then be closed, and the birds would march back and forth in a stiff erect posture, with the breastband much in evidence. They would at times also go into a crouching posture, raising the far wing and drooping the nearer one. Often they would then return to the spread-eagle position, and they sometimes enter mock battle, striking to-

ward each other with their wings and calling continuously throughout the encounter. These displays certainly suggest a high degree of territoriality or other agonistic behavior characteristic of breeding birds.

Reproductive biology. Regrettably, almost nothing is known of this species' reproduction. Friedmann and Smith (1975) stated that breeding activity in northeastern Venezuela occurred from May through July and that flightless young were seen in June and July. Dr. Helmut Sick (in litt.) informed me that he photographed young birds on the big sand beaches of the upper Xingu River in central Brazil, Mato Grosso, in late June of 1949. The nest and eggs are still inadequately described, but the reported clutch is only 2 eggs (Maclean 1972a).

Status and relationships. Almost nothing can be said of this species' status, but it seems to be very uncommon over much of its range, though Friedmann and Smith certainly observed rather substantial numbers in the savannas of Venezuela. The form's relationships are indeed questionable. It was traditionally placed in a monotypic genus among the lapwings until Bock (1958) merged it with *Vanellus*, indicating that the species was probably derived from the African lapwing group; he noted, however, that the color pattern of the back is unique. Strauch (1976) took strong exception to this treatment, stating that the plumage pattern of *Hoploxypterus* is typical of *Charadrius* and almost identical to that of such species as *C. melanops*. He supported this position by noting that Zusi and Jehl (1970) had found that *cayanus* was the only presumed lapwing to lack certain struts on some cervical vertebrae, and that Burton (1974) observed that, unlike other lapwings, certain muscle variations exist that he believed might represent convergence toward the condition in *Charadrius*. Certainly the smaller body size and modified foraging opportunities would result in convergence of anatomical similarities with *Charadrius*, and the agonistic behavior described above seems to closely fit the lapwing social pattern of highly developed territoriality in species with wing spurs. None of the *Charadrius* species have wing spurs, and, while this is an obvious adaptive correlate of territorial defense, Strauch himself believed that wing spurs represent "true cladistic characters." Until further evidence is forthcoming, I believe *Hoploxypterus* should be given tribal status and placed between the lapwings and the more typical plovers.

Suggested reading. None.

Tribe Charadriini (Typical Plovers)

Semipalmated Plover
(male threat display)

Plovers are very small to small shorebirds that usually have no hind toe, pointed wings, a reticulated tarsus that is short to moderately long, and a bill that is shorter than the head. All species lack wingspurs, bare facial skin, and iridescent plumage. The downy young usually have a white nape band and a "pebbly" dorsal pattern. The sexes are alike as adults and usually have a white nape band, a black neck collar, and a dark terminal or subterminal band on the tail. Seven genera, including 38 species, are recognized here.

KEY TO GENERA OF CHARADRIINI

A Bill bent laterally . . . . *Anarhynchus* (1 sp.)
A' Bill not bent laterally
  B Underparts finely barred, wing relatively short (under 120 mm), head mostly blackish, and sides of neck and upperparts rufous . . . . *Phegornis* (1 sp.)
  B' Not with above combination of traits
    C. Tail barred, wing at least 150 mm . . . . *Pluvialis* (3 spp.)
    C' Tail not barred, usually white-tipped, wing usually under 150 mm
      D Underparts tawny to creamy buff
        E Black V-mark present on throat . . . . *Peltohyas* (1 sp.)
        E' Throat lacking black, but black median patch on abdomen . . . . *Oreopholus* (1 sp.)
      D' Underparts usually white, never tawny to creamy buff
        E Four toes, head mostly black with white throat . . . . *Erythrogonys* (1 sp.)
        E' Usually three toes, if four toes, then head not mostly black . . . . *Charadrius* (25 spp.)

Genus *Pluvialis* Brisson 1760 (Greater Plovers)

Greater plovers are small to medium-sized arctic-nesting shorebirds that have only a minute hind toe or none, pointed wings that are nearly four times as long as the tarsus, and a plumage with extensive pale spotting on the upperparts. The tail is relatively short, square-tipped, and barred. The bill is relatively slender and more than half as long as the tarsus. The front toes are webbed at the base, particularly between the middle and outer toes. Three species are recognized.

KEY TO SPECIES OF *Pluvialis*

A Hind toe present but minute, culmen at least 27 mm, upperparts spotted with white . . . . *squatarola*
A' Hind toe absent, culmen under 25 mm, upperparts spotted with yellow
  B Axillaries and under wing coverts white . . . . *apricarius*
  B' Axillaries and under wing coverts brownish gray . . . . *dominicus*

Greater Golden Plover

Pluvialis apricaria (L.) 1758

Other vernacular names. Western golden plover; pluvier doré (French); Goldregenpfeifer (German); chorlito dorado común (Spanish).

Subspecies and ranges. See map 40.
  P. a. apricaria: Southern greater golden plover Resident in the British Isles, Denmark, western Germany, southern Norway, and southern Sweden.

  P. a. altifrons: Northern greater golden plover. Breeds in Iceland, the Faeroes, northern Scandinavia, and east to the Taimyr Peninsula. This subspecies, or both, winters from the southern end of the North Sea to the Iberian Peninsula and in the Mediterranean Basin, and in small numbers on the southern Caspian coast, rarely beyond.

40. Breeding distributions of the northern (N) and southern (S) greater golden plovers.

Measurements and weights. Wing: males 172–98 mm; females 176–97 mm. Culmen (from feathering): males 20–26 mm; females 21–26. Weights (breeding season): males 140–218 g, females 157–94 g (Glutz et al. 1975). Eggs ca. 52 x 35 mm, estimated weight 32.5–33 g (Schönwetter 1963).

DESCRIPTION

Breeding adults of both sexes have the sides of the head, chin, throat, foreneck, chest, breast, and abdomen dull black, broadly margined laterally with white from the forehead and superciliary area down the sides of the neck to the breast. The sharpness of the white and black is considerably greater in *altifrons* than in *apricaria,* which lacks uniform black on the sides of the head and has an intermixture of sepia-spotted white and golden feathers on the throat, breast, and abdomen, obscuring the sharply defined patterning present in *altifrons.* The uppermost parts are dusky, with bright ochre yellow speckling that expands on the scapulars and larger

wing coverts to form bars along the feather margins. The wing coverts are grayish brown with yellowish markings, the primaries and their coverts are dusky with whitish margins, and the tail is grayish brown with pale yellow and whitish or pale grayish brown barring. The under wing coverts and axillaries are white, as are the under tail coverts; the latter two have some darker markings. The iris is dark brown, the bill is black, and the legs and toes are grayish black to greenish gray. *Adults in winter* lack black on the sides of the head and underparts and have a white throat and abdomen, and the upperparts are less conspicuously marked with yellow. *Juveniles* have paler and duller yellow spotting than adults but are distinctly spotted on the breast, mantle, and scapulars, while the abdomen and flanks are white but extensively tipped with pale brown. *First-winter* birds are hard to tell from adults in winter but may retain dark-tipped feathers on the abdomen and flank until midwinter (Prater et al. 1977).

In the hand, the relatively large size (wings over 170 mm), lack of a hind toe, and ploverlike bill separate this from most species, and the white axillaries distinguish it from *dominicus*.

In the field (11 inches), this species is unlikely to be seen in company with lesser golden plovers but differs from them in all plumages by its white axillaries. It is more likely to be seen with the gray plover. In winter these two species are rather similar, but the gray plover has black axillaries and a white rump patch, rather than a brownish rump and white axillaries. Its usual call is a melodious, single-note whistle. In flight, it does not show a white wing stripe as does the gray plover.

NATURAL HISTORY

Habitats and foods. The typical breeding habitats of this species are swampy heaths having an abundance of sphagnum and heather, highland bogs and moors, barren heathers in high-latitude mountains, and low-lying marshes in moss tundra (Voous 1960). It is more temperate-adapted than the New World species and it still exists on some glacial relict areas as far south as central Europe (Germany, Denmark, and the Netherlands). In Great Britain the birds are most common on moors between 300 and 610 meters in elevation, especially those that have been burned and recovered to the point that grasses are mixed with short heather (Ratcliffe 1977). Breeding populations are usually well dispersed and often are less than a pair per 100 hectares (Glutz et al. 1975). On their wintering areas the birds typically occupy tidal shores, rocky outcroppings along the coast, salt swamps and marshes, and shortgrass fields, with some birds reaching inland lakes or damp areas (Frith

Distraction display, male

1976). In migration the birds often occur on inland areas of pastures, stubble, and plowed fields. Foraging is done by day or night, and the birds are relatively mobile, typically running a few yards then pausing in an upright posture and sometimes pecking. Some of the pauses are probably for listening, while others are followed by dashes to capture prey found visually. During the breeding season the birds eat a wide spectrum of invertebrate foods common on upland moors, especially insects and their larvae, particularly beetles. They also consume worms, small snails, spiders, and crustaceans at various times of the year and eat a very small amount of plant material (Glutz et al. 1975; Ratcliffe 1977).

Social behavior. Golden plovers are apparently strongly territorial, and males may actually begin advertisement displays as early as mid-February while still in flocks. They arrive on the breeding grounds singly, in pairs, or, as is most usual, in larger groups. Pairing begins in flocks, and males chase females well out over the moor before the two return to the flocks. Probably most pairs are formed by the male's establishing a territory, advertising it by aerial display and calls, and attracting or leading a mate into it. The male performs the advertisement flight at heights of 50–1,000 feet, while slowly flapping the wings and uttering a plaintive *per-pee-oo, per-pee-oo* call. Fights among the males over females are quite common, and unmated males sometimes challenge and fight paired territorial males. Such behavior tends to produce considerable spacing of territories and nests, though it is likely that individual pairs tend to maintain much the same territory from year to year, and in one case the same nest scrape was used in six successive years (Ratcliffe 1977). When the female approaches breeding condition the male often chases her on the ground or rises and hovers above her as she stands or runs below him. Sometimes the male performs a precopulatory display of raising and violently shaking his wings, but more often copulation occurs without any specific display after the female squats. At times the male will fly low from a distance to land on the female's back, and he flaps his wings for balance while he is mounted. Males apparently make the initial nest scrapes, although unlike the situation in *Charadrius* these do not serve as copulation sites. Later the female joins the male and completes one of the scrapes as a final nest site (Bannerman 1961). Nest-building may occur as early as 3 weeks before the first egg is laid, which in Great Britain occurs over a surprisingly long span of about 4 months (Ratcliffe 1977).

Reproductive biology. The nests are in exposed, but nondescript, sites that combine ground and mixed plant cover where there is a lack of uniformity or pattern in the substrate. The birds apparently avoid placing the nests near conspicuous objects, and they also avoid areas that collect water during wet weather. Nests are well spaced, averaging in various areas from about 200–240 meters apart, with surprising year-to-year constancy in spacing patterns. The eggs are usually laid at intervals of 48 hours in Great Britain, but 60-hour intervals are also frequent, and gaps of as long as 120 hours have been reported. Replacement clutches are very common, and pairs have been known to produce as many as three replacement clutches. The usual clutch is 4 eggs, but 3-egg clutches are fairly common in replacement nests, and very rarely there are 5-egg clutches. Of 231 clutches in Great Britain, 207 were of 4 eggs, 15 were of 3, and 9 were of 2, and some of these were certainly depleted or incomplete clutches. Both sexes incubate, and the role of the male seems to increase as incubation progresses. The incubation period varies from 27.5 to 34 days, and in one study it averaged 30.7 days (Bannerman 1961). For the first week both parents brood the young, with the brood sometimes being divided and each parent looking after one or more chicks. They fledge in less than 5 weeks, probably about 32–33 days. In spite of the rather long breeding season, there is no evidence of double brooding in Great Britain (Ratcliffe 1977).

Status and relationships. In Great Britain this species probably reached a peak population about 1850–60, and since that time it has generally declined. About 31,000 pairs now breed in Great Britain (Ratcliffe 1977), while on the Continent breeding has also declined, with the last Netherlands breeding record being from 1937. Only a few pairs survive in Germany, and the species is also extremely scarce in Denmark (Voous 1960). It has probably also retracted from some part of its breeding range in Norway and Sweden, but it is still fairly common in Iceland (Bannerman 1961). The species is obviously a very close relative of *dominica,* and at least one wild hybrid has been reported between these forms. Mayr and Short (1970) considered them as constituting a species group, but I believe they can be considered a superspecies. Bock (1958) also regarded them as a superspecies, indicating that the area of limited sympatry of breeding ranges requires that they be regarded as distinct species.

Suggested reading. Bannerman 1961; Ratcliffe 1977.

Lesser Golden Plover

Pluvialis dominica (P. L. S. Müller) 1776

Other vernacular names. American golden plover; eastern golden plover (Australian); pluvier faure (French); Sibirischer Goldenregenpfeifer (German); chorito dorado chico (Spanish).

Subspecies and ranges. See map 41.

P. d. fulva: Asiatic lesser golden plover. Breeds on the tundra of northern Siberia from the Yamal Peninsula eastward, on the west coast of Alaska from Cape Lisburne to the Kuskokwim River, and on the Bering Sea islands. Winters from India, China, Taiwan, and the Hawaiian Islands south to Australia, New Zealand, Samoa, and Oceania.

P. d. dominica: American lesser golden plover. Breeds in Alaska from Point Barrow east through arctic Canada to Baffin Island and south to northeastern Manitoba, central Mackenzie, and southwestern Yukon. Winters in South America south to northern Argentina and Uruguay.

Measurements and weights (dominica). Wing: males 175–88 mm; females 172–83 mm. Culmen (to feathering): males 22–25 mm. Weights (*fulva* in breeding season): males 106–34 g; females 120–30 g (Glutz et al. 1975). Summer males of *dominica* 127–69 g, average of 36 was 144 g; females 126–69 g, average of 24 was 146 g (Irving 1960). Eggs ca. 48 x 33 mm, estimated weight 26–27.8 g (Schönwetter 1963).

DESCRIPTION

Breeding adults of both sexes have the sides of the head, chin, throat, chest, breast, and abdomen uniformly black, and the sides of neck and chest, under tail coverts, and thighs white. The axillaries and under wing coverts are brownish black, thickly marked with spots of golden yellow and grayish white. The upper tail coverts are barred rather than spotted with yellow, and the tail is dusky with barring of grayish white or pale gray and yellowish. The primaries and their coverts are dusky to blackish. The iris is dark brown, the bill is black, and the legs and toes are grayish black. The race *dominica* tends

41.   Breeding distributions of the American (Am) and Asian (As) lesser golden plovers.

to show more white on the forehead and sides of the breast than does *fulva,* and the gold spotting on the back is paler. *Adults in winter* lack black on the sides of the head and underparts, which are brownish gray, becoming white posteriorly. The foreneck and chest are spotted with darker markings, and the upperparts are speckled mostly with grayish white rather than gold. There is also no pure white on the forehead, superciliary area, or sides of the neck (Ridgway 1919). *Juveniles* resemble winter adults but have brown-tipped white feathers on their flanks and abdomen (Prater et al. 1977).

In the hand, this plover can be recognized by its relatively large size (wing 170–90 mm), its lack of a hind toe, its dark and banded tail, and its brownish gray axillaries, which separate it from *apricaria* in all plumages. In breeding plumage it also has more extensive black on the underparts, including the under tail coverts, and in nonbreeding plumage has a more distinct superciliary stripe than that species. The North American race is also grayer than *apricaria,* while the Asian race tends to be more buffy.

In the field (10 inches), the distinctive black underparts and white lateral markings readily distinguish this from other North American birds. In winter it can be separated from the similar gray plover by its smaller and less chunky configuration, the absence of a white rump, and the lack of black axillaries when in flight. When in flight it also lacks the definite white wing stripe that occurs on the gray plover. The many calls include a two-syllable whistle, the first note ending with a quaver and the second falling in pitch. The birds are frequently found well away from water, often on plowed or recently burned fields during spring migration, or even on lawns throughout the winter on Pacific Islands (C. F. Zeillemaker, pers. comm.).

NATURAL HISTORY

Habitats and foods. During the breeding season this species is mostly associated with high arctic tundra, but locally it extends into the low arctic, being largely limited to dry but well-vegetated slopes and uplands, or to knolls on lowland tundra. Areas having a disrupted surface pattern, strewn with lichens, gravel, rocks, or other discontinuities, are favored over uniform landscapes. Gentle peaty slopes, with exposed boulders and adjacent to hummocky wet tundra or fresh water, are favored breeding habitats on Victoria Island, and the density there on low-lying tundra within 5 miles of the coast is about one pair per square mile (Parmelee et al. 1967). In conjunction with this sparse density, the territories are rather large and the birds are highly mobile, feeding both inside and outside the territory. During the breeding season the birds feed mostly on insects, including beetles, the larvae of lepidopterans, and the like. They also eat small mollusks, crustaceans, spiders, small worms, and berries such as crowberries. Very large quantities of terrestrial insects such as grasshoppers, crickets, and locusts are eaten by migrating birds, which often gather in plowed fields, meadows, stubble fields, and similar upland habitats. In coastal South America the birds feed mostly in shortgrass fields and tilled farmlands, but also occur in wetlands, along slow-moving streams, and in sloughs. Littoral areas are used only infrequently. Many birds defend winter foraging territories, which range up to 0.3 hectares, and thus there is considerable population dispersion over suitable upland areas or along streams (Myers and Myers 1979). In Hawaii the birds feed mostly on lawns, pastures, and taro paddies.

Social behavior. The first birds begin to arrive at Cambridge Bay, Victoria Island, about the first of June. They arrive in singles, pairs, and small flocks of up to about 6 birds, and within a few days some of them begin to act as if they are paired, suggesting that at least some arrive already paired or perhaps quickly reestablish pair bonds with mates of the previous year. Much of the tundra is then still covered with snow, but the birds soon become territorial and begin making aerial display flights over the snowy ground, as early as 3–4 days after initial arrival. The usual flight display, performed 100–150 feet in the air, is a long, circling flight that may cover a quarter- to a half-mile of area. The wingbeats are slow and hesitant, with a jerky aspect, and a repeated *chu-leek* or *too-lick* is uttered. The gliding descent to earth is usually fairly abrupt, but it typically is less dramatic than that of the gray plover. These quick descents may also be followed by a rapid climb and a continuation of the flight (Parmelee et al. 1967). Sometimes one or more males will fly directly above the female during such flights, and it may be that one purpose is to announce the male's arrival on the breeding grounds, in addition to proclaiming territorial "ownership." Territories are quite large, covering up to about half a square kilometer of area, and they always include dry, gravelly tundra on the mountain slopes and also have access to open water. Pair-formation has not yet been studied in the wild, but in captive birds it seems to involve a male's advertising his presence with "long calls" and performing lateral displays toward the female. The displaying male keeps his wings closed, his body low and declined, with sleeked plumage, the tail pointing upwards and the neck stretched downward. A long trill is uttered, and the bird usually displays in full view of the female, sometimes running in small circles in front of her (Sauer 1962). In one observation of a copulation, the male of a pair standing close together finished preening, stretched his wings high and lowered them again, then ran several yards to overtake his mate. A swift copulation then occurred (Parmelee et al. 1967). Thus it appears that copulation is independent of nest-scrape ceremonies in this species.

Reproductive biology. Nests are in relatively high and dry areas, where gravel and lichens cover the ground. Though sometimes placed beside a rock, they are fully exposed and lined with dry grasses, lichens, and dead leaves. The nest is typically situated where cushions of bright and varied lichens surround it, and materials added to the nest, such as white lichens and similar items, help provide a color mosaic that closely matches the dorsal coloration of the bird. The clutch size is almost invariably 4 eggs; all of 18 clutches in the Churchill area of Manitoba were of this number (Jehl and Smith 1970). At least in one instance, eggs were laid at 2-day intervals, and incubation began a day before the last egg was laid. The incubation period of the last egg in this clutch was slightly over 26 days, and it was approximately 27 days for the other eggs. Both sexes incubate the eggs throughout the incubation period, with the female perhaps taking the greater share. Both sexes also attend the young for the entire fledging period and sometimes remain with them even after they can fly. Young are able to fly by about 22 days of age. Adults begin to leave the area at about this time, well in advance of the departure of juveniles and of adult gray plovers (Parmelee et al. 1967).

Status and relationships. There is no numerical information on this species' status, but there is also no evidence that it is in a state of decline. Its relationships to the larger Eurasian species are obviously close, and were it not for a limited area of apparent sympatry, the two could probably be considered conspecific (Bock 1958). It thus seems most realistic to consider them as constituting a superspecies.

Suggested reading. Sauer 1962; Parmelee et al. 1967.

Gray Plover

Pluvialis squatarola (L.) 1758
(*Squatarola squatarola* of Peters, 1934)

Other vernacular names. Black-bellied plover; pluvier argénte (French); Kiebitzregenpfeifer (German); chorlito gris (Spanish).

Subspecies and range. No subspecies recognized. Breeds in the arctic tundras of North America from Alaska to the Melville Peninsula and on adjoining islands east to Baffin Island, and in Russia and Siberia from the Kanin Peninsula to the Chukotski Peninsula and Anadyrland and adjoining islands. Winters over a widespread area of North, Central, and South America south to southern Brazil, and in the Old World south to South Africa, Mada-

42. Breeding distribution of the gray plover.

gascar, India, Australia, and the central Pacific islands. See map 42.

Measurements and weights. Wing: males 180–91; females 182–99 mm. Culmen (to feathering): males 27–32 mm. Weights (autumn, various ages): males 183–280 g; females 176–203 g (Glutz et al. 1975). Eggs ca. 52 x 36 mm, estimated weight 34.2 g (Schönwetter 1963).

DESCRIPTION

Breeding adults of both sexes have the lores, sides of face, chin, throat, foreneck, and underparts as far as the lower abdomen black, while the forehead, superciliary region, and sides of neck are nearly pure white, as are the under tail coverts. The rest of the upperparts are pale gray to grayish white, the back, scapulars, and wing coverts have brownish black spots, and the rump often is barred with blackish, as is the tail. The primaries are dusky, with the inner ones having white stripes on the outer webs, and the secondaries are pale grayish brown, tipped with white, while the outer greater coverts are broadly margined with white. The axillaries are black. The iris is brown, the bill is black, and the legs and toes are black to grayish black. *Winter adults* have the sides of the head and underparts white, the former streaked with dusky, the foreneck and chest shaded with brownish gray also marked with dusky, and the upperparts without black spotting, the general color being brownish gray to grayish brown with darker and whitish feather markings (Ridgway 1919). *Juveniles* have grayish brown underparts and coverts, their margins heavily spotted with pale gold or yellowish white. The underparts are white, with extensive buff brown barring on the breast and flanks. *First-winter* birds remain distinctly spotted above and have white underparts, with some juvenile inner median coverts retained that show dark central wedges at the tips (Prater et al. 1977).

In the hand, the large size (wing 180–200 mm), ploverlike bill, presence of a small hind toe, and black axillaries readily identify this species.

In the field (11 inches), this plover can be separated from the two related *Pluvialis* forms by its larger and more chunky build, its lack of golden spotting in the breeding plumage, and its whitish rump and black axillaries in winter plumage. In flight it shows a definite white wing stripe, and it often utters a plaintive whistle of three notes, *pee-u-wee.* The birds are often found on coastal mud flats, along lakeshores, and in freshly plowed or flooded fields.

NATURAL HISTORY

Habitats and foods. In the breeding season this species generally occupies dry stony moss-, grass-, or lichen-tundra, preferably on rocky slopes, but does

17.   Breeding-plumage heads and axillaries of gray (*top*), greater golden (*middle*), and lesser golden (*bottom*) plovers. After Glutz et al. 1975.

not utilize low-lying coastal tundra (Voous 1960). Pairs are often associated with the drier and more exposed ridges, riverbanks, or raised beaches that can be found within a mile of the seacoast. These areas are among the first to become snow-free, and they usually are covered with sand or gravel, with scattered cobble, and with gray or black lichens and clumps of grass rush (*Luzula*). Peat ridges in tundra marshes are sometimes also used for breeding (Drury 1961). Population densities on breeding habitats tend to be very sparse, and on Victoria Island the maximum density was about 1–2 pairs per square mile (0.4–0.8 pairs per 100 hectares) on low-lying tundra within 4–5 miles of the coast, while higher populations of from 1.1 to 2.3 pairs per 100 hectares were found on Jenny Lind Island (Parmelee et al. 1967). In Hooper Bay of Alaska about 4–6 pairs per square mile is a typical density (Brandt 1943). On their wintering areas the birds occupy tidal sands or mud flats of ocean coastlines, bays, estuaries, and sometimes also inland habitats. In coastal Argentina they sometimes defend foraging territories, but usually they forage either alone or in small, loose flocks (Myers and Myers 1979). They feed on a variety of marine invertebrates, including crabs, sand shrimps, and marine worms. They also are known to eat mollusks, insects, and some plant seeds. Their foraging methods are much like those of golden plovers, although they tend to be less mobile and to forage more rapidly. They feed by alternating pecking with pausing or running movements, and they often combine pecking with a sideways flicking movement, probably to help expose prey (Burton 1974).

Social behavior. Sexual maturity is probably attained in the second year of life, with some courtship display occurring during the spring migration. The first birds to arrive on the breeding grounds are males or pairs, suggesting that some birds arrive already mated. Localization of pairs begins within about a week of spring arrival, and on Victoria Island flight displays may be seen as early as June 8, less than a week after first arrival. The male typically rises to 75–100 feet or more, then flies on shallow wingbeats that are slow and maintained above the body plane. This "butterfly flight" often covers several acres of ground but is not necessarily confined to the territory. During it, the male utters a loud three-syllable call sounding like *whee-li-ee,* with the first syllable longer and lower than the middle one. The terminal phase is usually a spectacular glide to the ground with the wings held straight out or somewhat upturned. On the ground the bird may display again, running stiffly toward the female, which may stand from a few feet to as far as 30 yards away from him. Sometimes he will rush right by her, but at other times he will stop and stand perfectly still, with tail up and the bill nearly touching the ground. Copulations may follow such a rushing display, and in two cases the display was followed by the male's flying away 10–40 meters (Hussell and Page 1976). Nest-scraping behavior has apparently not been observed but almost certainly is a part of pair-forming and early nesting behavior.

Reproductive biology. Nesting sites are usually in the areas that first become snow-free, but in spite of this apparent advantage the birds tend to nest relatively late. The sites are typically dry, exposed, and stony, much like those of golden plover except that the latter choose more peaty substrates. The nest is a simple scrape in mottled black-and-white lichen and moss tundra, usually with an unimpeded view of the surrounding area. The eggs closely resemble their background, as does the dorsal plumage pattern of the adult. There are normally 4 eggs; 19 of 21 clutches found on Victoria Island and Jenny Lind Island were of this size, while the exceptional clutches contained 3 and 5 eggs. The eggs are laid at intervals of 1–2 days, and both sexes incubate, though at times the male apparently does most of the incubation. The incubation period is 26–27 days (Drury 1961; Hussell and Page 1976). Apparently both sexes participate

equally in brood care, but the males appear to be bolder in brood defense. The fledging period is 23 days, with the female deserting the brood before the male. The entire laying, incubation, and fledging period thus occupies about 55 days for a single nesting, and it occurred within a 68-day period for a population observed on Devon Island (Hussell and Page 1976).

Status and relationships. There are no available estimates of this species' worldwide population, or even any major components of it. The birds are apparently not extremely abundant anywhere on their wintering areas, but these are so widespread that the total population may nevertheless be quite high. This species is obviously a close relative of the two "golden" plovers, and Bock (1958) included not only these two species but also *obscura* in the genus *Pluvialis*. This merging of *Squatarola* with *Pluvialis* has also been followed by several other workers (Vaurie 1965; Voous 1973), and I fully support the position. Differences in the structure of the skull and the presence or absence of the hind toe were shown by Bock to be of minor taxonomic value, and differences in the nuptial plumage are probably associated with need for effective concealment of different nesting substrates, with this form nesting on drier and more grayish or whitish substrates than is true of the "golden" plovers.

Suggested reading. Parmelee et al. 1967; Hussell and Page 1976.

Genus *Charadrius* Linné 1758 (Lesser Plovers)

Lesser plovers are small, short-billed shorebirds that lack hind toes (one exception) and have pointed wings that are under 150 mm. All species lack pale spotting on the upperparts, often have white facial or nape markings and black neck or breast markings, and usually have a long tail that has a black subterminal band. The downy young have a pebbly dorsal pattern and a white nape patch. Twenty-six species are recognized here.

KEY TO THE GENUS *Charadrius* (ADULT PLUMAGES)

A Hind toe present; adults with a broad cinnamon rufous or grayish brown breastband, becoming blackish posteriorly . . . . *modestus*
A' No hind toe
 B Head entirely black above the eyes; central upper tail coverts also nearly black . . . . *cucullatus*
 B' Head not almost entirely black above the eyes, usually with gray or white on crown or forehead
  C Dark subterminal band present on the tail, which is white tipped
   D White present at base of the outer web of innermost primaries
    E Mantle pale gray, black on head limited to a narrow crown band . . . . *melodus*
    E' Mantle grayish brown to dark brown

  F Most of the head and neck black, with a white stripe extending above the eye and across the crown, no separate black breastband . . . . *novaeseelandiae*

  F′ Head mostly brown and white, a black forehead bar and breastband present

   G Outer and middle toes connected by webbing . . . . *semipalmatus*

   G′ No webbing between outer and middle toes . . . . *hiaticula*

D′ No white at the base of outer web of innermost primaries

 E Axillaries gray or brownish

  F Belly white in summer, middle toe with claw shorter than bill . . . . *veredus*

  F′ Belly blackish in summer, middle toe with claw longer than bill . . . . *morinellus*

 E′ Axillaries white

  F Two blackish bands crossing the breast

   G Wing under 120 mm, throat and upper breast pale gray to brown . . . . *tricollaris*

   G′ Wing over 140 mm, throat and upper breast white . . . . *vociferus*

  F′ Only one band crossing the breast

   G Wing coverts and scapulars red, contrasting with the back . . . . *melanops*

   G′ Wing coverts and scapulars same color as the back

    H Larger (wing over 130 mm), outer tail feathers considerably (at least 12 mm) shorter than central ones . . . . *placidus*

    H′ Smaller (wing under 120 mm), outer tail feathers almost as long (less than 8 mm difference) as central ones . . . . *dubius*

C′ Tail lacking a black subterminal band; base of the innermost primaries white

 D Larger species (wing at least 120 mm, exposed culmen usually at least 18 mm)

  E Breast crossed by two bands (sometimes poorly defined) in breeding plumage

   F Both breastbands pale gray, grayish brown, or black; the upper one often incomplete . . . . *falklandicus*

   F′ Breastbands narrow black above and broader chestnut ventrally . . . . *bicinctus*

  E′ Breast crossed by one gray, rufous, or black band, or none in breeding adults

   F Larger (wing at least 150 mm), with a reddish breast and abdomen in breeding plumage (dark gray in winter) . . . . *obscurus*

   F′ Smaller (wing usually under 150 mm); if breast is reddish it does not extend to abdomen

    G Tarsus more than twice as long as middle toe without claw; no breastband in any plumage . . . . *montanus*

    G′ Tarsus less than twice as long as middle toe without claw; breastband present

     H Breastband black (warm brown in winter), legs and toes pink . . . . *wilsonus*

     H′ Breastband chestnut (buff to grayish brown in winter), legs and toes black, gray, or greenish yellow

      I Black on the lores and forehead in breeding plumage, outer tail feathers mostly white

       J Wing usually under 140 mm, tarsus usually under 35 mm, breastband broad in breeding plumage . . . . *mongolus*

       J′ Wing usually over 140 mm, tarsus usually over 35 mm, breastband narrow in breeding plumage . . . . *leschenaultii*

      I′ No black on the lores or forehead, outer tail feathers mostly brownish . . . . *asiaticus*

 D′ Smaller species (wing under 120 mm, exposed culmen usually under 18 mm)

  E Black or brownish breastband present in breeding adults, sometimes incomplete

   F Complete, pale chestnut breastband, legs olive colored . . . . *pallidus*

   F′ Breastband black or rusty brown, legs not olive colored

    G Black or rusty band, often incomplete frontally, legs gray to yellow . . . . *peronii*

    G′ Complete black breastband, legs not gray to yellow

     H Legs and feet blackish . . . . *thoracicus*

     H′ Legs and feet pink . . . . *collaris*

  E′ No breast collar present in any plumage, but sides of neck often blackish or brownish, forming semicollar

F Black line extending from the eye down the side of the neck
  G Primaries dark brown, exposed culmen over 20 mm . . . . *sanctaehellenae*
  G′ Primaries light brown, exposed culmen under 20 mm . . . . *pecuarius*
F′ No black line down the side of the neck from the eye
  G No black on the side of the neck
    H Legs blackish, no white stripe above eye . . . . *falklandicus alticola*
    H′ Legs yellowish, white forehead extends back above eye . . . . *marginatus*
  G′ Black semicollar on side of neck near front of wing, legs black or dark gray
    H White nape collar present . . . . *alexandrinus*
    H′ Hindneck brown, like crown . . . . *ruficapillus*

Red-breasted Dotterel

Charadrius obscurus (Gmelin) 1789
(*Pluviorhynchus obscurus* of Peters, 1934)

Other vernacular names. Red-breasted plover; New Zealand dotterel; pluvier roux (French); Neuseelandregenpfeifer (German).

Subspecies and range. No subspecies recognized. Resident in New Zealand, breeding on North Island from the northern tip to Kawhia on the west coast and the eastern Bay of Plenty. Sometimes present on South Island along Nelson, Westland, and Southland coasts to Foveaux Strait, but not known to breed there. Breeds locally at Mason Bay and on the bare hilltops of Steward Island. See map 43.

Measurements and weights. Wing: both sexes 150–65 mm. Culmen: both sexes 28–30 mm. Weights: 1 breeding male 160 g; 5 breeding females 100–150 g, average 110 g (Sylvia Reed, pers. comm.). Eggs ca. 45 x 32 mm, estimated weight 23.2 g (Schönwetter 1963).

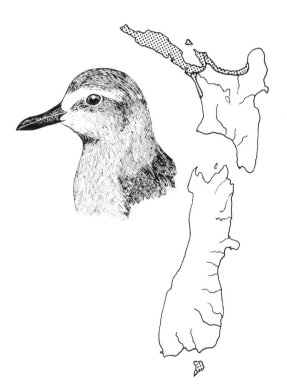

43. Breeding distribution of the red-breasted dotterel.

DESCRIPTION

Adult males in breeding plumage have a white band through the forehead and a narrow white line extending backward above the eye, the chin white, and these two areas separated by a blackish line extending from the base of the bill backward through the eye and merging with a brown hindneck and crown. The entire upper surface is brown, the feathers margined with pale chestnut, while the upper wing coverts are grayish brown with white margins. The primaries and central tail feathers are dark brown, while the outer tail feathers are much paler, with white tips. The underparts vary from pale cinnamon to chestnut brown, varying considerably in degree of underpart coloration. The iris is brown, the bill is black, and the legs and feet are gray. *Females* are usually paler on the underparts than males. *Adults in winter* lack the chestnut margins on the upper body surface and are mostly white below, the breast being crossed by a band of dark gray to rufous. *Juveniles* are similar to adults in winter, but they sometimes have rufous on their sides, and there is a band of spotted brown and fulvous on the breast.

In the hand, the combination of large size (wing at least 150 mm) and a rather stout (28–30 mm), plover-like bill separates this species from other *Charadrius* forms, and its lack of a hind toe separates it from the only species of *Pluvialis* with a culmen length in this range.

In the field (10½ inches), this species' large size separates it from the other New Zealand plovers, and it is likely to be confused only with wintering golden or gray plovers. However, these species are likely to show some black on the underparts and are conspicuously spotted with yellowish above, while this species is only faintly marked with lighter colors above and usually shows some rufous on the underparts. Its calls include a cricketlike *krik* or *kriki,* and when in flight breeding birds utter a variety of sharp notes.

NATURAL HISTORY

Habitat and foods. This species is mainly limited to sandy habitats, mostly associated with ocean beaches, lakeshores, and riverbeds, but it also occurs along mud flats of harbors and estuaries during the nonbreeding season. Besides its coastal nesting areas, it has been found breeding at altitudes of 6,000–8,000 feet on mountains along the south coast of New Zealand. Its food consists of crustaceans and small mollusks along the coast, and of insects in other localities (Oliver 1955). It has a relatively heavy bill, which is presumably well suited to crushing and digging.

Social behavior. Relatively little has been written of the social behavior of this species, but Phillips (1980) has provided some useful information. In the winter the birds that breed at higher elevations move to the coast, and at such times flocks of 200 birds or more have been reported. First-year birds probably wander somewhat, but in general the birds are believed to be rather sedentary. Males may become territorial as early as May, but they often establish territories as late as July or August. About 9 or 10 weeks elapse between the onset of territoriality and the laying of the first egg (Edgar 1969). Breeding pairs are rather aggressive, and ground or aerial chases are common. In some cases aerial chases that do not reach the opponent end in a glide, with the wings held above the back, in a manner similar to that described for the little ringed plover. Nest-scraping displays are well developed, and when the female approaches such a displaying male he steps out of the scrape, bows forward, and performs picking or choking movements at the sand or raises the farther wing while spreading or folding the tail. This is similar to the precopulatory display of several *Charadrius* species. In one observation of copulation the female walked stiffly beyond the displaying male, who followed her slowly while "parade marching." She gradually stopped and became horizontal, and the male finally hopped onto her back. Treading lasted a full minute before he suddenly grasped her nape feathers and pulled her backward. The behavioral sequence of breeding is thus quite similar to that

Distraction display

reported for the sandplover, mountain plover, and double-banded dotterel (Phillips, 1980).

Reproductive biology. These birds invariably nest in sandy situations, often on coastal dunes and usually near running water. Often the nest is on a slight mound or elevation that provides safety from flooding and improved visibility. The nest is usually also among jetsam, which helps conceal the bird. The normal clutch is 3 eggs, and although the males spend much time near the nest during the egg-laying period the females drive them away when incubation begins. However, males remain in the vicinity and protect the nest when danger threatens. The incubation period is 28–32 days, and both sexes exhibit intensive injury-feigning behavior when the young are threatened by terrestrial predators. The fledging period has been estimated to range from 38 to 50 days (Oliver 1955; McKenzie 1953). There is one record of a female tending young a year after she had been banded as a chick, indicating as least occasional breeding in the first year of life. More surprisingly, one male has been reported breeding at the age of 26 years. Two banded pairs were found together in two or more consecutive seasons, but such mate constancy may not be the general rule (McKenzie 1978).

Status and relationships. This species has apparently suffered somewhat from human activities, and according to Oliver (1955) it no longer nests in the mountains of South Island. It is now relatively common only in the northern part of North Island. Edgar (1969) reported that in 1968 there were slightly more than 1,100 birds on known breeding areas of the North Island. He indicated that the birds are increasing slowly in the Auckland area, and that a small breeding population may still exist in the high country of South Island. The species' relationships are still somewhat uncertain. Bock (1958) and Oliver (1955) believed that its affinities are with *Pluvialis,* while Peters (1934) considered it to constitute a monotypic genus *(Pluviorhynchus).* Although I have not seen the species alive, it seems to me to be somewhat intermediate between *Pluvialis* and the group of plovers that Bock referred to as mountain or plains plovers and that he suggested might be transitional between *Charadrius* and *Pluvialis.* Phillips (1980) believes that *obscurus* is clearly a member of the genus *Charadrius* and probably is fairly closely related to such forms as *bicinctus* and *frontalis.*

Suggested reading. Oliver 1955; McKenzie 1953; Phillips, 1980.

Mountain Plover

Charadrius montanus (Townsend) 1837
(Eupoda montana of Peters, 1934)

Other vernacular names. None in general English use; pluvier montagnard (French); Prärieregenpfeifer (German).

Subspecies and range. No subspecies recognized. Breeds in western North America from Montana south to New Mexico and western Texas; the range is declining but once extended into the Dakotas *(Condor* 78:265). There is a recent breeding record from northwestern New Mexico *(Wilson Bull.* 88:358). Winters from central California and Texas south to southern Lower California. See map 44.

Measurements and weights. Wing: males 138–54 mm; females 143–55 mm. Culmen: males 19–22 mm; females 21–23 mm. Weights: average adult

44. Breeding distribution of the mountain plover.

weight 106.8 g (Graul 1973); 4 males 67–104 g, average 93.8 g; 7 females 89–112 g, average 100.2 g (M.V.Z. specimens). Eggs ca. 37 x 28 mm, estimated weight 16.5 g (Schönwetter 1963).

DESCRIPTION

Breeding adults of both sexes have a white forehead and superciliary stripe, a narrow black stripe from the beak to eye, and a black forecrown (sometimes the entire crown), while the rest of the upperparts are buffy grayish brown, often with a buffy or ochre tinge, and the feathers are margined with the same. The greater wing coverts are tipped with white and the primaries are dusky with whitish shafts and white at the bases of the outer vanes, especially on the inner feathers. The tail is grayish brown tipped with whitish and crossed with a darker subterminal band. The entire underparts are white except for a grayish buff chest area. The iris is dark brown, the bill is black, and the legs and toes are light brownish yellow. *Adults in winter* are similar, but the black markings on the head are replaced with grayish brown (Ridgway, 1919). *Juveniles* have their upperparts grayish brown and the feathers with broad buffy tips and dark subterminal bands and have no black on the face. *First-winter birds* have inner median coverts with buff tips and the outer ones with rufous brown fringes (Prater et al. 1977).

In the hand, this species may be identified by its moderately large size (wing 135–55 mm), a relatively long tarsus (at least twice as long as middle toe without claw), and the absence of black or rufous on the underparts.

In the field (7½ inches), these pale-bodied birds are usually associated with arid grasslands on the breeding grounds and with cactus deserts on the wintering areas. They run well, tending to flee on foot rather than by flying, and when in flight they remain close to the ground and sometimes crouch immediately after landing. The white under wing lining is evident in flight, and contrasts with the sandy brown upper wing surface, which has a distinct wing stripe. The calls include whistled notes ranging from musical and plaintive to shrill and harsh.

NATURAL HISTORY

Habitats and foods. The shortgrass plains of the Great Plains are the primary habitat of this species. Nests are typically placed in grama grass, buffalo grass vegetation, which tends to be less than 8 centimeters (or 3 inches) high in April. They sometimes also nest in areas with scattered clumps of cacti, higher grasses, and scattered shrubs (Graul 1975). In the nonbreeding season they also occupy areas of semidesert and often occur far from water. Their major food consists of insects, probably primarily grasshoppers, from which they also presumably obtain much of their needed water.

Social behavior. In northeastern Colorado these birds arrive on their nesting areas in late March, usually in flocks of 3–10 individuals. Both courtship and copulation occur in these early flocks, but they soon disperse, and at least some of the adults return to the same areas they occupied in previous years. Males often reoccupy their old territories and seem to show a somewhat higher fidelity to previous nesting sites than do females. Territories are quite large, perhaps averaging about 16 hectares, but are not mutually exclusive. Many birds also forage on undefended neutral grounds. Territories are advertised by calling from the ground and also while in flight. One display flight is the "falling-leaf display," in which the bird descends from 5 to 10 meters with the wings held in a sharp V while rocking back and forth, and another display is the "butterfly flight," performed with very slow and deep wingbeats. Courtship behavior toward females includes several displays, of which the most common is scraping that appears identical to the ritualized nest-scraping behavior of other *Charadrius* species. A "bowing" display, with a soft mooing call, is sometimes also performed toward females and rarely by females toward a male performing the same display. Copulation sometimes occurs after a nest-scraping ceremony, with the male approaching the female in an upright posture while calling softly. If the female stops, he then performs a series of rapid foot movements apparently like the "high-stepping" or "goose-stepping" displays of other *Charadrius* forms. Copulation lasts for remarkably long periods, averaging about 5 minutes but sometimes lasting up to 13 minutes. Evidently pair bonds are formed only gradually, and females may visit the territories of several males before developing a preference for one. Second clutches are usual in this species, and some females switch mates before laying the second clutch (Graul 1973).

Reproductive biology. Before nesting, it is common for several nest scrapes to be constructed on a territory. However, the actual nest is usually on a nearly flat area, either on the top of a hill or in a valley. More than half the nests in one study were also situated very close to an old cow manure pile. Within populations, the nests are well spaced, and in one case the average distance between 21 active nests was 140 meters. Of 154 clutches, 134 had 3 eggs, with the extremes being 1 and 4 eggs. Although eggs are sometimes laid at daily intervals, as many as 96–108 hours may elapse between successive eggs. Incubation usually begins as soon as the clutch is complete and typically is performed by the male alone. The usual incubation period is 29 days, with extremes of 28 and 31 days. At least in some cases, the female begins laying a second clutch about 11–13 days after the completion of her first clutch, and she incubates it herself. In one case a third clutch was begun 9 days after the removal of the second clutch. Hatching of the young is usually synchronous, and the adults remove the eggshells from the nest, dropping them 20–50 meters away. The young are moved considerable distances from their nest during the fledging period, sometimes as far as 800 meters. Fledging occurs 33–34 days after hatching, and thereafter the birds begin to form fall flocks. Sometimes these flocks become very large, with on observation of about 2,500 in a single flock (Graul 1973, 1975).

Status and relationships. The status of this species is deteriorating (Graul and Webster 1976), and there is no doubt that its range has distinctly contracted in historical times. It no longer breeds in North Dakota or South Dakota and is probably accidental in both Nebraska and Kansas (Johnsgard 1979). Bock (1958) placed this species in his group of "mountain or plains" plovers, which also included *mongolus, leschenaultii, asiaticus* (including *veredus*), and *modestus*. I would also include *obscurus* in this general group, and I suspect that in spite of the zoogeographic difficulties it is possible that *montanus* and *obscurus* may be fairly closely related. More probably, however, *leschenaultii* represents the nearest relative of *montanus*.

Suggested reading. Graul 1973, 1975.

Greater Sand Plover

Charadrius leschenaultii Lesson 1826

Other vernacular names. Geoffrey plover; larger sand dotterel; large-billed dotterel; pluvier de Leschenault (French); Wüstenregenpfeifer (German); chorlitejo de Leshenault (Spanish).

Subspecies and range. No subspecies recognized here. A smaller form called *columbinus* has been proposed for western areas, but Vaurie (1964) has rejected it. Breeds from Armenia west through Transcaspia and Russian Turkistan to the steppes east of the Aral Sea. Also breeds in the steppes of southeastern Russian Altai, and in Mongolia east to about 110° E longitude. Recent local breeding has occurred in Afghanistan, Anatolia, and Jordan (Nielsen 1971). Winters from the Persian Gulf east to the Indo-Chinese countries, Taiwan, and the Philippines south to southern Africa, the Sundas, and northern Australia. See map 45.

Measurements and weights. Wing: males 137–51 mm; females 142–55 mm. Culmen (from feathering): males 22–26 mm; females 22–25 mm. Weights: 8 males 65–86 g, averaging 74 g; 6 females 77–90, averaging 82 g (Shaw 1936). Six Australian wintering birds ranged 65–98 g, averaging 77 g (D. Purchase, pers. comm.), while 6 African wintering birds averaged 96.8 g and ranged from 73 to 130 g (Summers and Waltner 1979). Eggs ca. 38 x 28 mm, estimated weight 15.6 g (Schönwetter 1963).

DESCRIPTION

Breeding adults of both sexes are very similar to *mongolus*, having rufous on the crown, hindneck, and sides of the head, with no definite pale superciliary streak. Instead the crown grades to a black facial stripe extending from the ear coverts forward through the eye, down the lores to the beak, and upward across the front of the crown. In the front of the forehead a median black line extends down to the base of the upper mandible, restricting the white on the forehead to two small patches on either side of this median line. The malar region, chin, throat, and upper foreneck are white, which is followed by a relatively narrow band of chestnut, the two sometimes separated by a narrow black stripe. The upperparts are generally a warm brown, including the upper wing coverts, while the tail is brownish centrally with white lateral feathers and white tips. There is a greater amount of white tipping on the wing coverts and bases of the flight feathers than in *mongolus*, and likewise more white on the outer tail feathers, with brownish on the outermost pair restricted to a small subterminal area. The underparts are entirely white. The iris is brown, the bill is black, and the legs and toes are dusky greenish to olive slate. *Adults in*

45. Breeding distribution of the greater sand plover.

winter have a pale brownish white forehead and superciliary stripe, with a broad dark brownish band from the base of the bill through the eye, and the crown as well as the other upperparts brownish gray. The breastband is composed of grayish brown lateral patches usually joined by a thin gray line across the breast. *Juveniles* have their upperparts and coverts extensively fringed with buff, a poorly defined head pattern, with no black present, and an indistinct superciliary line. The breast patches are grayish brown on a buffy background. *First-winter birds* resemble adults in winter but retain some buff-fringed inner median coverts (Prater et al. 1977).

In the hand, the major measurements (wing usually at least 140 mm, bill at least 20 mm, tarsus at least 35 mm) separate nearly all birds from *mongolus.* It is separated from similar-sized *asiaticus* by having black rather than white lores (breeding males), or by the dark central tail and whitish rump rather than the rump and tail of the same color (females), and by the slightly longer bill (exposed culmen usually 23 mm or longer, versus a maximum of 23 mm in *asiaticus*).

In the field (8½ inches), this species is found on sandy coastlines, mud flats, estuaries, and salt pans in wintering areas. It is easily confused with the smaller Mongolian plover, especially in winter plumage, when the differences in breastbands are not evident. At that time the slightly longer and heavier black bill provides a comparative means of distinction when both might be present, as well as the somewhat larger body size. There is more white in the tail, a more evident wing stripe is visible when the bird is in flight, and the legs have a more greenish or yellowish green tinge than in that species. The call in flight is a short *drrit,* and a quiet, dry trill, *chirrirrip,* longer than that of *mongolus,* is sometimes uttered.

NATURAL HISTORY

Habitats and foods. The breeding habitats of this species consist of flats of clay or solonetz soils, overgrown with wormwood (*Artemisia*) and other halophytic plants. The birds sometimes also breed in totally vegetation-free areas and occasionally also nest in sandhills, gravelly areas, or other barren substrates. In some very arid areas these are the only breeding birds, and apparently young can be reared in the absence of water. On migration the birds favor sandbars close to water, and in the wintering grounds they are rarely found far from coastlines, where they concentrate on tidal sands and mud flats. During that period they feed mainly on small crustaceans and mollusks, and the heavy bill is probably very useful at such times; they seem to be remarkably good predators compared with other similar-sized *Charadrius* forms. In the breeding areas the birds specialize on terrestrial insects, particularly weevils, but various beetles, ants, and other insects are also eaten (Glutz et al. 1975; Dementiev and Gladkov 1969).

Social behavior. Although rather little is known of the social behavior, it is probable that breeding occurs in the second year of life. First-year birds may spend the summer on wintering areas. The birds are territorial during the breeding season, and the male performs a territorial flight by circling about 30–50 meters above the ground. The territorial song is uttered as the bird flies with slow, owllike wingbeats, accompanied in rhythm with melodic *pipruirr-pipruirr-pipruirr* notes similar to those of a golden plover. This is alternated with more rapid wingbeats and a renewed ascent. The pairs are separated by distances of 150–300 meters, with no tendency for coloniality. Nonbreeders, however, are quite social during the breeding season, as is also true of wintering birds (Glutz et al. 1975). There are only a few observations on courtship behavior. Lehmann (1969) described a flight display of these birds, which consists of a flight about 3–4 meters about the ground, and a soft *drui-drui-drui* call. The ground display was seen only once, during which a male ran up to a female and postured with a level body and raised wings and feathers. As the female crouched somewhat, the male performed a typical "parade march," but no actual copulation followed.

Reproductive biology. The nesting biology is only rather poorly known. The nest site is a simple scrape, often thinly lined with stems and grasses and sometimes lined with the fragments of dried sheep or goat droppings. The usual clutch is 3 eggs, but occasionally 4 and rarely 2 have been reported. Both sexes have brood patches and both also incubate, though the incubation period is still undetermined, as is the fledging period. In at least one area, covering the eggs with sand during the hottest part of the day has been observed. Shortly after hatching, the young are led away to moister areas. Sometimes the adults divide the care of the young between them, and the young soon develop so that they are very difficult to capture (Lehmann 1969; Glutz et al. 1975).

Status and relationships. This species is considered quite common in many of its nesting areas, and the large numbers of birds seen on the diverse wintering grounds suggest that the total population of the species must be considerable. Its evolutionary relationships are obviously closest to *mongolus,* and the two clearly constitute a superspecies. Dementiev and Gladkov (1969) indicate that the two forms may be slightly sympatric near Lake Balkhash and the Aral Sea, though Nielsen (1971) shows no such sympatry. Since the southern limit of breeding in *leschenaultii* is still very uncertain, it is probably safest to consider these as separate species for the present. Two other Asian species, *asiaticus* and *veredus,* have certain features in common with these forms that indicate affinities, and the four forms probably constitute a single evolutionary complex (Nielsen 1971). Slightly less close affinities with the arid-adapted North American species *montanus* and the New Zealand species *obscurus* also seem likely. Bock (1958) believed that *mongolus, leschenaultii, asiaticus* (including *veredus*), *montanus,* and the South American *modestus* represent a group of species that probably descended from an *alexandrinus*-like ancestor and may help to connect the genera *Pluvialis* and *Charadrius.*

Suggested reading. Dementiev and Gladkov 1969; Lehmann 1969.

Mongolian Plover

Charadrius mongolus Pallas 1776

Other vernacular names. Lesser sand plover; Mongolian sand dotterel; pluvier Mongol (French); Mongolenregenpfeifer (German); chorlitejo Mongol (Spanish).

Subspecies and ranges. See map 46.
   C. m. pamirensis: Western Mongolian plover. Breeds in the central and eastern Pamirs, north to western Sinkiang.
   C. m. atrifrons: Southern Mongolian plover. Breeds in Kashmir north to the Karakoram range, and from Ladakh eastward through Tibet to central China (Kansu). This form and *pami-*

46. Breeding distributions of the interior (I), southern (S), western (W), and eastern (E) Mongolian plovers.

rensis winter from the Persian Gulf south to the Greater Sundas, and on the coast of eastern Africa.

C. m. mongolus: Interior Mongolian plover. Breeds in eastern Siberia (Stanovoi and probably Verhoyansk ranges), possibly also in the Dzhugdzhur Range. The northern coast of the Sea of Okhotsk is an intergrade area with *stegmanni.*

C. M. stegmanni Eastern Mongolian plover. Breeds in the Koryak Highlands, the Chukotski Peninsula, and on Kamchatka, Commander, Koraginski and northern Kurile Islands. This form and *mongolus* winter from China, Taiwan, and the Philippines south to the Solomons and Australia.

Measurements and weights. Wing: males 120–42 mm; females 125–40 mm. Culmen: both sexes 16–17 mm. Weights: 53–95 g; average of 12 was 69 g (Shaw 1936). Ranges of both sexes: *stegmanni* 52–68 g; *pamirensis* 55–70 g; and *mongolus* 58–71 g; but fall weights of *atrifrons* are only 39–50 g (various sources). Weights of 200 wintering *mongolus*

in Australia ranged from 56 to 102 g, averaging 69 g (D. Purchase, per. comm.). Eggs ca. 36 x 26 mm, estimated weight 12.2–12.6 g (Schönwetter 1963).

DESCRIPTION

Adult males in breeding plumage have a white forehead (*mongolus* and *stegmanni*), variably obscured or divided by a median black line or black spots (especially in *atrifrons* and *pamirensis*) behind which is a black bar running from eye to eye and continuous with a black stripe extending from the eye to the beak, continuing below the eye and expanding to include the ear coverts. There is usually also a small white crescent on the lower eyelid and a small whitish or buffy area between the ear coverts and the crown. The top of the head is brownish gray or grayish brown, becoming pinkish to cinnamon anteriorly, laterally, and posteriorly, grading to deep cinnamon on the sides of the neck and on the entire chest. The malar region, chin, throat, and foreneck are white, sharply separated from the cinnamon chest (in *mongolus* and *stegmanni*) by a narrow line of dusky. The anterior side feathers are cinnamon, tipped with white, and the more posterior ones are brownish gray, with broader white tips. The rest of the underparts are white, and the upperparts are generally grayish brown, with the upper tail coverts paler and margined with white. The middle tail feathers are deep grayish brown, and the outer ones paler except subterminally, with the outermost ones entirely white. The greater wing coverts and secondaries are tipped with white and the primaries are dusky, the inner ones becoming more whitish basally. The bill is black, the iris is dark brown, and the legs are grayish, with darker toes. *Females* are duller than males, with the black head markings variably replaced by grayish brown and the cinnamon coloration less pronounced. *Adults in winter* lack cinnamon on the chest, neck, or crown, with the chest being crossed by an indistinct narrow band of grayish brown, especially laterally, and the black head markings replaced by grayish brown (Ridgway 1919). *Juveniles* are extensively fringed with buff on the upperparts and coverts, especially on the scapulars, and the breast patches are grayish with buff edging. *First-winter* birds resemble adults in winter but retain buff-tipped inner median coverts (Prater et al. 1977).

In the hand, this species is intermediate in size between the smaller "ringed" plovers and the larger brown-banded species. The rather short (16–17 mm) black bill, lack of a white nape collar, chestnut breastband (in breeding plumage), and wing no more than 142 mm should identify this species.

In the field (8 inches), this species is found mostly along tidal flats and estuaries in winter, and on elevated stony or sandy plains in summer. It very closely resembles the somewhat larger greater sand dotterel but has a broader breastband and a smaller and noticeably weaker bill. In flight this species shows a narrow white wing bar and often utters a *pip-ip* or *drrit* call, as well as a short, quiet, and dry trill like that of the greater sand dotterel but shorter. Separation from the greater sand dotterel in winter is best made by relative bill size and shape (*British Birds* 73:206–13).

NATURAL HISTORY

Habitats and foods. This is primarily an alpine species that inhabits mountain steppes and deserts, sometimes occurring on alpine tundra as high as 4,000–5,000 meters above sea level and as high as 5,500 meters in Tibet. Sandy areas, or slaty ones, often with scant vegetation, are preferred breeding habitats but usually when close to flowing water. Somewhat more vegetated locations are sought out by pairs leading young. On the Commander Islands, however, sandy dunes near the sea are also used for *stegmanni* breeding. Terrestrial insects appear to be the primary summer foods, including weevils, various kinds of beetles, and the like. Wintering birds have been reported to eat small mollusks, amphipods, small crabs, and marine worms. Wintering habitats are mostly tidal sands and mud flats associated with bays, inlets, and estuaries. At that time of the year the birds are usually in twos and threes, but they sometimes occur in flocks of up to 100. Like related forms, this species forages by making quick runs over the sand or mud, with sudden stops to capture food (Kozlova 1961; Glutz et al. 1975).

Social behavior. Social behavior has been little studied but apparently is much like that of *leschenaultii*. Courtship begins with the arrival of the birds on their breeding grounds and lasts until about the middle of June. Territorial males call in flight as well as on the ground, with a melodious and repeated *kruit-kruit* call, and in territorial flight the bird flies in broad circles at great heights, with rapid wingbeats and uttering a two-syllable *tekr-ryuk* call. Territorial flights have also been described as slowed-down gliding flights similar to those of gulls. During courtship there is intensive building of sham nests and also vertical wing-lifting displays. Courtship occurs not only during daylight but also at night, especially when there is moonlight. Territoriality is apparently well developed, with little or no tendency for coloniality (Kozlova 1961; Glutz et al. 1975), although in Pamir colonies of 3–10 pairs have been reported (Dementiev and Gladkov 1969).

Reproductive biology. The nest site is a shallow cavity in the ground, lined with dried stems and leaves and sometimes plant seeds. On the Commander Islands nests have been found on sand, among *Angelica* plants and close to water, while in the western Himalayas a nest was found in a rocky desert area among gravel. The nest is also frequently lined with fragments of dried cow droppings as well. A full clutch consists of 3 eggs, and incubation is performed by both sexes, according to some authorities, since both sexes have well-developed brood patches. More recent observations by I. A. Abdusalyamov, however, have suggested that possibly only the female *pamirensis* incubates the eggs and broods the young. But males perform intensive nest-distraction behavior with wing-dragging and tail-lowering and A. A.

Kistchinski (in litt.) believes that males do most of the incubating and usually also normally raise the young as well. The incubation period lasts 22–24 days and there is a fledging period of 30–35 days (Glutz et al. 1975; Kozlova 1961).

Status and relationships. Apparently all the races of this species are not very numerous, which probably is at least in part a reflection of the rather rigorous environmental conditions associated with breeding. This species is obviously a very close relative of *leschenaultii* and is evidently allopatric with it, although there is considerable overlap on coastal habitats of the wintering ranges (Nielsen 1971). Other than being slightly larger and having a more massive bill, *leschenaultii* is remarkably similar to *mongolus*, and probably only their geographic allopatry on breeding areas serves to maintain species distinction.

Suggested reading. Nielsen 1971; Dementiev and Gladkov 1969.

Lesser Oriental Plover

Charadrius asiaticus Pallas 1773
(Eupoda asiatica of Peters, 1934)

Other vernacular names. Caspian sand plover; pluvier Asiatique (French); Kaspischer Regenpfeifer (German); cholitejo Asiatico (Spanish).

Subspecies and range. No subspecies recognized here. Breeds from the region north of the Caucasus and the southern Volga-Ural steppes eastward to the Aral Sea and the vicinity of Lake Balkhash. Winters in eastern Africa. See map 47.

Measurements and weights. Wing: males 142–53 mm; females 140–55 mm. Culmen (from feathering): males 19–22 mm; females 19–23 mm. Weights (breeding season): males 63–88 g; females 65–91 g (Glutz et al. 1975). Eggs ca. 38 x 27 mm, estimated weight 13.8 g (Schönwetter 1963).

DESCRIPTION

Adult males in breeding plumage have a white forehead, lores, superciliary stripe, malar region, chin, and throat. The crown and entire upperparts are sandy brown, and the lores and ear coverts are sim-ilarly colored. Below the white throat is a rather broad rufous breastband, terminated posteriorly by a narrow black band, with the rest of the underparts white, including the axillaries, which are slightly suffused with brown. The flight feathers are blackish brown, with the outermost primary having a white shaft, while the others are marked with whitish toward the outer end. The central tail feathers are the same color as the back, but the outer ones are considerably lighter. The tips of the outer greater coverts and the outer webs of the primaries are white, producing a small but definite wing stripe. The iris is brown, the bill is black, and the legs and toes are dull greenish gray to yellowish gray. *Females* differ from males in having a less extensive and less colorful breastband, which lacks a black lower border or has one that is indistinct. *Adults in winter* have a breastband that is grayish brown, and the white portions of the head become more buffy. *Juveniles* have extensive pinkish buff fringes on the edges of the upperpart feathers, and the breastband varies from buffy to grayish brown. Their legs may also be more yellowish than in adults. *First-winter birds* resemble adults in winter but retain some buff-edged median coverts (Prater et al. 1977).

In the hand, this species is most likely to be confused with *veredus,* but it is smalller (the wing usually under 155 mm, the tarsus usually under 42 mm), has whitish rather than brown axillaries, and has a small white wing stripe and more definite white superciliary stripe in spring plumage. It differs from *leschenaultii* in having less white on the wing and tail and in lacking dusky on the lores and forehead in spring.

In the field (7½ inches), this species is usually found in dry semidesert areas during the breeding season and in coastal areas and mud flats during winter. In spring plumage the extensive reddish breastband separates it from all species except the very similar but larger *veredus* (which has a broader black bar below the breastband), and the more short-legged *leschenaultii* (which has no black below the breastband but does have black on the face). Separation in the winter is more difficult, but this species has a small but definite wing stripe, while *veredus* lacks a wing stripe and *leschenaultii* has a conspicuous white wing stripe. Its calls include a sharp whistled *ku-wit* or *hweet,* and a piping *klink.*

NATURAL HISTORY

Habitats and foods. This species breeds on the dry steppes and semideserts of central Asia, apparently

47. Breeding distributions of the lesser (*shaded*) and greater (*hatched*) oriental plovers.

overlapping in range and also in breeding habitats with *C. mongolus* and *C. leschenaultii.* The usual breeding habitats include *Artemisia* and other halophytic shrubs, usually fairly close to water but sometimes at considerable distance from it. The species is especially associated with highly saline soils, and salt marshes are also listed as breeding habitats. Unlike the two other species just mentioned, it is not associated with coastal habitats on the wintering ground but instead inhabits arid inland plains that are dominated by grasses. The species' foods are only slightly known; the stomachs of some birds collected on the breeding grounds were tightly packed with insects. Apparently insects and their larvae constitute the major part of the food for the entire year, since birds collected on their wintering grounds in Zaire were also found to contain insect remains. Recently burned grasslands, which are rich in grasshoppers and other insect life, are favored winter habitats in Africa (Bannerman 1961). The birds are evidently rather independent of surface water, and wintering birds were not observed to drink (Glutz et al. 1975). However, breeding birds have been observed flying daily to distant drinking places (Kozlova 1961).

Social behavior. Very little is known of the social behavior of this species. The birds are known to be territorial on the breeding grounds but apparently breed in loose groups or colonies of up to 25 pairs. The territorial display flight is marked by a three-syllable call, and the "song" is said to be reminiscent of that of a wheatear (*Oenanthe oenanthe*). The male's call has also been described as "buzzing," and after alighting the male is said to run around the female with his feathers fluffed out. A group display activity involving about 20 birds has been reported on the breeding grounds, but details of pair-forming and copulatory behavior remain completely unstudied (Glutz et al. 1975; Dementiev and Gladkov 1969; Bannerman 1961).

Reproductive biology. Nests are in exposed sites and usually have a few twigs around the edges, as well as grass stalks and dried horse droppings in the scrape itself. The eggs are typically arranged among these objects and pieces of soft, crumbling clay; 3 eggs constitute the usual clutch. Some breeding begins very early, and the season is very irregular, with records extending from early April to the latter part of June.

Apparently both sexes perform incubation, since males possess brood patches, and incubating birds reportedly sometimes leave their nests to fly in small groups to distant areas for foraging and drinking. Distraction behavior has been reported by birds disturbed near the nest, but both the incubation period and the fledging period remain unreported. After the young have grown somewhat and become fairly mobile, the birds begin to concentrate near livestock watering places and sometimes also near villages. Soon they begin to assemble in large flocks before the fall migration (Glutz et al. 1975; Dementiev and Gladkov 1969). The spring and fall migrations of this species are apparently more concentrated in time and space than those of *mongolus* or *leschenaultii* (Nielsen 1971).

Status and relationships. This species is rather uncommon at the western edge of its breeding range but is rather common near the Aral Sea and around Lake Balkhash. Thus there is no apparent reason for present concern over its status. The species is clearly a very close relative of *veredus,* and the two forms are commonly considered conspecific, as for example by Bock (1958) and Dementiev and Gladkov (1969). They are certainly highly allopatric, but Vaurie (1964) and Kozlova (1961) both strongly objected to the conclusion that they should be regarded as the same species. Kozlova reported differences in molting sequences that she regarded as significant, and Vaurie pointed out nonoverlapping differences in the length of tail, wing, bill, and tarsus. There are also marked differences in dorsal plumage color and leg coloration, as well as in that of the axillaries, all of which seem to argue for species separation. As I noted in the account of *leschenaultii,* these two species are part of a rather large group of arid-adapted and mostly Asian-breeding forms that constitute a distinctive evolutionary complex.

Suggested reading. Bannerman 1961; Elliott 1956.

Greater Oriental Plover

Charadrius veredus Gould 1848
(*Eupoda veredus* of Peters, 1934)

Other vernacular names. Eastern sand plover; eastern dotterel; pluvier Asiatique (French); Hufeisenregenpfeifer (German).

Subspecies and range. No subspecies recognized here; this form is often considered a subspecies of

asiaticus. Breeds in Outer Mongolia from the valley of the upper Dzabkham River and steppes of Khangai east to northwestern Manchuria, and Inner Mongolia north of the Hwang Ho River. Winters from the Sunda Archipelago to Australia. See map 47.

Measurements and weights. Wing: males 161–76 mm; females 157–69 mm. Culmen: both sexes 29–34 mm. Weights: 5 males 77–88 g, average 80 g; 2 females 78 and 80 g (Shaw 1936). Two wintering birds weighed 86 g (D. Purchase, pers. comm.). Eggs ca. 38 x 27 mm, estimated weight 14.0 g (Schönwetter 1963).

DESCRIPTION

Breeding males are colored very much like those of the Caspian plover (with which it is often considered conspecific) but have a cream or yellowish ochre superciliary stripe and lighter brown ear coverts, with the rest of the head and neck white except for a smoky brown crown. There is a broad chestnut breastband, bordered posteriorly with a black band. The back plumage is rather uniformly brown, while the upper coverts are grayish brown. The outer greater coverts are narrowly tipped with white, and there are sometimes traces of white on the outer webs of the inner primaries, producing a very weakly defined wing stripe. The underparts are white, but the axillaries are brown with narrow white fringes, and under wing coverts are also more brownish than in *asiaticus.* Likewise, the legs and toes are yellowish, rather than greenish gray; the iris is brown; and the bill is black. *Breeding females* lack the black posterior border to the breastband present in males and have a pale buff rather than white throat, becoming darker brownish buff rather than chestnut, contrasting but little with the upperparts. *Adults in winter* have broad breastbands that are buffish brown in females and pinkish brown in males. *Juveniles* have the upperpart feathers extensively fringed with pale buff and pale chestnut, and the breast mottled with darkish brown and buff. *First-winter* birds resemble adults but have buffy-fringed median coverts that contrast with newly grown and chestnut-fringed outer median coverts (Prater et al. 1977).

In the hand, the differences mentioned above (especially axillary color and leg color) separate this species from *asiaticus,* which is also smaller (the wing of *veredus* is usually over 155 mm, that of *asiaticus* usually under 155 mm). None of the other Asian *Charadrius* species have measurements in this range.

In the field (9–10 inches), spring-plumaged birds can be distinguished from *asiaticus* by the broad black border to the chestnut breastband and the paler head plumage, especially around the ear coverts. The yellowish leg color also helps separate it from *asiaticus*, in which the legs are greenish gray, and from the smaller *leschenaultii*, in which they are much shorter and greenish to olive slate. In flight, there is no definite wing stripe evident, and it utters a piping *klink* note as well as loud melodious trilling calls in flight. Wintering birds are usually found in inland locations, especially on bare ground such as mud flats or at the edges of lakes and ponds. They are remarkably long-legged and slender compared with *leschenaultii*.

NATURAL HISTORY

Habitats and foods. Virtually unstudied on their breeding grounds, these birds are sometimes seen in northern Australia in winter, where they occupy inland plains and the flat borders of lakes and lagoons (Frith 1976). The birds are remarkably long-legged and long-winged, giving the impression of a pratincole rather than a typical *Charadrius*, and they evidently are well adapted to foraging in rather dry environments. They often occur on mud flats, clay pans, and airfields, sometimes in association with pratincoles or coursers. They sometimes also forage along the edges of roadways in Australia, especially those having little traffic, and in Malaya they have been observed on such disturbed sites as tin tailings.

Social behavior. In the wintering areas of Australia, this species sometimes occurs in very large flocks. However, details of its social behavior are unreported.

Reproductive biology. Reproductive biology is almost completely unknown. The eggs are still essentially undescribed, though Schönwetter (1963) provides the measurements of one reported egg. Presumably the biology of the species is much like that of the smaller *asiaticus*, but this is conjectural.

Status and relationships. Apparently uncommon to scarce on the Malay Peninsula and rather local in Borneo, this species is evidently not common anywhere on its wintering grounds except in northern Australia and possibly in the Sundas and the Moluccas. The form is obviously a very close relative of *asiaticus*, but it was considered specifically distinct by Kozlova (1961), Vaurie (1965), and others. Niel-

sen (1971) pointed out that *asiaticus* and *veredus* show an absence of winter overlap ("allohiemy"), but *veredus* and *leschenaultii* overlap somewhat on both their breeding and their wintering grounds, where they occupy similar habitats and perhaps compete. Whether regarded as full species or not, *asiaticus* and *veredus* are the products of a common evolutionary past.

Suggested reading. Nielsen 1971.

Rufous-chested Dotterel

Charadrius modestus Lichtenstein 1823
(*Zonibyx modestus* of Peters, 1934)

Other vernacular names. Chilean plover; winter plover; pluvier d'Urville (French): Rotbrustregenpfeifer (German); chorlito pechicolorado (Spanish).

48.   Breeding distribution of the rufous-chested dotterel.

Subspecies and range. No subspecies recognized. Breeds in southern Chile and on Tierra del Fuego; also on the Falkland Islands. Winters to northern Chile and to Uruguay and northern Argentina. See map 48.

Measurements and weights. Wing (flattened); males 131–45 mm; females 133–46 mm. Culmen (to feathering): males 19–25 mm; females 21–26 mm (Blake 1977). Weights: both sexes 71–89 g, average of 5 was 78 g (various sources). Eggs ca. 45 x 30 mm, estimated weight 22.0 g (Schönwetter 1963).

DESCRIPTION

Adults in breeding plumage have the forehead and a broad stripe above the eye to the side of the crown white, above which there is a dark grayish brown crown and below which the sides of the head are light gray, darkening to dusky at the base of the bill and lightening to grayish white on the chin and upper throat. The throat is posteriorly bordered by a bright cinnamon rufous breastband that grades to black ventrally and by rather uniform grayish brown, darkening on the rump and inner tail feathers, while the outer tail feathers are progressively more white. The upper wing surface is also uniformly brownish, although the shafts of the primaries are white and there are pale margins on some of the coverts. The iris is brown, the bill is black, and the legs and feet are greenish gray. *Adults in nonbreeding plumage* are darker above and have a white rather than a gray throat; the entire breast area is grayish brown. *Juveniles* resemble nonbreeding adults but have black upperparts, finely streaked with buffy white; the wing coverts are mottled with buffy, and the breast is buffy brown with fine dusky barring (Blake 1977).

In the hand, the combination of a small hind toe, a grayish throat, and moderate size (wing under 150 mm) should identify this species.

In the field (8 inches), this species appears as a medium-sized plover with a broad cinnamon-colored breastband. It might be confused with the tawny-throated dotterel, but that species has a brownish gray breastband and an orange-tinted throat. It has numerous calls, the most characteristic being a melodious and melancholy whistle, uttered in flight or while on the ground.

NATURAL HISTORY

Habitats and foods. On the Falkland Islands, this rather common species occurs in many habitats dur-

ing the nonbreeding season, ranging from coastline beaches of boulders, sand, or mud flats to flooded grasslands and eroded slopes. However, when it is breeding it occurs on dry slopes of open country, where stands of ferns and diddle-dee (*Empetrum*) are abundant and provide nesting cover. On Tierra del Fuego the birds breed on dry grasslands at high altitudes, such as in tussock grasses, but they also have been seen nesting in open peat bogs. Evidently they consume many kinds of insects, insect larvae, crustaceans, mollusks, and some plant materials such as algae (Humphrey et al. 1970). In Argentina the birds forage mostly in upland grassy areas, but they also use marshes, streams, sloughs, and tidal flats. They forage at the waterline, and use typical plover movements to capture prey (Myers and Myers 1979).

Social behavior. Wintering birds are often seen mingling with two-banded plovers on the Falkland Islands, forming flocks of up to a hundred or more individuals. Pair-formation probably occurs during winter, since birds returning to Tierra del Fuego in spring were noted to be already in pairs. One description of courtship states that the male (probably) flies with an unusual slow-motion wingbeat, glides with his wings raised, as high as 100 feet above his territory, and utters a low repeated *tik* call, alternated with a loud whirring or rattling call. However, the most typical call is a loud whistle, *peeoo,* or *piiru,* variously described as melancholy and as having tremulous quality (Woods 1975; Humphrey et al. 1970). Wintering birds often defend foraging territories, especially along streams and sloughs, which range in length from 10–100 meters (Myers and Myers 1979).

Reproductive biology. Nests on the Falkland Islands are placed in clumps of grass, ferns, or *Empetrum* on dry slopes, between late September and mid-January, with a peak of laying in October. On Tierra del Fuego the nesting period is probably from October to January. Evidently the usual clutch is 2 eggs, but there is at least one record of a 3-egg clutch. Apparently both parents incubate and exhibit anxiety by calling when disturbed near their nest or brood, but they seem to have no injury-feigning display (Woods 1975; Humphrey et al. 1970).

Status and relationships. This species is said to be common on the Falkland Islands as well as throughout Tierra del Fuego, and it is probably also fairly common on the mainland of southern South America, mainly along the Chilean coastline. Bock (1958) regarded the species as a close relative of *asiaticus*, the two forming a superspecies, and he considered both part of the group he called mountain or plains plovers. It seems to me that an equally likely close relative is *morinellus*, which Bock places in *Eudromias*, but which I have included in *Charadrius*. The two species nest in similar habitats, and perhaps some of their general similarities are the result of convergence rather than reflecting actual relationships. In any case, I see no reason for maintaining a separate genus (*Zonibyx*) for this form, which seems to fall easily within the limits of *Charadrius* as used here.

Suggested reading. Woods 1975; Humphrey et al. 1970.

Dotterel

Charadrius morinellus L. 1758
(*Eudromias morinellus* of Peters, 1934)

Other vernacular names. None in general English use; pluvier guignard (French); Mornellregenpfeifer (German); chorlito carambola (Spanish).

Subspecies and range. No subspecies recognized. Breeds discontinuously in Scotland, the Scandina-

vian Peninsula, northern Finland, the Kola and Kanin peninsulas, the Urals, the Taimyr Peninsula, the Lena, Yana, and Anadyr estuaries, and also in southern Siberia, northwestern Mongolia, and the Russian Altai. Sporadically breeds in the mountains of central Europe and Italy, and probably also in Alaska (Kessel and Gibson 1979). Winters from northwestern Africa and the Mediterranean Basin east to Iraq and the Persian Gulf. See map 49.

Measurements and weights. Wing: males 143–52 mm; females 151–61 mm. Culmen (from feathering): male 14–17 mm; females 15–18 mm. Weights (breeding season): males 88–116 g; females 99–130 g (Glutz et al. 1975). Eggs ca. 41 x 29 mm, estimated weight 17.0 g (Schönwetter 1963).

DESCRIPTION

Adults in breeding plumage have a blackish brown crown and upper nape and a broad white superciliary stripe extending from the lores to the nape. The face and sides of the neck are cream to pale pinkish cinnamon, with sepia streaking on the lores and cheeks, and the ear coverts are ash brown. The chin and throat are creamy to white, with darker streaking. The upper breast is ashy brown, followed by a narrow sepia line and a broader white band separating the upper breast from orange cinnamon on the lower breast and flanks. The abdomen is black, and the under tail coverts are white tinged with pale cinnamon outwardly. The lower nape, mantle, scapulars, back, and rump are ashy brown, the upper tail coverts are the same with warm buff fringing, and the tail is mostly ashy brown, with a darker subterminal band and white tipping that broadens outwardly. The wing coverts are ashy brown with lighter edges, the primaries are sepia with white tips on the inner feathers, and the secondaries are ashy brown with white tips. The iris is brown, the bill is black, and the legs and toes are dull yellow. *Females* are usually slightly brighter in color than males but not consistently so. *Adults in winter* are similar in general pattern to breeding birds but have paler upper parts, a white belly, and a grayish brown breast, with a white line across it. *Juveniles* have broad buff fringes on their upperpart feathers, interrupted by a central dark wedge extending to the tip of the feathers. They also have a brownish buff breast with an indistinct white line. *First-winter birds* resemble pale juveniles, and their inner wing coverts have distinctive dark brown central areas (Prater et al. 1977).

49. Breeding distribution of the dotterel.

In the hand, the distinctive breeding plumage, with the black abdomen and the white line crossing the lower breast between ashy brown and cinnamon areas, provides a ready means of recognition.

In the field (9 inches), these birds are found in tundra situations in the breeding season and along sandy or muddy shorelines in winter. The white breast stripe is evident in all plumages, and a broad white superciliary stripe is also conspicuous, forming a V-mark at the back of the nape. In flight, no wing stripe is present, and the usual call is a soft trill, *wit-e-wee, wit-e-wee . . . ,* as well as a soft peeping note and a sharper *ting.*

NATURAL HISTORY

Habitats and foods. In general terms this species is associated with northern tundra, heaths, and sparsely vegetated barrens. It is generally found where there is a combination of luxuriant moss vegetation and scattered boulders, not far from tree line. Although northerly in its distribution, it is perhaps more a mountain species than an arctic one. Its foods

are mostly the larvae of insects such as dipterans, beetles, and weevils, but it sometimes also eats spiders, worms, and occasionally even seeds and berries of *Empetrum.* The birds are evidently good beetle-hunters and sometimes lift spiders from between stones. They also probe into wet moss beside streams while searching for the larvae of craneflies. The birds have rather large eyes, sometimes foraging at night or very early in the morning (Nethersole-Thompson 1973).

Social behavior. Dotterels arrive in Scotland in early May, and soon females can be seen calling and displaying from ridgetops newly free of snow. They raise their wings while chasing and trying to isolate males, and they sometimes squat or perform scraping behavior. Often a pair will perform rapid display flights in close synchrony or fly slowly, with legs sometimes dangling, above the territory. Between the time of initial pairing and egg-laying, scraping ceremonies and associated "nest dances" help hold the pairs together. Mating often occurs quite close to the nest scrapes and is preceded by the female's squatting and upending her tail in a seemingly subordinate display. Pair bonds probably last only until the clutch is completed and the male begins incubation; thereafter the females begin to form unisex groups that roam about foraging and occasionally fighting. A few instances of probable polyandry have been reported in which the female seemed to be associated with males incubating eggs or brooding young about 11–12 days apart in age. There is also one record of a female with three probable mates, and one of a female that shared the incubation duties of her second clutch with a new mate. However, Nethersole-Thompson believes that these are exceptional cases and that the birds are normally monogamous.

Reproductive biology. Nests are usually in small hollows of moss, tussocks, sedges, or grasses, often beside or among stones or rocks. Further, the nests are usually on ridges or hilltops rather than on sloping land, perhaps because these become snow-free quite early. The eggs are usually laid at intervals of from 24 to 30 hours, apparently at any time of the day. Throughout the species' range, the usual clutch is 3 eggs. A very small proportion of the clutches contain 1 or 2 eggs, and rarely there may be 4 present. Only about a fourth of the females whose clutches were robbed laid replacement clutches, which suggests that multiple brooding by polyandry is also fairly infrequent in this species. The incubation period averages 26 days, and the chicks normally hatch synchronously. The parent removes the eggshells almost immediately, sometimes hiding or eating the smaller fragments. Normally the male alone tends the young, apparently even driving his mate from the brood when she tries to approach it. Fledging occurs when the young are 26–30 days old, although the birds can take short flights for about a week before this time (Nethersole-Thompson 1973).

Status and relationships. While there are no estimates of actual numbers, the highly discontinuous breeding range of this species is well known, and in general it seems fairly secure. In recent years the birds have begun breeding in Holland, England, Austria, Poland, and the south Caucasus. There is periodic nesting in Italy, in Sudetenland on the Czechoslovakian border, and in the Carpathians, on the Hungarian and Romanian border. Occasional breeding in Alaska is probable, though proof is lacking (Kessel and Gibson 1979). The relationships of this species are uncertain and intriguing. It has often been placed in the monotypic genus *Eudromias*, but Bock (1958) expanded that genus to include *Oreopholus ruficollis*.

He admitted that this might be an artificial grouping, and I personally doubt that these species are close relatives. I believe that the dotterel is best maintained in the genus *Charadrius*, probably within the "mountain and plains" plovers and perhaps fairly close to the South American species *modestus*. This species is a somewhat montane, moss-nesting form with distinct similarities to *morinellus* in plumage and flight call, and it seems a much more likely near relative than *ruficollis*.

Suggested reading. Nethersole-Thompson 1973.

Double-banded Dotterel

Charadrius bicinctus Jardine and Selby 1827

Other vernacular names. Banded dotterel; mountain plover (New Zealand); pluvier à deux bandes (French); Doppelbandregenpfeifer (German).

Subspecies and ranges. See map 50.

C. b. bicinctus: New Zealand double-banded dotterel. Resident in New Zealand, on North and South Islands, as well as Great Barrier, Kapiti, Mercury, and Chatham Islands. Migratory, moving to eastern Australia in the nonbreeding season.

C. b. exilis: Auckland Island double-banded dotterel. Resident on the Auckland Islands, possibly also on Campbell Island (Falla 1978). Birds have been reported from Enderby, Adams, Rose, and Ewing islands of the Auckland group.

Measurements and weights. Wing: 125–35 mm. Culmen: both sexes 16–19 mm. Both sexes of the Auckland Islands race have wing lengths that average slightly larger than those of *bicinctus* (Falla 1978). Weights: both sexes 46–75 g, average of 102 was 57 g (D. Purchase, pers. comm). Eggs ca. 34 x 25 mm, estimated weight 10.8 g (Schönwetter 1963).

DESCRIPTION

Adults in breeding plumage have a white forehead, bordered above and below with black, and a black band from the bill to the eye, continuing as an edging to the brown of the upper surface. The white forehead extends back above and sometimes also below the eye, terminating behind the eye and grading into

50. Breeding distributions of the double-banded dotterel (*hatched*) and Malaysian sandplover (*shaded*). The Auckland Island race of the double-banded dotterel is indicated by a dot.

the grayish brown of the crown, hindneck, and upper body coloration. The lower cheeks, chin, and upper throat are white, terminated by a narrow black neck band, which in turn is followed by a white upper breast and a broader chestnut band on the lower breast. The flanks and the rest of the underparts are white, the tail is brown, the outer feathers are paler, with white edging, and the flight feathers are brown with white shafts. The iris is dark brown, the bill is black, and the legs and feet are yellowish gray to greenish yellow. *Females* are slightly smaller and less brightly colored than males and have a narrower and less deeply colored chestnut band. *Adults in winter* are uniformly grayish brown above, and white below, with a dull brown instead of black lower neck band and little or no trace of the lower breastband. There is a dark sandy line from the bill to the eye and a pale superciliary stripe that extends to the forehead. *Juveniles* have the dorsal feathers margined with rufous, the forehead and undersurface tinged with rufous, and a dark mottled gray band where the black band later develops.

In the hand, the double breastband, of brown and black, is diagnostic, but it may be absent in immature or nonbreeding birds. In that case, the combination of moderately long (122–32 mm) wings and a relatively short bill (no more than 18 mm, with a terminal arch of 7 mm or less) will help identify this species.

In the field (7 inches), birds in breeding plumage can be readily identified by the double breastband. Immature or nonbreeding birds are much more difficult and closely resemble Mongolian plovers, in winter plumage, but they are browner dorsally, with greenish yellow rather than dark green legs. The birds are found on beaches and areas of bare ground and rather rarely around inland swamps and lagoons. In flight they show only a faint white wing stripe at the base of the flight feathers, and they usually fly swiftly but close to the ground and usually not for great distances. The calls include *chip, chip* and high-pitched *pit-pit* notes, as well as a trilled *whee-o-whit* display call and a scratchy aggressive call sounding like *che-rec-a-rec*.

NATURAL HISTORY

Habitats and foods. The prime breeding habitats of this species consist of riverbeds and lakeshores with sandy or shingle beaches, either along the seacoast or inland. During the winter, the birds occur along coastlines in bays, estuaries, and inlets, and also well inland on flats surrounding large lakes, especially those with salty or brackish water. The usual foods in New Zealand are insects and crustaceans, although the seeds of a grass (*Muehlenbeckia axillaris*) and the succulent fruits of *Coprosma petriei* have also been reported as foods along river flats of Canterbury (Oliver 1955; Frith 1976). The birds forage by the usual plover method of running and pecking, feeding mostly at shingle edges but rarely wading in water more than a few millimeters deep. Probably they consume relatively large and scattered prey compared with black-fronted dotterels feeding in the same areas. Much of the feeding is done on neutral grounds up to 800 yards from defended territories. Feeding areas are not defended (Phillips, 1980).

Social behavior. During the winter period in Australia, these birds often occur in flocks, usually of from 2 to 20 birds, but they also occur in numbers as high as 1,000. In northern New Zealand large flocks may be seen in some localities during the winter. During the breeding season the birds are distinctly territorial, and they defend areas of about 50–70 meters in diameter. Breeding densities of about one pair per hectare are typical of the species. Territorial males perform a rather large number of visual and vocal displays, including a circling "butterfly flight" that they perform a few meters above the territory while uttering repeated calls. The birds also call at birds passing overhead, sometimes chasing them aggressively regardless of sex. Threat displays on the ground include puffing or spreading the breast feathers. When responding sexually to a female, the male typically runs toward and past her in a horizontal bulged-breast posture, then begins to perform nest-scraping, assumes a bowing or choking posture, and sometimes utters a mooing sound. The female may approach and replace him in nest scrape, after which the male stands at right angles to her, bowing, uttering mooing calls, and tilting the body toward her while raising the more distant wing. Such encounters often lead to copulation. Typically the female walks slowly away from the scrape, stopping about a meter away in a hunched posture. The male approaches and begins a "parade march" display of walking in place. Treading follows, with the male grabbing the female's nape and finally pulling her over backward so that both birds land on their backs. As pair-formation occurs, the females spend more and more time on the territories and begin to join males in territorial disputes (Phillips, 1980).

Reproductive biology. Most of the birds arrive on the breeding grounds in July, primarily establishing territories on sandy or shingle areas, but at times placing their nests in open fields or on tilled ground. The egg-laying period is from August to November, and the nest is a simple scrape in the sand or shingle, with little or no vegetational lining. The clutch is normally 3 eggs, and the incubation period is at least 28 days. Both sexes participate in incubation and have well-developed distraction displays, including a "broken-wing trick." Renesting is frequent after clutch loss. Although the fledging period has not been reported, the birds are well feathered in 19 days, and breastbanding begins to appear in about 8 months, with females showing their breastbands about a month later. Evidently the birds breed when less than a year old and return to the winter plumage when about 13 months old (Oliver 1955; Cunningham 1973).

Status and relationships. The population of the newly described subspecies from the Auckland Islands and Campbell Island must undoubtedly be rather small and probably numbers no more than 100–200 birds (Falla 1978). The mainland population is evidently rather large and in no obvious danger. Bock (1958) considered this species one of the typical sandplovers and regarded it as a close relative of *falklandicus,* the two forms constituting in his view a superspecies. He noted that the birds probably evolved from an *alexandrinus*-like ancestor, a view supported by a known case of natural hybridization between *ruficapillus* and *bicinctus.* I have some difficulty accepting the idea of a superspecies consisting of *bicinctus* and *falklandicus,* since, though *bicinctus* seems to have the typical conformation of the small sandplovers, this is not the case with *falklandicus,* based on captive individuals I have seen. Nonetheless, a tentative alliance between these two forms is perhaps better than admitting total ignorance. Phillips (1980) has recently argued that the similarities in courtship among this species, *obscurus,* and *frontalis* especially in their choking and bowing displays, suggest that they are congeneric and closely related.

Suggested reading. Cunningham 1973; Phillips, 1980.

Two-banded Plover

Charadrius falklandicus Latham 1790
(*Charadrius falklandicus* and *C. alticola* of Peters, 1934)

Other vernacular names. Puna plover *(alticola)*; Patagonian sand-plover; double-banded plover; pluvier des Falkland, pluvier du puna *(alticola)* (French); Falklandregenpfeifer (German); chorlitejo doble collar, chorlitejo serrano (Spanish).

Subspecies and ranges. See map 51.

C. f. falklandicus: Patagonian two-banded plover. Breeds on the coasts of Chile and Argentina south to Tierra del Fuego. Winters north to extreme northern Chile, Uruguay, and southern Brazil. Also resident on the Falkland Islands (possibly representing a recognizable race).

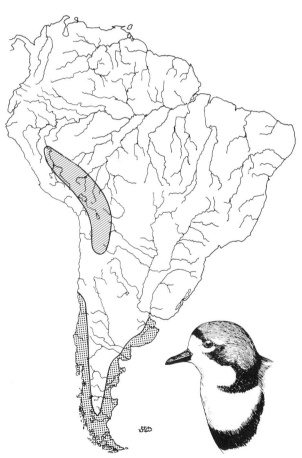

51. Breeding distributions of the Patagonian *(hatched)* and puna *(shaded)* two-banded plovers.

C. f. alticola: Puna two-banded plover. Breeds on the coast of northern Peru and in the puna zone of Peru, northern Chile, western Bolivia, and northwestern Argentina. Sometimes considered a separate species.

Measurements and weights. Wing (flattened): males 115–28 mm; females 106–28 mm. Culmen (from base): males 19–23 mm; females 18–24 mm (Blake 1977). Weights: 6 adult *falklandicus* of both sexes, 62–72 g, average 65 g; 5 *alticola* of both sexes, 41–49 g, average 45 g (various sources). Eggs ca. 37 x 27 mm, estimated weight 13.4 g (Schönwetter 1963).

DESCRIPTION

Adults in breeding plumage have a white forehead, terminated behind with a black band extending from eye to eye, and behind this the crown and sides of the head behind the eye are buffy cinnamon except for a more dusky area on the ear coverts and sides of the neck. The lower cheeks and throat are white, as are the underparts except for two pale grayish *(alticola)* to brown or black *(falklandicus)* bands on the lower neck and breast, the upper band less well developed and often incomplete on Falkland Island birds. The rest of the upperparts are pale grayish brown, while the flight feathers are fuscous to grayish brown, the primaries have white shafts, and the tail feathers are sooty centrally and white with dusky tips laterally. The iris is brown, and the bill, legs, and feet are black. *Females* closely resemble males but are less blackish on the face and are duller brown on the crown and hindneck. *Nonbreeding adults (of falklandicus)* have gray rather than black ventral bands on the lower neck and breast and are essentially uniform brown on the upperparts, including the forehead and sides of the head. *Juveniles* (at least of *alticola*) have the feathers of the back and the wing coverts narrowly edged with whitish and show little if any cinnamon or black on the head and neck (Blake 1977).

In the hand, the double black breastbands are unique to the smaller species of *Charadrius* (those with under 130 mm), and where these are lacking the absence of black on the lores and the entirely black bill, leg, and foot coloration helps distinguish this species.

In the field (6½–7 inches), this small plover is associated with sandy beaches, saltwater and freshwater ponds, and other shoreline habitats. It is most likely to be confused with the collared plover, which

18. Adult heads of Patagonian (*above*) and puna (*below*) two-banded plovers.

has only one black breastband, or the larger killdeer, which has yellowish legs and a rusty, banded tail.

NATURAL HISTORY

Habitat and foods. The two races of this species occupy two distinctly different habitats. *Falklandicus* occurs on the sandy beaches and muddy creeks of the Falkland Islands during the nonbreeding period, but moves to dry slopes covered with grasses and diddle-dee (*Empetrum*) for breeding (Woods 1975). It also breeds on Tierra del Fuego, where it has been found nesting around flat, sandy areas with numerous lagoons and ponds. In Argentina it forages on muddy, drying surfaces above the waterline, usually in vegetation-free areas, and on coastal areas it occurs on sand and mud flats, sloughs, and sandy beaches. Winter defense of foraging territories is common, with birds defending areas 10–70 meters long and attaining densities that average about 16 birds per kilometer (Myers and Myers 1979). Its foods seem to consist of small insects and worms obtained along the edge of the surf, in rocky pools or around pools of fresh water behind beaches. However, the race *alticola* has been found nesting in Argentina on a grassy plain near a salty lagoon at an elevation of about 13,000 feet (Hoy 1967). In Chile it probably breeds around freshwater and salt lakes between 12,000 and 16,000 feet elevation (Johnson 1965).

Social behavior. In the Falkland Islands, flocks of adults and young can often be seen during winter, when as many as 150 may occur together, usually resting or foraging along the beach. There seems to be no specific information on the timing of their pair-formation or its details. On the Falkland Islands it seems that the birds are quite sedentary and perhaps do not stray more than a mile or two in the course of a year (Woods 1975). Chases sometimes occur among males within flocks, accompanied by twittering whistles, and probably are a reflection of pair-forming processes. In Argentina winter flocks are rather loose and usually have fewer than 5 birds, with territorial behavior very common and the birds apparently roosting near their foraging areas.

Reproductive biology. Nesting on the Falkland Islands occurs from late September to mid-January, though clutches found after November are probably replacements rather than second nestings. Nests or downy chicks have been seen in Tierra del Fuego in November and December (Humphrey et al. 1970), and a nest of *alticola* was found in early January in Argentina. The nest is a simple scrape in short grass or other vegetation, or on the Falkland Islands may be among dry kelp near the beach. The usual clutch of *falklandicus* is 2–3 eggs, but occasionally there are as many as 4 (Woods 1975). One nest of *alticola* had a clutch of 2 eggs (Hoy 1967). However, in the only breeding of this species in captivity, 3 eggs were laid. These were found on May 22, 1970, and hatched on June 16, indicating a minimum incubation period of 26 days. The one chick that survived was nearly full grown within a month (*Animal Kingdom*, August 1970). It appears that these birds are unlike typical sandplovers in that they do not seek out areas of exposed sand for nesting, and their plumage pattern seems to blend more with pebbles and boulders than with a sandy substrate.

Status and relationships. This species is only locally common on the Falkland Islands and also is common on Tierra del Fuego. The race *alticola* is doubtless less abundant than *falklandicus*, but Johnson (1965) considered it comparatively common in its limited puna habitat. Bock (1958) believed that the species is a close relative of *bicinctus* and that both represent typical sandplovers that have been derived from an *alexandrinus*-like ancestor. The plump body configuration and seemingly large-headed shape of this bird do not conform to the usual sandplover "design," and I am not at all sure that it can be readily written off as simply one of the sandplovers. Strauch (1976) considered it (and *alticola*) as part of a "basal" group of *Charadrius* forms that also included *bicinctus, collaris,* and *ruficapillus*. Bock's assessment seems most reasonable.

Suggested reading. Woods 1975; Myers and Myers 1979.

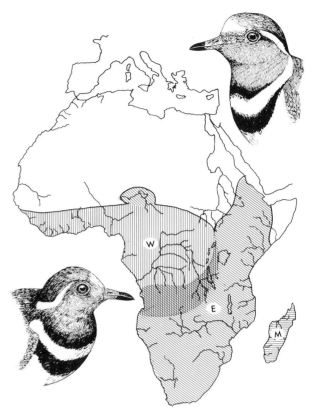

52. Breeding distributions of the eastern (E), Madagascan (M), and western (W) three-banded plovers. An area of possible sympatry or intergradation is shown by crosshatching.

Three-banded Plover

Charadrius tricollaris Vieillot 1818

Other vernacular names. Forbes's plover *(forbesi)*; pluvier de Forbes (French); Braunbänderregenpfeifer (German).

Subspecies and ranges. See map 52.
- *C. t. forbesi:* West African three-banded plover. Resident from Portuguese Guinea to Gabon, western Uganda, western Tanzania, Zaire, Angola, and Zambia. Possibly should be considered a distinct species.
- *C. t. tricollaris:* East African three-banded plover. Resident from Angola, South Africa, and western Africa north to southern Egypt, Sudan, and Nigeria.
- *C. t. bifrontatus:* Madagascan three-banded plover. Resident on Madagascar.

Measurements and weights. Wing: both sexes 103–33 mm; wing measurements of *forbesi* average larger (121–33 mm) than those of *tricollaris* (103–18). Culmen: both sexes 15–17 mm. Weights *(tricollaris)*: adults 25–38 g, average of 11 was 31.2 g (Skead 1977). Britton (1970) indicates similar figures for East African birds, while Summers and Waltner (1979) indicate an average of 34.0 for 55

postjuvenile birds. Eggs ca. 30 x 22 mm, estimated weight 7.3–7.5 g (Schönwetter 1963).

DESCRIPTION

Adults of both sexes have a white *(tricollaris)* or light brown *(bifrontatus* and *forbesi)* forehead and a dark brown crown, separated from the sides of the head by a white streak that extends from eye to eye around the back of the head and terminates at the eye *(forbesi)*, continues around the forehead *(bifrontatus)*, or joins the white of the forehead *(tricollaris)*. The back and mantle are olive brown, and the wings are blackish brown except for white tips on the secondaries. The tail is brown centrally and more whitish laterally, with nearly all the feathers tipped with white and with a subterminal black bar. The sides of the face and throat are pale grayish *(tricollaris)* to light brown, this color gradually changing ventally to dark brown, below which there is a white breastband, and behind this there is a second band of dark

19. Adult heads of East African (*above*) and West African (*below*) three-banded plovers.

In the field (7 inches), this species is closely associated with muddy rather than sandy or rocky areas and is often found near rivers. However, it often breeds on rocky outcrops well above savanna forest. It is usually found near the edge of the water, where it bobs its head and body when alarmed and rocks its tail when landing. The double breastband is the best field mark, and in flight there is a conspicuous white wing bar, as well as a barred tail, that is white laterally. The calls include repeated high-pitched whistles, including shrill *wick-wick* notes and more plaintive notes.

NATURAL HISTORY

Habitats and foods. The habitats of *tricollaris* may differ somewhat from those of *forbesi*, judging from the limited literature. The nominate form is said to frequent muddy pools, salt marshes, and lakeshores, as well as rivers having sandy banks. Probably muddy areas are most favored for feeding (Blaker 1966). The presence of a clear sandy or muddy shoreline may be important to this bird, and it is evidently not often found far from water. In Zambia it occurs along the edges of mud flats, on sandy beaches and dams, and sometimes on streams with shingle, flooded woodlands, or temporary pools on roads. However, *forbesi* not only is found around pools or on open mud flats, but also frequents burned areas and dirt roads (Benson et al. 1971). Generally, *forbesi* seems to be associated with drier habitats than *tricollaris*, including short grasslands, forest clearings, and, during the breeding season, rocky hills and slopes (Mackworth-Praed and Grant 1970). Foods of both probably consist of terrestrial and aquatic insects, insect larvae, worms, crustaceans, and small mollusks. Evidently a good deal of foraging occurs at night (Blaker 1966), as the eyes of this species are unusually large.

Social behavior. Apparently these birds are usually found in pairs, family groups, or small loose flocks up to as many as about a dozen or rarely up to 40 individuals. In Nigeria, flocking of *forbesi* occurs during the dry season, and it begins breeding with the first rains of March, with the flocks breaking up and the birds quickly moving to treeless granitic outcrops to begin breeding. In southern Africa *tricollaris* breeds in the dry season, mostly from July to September (Blaker 1966). In Madagascar, clutches have been found in July and September, and young birds

brown. The rest of the underparts are white. The iris is hazel brown with a large red eye ring, the distal half of the bill is black with a reddish pink base, and the legs and feet are purplish to pink. *Juveniles* have an incomplete brown upper chest band and are more uniformly colored on the head with dull white and dark brown, and their upperparts have buff-edged feathers.

In the hand, adults can be readily recognized by the distinctive tricolored breast pattern, but in both adults and immatures the dark-banded tail, white axillaries, and gray throat pattern should serve to separate this species from other *Charadrius* types. Where *forbesi* and *tricollaris* have to be separated, *forbesi* may be recognized by its more brownish throat, its slightly larger wing measurements, its lack of white on the forehead, and its breastbands, which are broader and are dark brown rather than black.

are seen between August and October. In *bifrontatus* and *tricollaris* the clutch seems to consist of only 1 or 2 eggs, while in *forbesi* 2 or 3 seem to be characteristic (Blaker 1966; Mackworth-Praed and Grant 1970). At Gordon's Bay, in the southwestern Cape area, where this form has been found breeding from July to October, pair-forming behavior begins in early July, when chasing may be seen among the small groups of birds foraging on shingle beaches. Somewhat later, nest-scraping behavior begins, which appears to be the same in this species as in other *Charadrius* species (Martin 1972).

Breeding biology. The nest of the *forbesi* form of this species was first discovered in Nigeria, when a family including a newly hatched chick was discovered on the top of a 500-foot-tall granitic outcrop emerging well above the surrounding savanna forest. Chicks were found in both July and August, and a clutch of 2 eggs was discovered in August, suggesting double brooding. It was determined that both sexes participate in incubation and that in *forbesi* the incubation period is about 20–22 days (Bannerman 1951). However, a recent estimate for *tricollaris* is 27–28 days (Blaker 1966). Observations on *tricollaris* in South Africa indicate that it too usually nests in gravelly or rocky locations, with the scrape sometimes lined with broken shells and the like (Martin 1972). However, unlike *forbesi*, nests of *tricollaris* are always close to water (Blaker 1966). Recent observations on this form (Tyler 1978) suggest that it too may exhibit multiple brooding.

Status and relationships. The several forms in this group are of uncertain taxonomic status. Bock (1958) treated them as conspecific and regarded them as a derivative of the ringed plover group. Maclean (1977) considered *tricollaris* a typical *Charadrius* form and a probably close relative of *melanops*. Although Peters (1934) considered *forbesi* conspecific with *tricollaris*, it has often been separated, most recently by Snow (1978), who considered the two forms to be clearly sympatric and distinct species. I hope that future research will clarify this interesting problem.

Suggested reading. Tyler 1978.

Black-fronted Plover

Charadrius melanops Vieillot 1818
(*Elseyornis melanops* of Peters, 1934)

Other vernacular names. Black-fronted dotterel; pluvier à face noir (French); Maskenregenpfeifer (German).

Subspecies and range. No subspecies recognized. Resident throughout Australia, including Tasmania, mainly in the interior. Also widely established in New Zealand, breeding on Hawkes Bay, on the Manawatu and possibly Ruamahang rivers, and on the Wairau River. See map 53.

Measurements and weights. Wing: both sexes 104–10 mm. Culmen (to base): both sexes 18–19 mm. Weight: females 30–33 g; males 28–29 g (Hall 1974). Eggs ca. 28 x 21 mm, estimated weight 6.3 g (Schönwetter 1963).

DESCRIPTION

Adults of both sexes have a black forehead and facial band that extends from the gape backward through the eyes to the nape, forming a narrow nape collar, and also continues downward along the sides of the neck to expand into a broader black breastband, the area of the cheeks and throat thus enclosed being entirely white, as is a stripe extending from in front of the eye backward around the back of the neck, setting off a brownish crown patch. The back, primary coverts, and smaller wing coverts are brown, while the upper tail coverts and scapulars are chestnut, the primaries are black, and the secondaries are white with black tips. The larger wing coverts are also tipped with white. The central pair of tail feathers are brown, tipped with black, the next three pairs are blackish brown with white tips and bases, and the outermost feathers are entirely white. The undersurface other than the breastband is white. The iris is brown with a bright red eye ring, the bill is orange with a black tip, and the legs and feet are orange pink. *Females* are virtually identical to males. *Juveniles* have a brown forehead, a mottled brown-and-white crown, no black breastband or facial streaks, brown bill, and brown legs and feet. The scapulars gradually become more chestnut, the bill becomes pinkish at the base, and the legs become more orange (Heather 1973).

In the hand, the chestnut scapulars serve to separate adults of this species from other small dotterel-

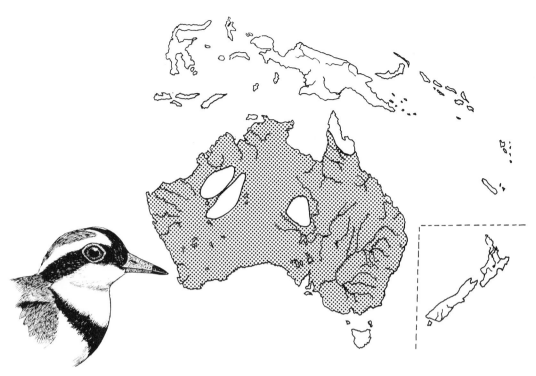

53. Breeding distribution of the black-fronted plover.

like birds, but in immatures the chestnut coloration is paler and extends over the wing coverts as well as the scapulars.

In the field (7 inches), this species is usually found near the water's edge, where it runs very fast, sometimes bobbing its head. When airborne it flies with a distinctive jerky and dipping motion. The chestnut scapulars are evident when it is on the ground or in flight; in contrast with the somewhat similar red-kneed dotterel, white is evident in the middle of the wing rather than at the tailing edge of the inner flight feathers. Unlike the hooded dotterel, it has no white evident on the primaries. Calls include a high-pitched metallic whistle, *tink-tink,* and also repeated clicking *tik* notes.

NATURAL HISTORY

Habitats and foods. The favored habitat of this species consists of rather small bodies of water, such as tanks and stock dams, having firm shorelines and usually a scattering of stones or gravel. Dry watercourses with shingle are sometimes also used by these birds, but they evidently avoid very sandy areas. They feed by pecking at the mud or water surface, usually at the edge of the water, standing on the shore or at most in very shallow water. The birds feed on crustaceans and on such insects as ants, weevils, water beetles, and midge flies and their larvae, and they sometimes also eat seeds (Frith 1969; Maclean 1977).

Social behavior. This is a relatively nongregarious species, apparently rarely occurring in numbers larger than those of a family group. In New Zealand the birds are usually in pairs, suggesting that mates may stay together throughout the entire year. The birds defend rather large nesting territories, so that pairs tend to be well spaced. One possible advertisement display is a circling flight over an area about 100 meters in diameter, with a deeply flapping wingbeat quite different from that of normal flight. This flight is accompanied by a series of loud calls. Hostile encounters usually occur between pairs, and the same loud calls used during aerial display are uttered as the pairs periodically chase one another. However, almost nothing is known of pair-formation or territorial advertisement. Copulation is preceded by various activities and does not always occur at a nest scrape. However, it sometimes occurs without specific preliminary display, and in contrast to many plover species lasts less than 10 seconds. Nest-scraping has been observed, and in one case the associated behavior showed similarities to the scraping ceremonies of *alexandrinus* and *hiaticula* (Phillips, 1980; Maclean 1977).

Reproductive biology. In eastern and southern Australia this species breeds from August to January during spring and summer, but in the interior breeding can occur at any time following rains. Nests are usually on gravelly or stony shorelines, as little as 4 meters or as far as 300 meters from water. The usual clutch is 3 eggs, but 2-egg clutches are not uncommon. The eggs are mottled in color and blend well with a pebbly background. Both sexes incubate about equally, and the incubation period was determined in one case to be at least 25 days. Maclean (1977) suggests that it probably lasts 26 days, as in most other plovers.

Status and relationships. This species is widespread and fairly common over much of Australia, requiring no special attention from conservationists. Its relationships are clearly with the other *Charadrius* forms; Bock (1958) considered it an offshoot from the ringed plover group. However, Maclean's (1977) observations indicate that it is probably a very close relative of *tricollaris* but not closely related to the rather similar *cinctus*, which he has shown to be distinct and apparently not a member of the genus *Charadrius*. Phillips (1980) suggested that *melanops* may be loosely related to *novaseelandiae* and *cucullatus*.

Suggested reading. Maclean 1977; Phillips, 1980.

Killdeer

Charadrius vociferus L. 1758

Other vernacular names. None in general English use; pluvier kildir (French); Schreiregenpfeifer (German); playero gritón, chorlitejo gritón (Spanish).

Subspecies and ranges. See map 54.

   C. v. vociferus: Northern killdeer. Breeds over much of North America, from southern Alaska (rarely), southern Mackenzie, and Quebec south to Lower California, central Mexico, and the Gulf Coast. Winters south to the West Indies, Central America, and northern South America.

   C. v. ternominatus: West Indian killdeer. Resident in the West Indies.

C. v. peruvianus: Southern killdeer. Resident in coastal South America from Peru to extreme northwestern Chile.

Measurements and weights (of *vociferus*). Wing (chord): males 154–67 mm; females 147–70 mm. Culmen (to feathering): males 19–23 mm; females 19–22 mm. Weights: average adult weight 91.3 g (Graul 1973); the averages of two samples obtained in summer were 87.6 and 93.3 g (Glutz et al. 1975). Measurements of *ternominatus* and *peruvianus* average slightly less than those given for *vociferus*. Eggs ca. 38 x 27 mm, estimated weight 14.5 g (Schönwetter 1963).

DESCRIPTION

Adults of both sexes have the forehead and a line over the eye white, the latter interrupted above the eye by a black band across the forecrown. There is also a black to grayish brown streak from the base of the bill along the side of the throat and below the eye, frequently bordered by whitish above, and a short black line at the back of the eye. The crown, upper back, scapulars, tertiaries, and most of the wing coverts are plain grayish brown, and the lower back,

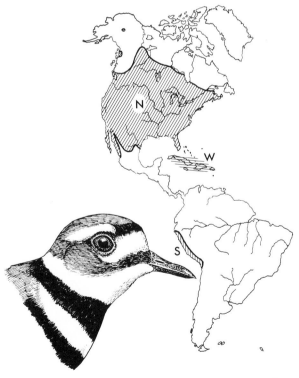

54. Breeding distributions of the northern (N), southern (S), and West Indian (W) killdeers.

rump, and upper tail coverts are pale rufous or tawny. Two black bands cross the lower throat and breast, separated by white, and the rest of the underparts behind the lower band are also white. The greater wing coverts are grayish brown tipped with white, while the primaries are blackish with increasing amounts of white at the base of the vanes proceeding inward. The primary coverts are blackish tipped with pale buffy. The outer tail feathers are largely white to pale cinnamon, barred and spotted to the base with blackish, and the rest of the tail is mostly pale grayish brown, the feathers crossed with a blackish subterminal band and tipped with white or pale buffy. The iris is dark brown, the eye ring is scarlet to orange red, the bill is black, and the legs and toes are pinkish gray to grayish yellow (after Roberts 1932). *Juveniles* have their mantle and covert feathers marked with buff fringes and darkish indistinct subterminal marks. *First-winter* birds retain some buff-fringed inner medians and scapulars; adult feathers are rufous-fringed in this area (Prater et al. 1977).

In the hand, the relatively large size and distinctive double black breastbanding are useful identification marks; this is the largest of the "ringed" plovers and the most widely known and recognized North American shorebird.

In the field (8 inches), the large size, rufous-colored rump, double breastband, and *kil-deee* call all provide easy recognition clues. In southern South America there is a chance of confusion with the similar-sized rufous-chested dotterel, but the bright chestnut breast area should serve to separate the two species.

NATURAL HISTORY

Habitats and foods. This species is extremely tolerant in its acceptable habitats and its capacity for human proximity, nesting in many open-land habitats, including pastures, golf courses, airports, gravel pits, roadsides, and sometimes suburban lawns. Frequently it is found well away from water, and it is especially attracted to gravelly substrates, sometimes even nesting on gravel rooftops. During the winter the birds are more closely associated with water, and they are frequently found along beaches, watercourses, and mud flats as well as in open fields. Throughout the year the birds are primarily insectivorous, with beetles especially prominent in the diet. Beetles and other mostly terrestrial insects made up nearly 80 percent of the foods eaten in one early analysis, while a variety of other invertebrates

constituted nearly all of the remainder. Grasshoppers and weevils are sometimes consumed in great quantities; individual birds have sometimes been found to have eaten hundreds of weevils or other pest insects (Bent 1929). Foraging is done in the usual plover manner of quickly running forward, stopping, and suddenly seizing a prey object from the surface, rather than probing for food. The birds have relatively large eyes and are relatively active during evening hours as well as during the daytime. As one of the largest of the American *Charadrius* species, the killdeer has evolved a niche that in many ways corresponds to that of various *Vanellus* forms in the Old World, and it is certainly the most widespread and familiar of all the North American shorebirds.

Social behavior. During the nonbreeding season killdeers are only semigregarious and are usually seen in numbers ranging from single birds to loose flocks of up to as many as 50 individuals. Apparently little pair-formation occurs before they arrive, either singly or in small groups, on the breeding grounds. Soon after their arrival the birds begin to disperse, and territorial advertising behavior begins. Territorial advertisement consists of the bird's standing in a conspicuous location and calling *di-yeet* or *di-yit* every few seconds for up to several hours. These calls are interspersed with runs or short flights, or with high circling flights that may cover an area a kilometer or more in diameter, and are accompanied by highly variable *killdeer* calls. Such flights are frequently marked by slow, deep wingbeats, and are usually performed only by males, though sometimes the female also participates. On the ground the birds often perform nest-scraping displays, accompanied by a variety of vocalizations, but typically by a trilling call, and by lowering the body until it nearly touches the ground. This display occurs in several situations, including territorial advertising by unmated males, precopulatory encounters, hostile encounters, and during nest-building. It is especially prevalent in the precopulatory situation, which can be initiated by either sex but usually is begun by the male. Behavior immediately following copulation is variable but typically consists of the male's standing very tall and still for as long as 50 seconds. During pair-formation the male digs a succession of scrapes, in the last of which the eggs are laid. Once a pair bond is formed the pair remains together, and together they vigorously defend their territory from conspecifics. However, the appearance of a human intruder or predator often stimulates a collective defensive behavior during which other killdeers are allowed on the territory. Males are usually more aggressive in their defense against humans, while females strongly attack conspecifics in the vicinity of the nest (Phillips 1972).

Reproductive biology. Nesting occurs within the territorial boundaries, and nest sites are usually characterized by a combination of little or no vegetation in the immediate site of the nest, the presence of nest materials (small stones and small plant stems), ground soft enough for digging, and an object beside the nest. Shallow water is usually available nearby, and stones that are flat, white, and between 5 and 10 mm in diameter are selectively used for nest lining. The egg-laying interval averages about 3.7 days, with the interval increasing between successive eggs. Nearly all nests have 4 eggs; the species is a determinate layer. Lost clutches are replaced about a week later. Both sexes participate about equally in incubation, with changeovers at the nest usually including the tossing of stones and plant chips by the departing bird. The incubation period averages 25.1 days, and hatching requires an average of 20 hours. Shortly after hatching the birds begin to leave the nest and forage for themselves. Initial flights occur at about 30 days, but in one case parental care of the first brood lasted 43 days, with the parents driving the young away after the second brood hatched. Double brooding is apparently common over much or all of the breeding range. By the time the young birds are 27–34 days old they have a body temperature typical of adults, but before that time they have subnormal body temperatures when exposed to cold (Bunni 1959).

Status and relationships. This is certainly one of the most common of North American shorebirds, and its capacity for exploiting agricultural habitats and tolerating the close proximity of humans results in a very favorable conservation outlook. The status of the West Indian and South American forms is not well known. The evolutionary relationships are clearly with the typical "ringed" plovers, which as defined by Bock (1958) include *melodus, wilsonius, dubius, placidus,* and *hiaticula* (including *semipalmatus*). Phillips (1972) has compared the social behavior patterns of most of these species as well as those of *alexandrinus* and *montanus* with that of *vociferus* and reports considerable similarity among them. The precopulatory behavior of *vociferus* is particularly close to that of *dubius* and *hiaticula* and less like those of *alexandrinus, montanus,* and per-

haps *wilsonius.* I believe that *vociferus* might have directly evolved from a *hiaticula*-like or *wilsonius*-like ancestor.

Suggested reading. Bunni 1959; Phillips 1972.

Thick-billed Plover

Charadrius wilsonius Ord 1814
(*Charadrius wilsonia* of Peters, 1934)

Other vernacular names. Wilson plover; pluvier de Wilson (French); Dickschnabelregenpfeifer (German); chorlitejo picogrueso (Spanish).

Subspecies and ranges. See map 55.
C. w. wilsonius: Northern thick-billed plover. Breeds in the southeastern United States from Virginia southward, and along the Gulf Coast south through Mexico, wintering from Florida to Honduras.
C. w. rufinucha: West Indian thick-billed plover. Resident in the West Indies.

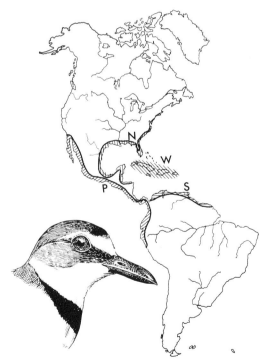

55. Breeding distributions of the northern (N), Pacific (P), southern (S), and West Indian (W) thick-billed plovers.

C. w. beldingi: Pacific thick-billed plover. Resident from Lower Califormia southward through Mexico and Central America to Peru. Possibly should be merged with *wilsonius* (Monroe 1968).

C. w. cinnamoninus: Southern thick-billed plover. Resident along the Atlantic coast of South America from Colombia to Surinam and perhaps Brazil; also resident on Trinidad, Aruba, Bonaire, and Margarita.

Measurements and weights. Wing (chord): males 106–21 mm; females 112–21 mm. Culmen (from feathering): males 19–29 mm; females 19–29 mm (Blake 1977). Weights: males 54–64 g; females 57–69 g (Haverschmidt 1968). Average adult weight 56.9 g (Graul 1973). Eggs ca. 36 x 26 mm, estimated weight 12.4 g (Schönwetter 1963).

DESCRIPTION

Adult males in breeding plumage have the forehead, sides of the rump, lateral upper tail coverts and underparts white except for a black breastband. There is also a black streak from the bill to the eye, becoming more brownish on the sides of the face and ear coverts, and the anterior part of the crown is black, becoming brown to cinnamon or rusty posteriorly. The upperparts are generally grayish brown, with white tipping on the greater wing coverts, the primaries and secondaries are dusky, with most of the shafts white, and there is white on the bases of some of the outer webs. The tail is light grayish brown, becoming darker subterminally and whitish at the tip and generally becoming whiter toward the outermost pair, which are entirely white. The bill is black, the iris is dark brown with a grayish eye ring, and the legs and toes are pale grayish to pink (Ridgway 1919). *Females* have warm brown to rufous breastbands and foreheads but otherwise resemble breeding males. *Adults in winter* also have the breastband and anterior crown pale brown. *Juveniles* have a very pale and indistinct breastband and anterior crown pattern, and *first-winter* birds are less black or dark brown in these areas than are adults (Prater et al. 1977).

In the hand, this species can be distinguished by the combination of a single large breastband of brown or black, a moderately long (19–29 mm exposed) culmen, black bill, and pinkish legs and toes. The larger but somewhat similar killdeer has two breastbands, and the smaller semipalmated plover has a yellow-and-black bill.

In the field (6½ inches), this species is found on coastal sandy beaches and mud flats, and its larger size and considerably larger bill size than *semipalmatus* are readily apparent, as are differences in the bill colors. This species also shows more white above the eye and on the forehead. In flight, both species show a considerable amount of white on the wings. The usual call is a single abrupt *wheet* or a deeper double whistle. The notes are generally sharp and unmusical, more like those of *collaris* than those of *semipalmatus,* with both of which it sometimes is associated.

NATURAL HISTORY

Habitat and foods. The favored breeding habitat consists of sand dunes and other barren or nearly barren habitat, usually close to salt water or brackish water. Open areas on sandy islands are favored; evidently the birds tend to select higher and drier portions of sandy beaches than do semipalmated plovers. Hard-packed sandy beaches are probably not suitable, but the birds sometimes occur on mud/sand flats near creek entrances. The birds feed largely by standing still until they sight food, then running quickly to it. Little if any wading is done. Although they doubtless consume a large variety of invertebrates, fiddler crabs are probably a major component in the diet (Tomkins 1944; Slud 1964).

Social behavior. Pair-formation apparently occurs before territorial establishment, though the two activities are not entirely separated. Courtship is initiated by the male and consists of chasing other males from the female, threat displays, and scrape-making. The displays include fluffing the throat so that the dark breastband stands out like a ruff, and spreading the breast feathers. The bird may stand erect or chase another in a running crouch. Initially the scrape-making displays are apparently done at various locations, but after a territory has been established it serves as a nest-site location mechanism. No specific aerial display flight has been described, but one may well occur. When the female is ready for copulation she postures and crouches before the male. He walks up behind her and performs a high-stepping display for a minute or so before mounting. Territories are usually no more than 100 feet in radius, rarely up to 300 feet. Nesting seems to be somewhat colonial;

as many as 18 birds have been noted by a person walking through a colony (Tompkins 1944).

Reproductive biology. Nest sites are apparently selected soon after the pair has settled on its territory, and as in other plovers are chosen by the nest-scraping ceremony. After the male has made several scrapes, the female begins to lay in one of them. The usual clutch is of 3 eggs, but sometimes there are only 2 and rarely there are 4. Incubation is by both sexes, with the female evidently performing the larger share. Incubation lasts 24–25 days, and the precocial young are able to begin feeding for themselves within 24 hours after hatching. They leave the nest soon after hatching, and are able to fly when about 21 days old (Tompkins 1944).

Status and relationships. This species is apparently still common enough that it has not yet appeared on the Audubon Society's "Blue List" of declining species. Its relationships are almost certainly with the other "ringed" plovers as defined by Bock (1958), although it has a relatively heavier bill than any of these. This bill size is probably directly related to its consumption of such relatively large arthropods as fiddler crabs and is of no great taxonomic significance.

Suggested reading. Tompkins 1944.

Long-billed Plover

Charadrius placidus Gray and Gray 1863

Other vernacular names. None in general English use; pluvier a long bec (French); Langschnabel-regenpfeifer (German).

Subspecies and range. No subspecies recognized. Breeds from the middle Amur Valley south through Manchuria and eastern China, and in Japan. Winters in Japan, China, the Indo-Chinese countries, and west to Nepal. Sometimes considered a subspecies of *hiaticula*, but recognized as specifically distinct by Vaurie (1965). See map 56.

Measurements and weights. Wing: males 132–44 mm; females 137–45 mm. Culmen: both sexes 18–21 mm. Weights: 4 males 41–70 g, average 58 g; 15 females 57–70 g, average 64 g (Shaw 1936). Eggs ca. 36 x 26 mm, estimated weight 12.3 g (Schönwetter 1963).

DESCRIPTION

Adults of both sexes are very similar to those of the ringed, semipalmated, and little ringed plovers, but there is no black stripe at the base of the upper mandible, so that the white forehead is broader and reaches the bill. Additionally, the ear coverts are less marked with black, and the black forehead band is the narrowest. The shafts of the primaries are almost uniformly brown, lacking white, except perhaps on the first, which is sometimes whitish or yellow. There is almost no white on the inner webs of the primaries, and it is present only to a limited extent on the secondaries. The outermost rectrices are similarly brown, with white only on the tip and base of the outer web. The iris is brown, and the eye ring is yellow. The bill is black, with only the base of the lower mandible orange yellow, and the legs and toes are slightly yellowish (Dementiev and Gladkov 1969). *Juveniles* resemble adults but have fringes of pale buff on the wing coverts (Prater et al. 1977).

In the hand, this species is separated from the similar ringed plover by its longer bill, which lacks yellow on the upper mandible, by the more extensive white on the forehead, and by the less distinct facial mask of the lores and ear coverts. Further, the eye ring is yellow rather than orange. Last, the black subterminal band of the tail extends to the outer web of the outermost feathers, and the wing stripe is poorly developed or absent.

In the field (8 inches), this species would be difficult to separate from the ringed plover, but the characteristics just mentioned might be visible at very close range or when the two are side by side. In flight, only the indistinct whitish wing stripe would be a useful field mark. The calls of this species are distinct from those of *hiaticula*. They include a repeated *piwii*, resembling that of the lesser ringed plover but louder and clearer, and a musical trisyllabic *tudulu*. The species is more associated with muddy than with sandy areas in winter, but it is associated with gravel and shingle areas during the breeding season.

NATURAL HISTORY

Habitats and foods. Shingle-lined rivers and mud flats are apparently favored wintering habitats. Vaurie (1965) stated that breeding occurs only on the edges of lakes or rivers covered with gravel, shingle, stones, and boulders. However, Dementiev and

56. Breeding distribution of the long-billed plover.

Gladkov (1969) describe the biotope as consisting of extensive and level pebble bars, but not including either sandy areas or areas with large pebbles or boulders. It has been reported (Kozlova 1961) that *placidus* differs considerably from *hiaticula* in its preferred habitats and feeding behavior, though at times they have been regarded as conspecific (Dementiev and Gladkov 1969). Its foods are still little known but apparently include flies and beetles. The longer and more slender bill of *placidus* as compared with *hiaticula* suggests that it is more likely to probe in mud or gravel than is that species.

Social behavior. In the winter this species is said to be usually solitary (Ali and Ripley 1969), and the only good observations of social behavior are for spring (April and May). By mid-April the birds are already paired and are locating nests on rivers with sandbars and areas of mixed pebbles and sand. Males perform nest-scraping displays in areas where there are pebbles with small sandy areas between the rocks. This ceremony appears to be much the same as in *hiaticula* except that the male utters three-syllable notes rather than repeated monosyllabic calls. Further, in the few cases so far observed the female approached the male from behind and came to a standstill with her head below the male's tail rather than crawling under the male from the side. Before copulation a high-stepping display is performed by the male that is evidently much like that of *dubius* and several *Charadrius* species. During display flight, which seems to be related more to territorial advertisement than to sexual display, the birds raise their wings almost vertically and utter a two- or three-syllable call. In Panov's (1963) opinion, these calls and displays are distinctly different from those of both *dubius* and *hiaticula* and do not suggest a closer affinity to one than to the other.

Reproductive biology. Little has been written on nesting biology, but the clutch is said to consist of 4 eggs, and the nest is the usual simple depression in the ground. Panov's (1963) observations suggest that the nest sites are usually on a higher portion of a sandy river shoreline, where there is a mixture of sand and gravel. Doubtless in most of its biology this species is much like *hiaticula*.

Status and relationships. Although Dementiev and Gladkov (1969) included *placidus* as a subspecies of *hiaticula*, Kozlova (1961) and Vaurie (1965) both regarded it as a distinct species, the former considering it close to *dubius* and the latter also listing it closer to *dubius* than to *hiaticula*. Eck (1976) regarded it as a close relative of *hiaticula*. Bock (1958) considered it as part of a superspecies with *hiaticula* (and *semipalmatus*). I suspect that *hiaticula* is the nearest relative and that its bill shape is associated with ecological foraging differences between these species.

Suggested reading. Eck 1976; Dementiev and Gladkov 1969.

Ringed Plover

Charadrius hiaticula L. 1758

Other vernacular names. None in general English use; pluvier à collier (French); Sandregenpfeifer (German); chorlito de collar, chorlitejo grande (Spanish).

Subspecies and ranges. See map 57.

C. h. tundrae: Eastern ringed plover. Breeds in northern Russia and Siberia from the Chukotski Peninsula west to about Novaya Zemlya, where intergrades with *hiaticula* occur. See *hiaticula* for winter range.

C. h. hiaticula: Western ringed plover. Breeds from the eastern coasts of Baffin and Ellesmere islands east through Greenland, Iceland, Spitsbergen, and the northern parts of Europe from the British Isles and France east to northern Scandinavia, the Kola Peninsula, and southern Novaya Zemlya. Breeds rarely on St. Lawrence Island. This form and *hiaticula* winter from the southern parts of their breeding ranges in Europe to southern Africa, the Persian Gulf, western India, and rarely farther east.

Measurements and weights. Wing: males 123–40 mm; females 123–40 mm. Culmen: males 16–23 mm; females 18–24 mm. Weights (*hiaticula*, breeding season): males 57–72 g; females 56–75 g (Glutz et al. 1975). Eggs ca. 34 x 25 mm, estimated weight 10.2–11.5 g (Schönwetter 1963).

57. Breeding distributions of the semipalmated plover (*shaded*), and the eastern (E) and western (W) ringed plovers.

DESCRIPTION

Breeding adults of both sexes are almost identical to those of the semipalmated plover but average slightly larger. The breastband is somewhat broader, the inner secondaries are darker, with the inner webs usually entirely brown, and the wing bar is slightly narrower and shorter. The soft-part colors are likewise essentially the same, although the eye ring is perhaps more orange. *Adults in winter* have the black of the facial mask and breastband replaced with brown, until midwinter. The coverts are grayish brown, with whitish fringes on the inner ones. The bill is dusky yellow at the base. *Juveniles* have a brown breastband, which is sometimes almost broken in the center. The head markings are also brown and are somewhat obscured. The upperparts and wing co-

verts are fringed with pale buff, with darker subterminal bars. The legs and toes are dull orange yellow, and the bill is blackish with a little yellow at the base of the lower mandible. *First-winter* birds retain some buff-tipped inner median coverts, the bill is only slightly yellowish at the base, and the primaries are usually noticeably pointed (Prater et al. 1977).

In the hand, this species may be separated from *semipalmatus* by its reduced webbing, which is present only between the outer and middle toes. Other distinctions are mentioned in that species' account.

In the field (7 inches), this species is most likely to be confused with the little ringed plover, which, however, has a black bill, a yellow rather than orange eye ring, a narrow white line behind the black forehead band, and pale pinkish legs. It also does not show a white wing stripe in flight. The ringed plover's calls include a *coo-eep* or *too-li,* and its song is a melodious trilling repetition of *quitu-weeoo.*

NATURAL HISTORY

Habitats and foods. This is essentially a coastal species during the breeding season, nesting on sandy, pebbly, or shell beaches, along coastal tundra ponds or lakes, on muddy plains with stones or pebbles, occasionally well away from water, and rarely along riverbanks (Voous 1960). However, unlike the lesser ringed plover, it is not river-adapted, and unlike that species it is rarely found at great distances from the coast. Smith (1969) found no significant differences in breeding habitat between *hiaticula* and *semipalmatus.* Both species were most numerous on coastal flats, followed in frequency by river terraces and lakeshores. The species is well dispersed during breeding, with most densities well below a pair per hectare, but with local concentrations of up to about a pair per hectare or 1.5 pairs per kilometer of coastline (Glutz et al. 1975). Outside the breeding season the species is also essentially coastal in its occurrence, using muddy, sandy, or pebbly coasts on tropical and subtropical climatic zones. Its foods are almost entirely animal matter and include a diverse array of mollusks, crustaceans, insects, annelids, and other invertebrates. It forages in a manner similar to that of most plovers and frequently uses foot-trembling movements during pauses between pecking, probably to cause prey to come to the surface.

Social behavior. These birds are strongly territorial during the breeding season. On Baffin Island they arrive in late May, often in groups of 3 birds that in-

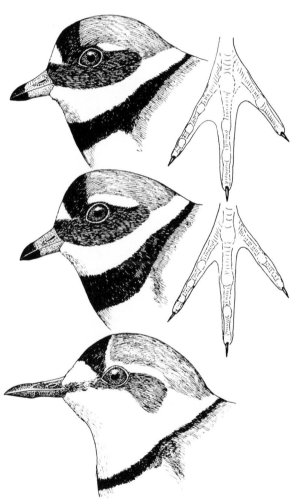

20. Breeding-plumage heads and feet of semipalmated (*top*) ringed (*middle*) and long-billed (*bottom*) plovers.

31. Downy young of (*left side, top to bottom*) black turnstone, eastern solitary sandpiper, northern cordillerian snipe,(*right side, top to bottom*), white-rumped sandpiper, bar-tailed godwit, and little curlew. *Painting by Jon Fjeldså*

32. Black turnstone, winter plumage. *Photo courtesy Kenneth Fink*

33. Ruddy turnstone, nuptial plumage. *Photo by author*

34.   Surfbird, winter plumage. *Photo courtesy Kenneth Fink*

35.   Eastern knot, adult incubating. *Photo courtesy Vladimir Flint*

36. Spoon-billed sandpiper, adult incubating. *Photo courtesy Vladimir Flint*

37. Rufous-necked sandpiper, adult incubating. *Photo courtesy Vladimir Flint*

38. Least sandpiper, nuptial plumage. *Photo by author*

39. Dunlin, winter plumage. *Photo courtesy Frans Lanting*

40.   Sanderling, winter plumage. *Photo courtesy Kenneth Fink*

41.   Stilt sandpiper, adult brooding. *Photo by author*

42. Ruff, male in nuptial plumage. *Photo by author*

43. Red phalarope, adult female. *Photo courtesy Ed Schulenberg*

44. Willet, nuptial plumage. *Photo by author*

45. Redshank, nuptial plumage. *Photo by author*

46. Lesser yellowlegs, nuptial plumage. *Photo by author*

47. Wandering tattler, winter plumage. *Photo courtesy Kenneth Fink*

48. Eurasian sandpiper, nuptial plumage. *Photo courtesy Kenneth Fink*

49. Spotted sandpiper, nuptial plumage. *Photo by author*

50. Upland sandpiper. Adult and young. *Photo courtesy Mary Tremaine*

51. Eskimo curlew and pasque flower. *Painting by James McClelland*

52. Whimbrel, adult and young. *Photo by author*

53. Long-billed curlew, adult and young. *Photo courtesy Mary Tremaine*

54. Eurasian curlew, adult. *Photo by author*

55. Black-tailed godwit, adult in nuptial plumage. *Photo by author*

56. Hudsonian godwit, adult brooding. *Photo courtesy Ed Schulenberg*

57. Marbled godwit, winter plumage. *Photo courtesy Kenneth Fink*

58. Long-billed Dowitcher, nuptial plumage. *Photo courtesy Frans Lanting*

59. Chatham Island snipe, adult. *Photo courtesy R. S. Phillips*

60. North American snipe, adult. *Photo by author*

clude 1 female and 2 males. However, many of the late-arriving birds are already paired. Males rapidly establish small territories that are often contiguous and that at least in two studies (Smith 1969; Mason 1947) averaged about 30 meters in diameter. In Smith's study, territories of ringed and semipalmated plovers were often adjacent but nonoverlapping, and contiguous territories between the species were more frequent than contiguous territories involving the same species. Aggressive behavior patterns are similar to those of other "ringed" plovers and tend to emphasize the facial pattern and the breastband. In this species the tail is often fanned and either depressed or elevated, and unlike the lesser ringed plover the head is held above the level of the body (Simmons 1953b). The territorial advertisement flight, or "butterfly flight," is performed about 3–10 meters above the ground and is relatively faster than in the little ringed plover or the Kentish sandplover. The associated call is a repeated *tche'rick* that speeds up to a rattle and ends on a descending slur, apparently much like that of the semipalmated plover (Drury 1961). These territorial flights are largely performed after the establishment of a pair bond and territory and are probably primarily aggressive rather than sexual in their significance. Besides the purely aggressive ground displays, nest-scraping behavior is well developed in this species. When the female approaches the nest she typically enters it by moving under the tail of the scraping bird, as is also typical of the lesser ringed plover, snowy plover, and killdeer. As Simmons (1953b) has pointed out, the courtship has some elements in common with aggressive behavior. When the female is approached by a displaying male, he may run toward her with his head held low and drawn in, while his tail and wings are held vertically upward in a posture comparable to that of strong threat display. As in the little ringed plover, the male gradually shifts from this low and horizontal attitude to a progressively more erect and intimidating one, marking time with his legs by stepping in place, sometimes actually striking his breast with the alternately raised feet. The male grasps the nape of the female while mounted, and copulation lasts 15–20 seconds. Unlike the snowy plover, copulation is not restricted to the nest scrape area. Pair bonds are renewed annually, but there is a tendency to mate with the previous season's mate as long as both birds are alive (Glutz et al. 1975).

Reproductive biology. The nest is a hollow in the sand, often quite open but sometimes sheltered by plants. It is often lined with small pebbles or shell fragments, and in one unusual instance it had a lining of about 2,000 pebbles that apparently had been transported at least 20 yards. The most favored locations seem to be near the high-water mark on shingle or sandy beaches, but some nests are placed as far as a half-mile from the nearest water. The eggs are laid at irregular intervals but usually between 24 and 48 hours apart, so that a 4-egg clutch requires 4–7 days (usually 6) for completion. Incubation is by both sexes, and as in other plovers a well-developed injury-feigning behavior is typical of both sexes, but especially females, when the nest is disturbed by humans. There is a frequent changing over of incubation duties, and the incubation period is typically 24–25 days, with extremes of 21 and 28 days reported. Both parents tend the young, which usually fledge in 21–23 days. Double brooding is typical, at least in some of the more southerly populations, and renesting is also regular. The eggs of the second clutch may be laid in the original nest or may be placed in a new site as much as 500 meters distant (Glutz et al. 1975).

Status and relationships. The population status of this species is apparently quite secure, and in some areas such as northern England and Scotland the breeding range has increased somewhat by gradual spreading inland. The species can be regarded as the central component of the "ringed" plovers as defined by Bock (1958). Its nearest relative is certainly *semipalmatus*, and, although the two forms locally hybridize (Smith 1969) and sometimes are considered a single species (Bock 1959), I prefer to follow Vaurie (1965) in maintaining them as separate species for the present. I agree with Mayr and Short (1970) that they constitute a superspecies, with *placidus* as their probable nearest relative.

Suggested reading. Prater 1974a.

Semipalmated Plover

Charadrius semipalmatus Bonaparte 1825
(*Charadrius hiaticula*, in part, of Peters, 1934)

Other vernacular names. Semipalmated ringed plover; pluvier semipalmé (French); Spaltfussregenpfeifer (German); chorlitejo semipalmado (Spanish).

Subspecies and range. No subspecies recognized. This form is sometimes treated as a subspecies of

C. hiaticula. Breeds in North America from Alaska east through Mackenzie, Victoria, Southampton, and southern Baffin Island (where it locally overlaps and hybridizes with *hiaticula*), the coastline of Hudson Bay, and east to Newfoundland and southern Nova Scotia. Winters from the southern United States south to southern South America and Hawaii. See map 57.

Measurements and weights. Wing: males 114–27 mm; females 115–29 mm. Culmen: males 15–18 mm; females 14–17 mm. Weights: wintering birds of both sexes 35–42 g (Haverschmidt 1968). Males in summer range 29–48 g, average of 11 was 42 g; females range 38–54 g, average of 5 was 42.4 g (Irving 1960). Eggs ca. 33 x 24 mm, estimated weight 9.6 g (Schönwetter 1963).

DESCRIPTION

Breeding adults of both sexes have the forepart of the crown, a narrow band at the base of the upper mandible, continuous with a stripe running back below the eye to the ear coverts, and a broad collar around the lower neck black or grayish brown. The forehead, lower eyelid, a more or less broken line over the eye, the chin, upper throat, a nape collar, and the underparts behind the chest are white. The rest of the upperparts are plain grayish brown except the greater wing coverts, which are tipped broadly with white,

the primaries, which are white on the inner webs, and the secondaries, which are inwardly and outwardly grayish brown with white tips. The tail is light grayish brown, the feathers darker near the tips and tipped with white, with the outer pair mostly white. The iris is dark brown, the bill is black distally and orange to orange yellow basally, the legs and toes are pale orange yellowish to pink, and the eye ring is orange to pale yellow (after Roberts 1932). *Adults in winter* have their black markings replaced by brown, and the legs and toes are more yellowish. *Females* tend to have a considerable number of brown feathers in the breastband and have paler bill and eye ring coloration than males. *Juveniles* have a brown breastband and extensive buff fringes and darker subterminal bands on the coverts and back and also have a blackish bill. *First-winter* birds resemble adults but retain some buff-tipped inner median coverts (Prater et al. 1977).

In the hand, this species is separated from *hiaticula* by its greater amount of webbing, which is extensive between the outer and middle toe and slightly present, reaching the first joint, between the middle and inner toes.

In the field (7 inches), this species can be readily separated from other North American *Charadrius* species by its small size, continuous breastband, and black-tipped yellow bill. It is very difficult to separate from the European ringed plover except in the hand, but when the two are side by side the black-and-white patch behind and above the eye tends to be narrower and more rounded and sometimes is almost absent. This species also appears smaller and shows less white on the head, secondaries, and tail. The vocalizations, however, are apparently identical. In flight the birds show a strong white wing bar, and the usual call is a *chur-wee.* The "song" is a series of short *ke-ruck* notes increasing in speed until they become a whinny or chuckle, followed by a rough *r-r-r-r-r-r,* and ending with a slurred, descending yelp or crow.

NATURAL HISTORY

Habitats and foods. Like its Old World counterpart, this species breeds preferentially in open sandy, gravelly, or pebbly environments, such as along the sandbars and gravel bars of rivers, stony or gravel ridges overlooking coastal beaches, or areas such as roadsides where the ground has been scraped by heavy equipment, exposing a gravelly substrate. Although it had earlier been suggested that *hiaticula*

Threat display, female

favors bare gravel flats and *semipalmatus* occurs on river terraces where the two forms are in breeding contact, Smith (1969) found that both species bred in essentially equal proportions along lakeshores, river terraces, and coastal flats of eastern Baffin Island. There was likewise no obvious segregation of breeding pairs within single uniform habitats, although territories of pairs representing the two species were more commonly in contact than were territories of the same species. It is likely that the foods and foraging methods of the two forms are essentially identical, considering the strong morphological similarities between them. The ecological significance, if any, of their differences in foot palmation has not been investigated.

Social behavior. There are apparently no obvious differences in the pair-forming behavior and vocalizations of *hiaticula* and *semipalmatus* on the breeding grounds, a fact that rather strongly supports the view that the two should be regarded as conspecific. Smith (1969) reported that on his study area pair-formation in the two forms apparently occurred at the same time but, for more than half of the individuals, not in the same region of eastern Baffin Island. In five of eight fjords, mixed pairings were observed, and in each of these five areas the incidence of mixed pairing was higher than would be expected by random chance. Yet the observed frequency of mixed pairing for the total population was less than would be expected from random mating, suggesting that other factors are also operative. Among these are the fact that about half the population of *hiaticula* was already mated on arrival, which of course reduced the probability of mixed pairing. Among the birds that did form mixed pairs, the subsequent average clutch sizes were as large as for unmixed pairs, and the average brood sizes were also at least as large, though not statistically larger. Smith suggested that such heterozygotic offspring might actually show superiority, favoring the maintenance of a polymorphic population, with a gradual shift toward the *semipalmatus* type during the past 80 years for which population data exist.

Reproductive biology. Nests in this form are placed on dry ground, usually on sandy or gravelly substrates with relatively sparse vegetation. Gravel flats on beach ridges or eskers are especially favored sites. The average clutch size in the Churchill area of Manitoba is nearly 4 eggs, with 19 of 23 clutches being of this size (Jehl and Smith 1970), while in eastern Baffin

Island there were 25 3-egg and 32 4-egg clutches reported by Smith (1969) for pairs that did not include *hiaticula* as one of the pair members. Four-egg clutches are typical on Baffin Island (Parmelee et al. 1967). Clutches are apparently completed within a week, and incubation is performed by both sexes. It lasts 23–24.5 days from the laying of the last egg, and the hatching of the last-laid egg may occur as long as 30 hours after the first 3 eggs have hatched, suggesting that some incubation may occur before the completion of the clutch. The fledging period is probably between 22 and 31 days (Sutton and Parmelee 1955).

Status and relationships. This is a relatively abundant species in arctic North America, and it certainly is in no apparent danger. It is obviously very closely related to *hiaticula,* and in some areas the two forms certainly behave as if they are a single species, while in other areas they reportedly are sympatric without interbreeding. Vaurie (1965) thus considers the two forms separate species, while Bock (1959) urged that they be considered conspecific although exhibiting circular overlap. Smith (1969) provided strong evidence on mating preference and chick survival supporting that position, but I believe that the situation should be further investigated rather than simply considered settled. For that reason I am retaining them as members of a superspecies, a procedure also followed by Mayr and Short (1970).

Suggested reading. Smith 1969; Sutton and Parmelee 1955.

Piping Plover

Charadrius melodus Ord 1824

Other vernacular names. None in general English use; pluvier siffleur (French); Pfeifregenpfeifer (German); chorlitejo picicorto (Spanish).

Subspecies and ranges. See map 58.
  C. m. melodus: Eastern piping plover. Breeds in coastal North America from Quebec and Newfoundland south to Virginia. Winters mainly from South Carolina to Florida.
  C. m. circumcinctus: Interior piping plover. Breeds locally in the prairie provinces of Canada, the Dakotas, and Nebraska; also breeds locally south of Lake Michigan and around Lakes

Huron, Erie, and Ontario. Winters on the Gulf Coast of the United States east at least to Alabama.

Measurements and weights (*melodus*). Wing: males 115–27 mm; females 110–22 mm. Culmen (to feathering): males 10–13 mm. Average adult weight 52.7 g (Graul 1973). Wilcox (1959) reported that breeding males range 46.5–63.7 g, average of 49 was 54.9 g; females 46.4–62.3 g, average of 38 was 55.6 g. Eggs ca. 31 x 24 mm, estimated weight 9.4 g (Schönwetter 1963).

DESCRIPTION

Adult males in breeding plumage have white on the forehead, the superciliary region, the cheeks, and the throat, extending back to form a collar at the nape, while the forecrown has a black bar extending from eye to eye, and the sides of the chest are also black,

58. Breeding distributions of the interior (I) and eastern (E) piping plovers.

continuing narrowly around the hindneck and sometimes also to the front of the breast. The rest of the upperparts including the top of the head are a very pale grayish brown, lightest on the rump and upper tail coverts, which are white laterally. The underparts behind the breastband are entirely white, the greater coverts are tipped with white, the secondaries are mostly white, and the primaries have dusky black tips and are variably white toward the base, the white increasing inwardly. The tail is white at the base, the feathers becoming darker toward the ends but tipped with white, and the outer feathers are increasingly white so that the outermost ones are wholly white. The bill is black terminally and orange yellow to dull orange at the base. The iris is brown, with a pale yellow eye ring, and the legs and toes are dusky reddish pink to orange yellow. *Females* have a blackish brown rather than black breastband and forecrown, with the breastband sometimes reduced to lateral bars, and the bill is more yellowish. *Adults in winter* lack dark forecrowns, and the breastband is reduced to lateral gray patches. The bill is generally blackish. *Juveniles* are sandy gray on the upperparts and coverts, with pinkish buff feather tips, and there is no breastband or dark forecrown. *First-winter birds* can hardly be told from adults in winter plumage except perhaps by differences in primary wear (Prater et al. 1977).

In the hand, this species is most likely to be confused with the snowy sandplover, but it is slightly darker, with a longer bill (the culmen is as long as or longer than the middle toe without the claw, or at least 13.5 mm). It also lacks a dark ear patch and has yellow to orange legs rather than slate gray ones (in life).

In the field (5½ inches), piping plovers are associated with sandy flats and riverbanks. They are hard to distinguish from snowy sandplovers in the field, but the more orange leg coloration and the more whitish ear coverts help separate them. In flight, both species show strong white wing stripes. The usual call is a clear, melodious *peep-lo,* the first note lower, and they also utter a series of short whistles on a descending scale.

NATURAL HISTORY

Habitats and foods. The favorite breeding habitat of this coastal population consists of dry, sandy areas, often outer beaches well above wave action or high-tide line. Grassless sandy areas are preferred over grassy ones, though openings in grassy dunes as small as 200–300 feet long may be used. The interior population favors the open shorelines of shallow lakes, especially salt-encrusted shorelines of gravel, sand, or pebbly mud. Outside the breeding season the birds feed on beaches, lagoon edges, and areas of rubble. They forage much like semipalmated plovers but do not use such abrupt actions. They also tend to run faster and farther between times when they stop and snatch their prey. The foods are not well studied but are known to include marine worms, fly larvae, and beetles. Crustaceans and mollusks have also been listed. The birds tend to separate when foraging, even though they may arrive at and depart from an area together (Bent 1929; Stout 1967).

Social behavior. These birds arrive on their nesting ground in Long Island in late March and soon spread out over the sandy beaches. Densities as high as 64 pairs along a 17-mile stretch of beach have been noted there. Territories are well spaced, and breeding birds will usually not allow others within 100 feet of their nest. Usually the nests are at least 200 feet apart. Adult birds return year after year to their same nesting areas, but young birds may nest from 5 to 18 miles from where they were hatched, or they may nest as near as 650 feet to where they were raised. When adults return and pair with their mates of the past year, the new nest is usually very close to the previous year's nest. But more frequently the mates are changed, even when the original mate is available. In one case, a male had 4 different mates in 4 years, though all 4 females were alive during one of the years. On the other hand, 2 pairs were known to remain intact for 3 years. When an adult takes on a new mate, the nest is likely to be closer to the nest of the previous year if the member of the original pair is a male. Eggs are laid at daily intervals, and the usual clutch is 4 eggs, but sometimes there are 3 and rarely only 2, probably as a result of egg losses. Both sexes incubate, apparently at about equal intensities, and incubation lasts from 27 to 31 days but usually is 27 or 28 days. The young birds leave the nest a few hours after hatching and are able to fly at 30–35 days of age. Brooding by the adult may occur until the young are about 20 days old, and injury-feigning behavior toward intruders is common (Wilcox 1959).

Status and relationships. There can be little question that this species is declining throughout its range and now is in rather serious trouble (*American Birds* 34:206–8). It has been on the Audubon Society's "Blue List" of declining species since the inception of that list in 1971. The interior race *circumcinctus* is probably especially in need of protection. It is still locally common in North Dakota (Stewart 1975) and eastern South Dakota (Harrell 1978), but it has become extremely local in Nebraska (Johnsgard 1979). In Canada its nesting range still includes parts of southeastern Alberta, southern Saskatchewan, and southern Manitoba (Salt and Salt 1976). The status of the Great Lakes population is less certain. Bock (1958) believes that *melodus* is a direct derivative of a *hiaticula*-like form rather than a typical sandplover. In my opinion this is the most likely pattern of evolution.

Suggested reading. Wilcox 1959; Cairns 1977.

Little Ringed Plover

Charadrius dubius Scopoli 1786

Other vernacular names. None in general English use; pluvier gravelot (French); Flussregenpfeifer (German); chorlitejo chico (Spanish).

Subspecies and ranges. See map 59.

C. d. curonicus: Eurasian little ringed plover. Breeds in Eurasia from Scandinavia east through Russia and Siberia to the Sea of Okhotsk and perhaps Sakhalin, and south to the Mediterranean and northern Africa, Asia Minor, central Asia, Mongolia, China, and Japan. Winters south to Sri Lanka, the Malay Peninsula, and the Sundas.

C.d. jerdoni: Indian little ringed plover. Resident throughout the Indian subcontinent from Sind through Burma, the Malay Peninsula, and southern China, and south to Sumatra and Borneo.

C. d. dubius: Pacific little ringed plover. Resident in the Philippines, New Guinea, and New Ireland.

Measurements and weights (for *curonicus*). Wing: males 111–19 mm; females 112–20 mm. Culmen (from feathering): both sexes 12–14 mm. Weights: males 34–38.5 g; females 35–45 g (Bub 1958). Eggs ca. 30 x 22 mm, estimated weight 6.0 (*jerdoni*) to 7.5 g (*curonicus*) (Schönwetter 1963).

DESCRIPTION

Breeding adults of both sexes have a narrow line on the forehead and a broad band on the front of the crown black, divided by a broad white band. The lores, a band under the eye, the ear coverts, and a narrow line over the eye are black, and above the latter there is a very narrow line that reaches the black forehead band and sometimes continues across the forehead. The crown is drab brown and there is a white nape collar followed by a black one that joins a black breastband. The upperparts are generally drab brown to ashy brown, but the sides of the back, rump, and upper tail coverts are white, as are the chin, throat, and sides of neck. The underparts behind the black breastband are also white. The central tail feathers are sepia, and the two outermost pairs are white with sepia markings, while the remainder are broadly tipped with white and banded subterminally with blackish. The primaries are sepia, with the tenth having a white shaft, but otherwise the shafts are brown. The secondaries are ashy brown with white tips, the primary and greater coverts are tipped with white, and the smaller coverts are also narrowly edged with white or buff. The iris is brown with a bright yellow eye ring, the bill is black with yellowish at the base of the lower mandible, and the legs and toes are pink to yellowish pink (after Witherby et al. 1941). *Adults in winter* have more brownish breastbands. *Juveniles* have extensive buff fringes on their sandy brown upperparts and have a brown breastband that is obscure and often incomplete, as well as obscure brown head markings. *First-winter birds* resemble adults but have buff-tipped inner median coverts, at least through midwinter (Prater et al. 1977).

59. Breeding distributions of the Eurasian (E), Indian (I), and Pacific (P) little ringed plovers.

In the hand, this species is most likely to be confused with the ringed plover, which is larger, has a white wing stripe, and an orange rather than a yellow eye ring. It might also be confused with *placidus,* which, however, has no black at the base of the upper mandible and also differs in the amount of white above and behind the eye. In this species, white is limited to the shaft of the outermost (tenth) primary, while in *hiaticula* it extends to the ninth and in *placidus* it is lacking.

In the field (6 inches), this species closely resembles the ringed plover, but lacks orange feet and an orange eye ring, has more white evident above the black facial mask, and lacks a white wing stripe. It is more likely to be found around fresh water than is the ringed plover, and it has a distinctive *pee-oo,* or *tee-u* note, as well as reedy notes and a trilling song.

NATURAL HISTORY

Habitats and foods. During the breeding season this species occurs on sandy and pebbly shores of slowly or rapidly flowing rivers, including river islands and dry, stony riverbeds. It also extends to about 8,000 feet above sea level in the mountains of Kashmir, where it occurs along the pebbly banks of mountain torrents, and it occurs locally on drying mud flats of fresh water and on exceptional occasions uses open arable land on clay soil. It is a southern ecological equivalent of the ringed plover, and in a few local areas is in ecological contact with it (Voous 1960). Breeding densities are greatly variable but at times are as high as about 4 pairs per hectare. Along rivers the density is generally less than 1 pair per kilometer of river distance. On African wintering areas the species occurs on extensive sandbanks in company with many other small shorebirds, but it occupies inland habitats rather than the coastal habitats normally used by the larger ringed plover. Feeding methods are the usual type found in the small plovers, consisting of short runs and stops, much as in the other species of "ringed" plovers. Foods consist of a wide variety of insects, especially beetles (primarily adults), dipterans (especially larvae and pupae), heteropterans,

Threat display, male

hymenopterans, and others. Much smaller numbers of snails, mussels, small crustaceans, and plant seeds are eaten (Glutz et al. 1975).

Social behavior. At least in Great Britain, the sexes typically arrive on the breeding grounds separately, with males generally preceding the females by about 1–3 weeks, but with a few birds arriving already paired. Almost from the first day of arrival the birds become localized, adopting "preferred areas" that soon become territories. Territorial defense is done mainly by the male, and advertisement is achieved by patrolling in "butterfly" flights. These are curving flights at heights of up to about 100 feet, with continuous singing. The wingbeats are usually fairly slow and arched, but at times they are quicker and the bird rocks from side to side as it utters its monotonous, deliberate, and rusty notes *cree-ah(k)*, *cree-ah(k)*. When near a flying intruder, the territorial male stalls before it in a raised-wing flight-threat display and utters a buzzing call. These flights apparently are aggressive rather than sexual in function. Surface intruders are threatened by ground threat displays, calls, chases, and attacks. Territories are usually not contiguous, and their limits are often rather vague since they include not only water areas but also a substantial amount of defended airspace. Some ground territories may be as small as about 600 square yards (1/8 acre), while in a Dutch study an

average territorial size of 1.2 acres was estimated. Much foraging occurs outside the territory on "neutral ground," but most of the courtship behavior occurs within the territorial limits (Simmons 1956). One of the important aspects of courtship is the scraping ceremony. The ceremony of this species is very much like that of the ringed plover, killdeer, and probably the other "ringed" plovers. That is, as the female approaches the nest scrape the male turns away while raising and spreading his tail, and the female then enters the nest by passing underneath the male's tail. Copulation occurs near the nest scrape and is preceded by the male's approaching in a horizontal posture and gradually becoming more upright and increasing the speed of his footsteps to a "parade march" or "goose-stepping" in place. The female assumes a horizontal posture and is soon mounted by the male; he remains on her back for 20–58 seconds before suddenly clutching her dorsal feathers and kicking strongly. Apparently in this species the male does not continuously grasp the female's nape during copulation (Simmons 1953b; Glutz et al. 1975). Over the course of several days, several nest scrapes are formed within the territory, and the female chooses one of these for depositing her clutch. The minimum distance that Simmons (1956) reported between nests of this species was 56 meters, compared with 21 meters between Kentish sandplovers and 30 meters between nests of the two species, which is a reflection of the relatively high level of aggression and social intolerance of this species.

Reproductive biology. Nests of this species are simple scrapes, usually on loose sand, but they sometimes are on dried mud or on flat and bare rocks outcropping from sand or mud. Nests are also often placed beside a large stone, or shell or underneath a scrubby plant. There are normally 4 eggs, but 3-egg clutches are more frequent late in the breeding season (presumably replacement clutches). However, 5-egg clutches are extremely rare and quite probably are laid by two females. Replacement clutches are certainly rather frequent, but at least in the British Isles double clutching is extremely rare. It may, however, be fairly common in the Balearic Islands, where the breeding season lasts for 4 months (Bannerman 1960). Known intervals between the loss of the first clutch and the initiation of a new one range from 4 to 11 days, and the highest known number of eggs produced by a female in a single season was 15, laid during four nesting attempts. Incubation is performed by both sexes. The incubation period ranges from 22 to 28 days, with most clutches hatching in from 24 to 26 days. In some cases the male may take over the care of the first brood, freeing the female to begin a second clutch, but the usual pattern is for both sexes to participate in brood care. Although some chicks may initially fly when only 21 days old, fledging probably usually occurs when the young are 22–25 days old, or at times even as late as 32 days after hatching (Glutz et al. 1975).

Status and relationships. This species has probably prospered as much as any European shorebird in recent decades, and in some areas such as Great Britain has very successfully colonized new breeding areas in the present century (Bannerman 1960). It has also increased locally in Europe, especially in the Low Countries, but also elsewhere where gravel pits or other excavations have provided open, gravelly environments. In its relationships it is clearly part of the "ringed" plover group as defined by Bock (1958), and it probably can be thought of as a southern offshoot of early *hiaticula* stock. Larson (1957) considered it part of a species group of temperate-zone forms that includes *leschenaultii, mongolus, melodus,* and perhaps *placidus.* Phillips (1980) includes it in a group that consists of *melodus, hiaticula, vociferus,* and perhaps *placidus* and *cinctus.*

Suggested reading. Simmons 1953a, b; 1956.

Malaysian Sandplover

Charadrius peronii Schlegel 1856

Other vernacular names. Malay sand plover; gravelot de Péron (French); Sundaregenpfeifer (German).

Subspecies and range. No subspecies recognized. Resident on Malaya, Thailand, Southern Vietnam, Borneo, the Lesser Sundas, and the Philippines. Specimens from Simeulue have been described as a separate race *chaseni* (Junge 1939). Of doubtful occurrence on Java (Hoogerwerf 1966). See map 50.

Measurements and weights. Wing: both sexes 94–105 mm. Culmen (to feathering): both sexes 14–15 mm. Eggs ca. 32 x 25 mm, estimated weight 9.6 mm (Schönwetter 1963).

DESCRIPTION

Adult males have a rufous crown and a hindneck collar of white, followed by a broad black band that extends to the mantle. The forehead is white, with a black band separating it from the crown except above the eyes, and there is also a black band extending from the base of the beak through the upper ear coverts, usually interrupted at the eye. The cheeks and underparts of the body are white except for a large patch on each side of the neck that sometimes extends across the breast to form a complete breast collar. The underparts are light ashy brown, becoming darker on the rump and central upper tail coverts, while the sides of the rump and lateral upper tail coverts are white, as are the outermost tail feathers. The wing coverts are mostly brown, with the marginal coverts darker and the greater coverts tipped with white. The inner primaries are white toward the bases of the outer webs, and the secondaries are tipped with white. The central tail feathers are dark brown, and the others are progressively lighter. The bill is black, with an orange base, the iris is brown, the eye ring is orange, and the legs and toes are gray. *Females* resemble males, but the black markings are replaced by rufous and brown. Their legs are usually more yellowish. *Juveniles* resemble adults but have no black on the face and lack black or rufous on the mantle or sides of the chest, and there are extensive sandy buff margins on the feathers of the upperparts (Sharpe 1896).

In the hand, this species closely resembles the sandplover but has more sandy-colored upperparts, and there is a definite black mantle stripe behind the white nape. The gray to yellow legs separate this species from the very similar *thoracicus,* and the bright orange eye ring and orange at the base of the bill separate it from the other small sandplovers.

In the field (6 inches), these birds usually are found on sandy beaches and might easily be confused with sandplovers or little ringed plovers. This species has a more sandy-colored back than the sandplover and also differs from the sandplover in that the black face streak is usually interrupted at the eye. It differs from the little ringed plover in that the breastband is often incomplete and the white forehead extends back above the eye, forming a superciliary stripe. In many ways it resembles a hybrid of the little ringed plover and the Kentish sandplover. The calls are apparently much like those of the sandplover, including a quiet *chit* resembling that of the Kentish sandplovers. It is usually found on sandy beaches, and at times it also occurs on mud flats.

NATURAL HISTORY

Habitats and foods. Apparently the primary habitat of this species is sandy shorelines, as suggested by its sand-colored upperparts, but at times it also has been seen on mud flats. It is said to be especially fond of coral sand beaches, where it usually forages in pairs and trios (Smithies 1968). No information is available on its foods or feeding behavior.

Social behavior. There does not seem to be any specific information on the behavior of these birds. They evidently occur as well-spaced pairs or family groups on sandy stretches of beach. The breeding season is said to include June in Borneo, and eggs have been seen from March to June on the Malay Peninsula (Medway and Wells 1976).

Reproductive biology. Although there are suggestions that a clutch of 3 or 4 eggs may occur, Medway and Wells (1976) indicate that the usual clutch is from 1 to 3 eggs on the Malay Peninsula. This generally conforms with the 2- to 3-egg clutch that seems typical of the sandplovers. No other information on their reproductive biology seems to be available.

Status and relationships. No specific information seems to be available, but apparently these birds are fairly common on sandy shorelines over much of their range. Bock (1958) suggested that *peronii* was derived from early *alexandrinus* stock and could al-

most be considered conspecific with it. I suggest that it may have come more directly from a ringed plover type such as *dubius,* which it closely resembles.

Suggested reading. None.

Collared Plover

Charadrius collaris Vieillot 1818

Other vernacular names. Azara's sandplover; pluvier d'Azara (French); Azararegenpfeifer (German); chorlitejo de collar (Spanish).

Subspecies and range. No subspecies recognized. Breeds in Central and South America, from coastal Sinaloa and Veracruz, Mexico, south to western Ecuador on the Pacific side and east of the Andes

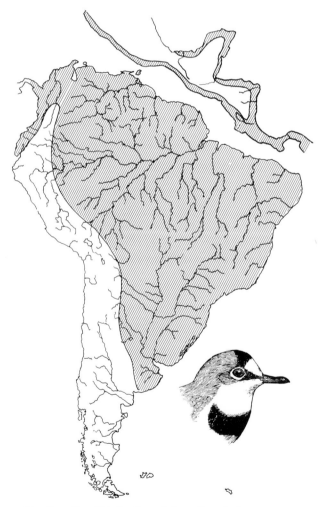

60.   Breeding distribution of the collared plover.

to central Argentina. Present in Chile in winter, sometimes also breeding there. See map 60.

Measurements and weights. Wing (chord): males 86–103 mm; females 89–102 mm. Culmen (to feathering): both sexes 13–16 mm. Weights: males 24–32 g; females 26–29 g (Haverschmidt 1968). Average adult weight 30.6 g (Graul 1973); 3 males 23–26 g, average 25 g; 3 females 29–30 g, average 29.3 g (M.V.Z. specimens). Eggs ca. 28 x 21 mm, estimated weight 6.8 g (Schönwetter 1963).

DESCRIPTION

Adults in breeding plumage are white on the forehead, below the eyes, and for a short distance behind the eyes; the lower part of the head is also white, separated from the white forehead by a black streak from the bill to the eye. The underparts are also white, except for a black band across the upper breast, and there is also a blackish band across the front half of the crown. Behind this the crown is cinnamon brown, grading to uniform grayish brown on the back and upper wing surface, except that the greater coverts, primary coverts, secondaries, and inner primaries are margined with white. The central tail feathers are grayish brown and the outer ones are progressively whiter. The iris is dark brown, the bill is black with orange at the extreme base, and the legs and toes are pale pink. *Adults in nonbreeding plumage* are very similar, but the feathers of the upperparts are margined with dull buffy. *Juveniles* resemble adults, but they lack the black crown patch, their lores are grayish brown rather than black, and the breastband is grayish to dull black and sometimes is incomplete (Ridgway 1919).

In the hand, the very small size (wings under 105 mm), single black breastband, and pink legs and toes serve to identify this species.

In the field (5½–6 inches), this species is usually found on coastal beaches, riverbanks, or sandy savannas in tropical areas. Its very small size, banded breast, and pink legs readily should identify it. Its calls have been described as *pec, peet,* and *peep-peep* (Haverschmidt 1968) and are generally sharp and unmusical, but they do include a cricketlike *chitit* and *tsick-tsilick* notes (Slud 1965).

NATURAL HISTORY

Habitats and foods. Undoubtedly this species is closely associated with sandy and gravelly habitats,

21. Adult heads of piping (*above*) and collared (*below*) plovers.

judging from its back coloration and what little has been written about it. Coastal beaches, sandy savannas, and the banks of rivers and ponds are common habitats of this form, according to Blake (1977). Wetmore (1965) reported seeing 2 on a gravel bed of a Panamanian river, and he also observed them feeding on a golf course after rains. He said that they often run quickly, stopping suddenly, in the general manner of the snowy sandplover, but rarely run as far without stopping as does that species. Their foods are little studied but probably are much like those of the other sandplovers. Haverschmidt (1968) mentions water beetles, dragonflies, fly larvae, and ants as foods in Surinam.

Social behavior. There is no specific information on the species' social behavior. Johnson (1965) reports seeing the birds in small flocks during the winter period in central Chile, but few remain in that country during the breeding period. Slud (1964) noted that the birds have a threat display in which the lower white feathers are fluffed laterally, so that from the front the bird resembles a starched shirt front.

Reproductive biology. Two nests and 2 pairs of birds with young have been reported for Chile, with all records occurring in January. Both nests had 2 eggs, as did a set from Brazil mentioned by Wetmore (1965). This latter clutch was found in late September. It is very probable that the breeding biology is very much like that of the other sandplovers, but no specific information is available.

Status and relationships. Apparently these birds are very common in much of eastern South America, mainly along the coasts and on such islands as Trinidad, wherever sandy habitats exist. The species occupies a habitat similar to that of *dubius* in the Old World and is probably derived from a *dubius*-like or *hiaticula*-like ancestor, though Bock (1958) indicates that it may be an offshoot of *alexandrinus* stock, which in turn was derived from a *hiaticula*-like form.

Suggested reading. Wetmore 1965.

Madagascan Sandplover

Charadrius thoracicus (Richmond) 1896

Other vernacular names. Black-banded plover; pluvier à bandeau noir (French); Madagaskarregenpfeifer (German).

Subspecies and range. No subspecies recognized. Resident in Madagascar. See map 61.

Measurements and weights. Wing: both sexes 102–15 mm. Culmen: both sexes 18–19 mm. Weights: no information available. Eggs 32 x 23 mm (Appert 1971), estimated weight 9.3 g.

DESCRIPTION

Adults of both sexes have white on the forehead, lores, cheeks, throat, sides, flanks, and under wing surfaces, as well as white lines over the eyes and ear coverts, with a black breastband and a white area below that grades into cinnamon buff on the abdomen and under tail coverts. There is also a black loral stripe from the bill to the eye that continues backward to include the ear coverts and extends across the nape. There is a black forehead band extending from eye to eye behind the white forehead and separating it from the white superciliary stripe. The crown and upperparts are generally hair brown, the feathers with paler edges, and this color extends to the central tail feathers, with the more lateral tail feathers progressively lighter, and the outer three feathers entirely white. The primary coverts are

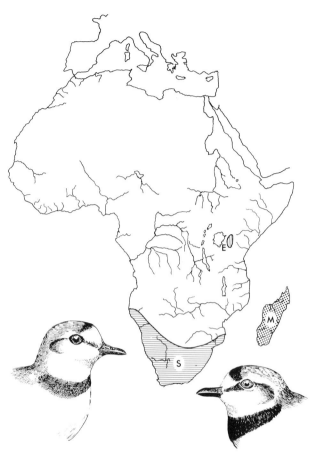

61. Breeding distributions of the Madagascan sandplover (M) and the eastern (E) and southern (M) chestnut-banded sandplovers.

tipped with white, the inner primaries are bordered and tipped with white, and the secondaries are also tipped with white, especially inwardly. The bill, legs, and feet are black (at least in dried skins). *Juvenile* plumages include a grayish brown breastband, with an area not yet completely white; the head markings are indistinct, and there is no reddish on the belly; rather, the undersides are grayish white, and the feet are gray (Appert 1971).

In the hand, this species most closely resembles *pecuarius,* but it differs in having a conspicuous breastband and in having the terminal shafts of the primaries tipped with white.

In the field (5¼ inches), these birds are perhaps most likely to be confused with the Kittlitz sandplover, which occurs in the same general area, but that species lacks a breastband. Its calls are also similar to those of *pecuarius* and include a rough *pit* or *pirds* and a rough *twitwitwi* (Appert 1971).

186 ❖ ❖ ❖

Habitats and foods. This species is often found in the same habitats as the Kittlitz plover, and in fact the two species are frequently seen together. It is not nearly so rare as was once believed, and it occurs not only along the coastline in southern and eastern Madagascar, but also around some lakes and brackish ponds, and sometimes several kilometers from water. It seems to prefer grazed areas of grasses and *Salicornia* frequented by zebus, and when in company with the white-fronted sandplover it seems to occupy drier habitats than does that species (Dhondt 1975). However, it is usually found rather near the coast, often feeding in silty areas at river mouths. Its foods are still unstudied but probably differ little, if at all, from those of other small plovers.

Social behavior. These birds seem usually to be found in pairs or at most in small groups, with the largest assemblage noted by Appert (1971) being a group of 33 birds sleeping on a vegetation-covered dune. Its social behavior often includes association with the Kittlitz plover, and its usual call, a *tui-tui-tui*, is similar to that of other small *Charadrius* species. Its displays and other aspects of its social behavior remain unstudied.

Reproductive biology. Rather few observations of breeding have been made, but 2 clutches (each with 2 eggs) have been found in November and January, and juveniles have been noted in August and December. One nest found by Appert in late November was on dry turf and consisted of a hollow containing a few dry grasses. The eggs closely resemble those of the Kittlitz plover and, like that species, the bird sometimes more or less covers its eggs with sand or dry twigs.

Status and relationships. This species is apparently not rare in its favored habitats, such as around Lake Tsimanampetsotsa, along the Mangoky River, and on the southern and eastern coastal areas. Its relationships are probably quite close to *pecuarius*, with which the earliest specimens were originally confused, and Bock (1958) believed the species probably originated from an invasion of pre-*pecuarius* stock, followed by a second, more recent invasion of *pecuarius*. I too believe that *pecuarius* is probably the species' nearest living relative.

Suggested reading. Appert 1971; Milon et al. 1973.

Kittlitz Sandplover

Charadrius pecuarius Temminck 1823

Other vernacular names. None in general English use; pluvier patre (French); Hirtenregenpfeifer (German).

Subspecies and ranges. See map 62.

C. p. allenbyi: Egyptian Kittlitz sandplover. Resident of Egypt from the Nile Delta south to Luxor. Clancey (1979a) considers *isabellinus* the proper name for this form, and extends its range to Zaire and Kenya.

C. p. pecuarius: Southern Kittlitz sandplover. Resident over most of Africa excepting the Sahara, the Congo Basin, and northeastern Africa; also resident on Madagascar. Birds from South-West Africa and northern Botswana have recently (1971) been separated by Clancey as *tephricolor*.

Measurements and weights. Wing: both sexes 96–107 mm. Culmen: both sexes 19–22.5 mm (Vaurie

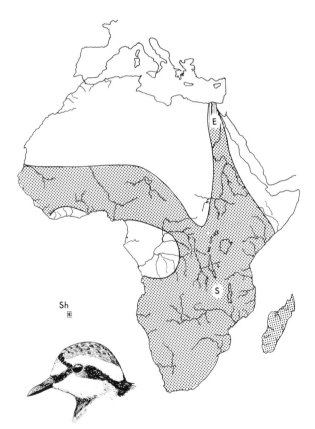

62. Breeding distributions of the Egyptian (E) and southern (S) Kittlitz sandplovers, and the St. Helena sandplover (Sh).

1964). Weight: 426 adults of both sexes 19–49 g, averaging 34.0 g (Skead 1977). A group of 91 post-juveniles averaged 42.3 g and ranged from 33 to 54 g (Summers and Waltner 1979). Eggs ca. 30 x 22 mm, weights 5.3–6.9 g (Conway and Bell 1969).

DESCRIPTION

Adults of both sexes have a white forehead, terminated by a black band, and a blackish streak from the base of the bill to the eye down the side of the neck and back around the base of the hindneck. The crown feathers are brown with sandy edging, and there is a white postocular stripe extending from the eye around the back of the neck to form a white collar. The upper parts are blackish brown, the feathers having sandy rufous edging. The tail feathers are white except for the central pair, which are blackish. The chin and throat are white, but the rest of the underparts are tinged with pale sandy rufous. The flight feathers are brown with white shafts. The iris is dark brown, the bill is black, and the legs and toes are greenish gray. *Females* closely resemble males, but the dark band on the front of the crown is narrower and may be brown rather than black. *Juveniles* resemble adults but have no black or white on the head. Instead the pattern is produced by buff and light brown markings.

In the hand, the distinctive black band from the eye down the side of the neck separates this species from other small *Charadrius* forms except for *sanctaehelenae*. Besides the differences in measurements in these two forms, *pecuarius* has more conspicuous buffy markings on the upperparts and has narrower and more brownish primaries, with their shafts white rather than brownish.

In the field (5¼ inches), this species is usually found on open, dry flats, sometimes well away from water. The black line down the side of the neck is a good field mark even in flight. When feeding, the birds often run at considerable speed, occasionally stopping suddenly to probe in the sand. The species does not show a white wing stripe in flight, and its calls include a plaintive, whistling *tu-wit*, a *tip-peep*, trilled notes, and an alarm *chirrt* or *prrrt*.

NATURAL HISTORY

Habitats and foods. At least in southern Africa, the primary habitats of this species consist of a variety of flat, exposed habits such as sandbanks, mud banks, and dry veld. It is rarely found on sandy or rocky seashores, but offshore islands and tidal mud flats are commonly used (Blaker 1966; Winterbottom 1963). Its foods are the usual array of insects and their larvae, crustaceans, and mollusks, and it often forages in small flocks in association with various wintering *Calidris* species.

Social behavior. Flocking in the nonbreeding season is regular and is partly the result of dispersal associated with summer flooding followed by concentration as the waters recede during winter. In Rhodesia and Zambia the winter flocks are usually fewer than 20 birds, but as many as 270 individuals have been seen during that season (Blaker 1966). Even during the breeding season the birds are said to be somewhat gregarious; Hall (1959a) noted that nests are often

22. Adult heads of Kittlitz (*above*) and St. Helena (*below*) sandplovers.

very close to one another, and he saw no obvious aggressiveness between pairs nesting only about 9 yards apart. However, Conway and Bell (1969) reported that under the crowded conditions of a zoo exhibit a single pair tended to try to dominate most of an entire 6 x 8 meter exhibit area, with the male undertaking the major part of this role and limiting his aggressiveness toward conspecifics. After hatching, the area defended depends on the location of the chicks.

Reproductive biology. In southern Africa, this species shows a summer breeding peak in the area of winter rainfall, after the rains have stopped. However, in the area of summer rainfall, the peak is in spring, before the rains begin (Winterbottom 1963). In central and western Africa breeding mostly occurs from June to September, and in Madagascar the records are widely spread through the year (Appert 1971). Nest locations vary greatly and include sand ridges or sand piles, sandy-soil patches in open grassland, areas of dried mud devoid of vegetation, and sometimes the droppings of cattle or horses, with the dung being partially broken up to form a lining. Even in clustered nesting areas the locations are usually at least 20 yards apart. During pairing, the birds often make several nest scrapes close together, with copulation usually occurring close to the scrape the female eventually chooses for egg deposition. Eggs are laid at intervals ranging from 2 to 4 days, and normally there are 2 eggs in a clutch. Hatching normally occurs from 23 to 27 days after the laying of the second egg, averaging about 25 days (Conway and Bell 1969). Both sexes incubate, and when disturbed from the nest by humans they quickly kick sand over the eggs before leaving. Although this seems to be an invariable response, the degree of actual covering of the egg is variable. This behavior does not occur during normal nest relief, though sideways-throwing movements by the departing bird frequently occur. Several kinds of distraction displays may be performed by disturbed birds, including "broken-wing" behavior, "rodent-run," "false-brooding," and combinations of these (Blaker 1966). As soon as the chicks hatch, one of the adults carries away the eggshells, and a few hours later they leave the nest site, to be brooded in various places and at diminishing rates for a rather prolonged period, sometimes as long as 42 days after hatching. However, the young are able to fly when about 30 days old, and in one case a pair was observed to begin a second nesting while they were still brooding young from the prior nesting (Conway and Bell 1968).

Status and relationships. This species is common over much of the open habitats of Africa and certainly is in no danger. Its relationships to the St. Helena form are discussed in the account of that species, so the only other question revolves around relationships to the other *Charadrius* species. Bock (1958) considered *pecuarius* and *thoracicus* to be closely related and both in turn to be derived from the "ringed" plovers such as *hiaticula*. He suggested that *thoracicus* may have evolved from pre-*pecuarius* stock in Madagascar, with a second invasion of *pecuarius* resulting in the present sympatry of the two. I agree that these two forms are probably a close-knit group and that they probably are offshoots of the "ringed" plovers. However, Strauch (1976) believes *pecuarius* and *sanctaehellenae* to be in a quite different assemblage that includes *novaeseelandiae*, *cinctus*, *melanops*, and *Hoploxypterus*.

Suggested reading. Hall 1958; Conway and Bell 1969.

St. Helena Sandplover

Charadrius sanctaehelenae (Harting) 1873

Other vernacular names. Wirebird; pluvier de Sainte Hélène (French); St. Helena-Regenpfeifer (German).

Subspecies and range. No subspecies recognized (this form is sometimes considered a subspecies of *pecuarius*). Resident on St. Helena Island. See map 62.

Measurements and weights. Wing: both sexes 112–18 mm. Culmen: both sexes 27.5–30 mm. Weights: No information available. Eggs ca. 34 x 25 mm, estimated weight 10.8 g (Schönwetter 1963).

DESCRIPTION

Adults of both sexes closely resemble those of the Kittlitz sandplover, but in *sanctaehelenae* the buffy margins of the feathers on the upperparts are less obvious than in *pecuarius*, and the feathers at the base of the upper bill are black instead of white. Further, the breast is white rather than buffy, the wing is rounder, and the webs of the primaries are almost twice as broad. The primary webs are darker, almost black, and their shafts are also black or very dark brown, except for the outermost primary, which has a white shaft (Vaurie 1964). *Juveniles* show whitish edging on the scapulars and back feathers, have a dusky wash on the breast, and lack black on the forehead, lores, and sides of the neck.

In the hand, the larger wing of at least 112 mm, bill of at least 27 mm, and tarsus of at least 36 mm (compared with maximums of 107 mm, 23 mm, and 30 mm in *pecuarius*), and plumage differences from *pecuarius* as noted above should serve to identify this species.

In the field (6 inches), the restricted distribution of this species should eliminate confusion with all other small sandpipers.

NATURAL HISTORY

Habitats and foods. These birds feed wherever suitable foraging conditions occur, such as in open grasslands, plowed fields, and even in large vegetable gardens, but mostly they are to be found in the more open plains of the island. The species' foods have not been specifically studied, but they are probably much like those of *pecuarius,* as are most of its "habits" (Pitman 1965).

Social behavior. This form is mostly seen in pairs, and the total population is so small that flocking seems unlikely. It breeds during the drier portion of the year, and, since the rainy season lasts from March to the end of August, this means nesting may begin in late September. No specific information is available on social or pair-forming behavior. Nonetheless, whereas *pecuarius* seems often to breed in small groups, this is evidently not true of *sanctaehelenae,* since none of 19 observed nests was found within 50 feet of another (Pitman 1965).

Status and relationships. The small size of St. Helena Island places limits on this form's population, and in the 1950s it was estimated that only about 100 pairs existed, confined to some 11 breeding grounds. More recently, the population has been estimated as slightly less than 1,000 birds (Pitman 1965). Although Bock (1958) argued strongly that this island form should be considered conspecific with *pecuarius,* Vaurie (1964) challenged this view and presented evidence for maintaining them as separate species. Pending further research, I am accepting Vaurie's conclusions.

Suggested reading. Pitman 1965.

Chestnut-banded Sandplover

Charadrius pallidus Strickland 1952
(*Charadrius venustus* of Peters, 1934)*

Other vernacular names. None in general English use; pluvier à bande chataine (French); Rotbandregenpfeifer (German).

Subspecies and ranges. See map 61.
  C. p. pallidus: Southern chestnut-banded sandplover. Resident on the coast of southern Africa from Angola to Port Elizabeth. Also breeds on temporarily flooded salt pans in the interior of South Africa.
  C. p. venustus: Eastern chestnut-banded sandplover. Resident in East Africa around the salt lakes of southern Kenya and northern Tanzania.

Measurements and weights. Wing: both sexes 90–105 mm. Culmen: both sexes 12.5–14 mm. Weights: 7 adults of *pallidus* 28–37 g, average 33 g (Skead 1977). Britton (1970) indicates slightly lighter weights for 9 *venustus,* and Summers and Waltner (1979) indicate an average of 36.0 g for 10 *pallidus.* Eggs ca. 29 x 22 mm, estimated weight 7.4 g (Schönwetter 1963).

DESCRIPTION

Adult males have a white forehead with a narrow black band behind and have a pale chestnut area on the anterior crown. There is also a narrow black line from the base of the bill to the eye, while the rest of the head is white. The entire upperpart coloration is pale grayish brown, and there is a narrow white band across the base of the primaries. The primaries are dark brown, and the primary coverts are the same, with white tips and white edging that merge to form a white wing bar with the broad white terminal bands of the gray secondaries. The tail is brown centrally and white outwardly, and the underparts are white except for a pale chestnut band extending across the chest and up the sides of the neck to meet the gray upperpart coloration. The iris is brown, the bill is black, and the legs and feet are olive. *Females* lack the black markings on the forehead and in front of the eyes. *Juveniles* also lack black on the head and the breast. The back feathers are light dove gray, with narrow lighter edging, and the front of the crown is buffy, becoming light gray posteriorly.

* The name *pallidus* has priority over *venustus* Fischer and Reichenow 1884.

In the hand, the small size (wings 90–105 mm) and tiny bill (culmen 12–14 mm) separate this species from all but the smallest plovers, and the chestnut breastband and bicolored (black and chestnut) forehead are distinctive.

In the field (7 inches), this species is most likely to be confused with the white-fronted sandplover, but even immatures show at least a partial light gray to grayish buff breastband. The birds are associated with salt flats that are often almost bare of vegetation, but they forage in a manner very similar to that of the white-fronted sandplover. In flight, the white wing bar is very evident. Few calls have been noted, one of which was a quiet *chuck* made when the birds take off or land.

NATURAL HISTORY

Habitats and foods. This species is distinctly associated with alkaline water; in southern Africa it is almost entirely restricted to salt pans, with a smaller number of occurrences on coastal lagoons and temporary vleis (Winterbottom 1963; Blaker 1966). It forages in a manner similar to that of *marginatus,* but that species tends to avoid highly alkaline areas, and thus competition between them is probably not severe. Its foods are unrecorded but presumably are much like those of other *Charadrius* forms.

Social behavior. These birds are highly local in distribution and tend to be resident, but they are subject to limited movements. They are usually to be found as singles or pairs, but they have been seen in flocks as large as 50 birds. Their social behavior is relatively poorly studied, but they are evidently quite territorial, since Jeffery and Liversidge (1951) noted that each pair in one area dominated a stretch of about 20 to 100 yards of the perimeter of the salt pan. The nesting season has not been well documented; most of the records from the western Cape area of South Africa are for November and December, appreciably later than for the other three *Charadrius* species breeding in the same general region (Blaker 1966). However, elsewhere in South Africa the season seems to be more prolonged (McLachlan and Liversidge 1957).

Reproductive biology. Few nests of this species have been found, partly because of the high degree of caution shown by nesting birds, and also perhaps because of their use of vegetation-free salt flat habitats that are of difficult access. In the study of Jeffery and Liversidge (1951), nests near Bredasdorp were found between the first of October and the middle of November, although breeding there apparently may begin as early as August and continue into December, with juvenile birds occurring as late as the middle of January. Both sexes participate in selecting and constructing the nest, with the scrape ceremony used as in other *Charadrius* species. In one case a scraping male was seen to open his beak and vibrate it while directing it downward and possibly calling when the female stepped on the nest, and in another instance the male left the nest at this time and began throwing small stone chips over his shoulder toward the nest. Both sexes perform incubation, with the female apparently incubating during the day and the male usually doing so during the evening. The length of the incubation period is unstudied, as is the fledging period. Both sexes care for the young, remaining very close to chicks under 10 days old and guarding the nesting territory even after the young birds are capable of flight. It is possible that the birds produce two broods, since a nest that had been used in October was found to be cleaned and prepared for a new use in November.

Status and relationships. This species no doubt has a rather small potential population, given its ecological constraints. So long as alkaline lakes and salt pans of eastern and southern Africa remain undisturbed, it should be able to persist. Without much question, the species is a derivative of the *alexandrinus* group, as Bock (1958) has already suggested.

Suggested readings. Jeffery and Liversidge 1951.

Sandplover

Charadrius alexandrinus L. 1758

Other vernacular names. Kentish plover (Great Britain); snowy plover (North America); pluvier à collier interrompu (French); Seeregenpfeifer (German); chorlitejo patinego (Spanish).

Subspecies and ranges. See map 63.

    C. a. nivosus: Snowy sandplover. Breeds in the United States from Washington and northern Utah south to Lower California. A separate population (sometimes recognized as *tenuirostris*) breeds in the salt plains of the southern Great Plains and on the Gulf coast, as well as in

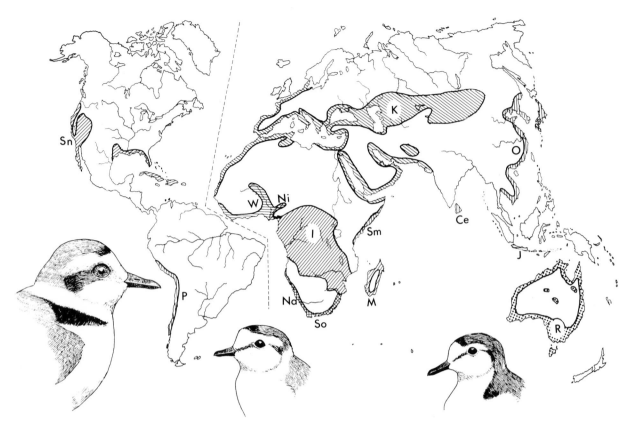

63. Breeding distributions of the Ceylonese (Ce), Javanese (J), Kentish (K), oriental (O), Peruvian (P), and snowy (Sn) sandplovers; the red-capped sandplover (R); and the interior (I), Madagascan (M), Namib (Na), Nigerian (Ni), Somalia (Sm), and western (W) white-fronted sandplovers.

the West Indies and islands off Venezuela. Winters south to Yucatan and northern South America on the Gulf drainage, and from central California to western Mexico on the Pacific drainage.

C. a. occidentalis: Peruvian sandplover. Resident on the coasts of Peru and Chile, south to Chiloé Island. Sometimes considered a distinct species.

C. a. alexandrinus: Kentish sandplover. Breeds in Eurasia, from southern Sweden to the northern Sahara, and east through the Mediterranean Basin and the coast of the Red Sea west to western India and through the steppes of central Asia to Transbaikalia, southern Ussuriland, and western China. Winters from the Mediterranean Basin south to tropical Africa and Angola, Arabia, and India to Sri Lanka.

C. a. javanicus: Javanese sandplover. Resident on Java and adjacent islands. Hoogerwerf (1966) reported that this race is of uncertain validity but probably includes birds from the Kangean Islands.

C. a. seebohmi: Ceylonese sandplover. Resident on Sri Lanka (Ceylon).

C. a. dealbatus: Oriental sandplover. Breeds in eastern China and Japan. Winters from Japan south to Taiwan and the Philippines, and from southeastern China to the Indo-Chinese countries and the Greater Sundas. Includes *nihonesis*, a poorly characterized Japanese race.

Measurements and weights. Wing: both sexes 101–19 mm. Culmen: both sexes 13–18 mm. Weights (of *alexandrinus*) vary considerably by sex and season; adults of both sexes average about 45 g and range from 32 to 69 g (cf. Glutz et al. 1975). Weights of *dealbatus* are very similar (Shaw 1936). Eggs average ca. 32 x 23 mm, estimated weight 7.8 (*nivosus*) to 9.0 g (*occidentalis*) (Schönwetter 1963).

DESCRIPTION

Adult males in breeding plumage have the forehead, the superciliary region, the sides of the face, and the throat entirely white, with black limited to a bar on the forepart of the crown (not reaching the eyes), a dusky patch on the upper ear coverts (extending in some races in front of the eye and reaching the base of the bill), and a transverse patch of black on the sides of the breast that sometimes extends narrowly around the back of the neck directly below the white

nape collar. Between the black forecrown and the white nape ring is a crown patch that ranges from light buff or brownish gray to brown (varying with subspecies). The rest of the underparts are white, and the upperparts are generally light brownish gray to grayish brown. The inner secondaries are mostly white, and the primaries are dusky with white shafts and increasing amounts of white on the inner feathers. The outermost tail feathers (up to three pairs) are white, and the others are progressively darker inwardly, with the central pair grayish brown to sepia, sometimes with paler tips. The iris is brown, the eye ring is blackish, and the bill is blackish, while the legs and toes are slate gray. *Females* in breeding plumage have brown breast and forecrown markings, rarely with a few black feathers. *Adults in winter* have brownish breast patches. *Juveniles* have pale brown breast patches, which often are indistinct, and their upperparts and coverts have distinct buffy fringes. *First-winter birds* resemble adults but retain buff-fringed inner median coverts (Prater et al. 1977).

In the hand, the small size (wing under 120 mm), pale dorsal coloration, short and black bill, black or brown half-collar on the breast, and up to three pairs of entirely white outer tail feathers help identify this species.

In the field (6 inches), identification of this species is complicated by the problems associated with uncertain species limits. In Africa, separation from the white-fronted sandplover is usually possible by the grayish legs and the black (or brown) markings on the sides of the breast that are absent in the African species. In Malaya, separation from *peroni* is difficult, but in that form a fairly broad black bar always extends around the base of the nape, and the lore stripe and ear covert patches are relatively separated rather than being joined by a dark eye stripe. In South America the similar species *collaris* has a complete breastband and pale pink legs and toes. In flight a conspicuous white wing stripe is evident. Reported calls include a low-pitched *poo-eet* or *chu-wee,* a soft *hwick,* a soft *wit-wit-wit,* and an alarm *kittup.*

NATURAL HISTORY

Habitats and foods. During the breeding season this species is essentially limited to sparsely vegetated and sandy areas, including sandy shores, sand dunes, salty steppes with scattered grasses, sand deserts, pebbly or muddy shorelines or plains, and sometimes tropical coasts of coral limestone (Voous 1960). Generally it is limited to the vicinity of seacoasts, but it

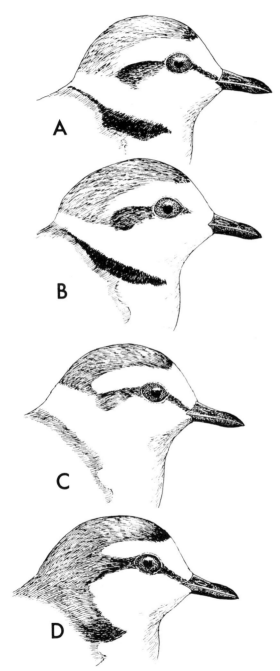

23. Breeding-plumage heads of *A,* Kentish sandplover; *B,* snowy sandplover; *C,* white-fronted sandplover; and *D,* red-capped sandplover.

occurs less frequently on inland sandy riverbanks, saline flats, and barren reservoir shorelines. Habitats during the nonbreeding season are essentially the same as during the nesting period, and many subtropical to tropical populations are essentially sedentary. Population densities are highly variable, but in general European populations probably range from about 0.5 to 20 pairs per hectare, with exceptional cases of up to about 40 pairs or even more than 100 pairs per hectare under conditions of extreme coloniality (Glutz et al. 1975). Foraging is done in the typical plover run-and-peck method of visual foraging. Like most, if not all, plovers, the birds often "patter" on the ground by vibrating one leg in such a way that it tends to make nearby prey move and thus reveal itself. Foods consist largely of insects and their larvae, including neuropterans, trichopteran larvae, dipterans, and beetles. Small mollusks and annelids, especially nereid polychaetes, bivalve and univalve mollusks, and gammarid crustaceans are also eaten locally or at various times, and brine shrimp have also been reported among foods consumed (Glutz et al. 1975).

Social behavior. Sexual behavior is attained near the end of the first year of life in both sexes, and thereafter the birds are either seasonally monogamous or maintain year-long pair bonds that may last as long as 6 years. However, in a Kansas study the spring arrivals were unpaired, and no definite case of a bird's pairing with its mate of the previous year was observed (Boyd 1972). On the other hand, Rittinghaus (1961) observed a substantial proportion of birds that remained with their mates of the previous year. He noted a total of 14 such pairs, while 42 birds mated with new mates even though their old mates were still available and 26 "widowed" birds mated with new mates. Territories are established by males and are advertised by "butterfly flights" and associated aerial calling. The flight closely resembles that of the lesser ringed plover but is performed less frequently than in that species or in the ringed plover. Territory size is rather variable, depending on population density, and in highly concentrated populations nesting pairs may be situated only a few meters apart. Once paired, both sexes participate in territorial defense and are often able to expel species considerably larger than themselves. Apparently the male has the primary responsibility for selecting the nest site, through his initiation of nest-scraping behavior. Often he will construct several scrapes only a few meters part. The nest-scraping ceremony is probably an important pair-forming and precopu-

latory display, since copulation is essentially restricted to the vicinity of the nest scrape. As the female approaches a scraping male at the nest, he typically stands on the edge of the nest with the bill pointed into the scrape, the tail raised and fanned away from the nest. The wing nearer the female is slightly drooped, and the farther wing is raised and slightly spread. Another male display associated with scraping is a stiffly horizontal position, with the throat feathers ruffled and a deep, gutteral *pikoor* note uttered. This posture may precede a "parade march" or walking-in-place, which probably normally precedes copulation. If the female is receptive she raises her tail, displaying the white-bordered cloaca. At that point the male immediately jumps on her back and gradually lowers himself until his belly touches her back. He may remain in this position for 1.5–2.5 minutes before cloacal contact is achieved, then he grasps the female's dorsal feathers and beats his wings, and the two birds tumble over backward (Boyd 1972; Rittinghaus 1961).

Reproductive biology. Nests are often near a conspicuous object such as a grass clump or shrub. The nesting pattern is one of semicolonial clumping, and in a Kansas study 17 nests were situated an average of 85 meters apart, with some nests as close as 15–20 meters to each other. Eggs are laid at approximately 2–day intervals, so that 5 days is normally required to complete a 3-egg clutch. Three eggs apparently represent the full clutch for this species throughout its range. Of 45 Kansas clutches, 32 had 3 eggs, 9 had 2 eggs, and 4 were abandoned nests with single eggs. Clutches of 4 eggs are found in only about 1 percent of the nests, but replacement nests often have only 2 eggs. The first egg of a replacement clutch may be laid as early as 4–5 days after the loss of the initial clutch. Although reportedly double brooded in some areas, the species is as a rule single brooded, but is a persistent renester. Both sexes incubate, and nesting changeovers are characterized by a symbolic nest-building ceremony. In many areas the eggs must be shaded from the heat and kept cool rather than warm, and egg-wetting behavior by incubating birds has been observed. Incubation lasts 24–28 days, with estimated averages of 25.5 and 26.3 days in two studies (Boyd 1972; Rittinghaus 1961). Both sexes also participate in brooding the young, and the young may fledge when only 28 days old (Boyd), or as much as 41 days after hatching (Rittinghaus).

Status and relationships. This species has regularly appeared on the Audubon Society's "Blue List" of declining North American birds since the list was initiated in 1972. In Europe it has greatly decreased in some areas, such as Sweden, Norway, Holland, and Great Britain (Bannerman 1960). However, the species as a whole has one of the most cosmopolitan ranges of any shorebird, and its collective population must be very large. It can be regarded as the central "type" of the sandplover group, which in Bock's (1958) classification also included *peronii, pallidus, collaris, bicinctus,* and *falklandicus.* However, he included *marginatus* and *ruficapillus* as part of *alexandrinus.* I tentatively prefer to list these three forms as members of a superspecies and believe that their probable nearest relative include *pallidus, collaris, thoracicus, pecuarius,* and *sanctaehelenae.* Relationships within this group of closely related and extremely similar species are clearly uncertain, and other grouping might be readily postulated.

Suggested reading. Boyd 1972; Rittinghaus 1961.

White-fronted Sandplover

Charadrius marginatus Vieillot 1818
(*Charadrius alexandrinus,* in part, of Peters, 1934)

Other vernacular names. White-fronted plover; pluvier à collier interrompu (French); Weisstirnregenpfeifer (German).

Subspecies and ranges. See map 63.
C. m. marginatus: Southern white-fronted sandplover. Resident in coastal South Africa.
C. m. arenaceus: Namibian white-fronted sandplover. Resident in coastal Namibia (South-West Africa); recently (1971) described by Clancey.
C. m. hesperius: Western white-fronted sandplover. Resident in western Africa from Liberia to Nigeria and the Central African Republic.
C. m. nigirius: Nigerian white-fronted sandplover. Resident on the Upper Niger from Gao to Bamako. Not recognized by Clancey (1975); included in *hesperius* or *mechowi.*
C. m. mechowi: Interior white-fronted sandplover. Resident on the Lonago coast, northern Angola, the Congo Basin, and in eastern Africa from Natal to Somalia.
C. m. tenellus: Madagascan white-fronted sandplover. Resident on Madagascar (Malagasy Republic).
C. m. pons: Somalia white-fronted sandplover. Resident in coastal Somalia.

Measurements and weights. Wing: both sexes 91–115 mm. Culmen: both sexes 15–18 mm. The average of 262 *marginatus* was 48.3 g, with minor monthly variations (Summers and Waltner 1979). Eggs ca. 29 x 21 mm (*hesperius*) to 33 x 24 mm (*marginatus*), estimated weight 6.8 g (*hesperius*) to 9.2 g (*marginatus*) (Schönwetter 1963).

DESCRIPTION

Adults of both sexes have a white forehead extending backward above the eyes and separated from the rusty brown collar by a black band. There is a black streak reaching from the base of the bill to the eye and continuing as a narrow line behind the eye to the ear coverts. The rest of the head, throat, and underparts are white, and there is a pale rusty patch on each side of the breast in front of the wings. The upper parts are warm brown and the feathers are edged with sandy brown, as are the wing coverts except for the greater coverts, which are tipped with white. The flight feathers are all brown, and the tail feathers are white outwardly and brownish centrally. The iris is brown, the bill is black, and the legs and toes are olive gray to buffish. *Females* closely resemble males but usually have less black above the forehead. *Juveniles* lack a black band on the forehead, and the streak from the base of the bill to the eye is brownish.

In the hand, this species is easily confused with several other small African *Charadrius* forms, but the absence of a white collar around the back of the neck, the lack of a black breast patch or black line down the side of the neck, pure white underparts, and a strong tinge of rusty on the upperparts will help identify it. Separation of immature or nonbreeding birds from *alexandrinus* is difficult, but individuals of *marginatus* tend to be paler and more scaly dorsally, have a longer tail (usually over 45 mm), and have yellowish legs and toes (Clancey 1975).

In the field (7 inches), this species might be confused with the Kentish sandplover, which, however, has blackish legs, lacks dorsal rusty tones, and has black patches in the breeding season, or with the Kittlitz sandplover, which is buff-tinted rather than pure white below. In flight this species shows a white wing bar, and its calls include a soft, liquid *woo-eet* and a more drawn-out *pirr,* given in alarm.

NATURAL HISTORY

Habitats and foods. Habitats used by this species in southern Africa include a wide ecological diversity;

Winterbottom (1963) lists the birds as occurring on all of ten designated shorebird habitats, but with the highest numbers on sandy seashores and progressively smaller numbers on tidal mud flats, rocky seashores, lagoons, and offshore islands. Blaker (1966) reported the maximum numbers on seacoasts, with saltwater lagoons, sand dunes, and tidal mud flats being of secondary usage. It feeds by running quickly, stopping suddenly to peck, then running on. It eats insect larvae, small crustaceans, and worms (McLachlan and Liversidge 1957).

Social behavior. Unlike many *Charadrius* species, this species is quite sedentary, with limited flocking outside the breeding season. However, it is to a large extent territorial, with birds tending to "guard" their own sections of beach or hollows between sand dunes (Blaker 1966). So far as is known, the behavior associated with pair-formation and nesting is virtually identical to that of *alexandrinus*.

Reproductive biology. Nests of this form are, as in other small sandplovers, simple scrapes with only a rudimentary lining. They are almost always on a sandy substrate, just above the high-tide line or in adjacent sand dunes. They sometimes also are placed in roadside gravel, quarries, and similar locations. Generally the nests are placed close to some conspicuous object, presumably to reduce the bird's own conspicuousness (Maclean and Moran 1965). The eggs are also relatively light in color and closely match the sandy substrate. The observed range is 1–4 eggs, with a modal clutch of 2 and a mean of slightly less than 2 eggs (including clutches that were probably incomplete or from which eggs were missing). Incubation is by both parents and probably normally ranges from 26 to 28 days, with one estimate of about 29 days. Interestingly, the birds probably perform much of the incubation by partly covering the eggs with sand during the daylight hours, whereas during the night the eggs are uncovered and brooded normally (Liversidge 1965). There is one estimate of 38 percent nesting success rate, based on 56 nests (Blaker 1955).

Status and relationships. This is one of the commonest of the small coastal shorebirds of southern Africa, and although the status of some of the subspecies with restricted ranges (such as *hesperius*) cannot be judged, the species as a whole is evidently in no danger. The birds are clearly very closely related to *alexandrinus*, and often (Bock 1958; Peters 1934) are considered conspecific with it. This may be the best ultimate taxonomic treatment, but it seems reasonable to maintain separate status for the present as Clancey (1975) has urged.

Suggested reading. Maclean and Moran 1965; Liversidge 1965.

Red-capped Sandplover

Charadrius ruficapillus Temminck 1822
(*Charadrius alexandrinus ruficapillus* of Peters, 1934)

Other vernacular names. Red-capped dotterel; red-capped plover; pluver a tete rousseau (French); Rotkopfregenpfeifer (German).

Subspecies and range. No subspecies recognized, but this form is sometimes considered a subspecies of *C. alexandrinus.* Resident throughout Australia and Tasmania along the coastlines, but locally also in the interior. See map 63.

Measurements and weights. Wing: both sexes 98–107 mm. Culmen: both sexes 13–14 mm. Weights: males 30–45 g, average of 108 was 35 g; females 30–47 g, average of 32 was 36 g (D. Purchase, pers. comm.). Eggs ca. 31 x 23 mm, estimated weight 8.2 g (Schönwetter 1963).

DESCRIPTION

Adult males have the top of the head, the hindneck, and the upper mantle rusty red. There is a white band on the forehead, separated from the rusty brown by a narrow band of black, and from the white chin and cheeks by a black band extending from the base of the bill through the eye to the side of the neck, where it ends in a patch of black and rufous that forms a semicollar. The rest of the upper surface is light brown, the feathers having rufous margins. The lateral upper tail coverts are white, and the outer tail feathers are also white, while the four central ones are dark brown. The marginal wing coverts are dark brown, and the other coverts are lighter brown except for the greater coverts, which are tipped with white. The primaries are dark brown, with the outermost ones more blackish and the innermost pair margined with white, and the outer secondaries are tipped with white. The sides of the face and all the underparts are white. The iris is dark brown, and the bill and legs are black. *Females* have brown rather than black on the forehead and a rufous band through the eye, and the rusty areas are paler. *Juveniles* have a dull brown crown and are brown where the adults are black or rufous. The bill is black, the iris is brown, and the legs and feet are black in males and olive gray in females.

In the hand, this species is easily confused with the other small sandplovers, but the rufous coloration on the head, the lack of a white nape collar, and the black legs and feet help identify it.

In the field (5 inches), this is the smallest of the Australian plovers and the only one with rusty brown on top of the head. Females lack the rusty color and are generally duller, but they do show a darker streak through the eye and have a brown semicollar. In flight the birds are rather pale brown above, with median and trailing wing stripes, and the rump appears dark centrally and white on each side. It has a shrill alarm call *kittup*, a long trill with a rising cadence, a fluty *poo-eet*, and other softer notes.

NATURAL HISTORY

Habitat and foods. Primarily coastal in distribution, but also present around salty lakes in the interior, this species has a patchy interior distribution and probably most birds are quite sedentary. They are typical sandplovers, with a pale brown back coloration that matches their background and an egg pattern that likewise blends with a sandy substrate. Their foods consist of midge larvae, weevils, and the larvae of water beetles, and they almost certainly also consume mollusks and crustaceans in coastal habitats. Like other sandplovers, they tend to forage while alternating between short, rapid runs and motionless postures.

Social behavior. Outside the breeding season the birds sometimes occur in small flocks, but most often they are seen in pairs. No specific information is available on pair-forming behavior, but it is known that typical nest-scraping displays occur at the onset of the nesting season and probably serve to determine nest sites as well as to form pair bonds. Copulation is typically preceded by the male's assuming a high and erect posture in front of the female, together with a "high-stepping" display that may occur in front of as well as behind her and that typically lasts for 3 or 4 minutes before actual mounting. There is no specific postcopulatory display, but the birds remain motionless for some time thereafter, with the male in the same high and erect posture (Davis and Reid 1964; Hobbs 1972).

Reproductive biology. In eastern and southern Australia the breeding season extends from August to March, and in the tropics and the interior it may occur at any time of the year. Hobbs (1972) reported that at Fletcher's Lake, New South Wales, the nesting season lasted from late June to late September, when he located 43 nests in an 800-meter stretch of shoreline. This area of localized nesting represented only about a twelfth of the total available area, and some of the birds would fly as far as 3 kilometers to feed. In spite of this seeming coloniality, the nests were well spread, with none closer than 28 meters, and they usually were 70 to 100 meters apart. The nests were often placed in a zone of *Juncus* growing around the shore, perhaps for concealment or shading, but as the season progressed and the waters receded they were often placed in loose sand near the *Juncus* zone. Most of the nests were on level sand, some were on dry sandy mud, some on drying mud, and a few on thin mats of weeds over dry mud. Those in damper sites had substantial vegetative platforms. There were usually 2 eggs, rarely one. Both sexes incubated, and at least at times the male would take the first chick to brood as soon as it hatched, while the female tended the remaining egg. Hobbs reported a rather high incidence of nest failure, but he obtained no definite information on renesting or multiple clutches. An 18-day incubation period has been estimated (Frith 1969), but this is appreciably shorter than the periods known for other sandplovers.

Status and relationships. This species is fairly common in sandy coastal areas and certainly does not seem to merit any special attention from conservationists. It is obviously a very close relative of *alexandrinus*, and some authorities (Peters 1934; Bock 1959) have regarded the two as conspecific. Considering the allopatry of breeding ranges, this is certainly reasonable, but the recent practice of Australian ornithologists is to maintain the two as separate species. As with *marginatus*, it is perhaps worthwhile maintaining them as distinct for the present, until some more definitive method of judging relationships becomes available.

Suggested reading. Hobbs 1972.

Hooded Dotterel

Charadrius cucullatus Vieillot 1818
(*Charadrius rubricollis* of Peters, 1934)*

Other vernacular names. None in general English use; pluvier à camail (French); Weissnackenregenpfeifer (German).

Subspecies and range. No subspecies recognized. Resident in southern Australia and Tasmania, mainly along the coast. See map 64.

Measurements and weights. Wing: both sexes 137–44 mm. Culmen (to feathering): both sexes 15–18 mm. Weights: both sexes 45–50 g (Hall 1974). Eggs ca. 36 x 27 mm, estimated weight 13.2 g (Schönwetter 1963).

DESCRIPTION

Adults of both sexes have the crown, face, and throat black, a broad white band on the nape, and a narrow black band on the upper back that continues around

*See *Auk* 36:279 for rejection of *rubricollis.*

to the sides of the upper breast. The rest of the back and scapulars is brownish gray, and the upper tail coverts and central tail feathers are black. The outer tail feathers and underparts are white, and there is a white stripe on the wing formed by the secondaries and the bases of the primaries; otherwise the wing coverts are brownish and the primaries are black. The iris is brown with a scarlet eye ring, the bill is orange with a black tip, and the legs and toes are pinkish brown. *Juveniles* lack black markings, and have a pale brownish crown and back, with a white nape, forehead, and throat, and a grayish streak through the eye. Their underparts are white, and their legs and feet are dull pinkish yellow.

In the hand, adults can be recognized by the uniformly black head and white nape, which are unique in the genus. Immatures are much more difficult, but the combination of pale brownish gray on the upperparts and pinkish yellow feet helps separate them from other immature Australian sandplovers.

In the field (8 inches), these birds are usually found along salt lakes, coastal beaches, and mud flats. The distinctive head pattern of adults serves as an excellent field mark, and their calls consist of a distinctive barking *fow-fow* as well as short whistled notes. In

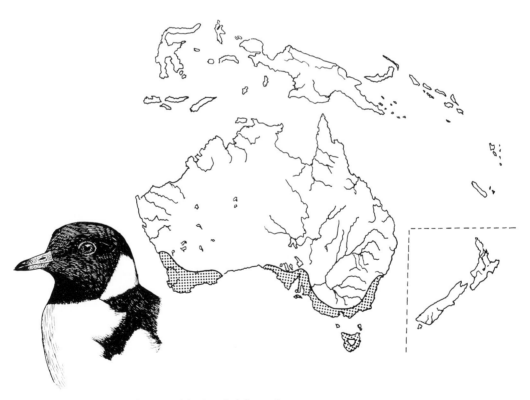

64. Breeding distribution of the hooded dotterel.

flight, the white wing stripe is strongly apparent, and the blackish central tail feathers and tail coverts are distinctive.

NATURAL HISTORY

Habitats and foods. According to Frith (1976), this species' favorite habitat consists of long, windswept beaches where low sand dunes are covered with tussock grasses and creepers. However, it also occurs abundantly around inland salt lakes. Although its back color is very pale and the color of dried sand, the strong black head pattern disrupts this concealing coloration and sets the species apart from the typical sandplovers. Its foods consist of aquatic insects and other "small animals," presumably invertebrates.

Social behavior. Family groups and small flocks are present during the fall and winter, but during the breeding season these birds are found as isolated pairs, which establish and maintain territories for long periods (Frith 1976). Perhaps the strong head patterning and bright bill and eyelid coloration are related to territorial signaling, but unfortunately there is no information on this aspect of the species' behavior.

Reproductive biology. Evidently these birds nest between August and January. They make a simple nest scrape in the sand above the high-tide mark, often in an open situation. The scrape is unlined except for a few scattered shells or bits of vegetation. The clutch is usually 2 or 3 eggs, but the incubation period and other aspects of the breeding cycle remain unreported.

Status and relationships. This is regarded as an uncommon species (Macdonald 1973), but probably its abundance is primarily determined by the distribution of open sandy habitats. It would be interesting to know the ecological relationships of this species and the other typical Australian sandplover, *ruficapillus*, which has a considerably broader distribution. Its evolutionary affinities are distinctly obscure. Bock (1958) included it in a loose group with *melanops, cinctus,* and *novaeseelandiae,* all of which he believed derived from the "ringed" plovers, but none of which seemed distinctly closely related to one another. I doubt that the species is closely related to any of these, and I can find no good evidence for judging its real affinities. A study of its behavior is greatly needed.

Suggested reading. Frith 1976.

New Zealand Shore Plover

Charadrius novaeseelandiae Gmelin 1798
(*Thinornis novae-seelandiae* of Peters, 1934)

Other vernacular names. New Zealand dotterel; pluvier de Nouvelle-Zélande (French); Kappenregenpfeifer (German).

Subspecies and range. No subspecies recognized. Now limited to South East Island, of the Chatham group, off New Zealand. Once more widespread and present on the New Zealand mainland. See map 65.

Measurements and weights. Wing: both sexes 113–28 mm. Culmen: both sexes 23–25 mm. Weights: no information available. Eggs ca. 36 x 26 mm, estimated weight 12.0 g (Schönwetter 1963).

DESCRIPTION

Adult males have the forehead, sides of the head, throat, forepart of the neck, and a narrow nape collar brownish black, separated from a grayish brown crown by a narrow band of white extending from the upper forehead backward to the nape. The remainder of the upper body surface is grayish brown, the wing coverts are brown with white tips on the greater coverts, the secondaries range partially white to entirely white, and the primaries are dark brown with a white central streak. The central tail feathers are dark brown, the outer ones are white on their outer webs, and the outermost ones are pure white. The under wing surfaces and underparts are white. The iris is dark brown with an orange eye ring, the bill is orange red with a black tip, and the legs and toes are pale orange. *Females* resemble males but are more brownish on the cheeks and neck and have more blackish on the bill. *Juveniles* have a tawny tinge on the upper surface, owing to the light fawn tips of the wing coverts, scapulars, and back, and have a light brown forehead. The white line above the eye is well developed, but the cheeks below this band are brown, becoming brownish white on the throat, and continuing as a light brown neckband rather than being black as in adults. The bill is blackish with a lighter base (Oliver 1955).

In the hand, the combination of an entirely black or dark brown head and neck, except for the brownish crown and the narrow white stripe around the crown, serves to identify adults of both sexes. Additionally, the bill is longer than either the middle toe

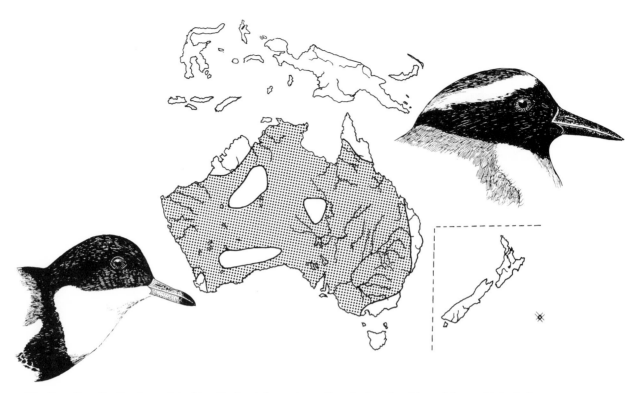

65. Breeding distributions of the New Zealand shore plover (*hatched*) and red-kneed dotterel (*shaded*).

and claw or the tarsus, and there is a good deal of white on the outer webs of the innermost primaries, characteristics that apply to all ages.

In the field (8 inches), this is the only New Zealand plover with so much black on the head and neck, though the locally established black-fronted dotterel has a black eye stripe and a separate black breastband. Both species show a good deal of white on the wings when in flight; this species is said to resemble a turnstone in its manner of flight and general appearance. Its calls include many loud and rising notes, an aggressive rattling call, and loud *kleet* or *splew* alarm notes (Phillips 1977).

NATURAL HISTORY

Habitat and foods. Phillips (1977) reported that the majority of birds observed in a post–breeding-season March census were on a narrow zone of wave-splashed rocks and platforms or in the sparsely vegetated zone just above high tide. A few others were observed at an inland area of fallen trees and exposed mineral soil in the middle of a meadow of *Salicornia.* Foraging was mostly done on rock plat-

forms, where the birds fed at low tide on barnacle-covered rocks and in shallow pools. Feeding is done in a succession of short steps and picking movements, rather than by the short runs and quick stops typical of many sandplovers. The birds often stand in shallow water and frequently pry under rocks or into crevices, and they sometimes catch small fish. They probably also eat small crustaceans, and they evidently often forage at night (Phillips 1977; Oliver 1955).

Social behavior. At present these birds are confined to a single small island and are completely sedentary. They evidently are not very gregarious, since Phillips (1977) observed that pairs not only chased other pairs from their apparent territories, but also chased juveniles that tried to follow and approach them. Banded individuals tended to remain on the same 100–200-meter stretches of shoreline, though foraging was done in common areas of shoreline. Terrestrial and aerial chases were frequently seen, and on several occasions birds were observed flying with the wings held in a V over the back during aggressive encounters, apparently comparable to the aggressive stall flight of the little ringed plover. Vocalizations of this species are loud and ringing, in contrast to those of other New Zealand plovers (Phillips 1977).

Reproductive biology. Nests are in many sites but almost always are well sheltered from above, with a lateral entrance. They are usually in rock crevices but sometimes are in deserted petrel burrows, in hollow logs, or in similar locations. The nest is bulky and surprisingly well lined for a shorebird, especially if it is on a moist substrate. The clutch size is usually 2 or 3 eggs, which are apparently laid daily. Incubation is by both sexes, and "broken-wing" distraction displays have been observed (Oliver 1955).

Status and relationships. Although at one time this species was rather widely distributed on North and South Islands, as well as on Mangare and South East Island, it is now limited to the last-named island, where in 1972 its population was determined to be only 82 individuals (Phillips 1977). It is thus one of the world's rarest shorebirds. Although it has often been placed in the monotypic genus *Thinornis,* this is based on its long and pointed bill, which seems certainly to be a foraging adaptation for probing and prying under rocks. Bock (1958) included it in the genus *Charadrius,* a view to which I subscribe, and suggested that it might be most closely related to *cucullatus* ("*rubricollis*"). I doubt that the plumage similarities of these species are very significant, and their vocalizations and ecologies are extremely different. However, I know of no possible closer relatives.

Suggested reading. Phillips 1977; Greenway 1958.

Genus *Erythrogonys* Gould 1838 (Red-kneed Dotterel)

The red-kneed dotterel is a small Australian plover that has a small hind toe and that compared with *Charadrius* has relatively long legs, toes, and bill. The downy young are much like those of *Charadrius* but have black-bordered brown patches on the back. One species.

Red-kneed Dotterel

Erythrogonys cinctus Gould 1838

Other vernacular names. None in general English use; pluvier ceinturé (French); Schwarzbrustregenpfeifer (German).

Subspecies and range. No subspecies recognized. Resident throughout Australia, except Tasmania, mainly in the interior. See map 65.

Measurements and weights. Wing: both sexes 108–116 mm. Culmen (to feathering): both sexes 21–22 mm. Weights: both sexes 46–64 g, average of 26 was 54 g (D. Purchase, pers. comm.). Eggs ca. 31 x 22 mm, estimated weight 8.1 g (Schönwetter 1962).

DESCRIPTION

Adults of both sexes have the entire upper part of head black, from the base of the bill back to the lower hindneck, while the lower cheeks, chin, throat, and neck are white. This area is terminated below by a broad black breastband that extends upward in front of the wing to join the black hindneck. The back, upper wing coverts, and central tail feathers are bronzy brown, while the outer tail feathers are white. The primaries and bases of the secondaries are dark brown, with the inner primaries increasingly white-tipped, and all the secondaries are broadly tipped with white. The underparts other than the breastband are black, except for some reddish brown streaks on the flanks. The iris is brown, the bill is pinkish red with a black tip, and the lower tarsus and toes are grayish blue, while the "knees" and upper part of the tarsus are pinkish red to violet. *Females* are essentially identical in plumage to males. *Juveniles* lack the black head and breastband, are a dull grayish brown above, lack brown flank stripes, and have brownish soft-part colors. However, they still show a bicolored head pattern, with no white behind, above, or in front of the eyes.

In the hand, a white throat and the presence of a hind toe separate this species from other small dotterels.

In the field (7 inches), adults can be readily recognized by their bicolored heads, which are black on the upper half and white on the lower half. The birds are usually found near margins of swamps or lagoons, especially those with grassy cover, and they run easily, frequently bobbing the head. The species flies swiftly, but usually for short distances, and is the only plover or dotterel that shows white on the trailing edges of the inner flight feathers. It has an alarm note, *chet-chet,* and some musical trilling notes.

NATURAL HISTORY

Habitats and foods. The margins of shallow swamps and lagoons, particularly those with canegrass (*Eragrostis*) cover, with a substrate of fine sand or mud, are the preferred habitat of this species. It typically runs rapidly over the mud and usually feeds by probing into mud with its rather long bill. It readily wades and sometimes even swims from one islet to another while foraging, which is virtually unique in the Charadiidae. It often forages among vegetation in the water, and its disruptive coloration affords good concealment. It consumes adults and larvae of aquatic insects, including midge larvae, and also eats seeds of various plants, especially legumes (Frith 1969; Maclean 1977).

Social behavior. This is a rather gregarious species and often occurs in loose flocks of as many as 60 birds. Although the pairs are territorial, the territories are quite small, and thus breeding is almost semicolonial. Threat postures are very different from those of typical *Charadrius* species, the most common being an upright posture that involves crown-flattening and a raised nape. This posture presum-ably exhibits the black breastband. Before copulation the male assumes a rather horizontal posture, with ruffled back feathers, and utters a trilling call. Injury-feigning is present but takes several forms, all rather different from those of *C. melanops.* Advertising flights were not seen by Maclean (1977).

Reproductive biology. Breeding occurs from September to December, and nests are typically under dead shrubs or tall canegrass on damp soil or mud. They are usually situated on tiny islets some distance from the nearest mainland. Clutch size ranges from 2 to 4 eggs, averaging slightly more than 3. The eggs are well blotched and seem to blend with a background of dappled sunlight. Both sexes incubate, but the incubation period has not been determined. The nests are highly susceptible to flooding or to drought, and the incidence of renesting is still unknown. Both adults and young swim readily, and when alarmed the chicks swim from the shore to the nearest water plant and hide beneath it. The rather short breeding season of this species may prevent second nestings, and Maclean (1977) has suggested that the large clutch size may compensate for this and for the rather high vulnerability of the eggs to various mortality sources.

Status and relationships. This species is considered uncommon to fairly common in Australia, and probably requires no special conservation efforts. It has at various times been considered a member of the genus *Charadrius;* Bock (1958), for example, thought it might be a rather aberrant derivative of the ringed plover group. However, Maclean (1977) has argued that it is behaviorally and ecologically distinctive, and although it does have some affinities with *Charadrius* he believed that recognition of a monotypic genus for it is justified.

Suggested reading. Maclean 1977.

Genus *Anarhynchus* Quoy and Gaimard 1830 (Wry-billed Plover)

The wry-billed plover is a small New Zealand plover with a slender, sharply pointed bill that is deflected to the right and is longer than the middle toe and claw. The hind toe is absent, and the front toes are narrowly margined and webbed at the base. The downy young are pale gray, with indistinct dorsal patterning. One species.

Wrybill

Anarhynchus frontalis Quoy and Gaimard 1830

Other vernacular names. Wry-billed plover; pluvier anarhyngue (French); Schiefschnabelregenpfeifer (German).

Subspecies and range. No subspecies recognized. Breeds on the middle reaches of some of the larger riverbeds between the latitudes of 42° and 45° S, on South Island, New Zealand. Winters mainly on the Firth of Thames, at Manukau Harbor, and on the Kaipara River, North Island. See map 66.

Measurements and weights. Wing: both sexes 114–20 mm. Culmen: both sexes 28–30 mm. Weights: males 47–70.5, average of 32 was 60.7; females 49–67 g average of 53 was 58.0 g (Rodney Hay, in litt.). Eggs ca. 35 x 26 mm, estimated weight 12.4 g (Schönwetter 1963).

DESCRIPTION

Adults in breeding plumage have a white forehead that continues back as a narrower line above the eyes, a band of dark gray that passes from the base of the bill backward below the eye to the ear coverts, changing to a lighter gray that extends forward to the crown, where it is separated from the white forehead by a thick blackish line. The rest of the upperparts are bluish gray, as are the wing coverts except for white edging, while the flight feathers are brown, the outer ones becoming pale gray with white edging. There is a broad black band across the upper breast, the feathers having white edging. The iris is brown, the bill is black, and the legs and toes are dark gray. *Females* closely resemble males but usually have a narrower and paler breastband, and the black streak above the forehead is less pronounced. *Adults in winter* lack the breastband. *Juveniles* resemble females but are paler and lack the breastband.

In the hand, the uniquely bent bill, which is curved to the right, identifies this species in all plumages.

In the field (8 inches), this bird resembles other small plovers and is found along the open seashore or large shingle-covered riverbeds, where its color blends well with sand and rocks. The laterally curved, sharp-tipped black bill is a good field mark, and the birds often stand on one leg, sometimes hopping about in this manner, or run rapidly over the open shoreline with the head held close to the body. In flight no white wing stripe is evident, and sometimes flocks perform complicated aerial maneuvers in

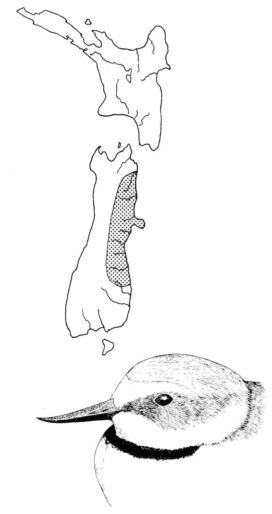

66. Breeding distribution of the wrybill.

synchrony. A variety of calls have been described, including staccato whistles, chattering notes, and trilled calls associated with sexual behavior. The most common alarm call is a short, clear *weet,* and a similar but harsher version is indicative of greater alarm (Rodney Hay, in litt.).

NATURAL HISTORY

Habitats and foods. The breeding distribution of these birds is closely associated with rivers, especially the middle to upper reaches of rivers flowing to the sea and the deltas of rivers feeding lakes. Breeding is almost entirely on shingle riverbeds, where there are large expanses of stones fairly close to water and free of vegetation. Outside the breeding season the birds typically occur on mud flats, fairly close to high-tide roosting sites. Ideal mud flats are those at middle to low tidal levels and those with soft, silty mud and a surface film of water. The birds forage opportunistically, feeding mostly on insects while on riverbeds or along stream margins. Aquatic foods include insect larvae, with occasional small fish and eggs, while riparian foods taken by pecking or probing under and between stones include beetles, saldids, craneflies, dipterans, and spiders. Rather few items are taken from sandy areas at this time. During winter,

worms, small bivalves, and tiny crustaceans are the primary foods, mainly obtained by probe and feeding in the mud. A distinctive sideways sweep-feeding behavior is also used, resulting in the capture of minute crustaceans occurring in the surface film of water, and the somewhat spoonlike bill shape is well adapted to this unique foraging method (Rodney Hay, in litt.).

Social behavior. Evidently pair bonds and family bonds break up outside the breeding season, whereas during the breeding period pairs are widely dispersed, maintaining exclusive breeding areas through aggressive and mutual avoidance behavior. Birds often pair with their mates of the previous year and frequently place their nests close to the location of the previous year. In one instance a nest site was used twice in one season and once in the following one. Intruders are chased by threatening runs in a horizontal posture, with the chest stripe expanded, and the run is terminated by a chirring call. Copulation is preceded by a series of horizontal runs by the male, "choke" displays (given while crouching in a hollow, moving the head and bill as if choking) by both birds, and conspicuous "parade marching" or "goose-stepping" by the male. Year-old birds apparently attain adult breeding plumage and sometimes visit the

breeding areas; 2-year-olds may form pairs but have not been proved to breed. Thus, breeding may be confined to birds at least 3 years old (Rodney Hay, in litt.).

Reproductive biology. Nesting may begin as early as late August in some locations such as the Rakaia River, but it occurs later farther south. Nests are typically built on the higher parts of bare islands or banks, where plants are sparse and there is little sand. The nest consists of a shallow scrape lined with small pebbles and is formed by the male during courtship display. The pebble lining may be produced by the birds' flicking them into the scrape or regurgitating them from the crop. The clutch size is almost invariably 2 eggs, laid at average intervals of about 48 hours. Incubation begins with the second egg and is performed by both sexes, with the female contributing the greater part in most cases, though sometimes the male incubates more. In one instance (the only proved case of polygyny in the species), two females and one male shared the incubation of 4 eggs. Incubation lasts 30–31 days, or rarely as long as 34 days. The first chick to hatch usually remains in the nest until the second one hatches, which may be as long as 36 hours later, but hatching is sometimes simultaneous. Both adults guard the chicks, which gather all their own food. Fledging occurs in about 28 days. Normally a second nesting occurs, with a second clutch laid shortly after the fledging of the first brood (Rodney Hay, in litt.).

Status and relationships. As of the late 1970s there were about 6,000 wrybills in existence, compared with about 5,000 in the early 1940s. They were then known to be breeding along the following rivers: Ashly, Waimakariri, Radaia, Ashburton, Rangitata, MacAulay, Godley, Cass, Tekapo, Tasman, Pukaki, Dobson, Hopkins, Ohau, Ahuriri, Waitaki, Hunter, Matukituki, and Makarora. A southern range extension to Otago has occurred in recent years, but the birds no longer breed on rivers north to the Wairau in Marlborough, where they were once recorded (Rodney Hay, in litt.). The taxonomy of the species is of some controversy. Although its unique bill shape has been the basis of a monotypic generic allocation, Bock (1958) found no unique muscle or skeletal specializations separating it from *Charadrius*. Rodney Hay (in litt.) notes that its behavior is also like that of the other small plovers, but the carrying of nest materials in the crop is "unusual" (probably unique), and the clutch size of 2 eggs is also unusually small. He attributed both of these factors to ecological adaptation. Thus, although *Anarhynchus* could perhaps readily be included in *Charadrius*, I believe it can for the present be retained as a separate but adjacent genus.

Suggested reading. Sibson 1943; Oliver 1955.

Genus *Peltohyas* Sharpe 1896 (Inland Dotterel)

The inland dotterel is a small, arid-adapted Australian plover that lacks a hind toe and has a short, black bill and a tarsus that is relatively long and is scutellated both in front and behind. The downy young are distinctive, lacking nape patches and resembling certain species of Glareolidae. One species.

Inland Dotterel

Peltohyas australis (Gould) 1840 (1841)

Other vernacular names. Australian dotterel; desert plover; courvite Australien (French); Ringrenn-vogel (German).

Subspecies and range. No subspecies recognized. Resident in the dry interior of Australia; nomadic and somewhat migratory. See map 67.

Measurements and weights. Wing: both sexes 113–44 mm. Culmen (to feathering): both sexes 15–18 mm. Weights: 1 male 100 g; 7 females 72–103 g, average 84.9 g (National Museum of Victoria). Eggs ca. 37 x 27 mm, estimated weight 13.8 g (Schönwetter 1963).

DESCRIPTION

Adults of both sexes have a broad black band across the crown, through the eye, and down the face and a broad black collar that becomes a Y-shaped band in the center of the breast. The forehead, front of the face, and throat are white, tinged to varying degrees

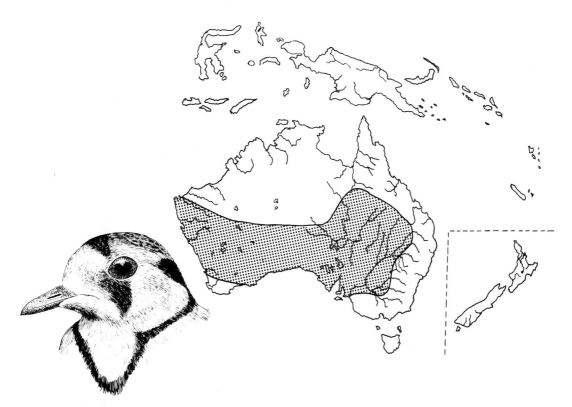

67. Breeding distribution of the inland dotterel.

with cinnamon, and the upperparts are generally cinnamon buff with mottled blackish brown. The sides of the breast, the flanks, and under wing surface are cinnamon brown, the center of the abdomen is chestnut, and the posterior abdomen and under tail coverts are white. The flight feathers are mostly blackish, with some cinnamon brown on the outer webs of some of the primaries. The iris is dark brown with a grayish eye ring, the bill is black, becoming gray at the base of the upper mandible and yellowish at the base of the lower mandible, and the legs and toes are brown, yellow, or gray. *Juveniles* closely resemble adults but have an incomplete or indefinite breastband, undeveloped facial markings, and (in museum skins) a more brownish bill.

In the hand, the distinctive cinnamon coloring throughout most of the plumage is unique, as well as the distinctive patterning of black on the head and breast.

In the field (8 inches), these birds are found in dry, sandy, or stony areas, and they prefer to run rather than fly when frightened. The distinctive Y-shaped black mark on the breast is readily seen, and when in flight the birds appear to be uniformly patterned with cinnamon and brown on the back and upper wing surface. The usual call is a low *kroot* sound, and a metallic *quoick* note is also frequently uttered several times in succession.

NATURAL HISTORY

Habitats and foods. This arid-adapted species occurs on many flat, open habitats, often far from water, which primarily include clay pans, gravel flats, and gibber plains, all of which have a sparse growth of low, shrubby plants but with few or no grasses. Unlike most shorebirds, the species is mostly herbivorous by day, consuming many succulent plants, from which it probably gets much of the water it needs. It also eats various insects, including beetles, ants, and grasshoppers, but it probably rarely takes water directly. Plants are eaten and swallowed in rather large pieces, and the bill is shaped to provide maximum mechanical strength for cutting and tearing plant materials (Maclean 1976).

Social behavior. These birds occur mostly in small flocks, rarely numbering as many as 50 birds, but they show considerable coordination of activity and movements. The birds are largely nocturnal in their activities, and surprisingly little has been learned

of their social behavior. Courtship has not been observed, in part because the birds tend to be very shy and run or fly when approached. The birds are known to have a well-developed injury-feigning or "broken-wing" display, which seems to exhibit the white areas of the wing coverts, tail coverts, and tail. They also have a "rodent-run" display much like that of various other plovers (Maclean 1973, 1976).

Reproductive biology. Rather surprisingly, this species has a distinct nonbreeding or "eclipse" plumage, which suggests that it has a regular breeding season rather than being an opportunistic breeder like most arid-adapted forms. Most of the breeding records are from February to October, suggesting that the birds avoid breeding in midsummer, and breeding is not necessarily timed by rainfall. The birds make scrapes in various open sites, sometimes on a mound, but not always. Usually the nest is placed where vegetation is scattered and where small stones are present to disrupt the substrate background. The clutch is invariably 3 eggs, and in common with a few other shorebirds (*Charadrius pecuarius, C. marginatus, Vanellus armatus,* and possibly others) it regularly covers its eggs with dirt when leaving the nest. There is no information on posthatching care or development of this species (Maclean 1973).

Status and relationships. This is a fairly common, but nomadic and erratic, species over much of Australia's dry interior. Its confusing, taxonomic history has been summarized by Bock (1958, 1964), who concluded on anatomical evidence that it is a member of the Charadriidae rather than the Glareolidae, with which it has often been aligned. Maclean (1973) agreed with this assessment, mostly on biological grounds. Others who have also concluded that *Peltohyas* is a plover are Burton (1974) and Strauch (1976). Other than the reasonable certainty that it is a member of the Charadriidae, it seems impossible to further define its relationships at present.

Suggested reading. Maclean 1973, 1976; Bock 1964.

Genus *Phegornis* Gray 1846 (Sandpiper-plover)

The sandpiper-plover is a small South American shorebird with a bill that is relatively weak and sandpiperlike but slightly swollen at the tip, no hind toe, and a unique plumage pattern that includes vermiculated sides and a brownish head with a white superciliary stripe. The downy young are mottled with brownish gray and black, lacking distinct dorsal patterning. One species.

Diademed Sandpiper-plover

Phegornis mitchellii (Fraser) 1845

Other vernacular names. Mitchell's plover; pluvier des Andes (French); Bänderregenpfeifer (German); chorlito Cordillerano, chorlito de las ciénagas (Spanish).

Subspecies and range. No subspecies recognized. Breeds in the puna zone of the Andes from Peru and Bolivia south to central Chile and Argentina.

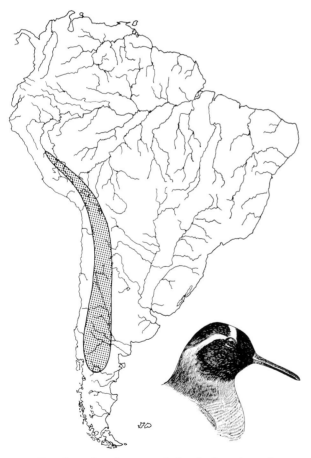

68. Breeding distribution of the diademed sandpiper-plover.

Apparently migratory in the south, occurring as low as 2,000 meters in October. See map 68.

Measurements and weights. Wings (flattened): males 110–14 mm; females 110–16 mm. Culmen (from base): males 28–31 mm; females 28–35 mm (Blake 1977). Weight: 1 adult of unknown sex 46 g (A.M.N.H. specimen); 2 males 28 and 32 g; 2 females 34.5 and 37 g (L.S.U. specimens). Eggs ca. 34 x 25 mm, estimated weight 11.7 g.

DESCRIPTION

Adults of both sexes have the entire head a deep chocolate brown with the exception of a white band extending from the middle crown back around the nape. The hindneck, sides of the neck, and anterior back are chestnut rufous, the scapulars, interscapulars, tertials, and primaries are deep fuscous, and the secondaries and wing coverts are grayish brown. The secondaries are edged and notched with white on the outer webs, as are the outermost tail feathers, while the inner rectrices are uniformly brown. There is a white crescent on the lower foreneck, below which the breast is finely vermiculated with brown and white, while the abdomen and sides are whitish with dusky barring. The iris is dark brown, the bill is black, and the legs and feet are yellowish or pink. *Juveniles* are similar to adults, but the upper surface is barred with rufous, the hindneck and anterior back are ashy gray, the throat is also ashy gray, and the breast is more coarsely barred (Blake 1977).

In the hand, the uniquely vermiculated breast pattern and the long and slender bill, which is slightly swollen at the tip and generally intermediate between that of plovers and sandpipers, provide ready means of recognition.

In the field (7 inches), this species is associated with Andean lakes and streams, and in size, general appearance, and manner of flight it somewhat resembles a painted snipe. The white line across the top of the head and the white crescent on the throat are useful field marks. The typical calls include a clear,

penetrating whistle, a much lower-pitched and more plaintive whistle, and soft, conversational notes.

NATURAL HISTORY

Habitats and foods. Johnson (1965) located breeding birds of this little-studied species in an Andean valley at some 8,500 feet elevation, where mountain streams cut through a U-shaped valley that was bordered on both sides by snow-covered slopes. The birds were observed feeding along the edge of the stream, which had gravel and sand spits and occasional grassy hillocks on each side. The food consists of insects, which the birds evidently pick directly off the water and from surrounding aquatic plants.

Social behavior. Johnson (1965) reports that he has never seen more than 2 of these birds associating at any one time. The species apparently occurs almost entirely as isolated pairs along stretches of mountain streams or along Andean lakeshores. Nothing specific has been written on its territorial or courtship behavior, but its clear, penetrating whistle may be an important territorial signal.

Reproductive biology. According to Johnson (1965), only 2 nests have ever been found. One clutch of 2 eggs was found more than half a century ago, but the eggs were broken and then discarded, and no details remain. The other clutch, which Johnson found in December of 1961, also consisted of 2 eggs. It was on a slight ridge, near the center of a sand and shingle area near a small stream tributary of the Yeso River. The male was observed sitting on the eggs, which were later collected. On the same day 2 newly hatched young were observed with a second pair. Two additional nests and one newly hatched chick were located in the same valley in the winter of 1971–72 by Brian Lavercombe and R. K. Templeton.

Status and relationships. This is apparently a rare but widespread species, the exact status of which will probably never be known because of its remote habitat. Its relationships are equally obscure. Seebohm (1888) suggested that it might be related to the painted snipes, and Johnson (1965) also commented on its similarity to this group. However, Bock (1958) considered its affinities still unknown and was unsure whether it should be included in the Charadriidae or the Scolopacidae. The most complete study (Zusi and Jehl 1970) concluded that the genus *Phegornis* should be retained in the Charadriidae, with most of its characters occurring in somewhat modified form among species of the genus *Charadrius*. Other persons providing evidence supporting the idea that *Phegornis* is a modified plover are Burton (1974) and Strauch (1976).

Suggested reading. Johnson 1965.

Genus *Oreopholus* Jardine and Selby 1835 (Tawny-throated Dotterel)

The tawny-throated dotterel is a small South American plover with a bill that is relatively weak and sandpiperlike, tapering toward the tip, no hind toe, a tarsus that is incompletely scutellated in front, and a pectinated (comblike) middle claw. The downy young have a whitish nape and indistinct dorsal markings. One species.

Tawny-throated Dotterel

Oreopholus ruficollis (Wagler) 1829

Other vernacular names. Slender-billed plover; pluvier oréophile (French); Klippenläufer (German); chorlito cabezón (Spanish).

Subspecies and ranges. See map 69.
> *O. r. pallidus:* Northern tawny-throated dotterel. Breeds on the arid coast of northern Peru; presumably resident.
> *O. r. ruficollis:* Southern tawny-throated dotterel. Breeds on the coast and in the mountains of central Peru and southward in the Andes through Bolivia, Chile, and western Argentina to Tierra del Fuego. Migratory in the south, wintering north to Ecuador, Uruguay, and southern Brazil.

Measurements and weights. Wing (flattened): males 146–77 mm; females 148–80 mm. Culmen (from base): males 33–40 mm; females 33–38 mm (Blake 1977). Weights both sexes of *pallidus* 111–18 g, average of 3 was 115 g; adults of *ruficollis* 120–45 g, average of 6 was 133 g (various sources). Eggs ca 42 x 31 mm, estimated weight 20.8 g (Schönwetter 1963).

DESCRIPTION

Adults of both sexes have a broad creamy white to buff superciliary stripe, below which is a black stripe extending from the base of the bill through the eye and backward past the ear, and above which is a pale brownish gray crown. The sides of the head and chin are whitish, grading on the throat to an orange rufous, and to brownish gray on the lower breast. The anterior back, rump, and upper tail coverts are

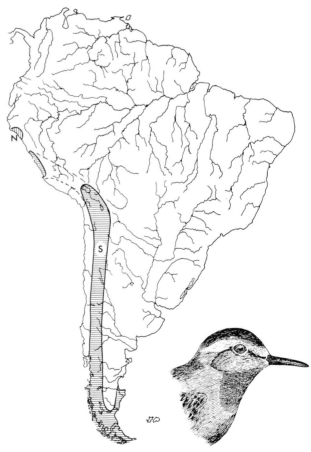

69. Breeding distributions of the northern (N) and southern (S) tawny-throated dotterels.

also brownish gray, while the scapulars, interscapulars, tertials, and wing coverts are heavily striped with black and tawny buff. The flight feathers are fuscous, the primaries are extensively white on their inner webs and have white shafts, and the secondaries have narrow white tips. The tail feathers are pale gray with a black subterminal spot or bar. The underparts are white, except for a large black patch in the middle of the abdomen. The iris is brown, the bill is black, and the legs are pinkish with black toes (Blake 1977). *Juveniles* develop the black abdominal patch quite early and thereafter closely resemble the adults.

In the hand, the unusually slender bill, almost resembling that of a sandpiper, the barred tail, and the black abdominal patch provide a unique combination of characteristics for identification.

In the field (10–11 inches), these birds are associated with grasslands and semiarid slopes as high as 13,000 feet. The tawny throat color, black abdomen, and black flight feathers with white edging should provide adequate field marks. In taking flight the birds call loudly with a tremulous and falling inflection, often repeated three or four times, but they are rather silent and inconspicuous on the ground, some-

times uttering a low and plaintive whistle *whees-tur-tur* when disturbed.

NATURAL HISTORY

Habitat and foods. This species occupies two quite distinct altitudes and habitats. The southern race *ruficollis* occurs both in the mountains up to 13,000 feet and along the coastal zone, in grasslands and other unforested areas, preferring semiarid hillsides, apparently often rather far from water. The northern race *pallidus* is associated with the coastal portions of Peru, where it nests on the sandy and nearly vegetation-free lomas zone near sea level. The species has an unusually long and thin bill for a plover and presumably probes for its food, but its feeding behavior and foods are virtually unreported. One young bird was reported to have insect remains (beetle) and grit in the stomach, but no descriptions of foraging behavior are available. Quite possibly the birds forage at night, since they have unusually large eyes (Johnson 1965; Humphrey et al. 1970).

Social behavior. During the nonbreeding season these birds occur in flocks of from 10 to 30 birds in the coastal zone and Andean foothills of Chile, and "great flocks" have also been seen migrating over Isla Grande, Tierra del Fuego, in March and April. Almost nothing has been written of their display or territorial behavior, but in mid-August flight displays have been noted in feeding flocks. In contrast to most plovers, this species seems to be very unaggressive, with wintering birds feeding in loose flocks and apparently not establishing foraging territories (Myers and Myers 1979).

Reproductive biology. Few records are available, but an incubated nest was found on Isla Grande in November, and a young bird was collected in December. Nesting in northern Chile starts considerably sooner; young have been seen in September and October in Aconcaqua, and an incubated clutch was found in mid-September in Atacama. One nest found by Johnson (1965) was on top of a clump of tussock grass. It had 4 eggs, as did a nest found in Atacama. No information is available on the roles of the sexes, incubation period, or other aspects of breeding biology.

Status and relationships. This species is heavily hunted in Chile and Argentina during the fall migration, and its numbers have reportedly declined considerably. Its relationships are definitely obscure. Seebohm (1888) regarded it as a member of the genus *Charadrius* and a close relative of *morinellus*. Likewise, Bock (1959) considered these two species closely related and placed them in the genus *Eudromias*, though he suggested that this might be an artificial grouping. Blake (1977) also included this species in *Eudromias*. I have not seen live specimens, but G. Maclean (in lit.) informs me that in life the species bears little similarity to *morinellus*, and he is unsure (1977) that it is even a plover. I thus am retaining it as a monotypic genus, with no strong opinions on its probable affinities.

Suggested reading. Humphrey et al. 1970.

Family *Scolopacidae*
(Sandpipers, Snipes, and Allies)

Tribe *Arenariini*
(Turnstones)

Ruddy Turnstone

This is the largest family of shorebirds and includes a wide range of mostly northern temperate to arctic-breeding species that vary greatly in size, bill length and shape, and feeding adaptations. However, in contrast to the Charadriidae, these birds tend to have bills that are longer than the head, often are longer than the tarsus, and are frequently pitted or ridged near the tip, in conjunction with tactile sensory endings. Feeding is more often by probing in soft substrates than by foraging visually, such as by running and pecking in the typical manner of plovers, and though some species have large eyes these are usually nocturnally active forms rather than visual foragers. The hind toe is usually present, and the tarsus is always scutellated in front and often behind also. The downy young are frequently distinctively marked, often with white-tipped powder-puff down, forming a concealing pattern well adapted to the lichen-covered tundra habitats in which many species nest. Nearly 90 percent of the species breed in the northern hemisphere, and only a few are adapted to tropical breeding. The sexes are alike in adult plumage, sometimes varying seasonally. Two subfamilies and 7 tribes, containing 83 species, are recognized here.

Subfamily Calidridinae (Sandpipers, Turnstones, and Surfbirds)

Members of this subfamily are very small to small, mostly arctic-breeding, shorebirds with relatively short legs, with scutellate scales on the anterior surface of the tarsus, with or without a hind toe, and with a bill that is usually relatively short, straight, and tapering and lacks terminal pitting or swelling. Three tribes containing 26 species are recognized here.

Tribe Arenariini (Turnstones)

Turnstones are small arctic-breeding shorebirds having a small hind toe, a short tarsus with scutellations both in front and behind, no webbing between the bases of the anterior toes, a pointed and somewhat wedge-shaped bill, and a long and slightly rounded tail. The downy young are gray, with a complex darker dorsal pattern and white-tipped powder-puff down. One genus and 2 species.

Genus *Arenaria* Brisson 1760

This genus of two species has the characteristics of the tribe Arenariini.

KEY TO SPECIES OF *Arenaria*

A   White present on the chin and throat . . . . *interpres*
A' Chin and throat gray to black, never white . . . . *melanocephala*

Ruddy Turnstone

Arenaria interpres (L.) 1758

Other vernacular names. Turnstone; tournepierre à collier (French); Steinwalzer (German); playero turco, vuelvepiedras (Spanish).

Subspecies and ranges. See map 70.
    A. i. interpres: Eurasian ruddy turnstone. Breeds on Ellesmere Island, Greenland, and in Eurasia from Scandinavia and Estonia eastward across

Russia and Siberia to Anadyrland. Also breeds in northwestern Alaska from Point Barrow south to St. Lawrence and St. Matthew islands and the Yukon Delta. Winters from California to Mexico and southern South America, and from Iceland, the British Isles, the Mediterranean coast, India, China, Japan, and the Hawaiian Islands south to southern Africa, the Malay Archipelago, Australia and New Zealand, and the Pacific islands including Samoa.
    A. i. morinella: American ruddy turnstone. Breeds

70. Breeding distributions of the Eurasian (E) and American (A) ruddy turnstones.

in North America from northeastern Alaska (where it intergrades with *interpres*) east across arctic Canada to Southampton and Baffin islands. Winters from South Carolina and the Gulf of Mexico south to southern South America.

Measurements and weights (*of interpres*). Wing: males 149–61 mm; females 153–65 mm. Culmen (to feathering): males 19–25 mm; females 19–26 mm. Weights: males 97–149 g, average of 16 was 110.3 g; females 99–141 g, average of 7 was 120.3 g (Morrison 1975). Eggs ca. 40 x 30 mm, estimated weight 15.5 (*morinella*) to 17.9 g (*interpres*) (Schönwetter 1963).

DESCRIPTION

Adult males in breeding plumage have white faces except for a black loral stripe and a black malar stripe extending back from the bill, the two connected by a black line extending downward from the eye to the lower cheek, and a third black line extending from

the posterior ear coverts forward and downward to the juncture of the other two lines. This black area extends forward to form a large breastband and continues laterally on the sides of the upper back to include much of the interscapular region. The crown is black, broadly streaked with white, and the mantle is mostly cinnamon rufous to hazel, with the posterior

scapulars becoming black. The central upper tail coverts are black, the tail is white basally and black distally but tipped with white, and the lower rump and lateral tail coverts are also white, as are the flanks and underparts. The primaries, primary coverts, and greater coverts are black, the coverts with white margins and the primaries white basally. The secondaries are white basally and dusky toward their tips, the amount of white increasing inwardly and finally replacing the dusky. The iris is brown, the bill is blackish, and the legs and toes are orange red. *Females* have a more extensively streaked brownish black crown, the nape is buffish white and streaked with brown, the mantle has a more limited area of reddish, and the black breast feathers have some light gray tips. *Adults in winter* resemble females in summer, but well-defined areas of black and white are lacking; the blackish feathers are obscured by whitish tips, and there is no cinnamon rufous on the upperparts (Ridgway 1919). *Juveniles* resemble adults in winter, but the scapulars, tertials, and wings coverts are fringed with pale chestnut buff, and the legs are dull yellowish brown to orange. *First-winter birds* retain some buff-fringed inner median coverts to spring (Prater et al. 1977).

In the hand, the short, pointed, and slightly upturned bill is distinctive and limited to the genus *Arenaria*. The presence of white on the chin and throat distinguishes this species in all plumages.

In the field (7–9 inches), turnstones are usually associated with rocky coastal areas, but they also occur on the shores and beaches of lakes. In spring plumage the harlequin facial pattern is unmistakable, but in winter the orange legs and dark brownish breast patch are useful field marks. In flight a broad white wing stripe is visible, as are the white rump and white-banded tail. The usual call consists of metallic rattles and a series of rapid and low-pitched slurred whistles.

NATURAL HISTORY

Habitats and foods. The breeding habitat of this high arctic species consists of a variety of substrates ranging from marshy slopes and flats in the lowlands and foothills to well-drained *Dryas*-hummocked slopes and tundra. In all cases, however, the birds breed fairly close to wet areas that remain moist until late summer, assuring them a supply of food in the breeding area. Breeding-season foods include some plant materials early in the season, before most invertebrates are available, with a gradual shift to these animal materials as they appear. For most of the

summer, dipterans (especially the adults and larvae of midges) are the prime food resource for both adults and young birds, and there is a close correspondence between the peak abundance of adult midges and the period of turnstone hatching. To a limited extent the birds also eat the larvae of lepidopterans, spiders, and hymenopterans, and at times they will resort to carrion and scavenging, especially when other foods are scarce. Probing and jabbing are the primary methods of feeding on the breeding areas (Nettleship 1973). Outside the breeding season the birds are primarily associated with rocky shorelines, but they also occur on sand or mud, especially where there are beds of mussels or cockles. They flip over objects such as stones and shells with a quick jerking movement of the head and beak and rapidly eat the organisms found underneath. They sometimes push larger objects with their breasts, and they also probe into mud and crevices for prey. They take a great diversity of foods at such times, including crustaceans, mollusks, carrion such as dead fish or mammals, discarded human food, and even partly incubated eggs, which they sometimes steal from incubating terns (Beven and England 1977). Some age-related differences in foraging rates have been observed in fall migrant birds, suggesting that the birds improve with experience in their foraging techniques, especially in their efficiency in searching for and mechanically handling their prey (Groves 1978). Immature birds usually remain on their wintering areas throughout their first summer of life, though they often molt into breeding plumage at that age.

Social behavior. Turnstones arrive on their breeding grounds in late May or early June with considerable regularity, apparently timing their arrival to coincide with the gradual emergence of tundra from its snow cover. They typically arrive in mixed-sex flocks within a short period each year, in spite of yearly weather variations. It is probable that pair formation may occur either before or after arrival, and in the latter case courtship occurs either on the males' territories or on coastal beaches. Courting parties of up to 30 birds have been observed on beaches of Ellesmere Island (Parmelee and MacDonald 1960). In any event, pairing is completed by early June, and males exhibit well-developed territorial behavior from the time of pair-formation until hatching. Associated behavior includes ground displays, calls, aerial displays, and territorial boundary patrolling. Although the size of individual territories is still unreported, one area of 240 hectares supported 13–14 pairs, and over a broader region the breeding density was estimated to be about 3.04 pairs per square kilometer (Nettleship 1973). Males frequently perch on rock outcrops or other elevated sites for territorial advertisement, crouching, vibrating their tails, and uttering metallic clicking notes. Noisy sexual chases have also been noted. In its song flight the male flies from 10 to 50 meters above the ground with slow, deep wingbeats and utters a series of *tjy-tjy-tjy* calls, about three calls every two seconds. Most displays and territorial encounters occur before incubation, and thereafter territorial boundaries are apparently recognized and respected by neighboring birds. Copulation is almost always preceded by the male's lifting both wings vertically upward from 1 to 3 times and holding them in this position 1–5 seconds. He then flies over the female's back, lands, and remains on her for 10–30 seconds, while holding onto her neck feathers (Bergman 1946).

Reproductive biology. Nesting areas of this species on Ellesmere Island were always in sites that were often wet and hummocky and had willow, *Dryas*, and heather vegetation. Some nests were built over wet mud and others over dry earth or dense vegetation, but none were built on stony ground or in areas distinctly sheltered by vegetation. On Victoria Island and Jenny Lind Island the birds breed on dry, peaty slopes, with the nests in shallow depressions in peat or low vegetation. At Hazen Camp, Ellesmere Island, they have been found on *Dryas*-covered hummocks or on open tundra. Most are found on well-elevated sites; on Ellesmere Island none was found less than a mile from the coast, and 2 were nearly 20 miles inland. There is some tendency for population clumping, and 2 nests found on Jenny Lind Island were only 32 paces apart. Nests almost invariably contain 4 eggs, as was true of all of 20 clutches found on Ellesmere Island and of 17 nests on Victoria Island and Jenny Lind Island. Some smaller clutches found at Hazen Camp were unusually early or late nestings. On Kandalaksha Bay, USSR, 90 percent of 81 nests contained 4 eggs, with extremes of 2 and 5 (Bianki 1977). At Kandalaksha Bay as well as in more southern parts of the breeding range, replacement clutches are sometimes laid (Vuolanto 1968). Both sexes assist in incubation, and some reports indicate that the male undertakes the major role in this behavior, while others suggest that the female does. The incubating bird is often warned of danger by its mate, and it leaves the nest well before an intruder gets

close. Injury-feigning in typical plover fashion, with wings dragging or flapping, has also been observed in birds flushed from the nest. The incubation period is normally 21–23 days, probably averaging 22 days. Although both sexes tend the young initially after hatching, the role played by the female soon diminishes, and she usually leaves the brood and the nesting grounds long before they have fledged. Fledging occurs at the age of about 19 days, and the young are independent thereafter. Males leave the nesting areas at about this time. The young soon gather in flocks on gravel beaches, feeding on midges, and by the end of August most of them have also departed from nesting areas (Nettleship 1973; Parmelee and Mac-Donald 1960; Parmelee et al. 1967).

Status and relationships. The relatively wide breeding distribution of this species favors its survival, and its extremely northerly limits (up to 83° N latitude), places it well out of the reach of most direct human disturbance. In spite of its extraordinarily long migration and high arctic breeding, the species seems to have a relatively low annual mortality rate, and although its total world population is completely unknown it must be substantial. The evolutionary relationships of the genus *Arenaria* are very controversial and speculative. The turnstone genus has traditionally been placed in a family close to the plovers, presumably because of its short, albeit distinctly shaped bill, and yet it differs from the plovers in nearly all major respects. The turnstones have almost always also been grouped in a common subfamily with the surfbird, but Jehl (1968a) has argued that these genera are not closely related, and while *Aphriza* almost certainly has calidridine affinities he believed that *Arenaria* represents an offshoot of the tringine sandpipers. However, Ahlquist (1974) suggested that *Arenaria* is also related to the calidridine sandpipers, and evidence provided by Burton (1974) and Strauch (1976) supports his view. Its downy plumage is not distinctly calidridine, but Fjeldså (1976) nevertheless suggested that it represents a side branch of the calidridine type, with degenerate "powder-puff" down. I am inclined to believe that *Arenaria* should be placed in the same subfamily as the calidridine sandpipers, but in a different tribe. I do not consider it to be an evolutionary link between the Charadriidae and the Scolopacidae, but rather agree with Fjeldså that it is probably a specialized side branch of the calidridine sandpipers.

Suggested reading. Nettleship 1973; Beven and England 1977; Parmelee et al. 1967.

Black Turnstone

Arenaria melanocephala (Vigors) 1928 (1829)

Other vernacular names. None in general English use; tournepierre noire (French); Schwarzkopfsteinwälzer (German).

Subspecies and range. No subspecies recognized. Breeds in western and southern Alaska, from Shismaref Inlet south to the Sitka district. Winters from southeastern Alaska south to southern Baja California and central Sonora, Mexico. See map 71.

Measurements and weights. Wing: males 138–49 mm; females 141–54 mm. Culmen (from feathering): males 21–24 mm; females 22–24 mm. Weights: 12 males 101–27 g, average 113.6 g; 9 females 98–148 g; average 124.2 g, (M.V.Z. specimens). Eggs ca. 41 x 29 mm, estimated weight 17.3 g (Schönwetter 1963).

DESCRIPTION

Breeding males are dark sooty brown on the head, neck, chest, and upperparts, with white lores, a white superciliary stripe, and white spotting on the sides of the neck and breast. The feathers of the lower breast are margined with white, which becomes continuous over the abdomen, sides, flanks, lower rump, and tail coverts. The wings are dusky, like the back, but the smaller inner coverts are white, as are the tips of the greater coverts. All the secondaries are white except for the outer ones, which are tipped with dusky, as are the basal parts of the outer webs of the primaries, especially the inner ones. The iris is brown, the bill is blackish, and the legs are dark purplish brown. *Females* average slightly smaller in size and have a less distinct head pattern, with a smaller white head patch. *Adults in winter* are much lighter on the throat, chest, and sides of the head, and there are no white markings on the head, neck, and chest. *Juveniles* resemble winter adults, but there are extensive pale buffish fringes on the upperparts, and the tail is tipped with light grayish brown rather than white. *First-winter* birds may retain some buff-fringed inner median coverts, but otherwise they closely resemble winter adults (Ridgway 1919; Prater et al. 1977).

In the hand, the distinctively pointed and uptilted bill identifies this as a turnstone, and the gray (winter) to black (summer) throat distinguishes it from the ruddy turnstone.

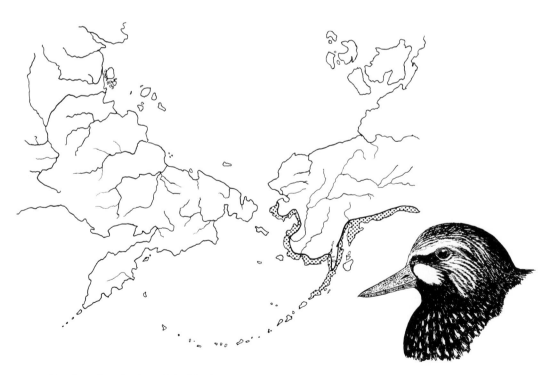

71. Breeding distribution of the black turnstone.

In the field (7–9 inches), black turnstones are found along rocky coastlines and in winter may be confused with ruddy turnstones, but they have darker and more uniform upperparts and brownish rather than orange legs. In flight they show extensive white on the upper wing surface and a strongly banded tail somewhat like that of a surfbird, but the black on the tail is a band rather than a triangular wedge, and the white of the rump does not extend up the back. Their calls are very similar to those of the ruddy turnstone but may be slightly higher in pitch.

NATURAL HISTORY

Habitats and foods. The breeding habitats of this species consist of grass-covered shorelines of brackish ponds, tidal sloughs, and similar habitats of the coastal tundra of western Alaska. Their foraging behavior and foods on the breeding grounds have not yet been studied, but no doubt they are very much like those of the ruddy turnstone. During the winter period the birds tend to inhabit rocky shorelines, jetties, and barnacle-covered reefs, especially those of

outlying islands and ledges. There they have been seen feeding by following receding breakers and gleaning from the surface, by foraging on wet sand just above the limits of the falling tide, and also by turning patches of seaweed up with the bill, sometimes running at the heavier objects and using their heads like battering rams. Small shrimps and prawns are apparently their primary quarry in such areas, but they probably also eat small mollusks, barnacles, sea slugs, and the like (Bent 1929; Brandt 1943).

Social behavior. Regrettably little is known of the social behavior of this species. Brandt's (1943) observations suggest that chases begin shortly after arrival on the breeding grounds, with the female dashing off on zigzag flights, chased closely by the male. Brandt also reports that the male performs an aerial display in which he ascends so high that he is completely out of sight, when he produces a "strange *zum-zum-zum* noise as made by the Wilson snipe." It seems unlikely that this species produces a sound comparable to that of the common snipe, and it is more probable that Brandt was mistaken on this point. Copulatory behavior has not been described. A clear, peeping *weet, weet, too-weet* call is produced during the nesting season, apparently associated with disturbance at the nest site, but injury-feigning has not been observed.

Reproductive biology. Nests are quite close to the water's edge along tundra pools, particularly on small inlets or points. The nest is simply a depression in the dead grass, or sometimes in almost bare mud, and there is little effort to line it. In some areas the birds nest very close to one another, producing a semicolonial distribution. Both sexes share incubation, and the clutch size is invariably 4 eggs. In the Hooper Bay area, fresh eggs were found over the period May 29 to June 21, suggesting that some renesting efforts may occur in that region. The incubation period of one clutch was determined to be 21 days, and the first downy young were seen on June 20. The fledging period is not exactly known, but when the young take wing in July they move from the tundra flats to the seacoast, where they begin to search for food along the tide line (Bent 1929; Brandt 1943).

Status and relationships. There is no good information on the status of this highly restricted species, but its breeding grounds are remote and free from disturbance, so there seems to be no cause for concern at present. This species is locally sympatric with the ruddy turnstone and obviously is its close relative, but it is much more closely restricted to coastal habitats. Mayr and Short (1970) consider the two forms as constituting a superspecies, with no evidence of hybridization. Relationships of the genus *Arenaria* are discussed under the *interpres* account.

Suggested reading. Brandt 1943; Bent 1929.

The surfbird is a small arctic-breeding shorebird with a small hind toe, a short tarsus with scutellations in front and reticulations behind, the anterior toes without webbing but with lateral membranes, the bill ploverlike, and the tail slightly forked. The downy young are *Calidris*-like. One genus and species.

Genus *Aphriza* Audubon 1839

This genus of one species has the characteristics of the tribe Aphrizini.

Surfbird

Aphriza virgata (Gmelin) 1798

Other vernacular names. None in general English use; bécasseau du ressac (French); Gischläufer (German); chorlito de las rompientes (Spanish).

Subspecies and range. No subspecies recognized. Breeds uncommonly in alpine tundra of central Alaska, including the Tanana-Yukon highlands, probably to the White Mountains, in the Alaska Range, and probably in the Wrangell Mountains. Also breeds at least rarely at the base of the Alaska Peninsula, and in the Kilibuck Mountains of western Alaska (Kessel and Gibson 1979). Known nesting in the Yukon is limited to the Richardson and Ogilvie mountains (Frisch 1978). Winters from

southeastern Alaska to the Strait of Magellan. See map 72.

Measurements and weights. Wing (chord): males 164–83 mm; females 169–81 mm. Culmen (to feathering): males 23–26 mm; females 23–26 mm. Weights: 25 males 121–92 g, average 155.5; 6 females 150–93 g, average 171.4 g (M.V.Z. specimens). Eggs ca. 43 x 31 mm, estimated weight 21.0 g (Schönwetter 1963).

DESCRIPTION

Breeding adults of both sexes have the entire head and neck streaked with blackish and grayish white; the scapulars and interscapulars are black with grayish white margins, becoming cinnamon rufous poste-

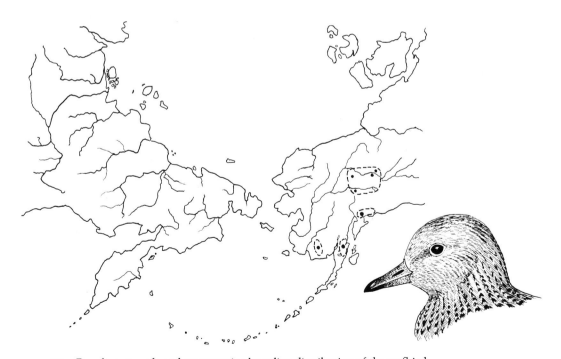

72. Breeding records and presumptive breeding distribution of the surfbird.

24. Winter-plumage heads of surfbird (*above*) and black turnstone (*below*).

riorly. The wing coverts are brownish gray, as are the secondaries, which are tipped with white and become mostly white inwardly. The primaries are similar but are more grayish inwardly and white basally. The rump is dusky, and the tail coverts are white, as is the tail, except for a blackish distal portion that is narrower outwardly and tipped with white. The underparts are white, extensively marked with V-shaped bars of blackish, especially on the sides and flanks. The iris is brown with black eyelids, the bill is black terminally and yellowish basally, especially on the lower mandible, and the legs and toes are olive green to yellow. *Adults in winter* are generally plain brownish gray above and on the breast, with a white spot on each side of the forehead, extending backward into a superciliary stripe, and grayish white on the chin. The breast and sides are variably streaked and spotted with dusky grayish. *Juveniles* have pale buff fringes on their wing coverts and a brownish breast. *First-winter* birds retain juvenile coverts, though the buffy fringes become well worn (Prater et al. 1977).

In the hand, the rather ploverlike bill, well-developed hind toe, and square tail (slightly forked when closed), which has a blackish triangle surrounded by white, provide for diagnostic identification.

In the field (8 inches), these birds are often found

on rocky promontories with turnstones. The two species appear rather similar, but surfbirds have yellowish legs and a more grayish than brownish appearance in winter. In flight, no white appears on the anterior wing coverts or lower back, and the black markings on the tail are wedge-shaped rather than forming a uniform band. The birds are much less noisy than turnstones but sometimes utter a shrill *ke-week* note. Wandering tattlers sometimes occur in the same areas but totally lack white on the tail and upper wing surface.

NATURAL HISTORY

Habitat and foods. The breeding habitat of this species consists of alpine tundra above 4,000 feet elevation, usually in the vicinity of rock slides and other rugged country occupied by mountain sheep. In that habitat the birds forage on open, sun-warmed slopes, catching insects either by stealth or by active chase. The food is predominantly flies, with beetles somewhat less important, and these two food sources constituted over 90 percent of the foods found in 8 specimens obtained on the nesting grounds (Dixon 1927). In the Richardson and Ogilvie mountains of the Yukon, breeding habitat occurs between 2,500 and 5,500 feet, and all apparent breeding sites were on sparse mountain heath consisting of mixed lichen-moss-avens-heather vegetation, interspersed with sedges and mosses (Frith 1976). By contrast, foods on the wintering habitats are quite different and consist of such marine invertebrates as barnacles, mussels, and the like. A few birds examined in California contained primarily mussels and periwinkles that were under 12 mm in length. The birds feed along the water's edge on wave-tossed rocky or stony beaches, generally foraging on smaller rocks and remaining closer to the water than do wandering tattlers, but otherwise they are probably dependent on much the same foods (Orr 1942).

Social behavior. On the wintering grounds these birds appear phlegmatic and sleepy, occurring in numbers ranging from single individuals to flocks of several hundred. During the summer, however, the birds are widely scattered over the alpine tundra; Dixon (1927) reported encountering surfbirds only 7 times during 72 days spent in the field, and the largest group seen during the entire period was a flock of 7 adults. There is no specific information on territoriality or display. Dixon does not mention any specific vocalizations associated with nesting, but he does

note that the birds were very shy and difficult to observe undetected.

Reproductive biology. Very few nests and young have been found. The first nest ever located by ornithologists was found on a southwest-facing slope about 1,000 feet above timberline, on a rocky ridge that was relatively snow-free. It was highly exposed and very near a well-traveled sheep trail. There was no fabricated nest as such; instead the 4 eggs were deposited in a natural depression that was lined with a few bits of lichen and caribou moss. This nest was found in late May, in Mount McKinley National Park, and subsequently a nest with 4 eggs was located in the Chugach Mountains in late June of 1971, and one with 3 eggs was found on the Seward Peninsula in mid-June of 1977 (Kessel and Gibson 1979). Young in Alaska have been seen as early as late June, and in the Yukon they have been found in early July. Although Dixon believed that only males develop incubation patches and do most or all the incubation, young have been found attended by paired adults, so that monogamy is the most likely breeding system. Further, Jehl (1968*b*) noted that both members of a pair collected in Alaska had well-developed brood patches, thus both sexes probably incubate. There is no information on the incubation or fledging periods. However, in the Yukon the birds leave by late July, suggesting that fledging and migration out of the breeding areas must occur less than a month after hatching. The protective coloration of the downy young is remarkable and closely matches the mottled and speckled pattern of the lichen-covered rocks (Gabrielson and Lincoln 1959).

Status and relationships. Little can be said of the status of this elusive bird that winters over a tremendous stretch of the Pacific coastline and breeds in some of the most remote areas of North America. Its evolutionary relationships are also somewhat obscure. Jehl (1968*b*) has reviewed the long and controversial taxonomic history of the genus, which at times has been placed in a monotypic family but more recently has usually been included in a separate subfamily with the turnstones. Together they have usually been assigned to the plover family Charadriidae, but, as Jehl has pointed out, the affinities of the surfbird, and probably also the turnstones, are much more likely to be with the calidridine sandpipers. Jehl believed that the great knot is probably a close relative of the surfbird, inasmuch as both have very similar downy, juvenile, and breeding plumages, and the bill and toes also have certain characteristics in common. In Jehl's view there is no good reason to believe that the surfbirds and turnstones are close relatives. Strauch (1976) agreed that the surfbird and the great knot might be close relatives and noted that these two species also share a pectinate claw on the middle toe. This character is of rather doubtful taxonomic significance, but it occurs in only a few other shorebirds (*Oreopholus, Limosa limosa,* and possibly others). Sibley and Ahlquist (1972) noted that *Aphriza* has an electrophoretic egg white profile that is like that of the Scolopacidae rather than the Charadriidae, and Bock (1958) also supported the idea that both turnstones and surfbirds are probably members of the scolopacid group.

Suggested reading. Dixon 1927; Jehl 1968*b*.

Tribe *Calidridini* (Typical Sandpipers)

Sanderling

Typical sandpipers are very small to small arctic-breeding shorebirds with the nostrils in a depression extending anteriorly as a groove that nearly reaches the tip of the mandible. A hind toe is usually present, and the anterior toes are usually unwebbed. The tarsus is scutellated anteriorly and posteriorly and varies from short to relatively long. The downy young are distinctively marked with white-tipped "powder puff" down, so that the plumage appears strewn with snowflakelike markings. The sexes are usually alike in adult plumage, but seasonal variations are frequent, and strong sexual dimorphism occurs in one genus. Five genera, including 23 species, are recognized here.

KEY TO GENERA OF CALIDRIDINI

A Bill and mantle snipelike, double dark stripe behind eye . . . . *Limicola* (1 sp.)
A' Bill and mantle not snipelike, no double dark eye stripe
  B Legs elongated (tarsus over 40 mm), tarsus longer than bill
    C Legs greenish, toes slightly webbed . . . . *Micropalama* (1 sp.)
    C' Legs varying from yellow to reddish, no webbing between inner and middle toes . *Philomachus* (1 sp.)
  B' Legs not elongated (tarsus under 40 mm), tarsus sometimes shorter than bill
    C Uniformly buffy on underparts, undersides of primaries and secondaries with black spotting
    . . . . *Tryngites* (1 sp.)
    C' Not buffy below, nor with black specks on undersides of flight feathers
      D Bill spoon-shaped . . . . *Calidris pygmeus* ("*Eurynorhynchus*")
      D' Bill not spoon-shaped . . . . *Calidris* (18 additional species)

Genus *Calidris* Anon. (=Merrem) 1804

Sandpipers and stints are small to medium shorebirds with a small hind toe (one exception), anterior toes that are usually unwebbed (two exceptions), and a bill that is straight or only slightly decurved, with slitlike nostrils in a long groove. The wings are pointed, and the tail is rather square, with the central pair of feathers longer and more pointed than the others. The downy young have irregular brown markings on the back and diffuse powder-puff feather tips. Nineteen species are recognized here.

KEY TO SPECIES OF *Calidris*

A Anterior toes with webs between them
  B Bill shorter (culmen no longer than middle toe with claw, maximum 23 mm) and gradually tapered, ratio of culmen length to wing length usually under 0.18 . . . . *pusilla*
  B' Bill longer (culmen longer than middle toe with claw, or at least 22 mm) and much heavier at base than toward tip, ratio of culmen length to wing length usually over 0.19 . . . . *mauri*
A' Anterior toes with no web
  B Hind toe absent . . . . *alba*
  B' Hind toe present
    C Middle pair of tail feathers not longer than rest, wing at least 150 mm
      D Wing under 180 mm, shafts of primaries brownish basally . . . . *canutus*
      D' Wing over 180 mm, shafts of primaries white basally . . . . *tenuirostris*
    C' Middle pair of tail feathers longer than rest, wing under 140 mm
      D Exposed culmen longer than tarsus
        E Bill stout and straight, scapulars and upper wing coverts margined with light gray
          F Scapulars and upper wing coverts of summer plumage with little if any rusty or cinnamon on upperparts, bill and feet yellowish . . . . *maritima*
          F' Summer plumage with much rusty or cinnamon present, bill and feet blackish . . . . *ptilocnemis*
        E' Bill more slender and decurved, scapulars and upper wing coverts not obviously pale edged
          F' Bill decurved over much of its length, white rump . . . . *ferruginea*
          F' Bill only slightly decurved near tip, dark rump . . . . *alpina*

D′ Exposed culmen shorter than tarsus
  E Larger (wing over 110 mm, tail over 48 mm)
    F Rump and median upper tail coverts sooty blackish
      G Shaft of outermost primary entirely white, middle tail feathers considerably longer than more lateral ones . . . . *melanotos*
      G′ Shaft of outermost primary partly white, tail feathers progressively longer from outer pair toward the middle ones . . . . *acuminata*
    F′ Rump and median upper tail coverts not sooty black
      G Median upper tail coverts dark grayish brown, like rump . . . . *bairdii*
      G′ Median upper tail coverts white . . . . *fuscicollis*
  E′ Smaller (wing under 105 mm, tail under 48 mm)
    F Middle toe and claw shorter than tarsus, culmen less than one-fifth as long as wing, which is usually over 90 mm
      G Legs and feet greenish yellow to olive, only outermost primary with white shaft, white outer tail feathers . . . . *temminckii*
      G′ Legs and feet blackish, all primaries with partly white shafts, outer tail feathers not pure white
        H Bill more slender, throat and foreneck white in summer . . . . *minuta*
        H′ Bill stouter, throat and foreneck cinnamon rufous in summer . . . . *ruficollis*
    F′ Middle toe and claw longer than tarsus, culmen more than one-fifth as long as wing, which is usually under 90 mm, legs and feet yellowish
      G All the primaries with shafts at least partly white, middle toes and claw under 18 mm . . . . *minutilla*
      G′ Only the outermost primary with white shaft, middle toe and claw over 18 mm . . *subminuta*

Summary of Field Marks of Confusing "Peeps"
(In part after Wallace 1974 and Prater et al. 1977; arranged from small to large)

Least Sandpiper. 5″, tiny body, bill short and needlelike, upperparts dark brown, with thin V-mark on edge of mantle, breast often forming a clear band, *legs short and yellow,* call a prolonged *kreet.*

Little Stint. 5½″, rufous upperparts, with buffy scaling and clear V-mark at edge of mantle, white throat above brownish breastband, *legs blackish,* short *chit* calls.

Long-toed Stint. 5½″, *upperparts spangled* with prominent black and chestnut, whitish V-mark at edge of mantle, legs and toes long and yellowish, call a purring *trerp.*

Temminck Stint. 5¾″, bill rather short, upperparts uniformly olive brown, throat and breast brownish, *white outer tail feathers,* legs brownish to yellowish, call a trilled *tit-tit-tit.*

Rufous-necked Sandpiper. 5¾″, *pale chest and whitish forehead,* grayish black upperparts with indistinct markings, legs blackish, squeaky *pit* or *chit* calls.

Semipalmated Sandpiper. 6″, *bill relatively stubby, with an expanded tip,* upperparts mostly dun gray with indistinct markings, legs black, call a low-pitched *chirrup.*

Western Sandpiper. 6½″, *bill heavy at base and drooping at tip,* upperparts marked with *bright rufous* and dark gray, breast strongly streaked and spotted, legs blackish, *cheet* call.

Baird Sandpiper. 7″, bill tapering to a sharp point, upperparts grayish brown and blackish, with *buffy scaling, wings reach well beyond tail,* legs blackish, call a guttural *kreep.*

White-rumped Sandpiper. 7″, upperparts streaked but not scaled with brown and black, flanks also streaked, *white rump,* wings reach well beyond tail, legs blackish, call a thin and squeaky *tzeet.*

Sharp-tailed Sandpiper. 8″, robust body, like pectoral but with *rufous* and black upperparts and more *rufous crown,* the breast lightly streaked and blending with sides and underparts, legs olive green, call a two-noted whistle.

Pectoral Sandpiper. 8½″, robust body, with brownish upperparts and a heavily streaked brownish *breast sharply set off from white underparts,* legs yellowish green, call a low and creaky *pritt.*

Great Knot

Calidris tenuirostris (Horsfield) 1821

Other vernacular names. Eastern knot; bécasseau d'Anadyr (French); Gross Knutt (German).

Subspecies and range. No subspecies recognized. Breeding evidently extends from the Verkhoyansk Range east to Magadan, Koryak Highlands, and the southern part of the Chukotski Peninsula. However, actual breeding records for the species are relatively few. Winters in Burma, Malaya, India, and the Sundas to Australia. See map 73.

Measurements and weights. Wing, males 174–90 mm; females 173–90 mm. Culmen (from feathering): both sexes 39–47 mm. Weight: 10 males 135–95 g, average 156 g; females 155–207 g (Shaw 1936). Eggs ca. 44 x 32 mm, estimated weight 22.0 g (Schönwetter 1963).

DESCRIPTION

Breeding adults of both sexes have the upperparts blackish brown, with whitish feather tips and many feathers with subterminal rufous patches. The rump is blackish brown, slightly lighter than the back and with less obvious white feather tips. The upper tail coverts are white with numerous black bars. The underparts, neck, cheeks, and sometimes the nuchal area are heavily spotted with blackish, with the abdomen somewhat lighter. The flanks, axillaries, and under tail coverts are white, with dark brown bars on the flanks and axillaries. The tail feathers are smoky gray, the flight feathers are blackish brown with white shafts, and the secondaries also have white borders on their outer vanes (Dementiev and Gladkov 1969). The iris is blackish brown, the bill is black distally and brown basally, and the legs and toes are dull greenish gray. *Females* average larger than males and may have less reddish on the upperparts as well as some white feathers in the rufous underparts. *Adults in winter* are smoky gray above and white below, with brown or smoky streaks on the breast, flanks, and under tail coverts. Their outer wing coverts are edged with paler gray, and the inner ones are edged with white. *Juveniles* have pale buff tips on their blackish brown upperparts, and their wing coverts are fringed with whitish buff. *First-*

73. Breeding records and presumptive breeding distribution of the great knot.

winter birds resemble adults but have some brownish wing coverts with pale buff tips, especially the inner medians (Prater et al. 1977).

In the hand, the large size (wings usually 175–90 mm), a fairly long exposed culmen (40 mm or more) that is longer than the tarsus, and the shafts of the primaries being entirely white basally usually distinguish this species from others, including the red knot, which has primaries with shafts that are grayish brown basally and terminally. Also, in breeding plumage the heavy breast streaking distinguishes it from the smaller species of knot.

In the field (12 inches), this species is likely to be seen in winter plumage except on its Siberian breeding grounds. In winter plumage it looks like a large, plump sandpiper that closely resembles the red knot. Yet it is larger, darker, and more mottled on the upperparts and has a more heavily spotted breast and sides, a whiter rump, and a longer as well as thicker bill. In flight it shows a narrow white wing stripe, with white otherwise limited to the tail coverts, whereas the red knot has lightly barred white feathers on both the tail coverts and the rump and has a less conspicuous wing stripe. Its usual call is a double-noted whistle, *nyut-nyut.*

NATURAL HISTORY

Habitats and foods. In the winter these birds assemble on muddy tidal flats, and they rarely occur along inland rivers of Australia. They typically forage in small groups on muddy banks, rarely staying more than a few days at any one time. Their winter and summer foods are unstudied, but some fall specimens were found to have been feeding on crustaceans (Gammaridae). The summer habitat is unusual, consisting of mountain tundra from about 300–420 meters elevation in some areas to about 1,000 meters in the upper Kolyma basin, where there is an abundance of bare gravelly areas covered by crustose lichens and occasional patches of herbs, heathers (*Empetrum, Dryas, Vaccinium*), and similar arctic vegetation (Kozlova 1962; Frith 1976). Summer foods include dipteran larvae, beetles, spiders, berries of *Empetrum nigrum,* and even seeds of *Pinus pumila* (A. A. Kistchinski, in litt.).

Social behavior. On their Australian wintering areas these birds have been seen in groups of as many as 50 birds. First-year birds may also spend their first summer of life on the winter range. However, adults return to their nesting areas in late May or early June, as thawed areas are appearing on the tundra. During courtship flights the male circles high above the ridges uttering a gutteral *kurru . . . kurru . . .* , and later *trry-ha . . . , trry-ha . . .* The birds are territorial, occupying areas of about 1 km² (A. A. Kistchinski, in litt.).

Reproductive biology. Very few nests have been found. One with 4 eggs was found in mid-June, 1917, on the Kolma lowlands, and was attended by both parents. In a nest studied by A. A. Kistchinski (in litt.), both sexes incubated, but the female apparently left the breeding area before the end of incubation. Only males have been observed accompanying broods, and they have been observed performing strong distraction behavior. Broods of 3 or 4 young have been observed in the Gorelov Mountains of the

Anadyr Basin in July (Dementiev and Gladkov 1969; Kozlova 1962).

Status and relationships. This is apparently an uncommon species, but nothing can be said of its actual population status. Its relationships are apparently with *canutus*, since its downy young are much like those of the red knot (Kozlova 1962), and the overall plumages and proportions of these two species are very similar. Kozlova attributes the short legs and toes of the red knot and great knot to their rocky habitats of arctic tundra and alpine tundra respectively, and Larson (1957) regards them as a species pair resulting from isolation in Tertiary times.

Suggested reading. Dementiev and Gladkov 1969; Uspenski 1969.

Red Knot

Calidris canutus (L.) 1758

Other vernacular names. Knot; European knot; bécasseau maubèche (French); Knutt (German); correlimos gordo (Spanish).

Subspecies and ranges. See map 74.

C. c. canutus: Eurasian red knot. Breeds on Greenland, Spitsbergen, the northern Taimyr Peninsula, the New Siberian Archipelago, and Wrangell Island. This form or *rufa* also breeds locally in northern Alaska (Brooks Range, Seward Peninsula, Barrow), on Ellesmere Island, and on adjoining arctic islands. Winters from the North Sea to the west coast of Africa and in Asia to New Zealand. Includes the eastern population sometimes separated as *rogersi*.

C. c. rufa: North American red knot. Breeds on Victoria Island, northern Melville Peninsula, and south to Southampton Island. Winters from Massachusetts south through Central and South America to Tierra del Fuego.

Measurements and weights. Wing: males 155–66 mm; females 158–72 mm. Culmen (from feathering): males 28–38 mm; females 28–33 mm. Weights (on breeding grounds): males 112–36 g, average of 13 was 125.5 g; females 135–69 g, average of 9 was 147.9 g (Parmelee and MacDonald 1960). Eggs ca. 43 x 30 mm, estimated weight 19.0–19.6 g (Schönwetter 1963).

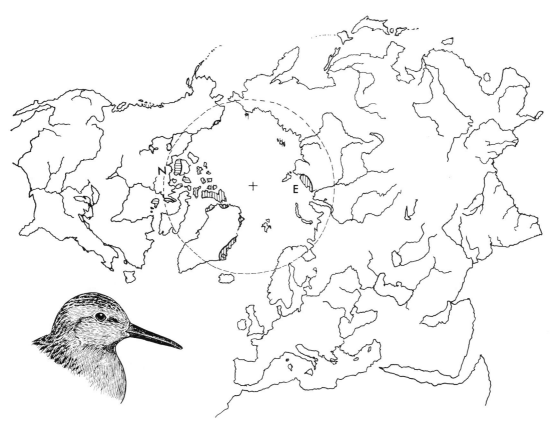

74. Breeding distributions of the Eurasian (E) and North American (N) red knots.

DESCRIPTION

Males in breeding plumage have the sides of the head, including the superciliary region, the throat, foreneck, chest, breast, and upper abdomen plain buffy cinnamon to chestnut brown, becoming paler posteriorly, with the sides and flanks mixed with white and dusky markings and the abdomen and under tail coverts almost entirely white. The forehead and crown are whitish, the latter with black streaking that becomes broader on the hindneck. The rest of the upperparts are mostly a mixture of light gray, pale cinnamon rufous, and black. The rump is pale gray with darker markings, the upper tail coverts are mostly white with similar markings, and the tail is pale brownish gray, the feathers margined with white. The wing coverts are light gray with whitish edge markings, the greater coverts are tipped with white, the secondaries are brownish gray with narrow white edges, and the primaries and primary coverts are dusky, the coverts edged with white, and the primaries with white shafts. The iris is dark

25. Winter-plumage heads of sanderling (*above*) and red knot (*below*).

brown, the bill is black, and the legs and toes are blackish (Ridgway 1919). *Females* are usually paler on the underparts, with the buffy more often broken laterally with whitish feathers, and they are more grayish and less chestnut on the upperparts. *Adults in winter* are light brownish gray above, with a white superciliary stripe and white on the underparts, including the lower cheeks and chin. The throat, foreneck, chest, and sides are streaked or otherwise marked with dusky. *Juveniles* have whitish underparts that are strongly washed with pinkish buff and have pale fringes and dark subterminal lines on grayish brown rather than grayish upperparts (Prater et al. 1977).

In the hand, the rather large size (wing 150–80 mm), a straight and fairly stout sandpiperlike bill that is longer than the tarsus, and the uniformly brownish gray tail without markings and with the central feathers not projecting beyond the others serve to identify this species.

In the field (10 inches), knots are usually found along coastlines, often probing in mud flats or standing in shallow water. Their chunky shapes and their behavior make them resemble dowitchers, but they have much shorter bills and only the rump is whitish. In winter the birds appear to be mostly white underneath, but in spring the bright cinnamon on the breast and sides helps to identify them. In flight they show a narrow white wing stripe and a white rump. Their calls include a low-pitched *knut* grunting note, and a low double-noted whistle, *wah-quoit* or *tlu-tlu*, is also sometimes uttered.

NATURAL HISTORY

Habitats and foods. The breeding habitat of this arctic species is apparently rather variable, but in general it consists of dry upland tundra, including weathered sandstone ridges, upland areas with scattered willows, *Dryas* and poppy vegetation, moist slopes and flats in foothills or sometimes lowlands, and well-drained slopes hummocked with *Dryas*. Evidently the most favored areas consist of hummocked slopes and tundra, either dry or moist, frequently dominated by *Dryas*. Population densities are generally only about one pair per square kilometer. On Wrangell Island the birds occur at a density of up to several pairs per square kilometer (Portenko 1972). On the breeding grounds the birds are by necessity relatively vegetation-dependent when they first arrive, owing to the lack of insect food. At that time they probe and jab in marshes and moist slopes, eating the seeds of sedges, horsetails (*Equisetum*), grass shoots, and the like. They also probe into snow-free slopes, probably for invertebrates, and caterpillars sometimes become important foods in middle to late June. As dipterous insects appear on the tundra, the birds begin to peck for them, and beetles, craneflies, and other insects are all eaten where they are available (Portenko 1972; Parmelee and MacDonald 1960; Nettleship 1974). The high incidence of plant food taken by adults during the early part of the breeding season is of special interest and is nearly unique among arctic shorebirds, and the availability of suitable invertebrate food, especially midges, for the young at the time of hatching may restrict the birds to the vicinity of water during nesting (Nettleship 1974). In the wintering grounds of coastal Argentina this species is almost entirely maritime in habitat, foraging on sandy beaches and mud flats, sometimes with sanderlings. It occurs in flocks of 5–100 birds, and does not exhibit winter territoriality (Myers and Myers 1979). In a German study, migrating red knots concentrated on small, mollusks (*Macoma, Littorina, Hydrobia*), while sanderlings in the same areas primarily ate insects (Ehlert 1964).

Social behavior. The birds arrive at the northern portion of Ellesmere Island in early June and waste little time in dispersing to inland breeding areas, which are at that time still largely snow-covered. After males become established on territories, they begin advertisement flights that peak in mid-June but last for nearly a month. The bird ascends quickly to 20–160 meters, where the song flight display is performed. There are three phases to the display, including an initial diagonal glide and a *whip-poo-mee* call uttered at the rate of about one per second, followed by a horizontal glide and shifting of the call to a flutelike *poo-mee*. Finally, there is a quivering flight during which the bird regains altitude with rapid but shallow wingbeats and often continues the *poo-mee* calls. Sometimes an area as wide as a mile in diameter will be covered during a single flight. Last, the bird glides swiftly to the ground with the wings held in a high V above the body. It lands in a wing-up display posture. This same posture typically occurs immediately before copulation, judging from the few available observations (Nettleship 1974; Hobson 1972). This wing-up display appears to be especially common in the red knot, but details of pair-formation are still essentially unknown. Apparently the pair bond is

strong, and the very short available breeding period suggests that pair-bonding must occur rapidly after territories are established. Hobson (1972) indeed implied that the pair establishes the territory after having either arrived in pairs or paired shortly after arrival.

Reproductive biology. There is considerable variance in the favored nest sites of this species, with some authors (Parmelee and MacDonald 1960) suggesting that the birds prefer to nest on hummocks surrounded by mud and water, while most other nests seem to have been found on stony or gravelly ground, often 500–800 feet above sea level and sometimes miles from the coast (Hobson 1972). Nine nests found on Ellesmere Island averaged nearly 500 feet above sea level, and all were in low-spreading vegetation on relatively hummocky ground. All contained 4 eggs or eggs and young. The interval between successive eggs is probably 24–28 hours. The incubation period is probably between 21.3 and 22.4 days, and the male evidently performs the majority of the incubation. The male is also believed to play the major role in the brooding and care of the young. The fledging period is probably 18 days, at which time the males probably abandon their broods and begin to leave the area. The young knots remain for another few weeks (Parmelee and MacDonald 1960).

Status and relationships. Although this species has an extremely scattered breeding range, it winters in high concentrations in only a few areas. For example, Great Britain supports winter populations of up to 100,000 on the Wash and similar numbers on Morecambe Bay, while another 130,000 regularly winter along the coast of Europe between Denmark and the Spanish border. These tidal mud and sand flat areas are potential locations for serious oil spills, and thus the status of the species may be less secure than one might imagine. The relationships of the species are quite obviously close to those of the larger eastern knot, and Larson (1957) regards *canutus* and *tenuirostris* as constituting a species pair. Beyond this immediate affinity, there are no other clear-cut relationships within the genus *Calidris*, and the knots seem to be a rather peripheral component of the genus. The downy young of the knots have relatively "primitive" powder-puff down characteristics that they share with the surfbird, and Fjeldså (1976) therefore regards these three forms as transitional between *Aphriza* and the typical *Calidris* forms.

Suggested reading. Nettleship 1974; Hobson 1972.

Sharp-tailed Sandpiper

Calidris acuminatus (Horsfield) 1821
(*Erolia acuminata* of Peters, 1934)

Other vernacular names. None in general English use; bécasseau à queue pointé (French); Spitzschwanzstrandläufer (German); correlimos acuminado (Spanish).

Subspecies and range. No subspecies recognized. Breeds in arctic Siberia, probably from the delta of the Yana to the Kolyma delta. Has also bred at Tiksi Bay, and has been collected during summer on the Lena delta. Winters in Melanesia from New Guinea south to New Caledonia, Australia, and New Zealand, and the Pacific islands to the east. See map 75.

Measurements and weights. Wing: male 131–43 mm; females 125–31 mm. Culmen (to feathering): males 24–28 mm; females 23–26 mm. Weights: 10 males 53–82 g, average 70.3 g; 10 females 39–81 g, average 63.5 g (U.S.N.M. specimens). Shaw (1936) provided weights for Chinese birds, with 8 males 63–114 g, average 80 g, and 8 females 43–93 g, average 72 g. Summer weights range from 69 to 102 g (Uspenski 1969), while wintering males and females average ca. 74 and 56 g respectively (Serventy and Wittell 1962). Eggs ca. 38 x 27 mm, estimated weight 13.7 g (Schönwetter 1963).

DESCRIPTION

Breeding adults of both sexes have a rufous brown crown broadly streaked with black, a white superciliary stripe that is slightly streaked with dusky, the sides of the head brown with dusky streaks, a white chin and throat, and the foreneck, chest, and sides of breast pale cinnamon to reddish buffy, with irregular dusky markings. The rest of the underparts are white, except for some V-shaped markings on the breast, upper sides, and flanks. The upperparts are mostly of feathers that are blackish centrally and edged with brownish, pale rufous, or whitish. The rump and upper tail coverts are brownish black, as are the central tail feathers, and the rest of the tail is grayish brown, narrowly tipped with white. The wing coverts are deep grayish brown with paler margins, the greater coverts have narrow white margins, and the flight feathers are of same color, the inner ones increasingly edged with white. The iris is brown, the bill is blackish distally, becoming yellowish toward the base, and the legs and toes are olive to yellowish brown. *Females* can often be sexed by their shorter wing length (males 136 mm or more, females

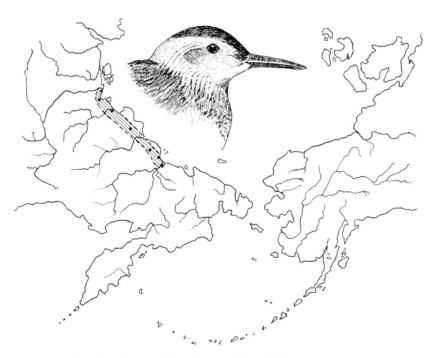

75. Breeding distribution of the sharp-tailed sandpiper.

26. Winter-plumage heads and half-tails of pectoral (*above*) and sharp-tailed (*below*) sandpipers.

135 mm or less). *Adults in winter* are similar to summer adults but are less rufous above and more grayish brown, with a chest of pale grayish buff, indistinctly streaked with dusky. The wing coverts are gray with paler edging. *Juveniles* have a blackish crown that is edged with bright chestnut, and the upperpart feathers are fringed with bright chestnut, buff, or white. The breast is bright buff with a few brown streaks. *First-winter* birds retain some inner median coverts with buffy edging (Prater et al. 1977).

In the hand, this species may be separated from *melanotos* by its rather graduated or wedge-shaped tail, with all the feathers rather than just the central ones relatively pointed. The size (wing over 125 mm) separates the species from the smaller *Calidris* forms,

and only *melanotos* has a comparable bill shape and culmen length (23–28 mm).

In the field (8–9 inches), this species is most like a pectoral sandpiper but tends to be generally more reddish, with a bright chestnut cap, especially in juveniles and breeding adults. The breastband is less abruptly terminated in this species, and the legs are more greenish and less yellowish in *acuminatus*. The calls of the two species are distinctive; this species utters a two-noted *krip-krip* or *chet-chet* on flushing and also has some soft grunting notes. It is found in coastal and inland sites in winter, often on mud flats or sandbars. For additional field identification information, see *British Birds* 73:333–45.

NATURAL HISTORY

Habitats and foods. Wintering habitats of this species in Australia are open and shallow coastal or inland swamps, tidal mud flats, and estuaries, particularly where there is sparse and low vegetation. There they eat mostly small crustaceans, mollusks, kelp flies, small insects, and small amounts of vegetation (Frith 1976). Breeding habitats consist of wet tundra with grassy moss cover and shrubs in elevated situations (Vorobiev 1963). Very damp, hillocky tundra and moss and sedge bogs are apparently optimum habitats for this species. Although it often nests in the same area as the pectoral sandpiper, and though their breeding habitats overlap, the pectoral sandpiper prefers somewhat drier areas that are devoid of shrubs, while the sharp-tailed sandpiper seems to prefer areas where shrubby tundra alternates with sedge bogs (Flint and Kistchinski 1973). Early in the breeding season its major foods are cranefly larvae (Tipulidae); later on the females continue to specialize in these larvae and small mollusks while the males begin to eat seeds in somewhat drier locations (Kistchinski 1973).

Social behavior. On the wintering grounds, these birds are seen in flocks that range in size from small groups or single birds to as many as a thousand or more individuals. However, they are highly territorial during the breeding season. Male territories are about 150–300 meters in diameter, and from 5 to 10 territories may occur in a square kilometer area. Courtship flights are performed above the territories, but there is only weak aggression toward neighboring birds, and none toward male pectoral sandpipers, whose territories sometimes overlap those of sharptailed sandpipers. The territorial flight is much like that of the pectoral sandpiper, the only other *Calidris* known to have an inflatable esophagus. The bird first rises at a steep angle, and at a height of about 30–40 meters it sets its wings and utters a short sound apparently associated with the intake of air into its throat. It then starts to glide and begins to utter a dry, crackling warble that lasts 5–6 seconds. About halfway through its descent the trilling ceases, and the bird continues to glide down silently. Precopulatory behavior consists of the male's approaching a female, squatting somewhat, fluffing his back and tail coverts, drawing in his neck so that the bill touches the inflated throat, and slightly spreading his wings. His tail feathers are also spread, and after initially being held down are lifted almost vertically and quivered slightly. Two distinct sounds are made at this time, a series of toneless *khruk-khruk-khruk* sounds, and a group of ringing, clicking *pot'* . . . *pot* . . . *pot'* notes. According to Flint and Kistchinski (1973), this differs in several respects from the display of the pectoral sandpiper, which spreads its wings back, not down, does not draw its neck in toward the breast, and raises its tail at the beginning rather than the end of the ceremony. Further, the exposed tail covert coloration is different, as are the accompanying sounds. Since the males do not participate in incubation, it seems quite possible that a nonmonogamous mating system may occur in this species.

Reproductive biology. Of a total of 25 clutches found by Flint and Kistchinski (1973), 21 had 4 eggs and the rest had 3. Nearly half were on dry moss and lichen hillows and alongside frost polygons covered with dwarf birch and osiers, 5 were in thick sedges in the middle of wet polygons, 4 were in the middle of dry, mossy polygons, and 4 were in thick sedges on dry, hillocky tundra. The nests were nearly always covered with sedge leaves of the previous year's growth and were very well hidden. The eggs average slightly heavier and larger than those of pectoral sandpipers. The eggs are incubated only by females, and males leave the breeding grounds during incubation.

Status and distribution. In its limited nesting range this species is fairly abundant; Flint and Kistchinski (1973) stated that in two areas studied the population (5–10 territorial males per square kilometer) was surpassed only by those of pectoral sandpipers, ruffs, and red and red-necked phalaropes. Its evolutionary relationships are obviously closest to the pectoral sandpiper, but these forms are obviously distinct species on the basis of information presented by Flint and Kistchinski. It is of interest that the parts of the plumage that are diagnostic of this species (the tail feathers and under tail coverts) are those that seem to be most strongly exhibited during courtship display. Vorobiev (1963) has pointed out that the coloration of the downy young is also different in the two forms and has urged that they be considered two species.

Suggested reading. Dementiev and Gladkov 1969; Uspenski 1969.

Pectoral Sandpiper

Calidris melanotos (Vieillot) 1819
(*Erolia melanotos* of Peters, 1934)

Other vernacular names. None in general English use; bécasseau tachete (French); Graubruststrandläufer (German); correlimos pectoral (Spanish).

Subspecies and range. No subspecies recognized. Breeds in Siberia from the east coast of the Taimyr Peninsula east to the Kolyma delta and probably sporadically to the tip of the Chukotski Peninsula. Also breeds in North America from northern Alaska (south to Goodnews Bay) east to Southampton Island and James Bay. Also breeds on Banks, Victoria, and Prince of Wales islands. Winters in South America and in smaller numbers in Australia, New Zealand, and the Central Pacific Islands. See map 76.

Measurements and weights. Wing (chord): males 138–49 mm; females 126–35 mm. Culmen (to feathering): males 26–32 mm; females 24–29 mm. Weights (on Siberian breeding grounds): 9 males 96–126 g, average 110 g; 5 females 60–97 g, average 84.2 g (Glutz et al. 1975). Males on Alaska breeding grounds average about 20–30 g heavier than females (Pitelka 1959). Eggs ca. 37 x 26 mm, estimated weight 13.1 g (Schönwetter 1963).

DESCRIPTION

Breeding adults of both sexes have a brownish crown streaked with blackish, a whitish superciliary stripe, becoming streaked posteriorly, and an indistinct loral stripe; the rest of the head is light brown to brownish white, with dusky streaking. The chin,

76. Breeding distribution of the pectoral sandpiper.

upper throat, and underparts are entirely white; these areas are separated by a broad brownish white breastband that is strongly streaked with dusky, and some dusky markings extend backward along the upper sides and flanks. The upperparts are mostly of feathers that are blackish centrally with margins of brown, cinnamon, or grayish. The wing coverts are grayish brown, edged and margined with whitish, and the primaries and their coverts are darker, with the shaft of the outermost quill entirely yellowish white, while the others are light brown. The rump and middle upper tail coverts are sooty black, and the lateral upper tail coverts are white. The middle tail feathers are dusky and the outer ones are light grayish brown, with narrow pale edges. The iris is brown, the bill is black distally and dull greenish yellow at the base, and the legs and toes are greenish to yellowish, predominantly the latter. *Adults in winter* lack the cinnamon to rusty tinge on the upperparts, and the black markings are less conspicuous. The wing coverts are grayish brown with paler edges. *Juveniles* resemble adults in spring, but their upperparts are brighter, with more whitish buff on the

scapulars, and the wing coverts have extensive buffy fringes. The breast is streaked with brown on a buffy background. *First-winter* birds have inner median coverts with buffy fringes that are held until spring (Prater et al. 1977).

In the hand, this moderately large *Calidris* (wing 125–50 mm) may be identified by the combination of a distinct breastband, yellowish legs, and a relatively short (culmen 24–29 mm) and straight bill.

In the field (8–9 inches), this is the largest of the common peeplike sandpipers, and has the most conspicuous "bib," sharply separated from the white abdomen. In this feature and in its lack of a rusty crown it differs from the sharp-tailed sandpiper, and it is also usually found near grassy cover rather than on open mud flats, as are the small "peeps." When flushed it utters a *kreek* or *pritt*, sometimes repeated in series, but of a low and creaky quality.

NATURAL HISTORY

Habitats and foods. During the breeding season the birds occupy a variety of tundra habitats, but they are usually associated with terrain that is flat, poorly drained, and wet. Areas used include low grass-sedge cover, and taller cottongrass tussock/dwarf shrub communities. Males prefer to establish territories where the wet tundra has promontories such as mounds or hummocks, or along low ridges adjoining ponds or marshy basins. Females nest in all kinds of tundra vegetation that provide continuous cover of grasses or sedges and well-drained nest sites (Pitelka 1950). During early portions of the breeding season the adults concentrate on cranefly larvae, gradually shifting in July to midge larvae and eating moderate numbers of adult insects, especially dipterans and craneflies, during early July. There is considerable overlap in foods eaten by this species and by other *Calidris* forms nesting in northern Alaska, with a fairly early departure of adult male pectoral sandpipers from the breeding areas during July (Holmes and Pitelka 1962). This is possible because male pectoral sandpipers do not participate in incubation and brood-rearing. Breeding densities are rather variable from year to year and between localities, ranging in good habitats from 5–12 males per 40 acres near Barrow, Alaska (Pitelka 1959), to about 10–15 males and 40 females per square kilometer (150 acres) in eastern Siberia (Kistchinski 1974). In another Alaskan study, summer densities ranged from 3.8 to 40.4 birds per square kilometer over a five-year period (Bergman et al. 1977). During migration the foods and habitats utilized change considerably, with crickets and grasshoppers often eaten by migrating birds during the spring and fall (Bent 1927). In its South American wintering areas it occupies many wet, grassy habitats, such as coastal marshes, temporary pools in grasslands, and freshwater ponds. Foods at that time of year have not yet been studied, but the birds usually forage in small flocks of 10–50 birds, sometimes with individuals being territorial and defending areas of 0.01–0.05 hectare (Myers and Myers 1979).

Social behavior. In the Barrow area of Alaska the first birds begin arriving early in June, reaching a peak by the end of that month. Territorial flights and pairing begins by the end of the first week in June, or only a few days after initial arrival. Females probably arrive a few days later than males on the average, but sometimes both sexes arrive together. In any case, males are soon well dispersed on territories that range in size from 10 to 15 acres. The most frequent display of territorial males is performed in flights that are only a few feet above the ground and cover distances of 400–500 feet. The flight is slow, with shallow or deep wingbeats, and the bird has a swollen chest, with the streaked "bib" extended below the white underparts. A hooting call, uttered several times per second and usually given 10–20 times in series, is produced during the flight, the hooting notes often being synchronized with the wingbeats. The flight may end with a direct drop to the ground, or with a soaring upward and a gliding descent. The same notes are uttered during a ground display, when the male expands his chest, droops his wings slightly, raises his tail, and follows a female in short runs. This is doubtless a courtship display, but Kistchinski (1974) indicates that a vertical wing-raising display may immediately precede copulation. Tape recordings of the male's flight display made near Barrow in June 1978 by J. P. Myers indicate that the male utters his booming notes at an average rate of about 4 per second, and in a series of from 17 to 37 (9 counts) notes in each song bout. The call has a nearly uniform fundamental frequency of approximately 500 cycles per second (cps), with a very slight rise at the end of each note. No harmonic development is evident in the sonograms. The female's response note is a highly modulated call ranging from about 3,000 to 5,000 cps and lasting about 0.25 second, with a sharp drop in frequency at the end. A synopsis of Mr. Myer's field notes reads as follows: "The male stood alert on the crest of a 40 cm polygon mound, downwind and adjacent to a shallow sedge-filled pond on which the female fed, some 8 m. from the male. As the male stood alert, he groomed his breast feathers with frequent bill strokes, surveying over the pond while not preening. Once every 5 to 10 minutes for at least ½ hr, he flew over the female in flight display. This began with him turning sideways to the wind, taking off, and flying downwind away from the female in a large loop. After traveling approximately 30 m. downwind, he circled and began to approach the female. Ten to 15 m. downwind from her he commenced the hoot, dropping to within 20–50 cm of the ground during the hooting phase of the flight, during which he pumps his breast sac and wings while making the sound. He passed directly over the female during each hoot and as he flew over, she more often than not gave a harsh *churr* note. After finishing the hoot the male changed his flight pattern: instead of the rapid hooting pump he alternately soared and fluttered in a strongly undulating flight. In this mode he turned in a large circle and finally landed back on the original polygon mound." Territories are almost certainly little more than mating stations, since females generally place their nests in locations other than those used heavily by territorial males. Since males have not been observed associating on territories with more than one female at any single time, a short-term pair bond may be present. However, males have been seen with different females during periods separated by intervals adequate for the completion of a clutch, which may indicate a polygynous mating system or possibly males associating with renesting females. Most likely males seek a succession of mates during their period of territorial occupancy, which lasts about a month (Pitelka 1959; Portenko 1968).

Reproductive biology. In the Barrow area, clutches may be started as early as about June 10 and as late as July 8, but most are laid during the last ten days of June. All clutches studied there by Pitelka (1959) consisted of 4 eggs. Females sometimes nest within male territories but often nest outside of them, and on Victoria Island two nests were found about a hundred yards apart, suggesting polygynous mating. Nests there were typically well hidden in grassy hummocks over wet ground or ground that had previously been wet. Some were hidden from above by willows, and the nests were lined with dried willow leaves and grasses or sedges. Males do not visit the nest but often join the female when she leaves the nest, perhaps to protect her from the attentions of other males. The incubation period is probably 21–23 days. Incubation is performed entirely by the female, but males do sometimes approach the nest and may display near it. Likewise, although males do not directly help in rearing the young, they will occa-

sionally participate in brood defense (Parmelee et al. 1967). The fledging period has not been established with certainty, but in the Cambridge Bay area of Victoria Island the fledged birds were seen in late July, 20 days after the earliest date of hatched young, suggesting an approximate 3-week fledging period. By early July most of the males had already left the area, and females evidently begin to leave the breeding grounds shortly after their young have fledged, while unattended juveniles may remain nearly a month longer. The reduction of the summer population is probably adaptive, allowing maximum food availability for the young birds (Pitelka 1959).

Status and relationships. The very broad geographic range of this species and the moderately high breeding densities attained in favorable habitats suggest that the total population of pectoral sandpipers must be substantial. Its evolutionary relationships are of special interest. Pitelka (1959) noted that there are some behavioral similarities with *fuscicollis*, and also some with *Philomachus pugnax*, and he doubted that a close relationship with *acuminatus* exists. However, Larson (1957) regarded *melanotos* and *acuminatus* as a closely related species pair, while Mayr and Short (1970) mentioned *ferruginea*, *acuminatus*, and perhaps *fuscicollis* as possible members of a species group. I have little doubt that *melanotos* and *acuminatus* should be considered at least a species pair and probably a superspecies, and though they are peripheral members of the genus *Calidris* they are probably connected to the more typical "peeps" through *fuscicollis* and *bairdii*. I certainly do not agree with Drury's (1961) suggestion that a separate genus be recognized for *acuminatus*, *melanotos*, and *fuscicollis*, though I believe all three are fairly closely related.

Suggested reading. Pitelka 1959; Parmelee et al. 1967.

White-rumped Sandpiper

Calidris fuscicollis (Vieillot) 1819
(*Erolia fuscicollis* of Peters, 1934)

Other vernacular names. Bonaparte's sandpiper; bécasseau de Bonaparte (French); Weissbürzelstrandläufer (German); correlimos de Bonaparte, correlimos lomiblanco (Spanish).

Subspecies and range. No subspecies recognized. Breeds in North America from northern Alaska

77. Breeding distribution of the white-rumped sandpiper.

streak and chin; the sides of the head are whitish with dusky streaking that extends down the sides of the neck and the foreneck and chest, becoming irregular on the sides and flanks. The rest of the underparts are white. The upperparts are mostly light grayish brown, the feathers black centrally and with cinnamon or light rusty edging, producing streaking on the upper back. The wing coverts are grayish brown, with paler margins and the greater coverts narrowly tipped with white. The flight feathers are dark brownish gray to dusky, becoming lighter inwardly, and the secondaries are margined with dull white. The rump is dusky grayish brown, contrasting with white on the upper tail coverts, while the central tail feathers are dark grayish brown, becoming much paler outwardly and narrowly edged with whitish. The iris is brown, the bill is blackish teminally and brownish or yellowish at the base, and the legs and toes are yellowish brown. *Adults in winter* are much more uniformly brownish gray dorsally, initially with a mixture of browner feathers. The underparts are white, with spotting and streaking of brown on the breast and flanks. *Juveniles* are blackish brown above, the feathers with chestnut, whitish, and buffy fringes, and the wing coverts have extensive buff and whitish fringes. The breast and flanks are buffy to grayish with pale brownish streaking. *First-winter* birds resemble winter adults but have a mixture of brown and gray feathers to midwinter and usually retain some brownish inner medians with buffy or creamy fringes until spring (Prater et al. 1977).

In the hand, this rather small (wing 118–30 mm) *Calidris* differs from others in its white upper tail coverts and its rather straight, short (22–26 mm) bill.

In the field (7 inches), this rather small "peep" closely resembles a Baird sandpiper in size and general color, but it is streaked rather than scaly on the upperparts, and the streaking on the head, neck, and breast is more distinct and extensive. The distinctive white band across the upper tail coverts may be visible only in flight. When flushed the birds utter a squeaky *tzeet* call that is quite different from that of the other "peeps." The birds are rather tame and often forage in water up to their bellies, feeding slowly and submerging the entire head.

NATURAL HISTORY

Habitats and foods. The breeding habitat of this species consists of wet tundra and the edges of ponds or streams, in the general vicinity of coastlines but usually somewhat inland in a local sense. It avoids the

and northern Mackenzie east to northern Keewatin, and on Banks, Melville, Prince of Wales, Bylot, Baffin, and Southampton islands. Winters in eastern South America south to Tierra del Fuego and the Falkland Islands. See map 77.

Measurements and weights. Wing (flattened): males 117–26 mm; females 116–25 mm. Culmen (from base): males 25–38 mm; females 29–33 mm (Blake 1977). Weights (on breeding grounds): males 31–45 g, average of 7 was 39.7 g; females 31.7–51 g, average of 6 was 45.8 g (Glutz et al. 1975). Eggs ca. 34 x 24 mm, estimated weight 10.8 g (Schönwetter 1963).

DESCRIPTION

Breeding adults of both sexes have a cinnamon crown with blackish streaking and a whitish superciliary strictly coastal tundra, and the ideal habitat consists of hummocky, well-vegetated tundra that remains almost permanently wet and often occurs around the edges of ponds and lakes. It thus occupies a breeding habitat similar to that used by the least sandpiper, but it has a more northerly distribution than does that species, and while overlapping broadly with the Baird sandpiper it occurs in much wetter habitats. Population densities are apparently quite sparse, with the highest available estimate being 7 pairs per 80 acres, but more commonly the birds are found in concentrations of from 0.5 to 12 per square mile (Parmelee et al. 1967). The birds forage by rather methodically moving about, picking and probing in moist areas. Their foods have been only very poorly documented, but they probably differ very little from those of other *Calidris* species with bills of similar length. On the wintering areas of South America the birds are usually found foraging on mud flats at low tide, and they also occur at inland lagoons. They usually are found on muddy substrates well away from emergent vegetation. They forage in groups of fewer than 10 birds, and in some areas individual birds defend foraging territories for varying lengths of time (Myers and Myers 1979). On the breeding grounds both adults and young feed by probing deeply into moss and wet vegetation, and adult and larval craneflies as well as beetle larvae have been noted among stomach contents (Drury 1961).

Social behavior. On Victoria and Jenny Lind islands this species arrives in late May or early June, and almost immediately after arrival males take up territories and begin displaying. Females evidently arrive at almost the same time as do males and are ready for immediate reproduction if conditions permit. Ground and aerial displays of the males include overt pursuits of females and attempted copulations on the ground and even in the air. The most typical posture is a "sharp-tailed grouse dance," during which the rump is exposed, the wings are held stiffly outward, and the tail is slightly raised, exposing the under tail coverts. Running about in this posture, the bird utters a low buzzing or growling sound. At least two aerial displays occur. One begins with horizontal flight 15–25 meters above the ground. The bird then begins a series of shallow wingbeats and utters a rattling sound similar to that of a fishing reel or a typewriter carriage, interrupted with a few piglike *ng-oik* sounds while the head is sharply extended and drawn back. At the end of the display the bird glides back to the ground, either silently, uttering the

fishing-reel call, or making *zip-zip* sounds (Drury 1961). Another display is a fast and low horizontal flight over the tundra while uttering the mechanical rattling call or *quo-ick* notes, apparently used to drive a male intruder from the territory. All females that enter the territory are, however, immediately courted in a manner very similar to that of pectoral sandpipers. Nesting by the female is done without regard to the male's territory. The male's territory is therefore apparently mainly a center for sexual advertisement rather than a resource-rich area, and the pair bond, if any exists at all, is of extremely short duration. It probably terminates with the completion of the clutch or soon thereafter. The fall departure of males from the breeding grounds probably occurs shortly after territorial abandonment, resulting in an extremely short period of summer occupancy of the breeding grounds by males. That probably only the female incubates (Sutton 1932) indicates that a polygynous mating system likely occurs in this species. In this aspect of its biology, as well as in the male's swollen-neck display and associated subcutaneous "glands" of unknown function, the species has some distinct similarities to the pectoral sandpiper.

Reproductive biology. Egg-laying begins as early as June 6 on Jenny Lind Island or as late as about June

20 on Bylot Island, but in both areas it probably begins only a few days after arrival. Although females often nest within a male's territory, this is not always the case, and they appear to be indifferent to its limits or to the presence of other nesting females nearby. Nests have been found as close as 13 yards apart and are typically very well concealed in deep depressions in hummocks and lined with willow leaves, mosses, and lichens. Eggs are laid at approximately 24–30 hour intervals, and the egg-laying period in a single population and year covers a span of only about 12 days, indicating a very high degree of reproductive synchronization. Of 47 nests on Jenny Lind Island, at least 46 contained 4 eggs. All the incubation is by the female, and it lasts 21–22 days. During this time the males apparently show no interest in the nest, nor do they participate in brood-rearing. The period between the hatching of the first and last eggs in a clutch is often quite short, as little as an hour, but it also may be as long as 17 hours. Pipping of the eggs may occur as long as 4 days before actual hatching, but usually only about 24 hours elapse between initial pipping and the departure of the brood from the nest. Females usually keep their broods in the vicinity of the nest for the first week or so after hatching, but they may move them as far as a mile and a half in 12 days. The fledging period is 16–17 days, and the parent-offspring bond dissolves shortly after fledging of the young (Parmelee et al. 1967; Drury 1961).

Status and relationships. This species occurs widely across the high arctic of Canada but apparently at rather low densities everywhere. So far as can be determined, it offers no special conservation problems. Its relationships are somewhat controversial. Drury (1961) urged that this species, together with *melanotos* and *acuminatus*, be removed from *Calidris* and included in the revived genus *Heteropygia*, a conclusion that Holmes and Pitelka (1962, 1968) criticized. The latter authors argued that *fuscicollis* resembles *Calidris* in its important characters, while *melanotos* is quite distinctive. Parmelee et al. (1967) reviewed the controversy and, while agreeing with Drury (1961) that *melanotos* and *fuscicollis* are probably closely related, considered it premature to begin splitting the genus *Calidris* in the way Drury suggested. I believe that some of the behavioral similarities of *melanotos* and *fuscicollis* are the result of their similar social structuring, which is primarily of ecological rather than taxonomic significance. It is interesting, however, that a pectoral sac exists in males of both forms, though it is less developed in *fuscicollis*. This sac presumably accounts for the piglike call of *fuscicollis* and the "booming" of *melanotos* by acting as a resonating chamber, and in the latter species it also adds a striking visual display component. This feature may also be of independent origin, but at least it suggests possible affinities. Like Larson (1957), I believe that *bairdii* and *fuscicollis* are probably very close relatives and may constitute a species pair.

Suggested reading. Drury 1961; Parmelee et al. 1967.

Baird Sandpiper

Calidris bairdii (Coues) 1861
(*Erolia bairdii* of Peters, 1934)

Other vernacular names. None in general English use; bécasseau de Baird (French); Bairdstrandläufer (German); correlimos de Baird (Spanish).

78. Breeding distribution of the Baird sandpiper.

Subspecies and range. No subspecies recognized. Breeds in eastern Siberia on the tip of the Chukotski Peninsula, and in North America from northern Alaska (south to Cape Romanzof) eastward to the Melville Peninsula, and on most of the arctic islands of Canada north to Ellesmere Island and adjacent Greenland. Winters in central and southern South America. See map 78.

Measurements and weights. Wing (chord): males 114–22 mm; females 119–26 mm. Culmen (to feathering): males 20–23 mm; females 21–24 mm. Weights (on breeding grounds): males 32–48 g, average of 30 was 39.0; females 34–45 g, average of 12 was 39.0 g (Irving 1960). Eggs ca. 33 x 24 mm, estimated weight 9.6 g (Schönwetter 1963).

DESCRIPTION

Breeding adults of both sexes have a white superciliary stripe, bordered above by a grayish brown crown streaked with sooty black, while the sides of the head are whitish with dusky streaking and the chin and throat are clear white. There is a buffy chest band, narrowly streaked with dusky, and the rest of the underparts are white. The scapulars and interscapulars are blackish centrally, with broad whitish and bright buffy margins, and the wing coverts are light grayish brown with paler margins, especially the greater coverts. The flight feathers are grayish brown, with the inner secondaries edged with whitish and with the primaries edged with grayish and the shafts whitish near their tips. The rump and upper tail coverts are sooty grayish brown, becoming whitish laterally, while the middle tail feathers are also sooty grayish brown and the outer ones are paler and margined with whitish. The iris is brown, the bill is black, and the legs and toes are blackish to greenish black. *Adults in winter* are grayish brown on the upperparts, and the sides of the neck and breast are tinged with dull brownish buffy but are not streaked. The wing coverts have very pale buff fringes. *Juveniles* have a scaly upperpart pattern formed by extensive whitish buff fringes and a buffy breast with indistinct brown streaking. *First-winter* birds retain some bright buff edging on the inner median coverts until midwinter (Prater et al. 1977).

In the hand, this species is separated from the other small (wing 114–26 mm) "peeps" by its distinctly buffy plumage, including a buffy breastband, and the relatively short bill and toes in proportion to the wings (exposed culmen up to 24 mm, middle toe to 19 mm).

27. Winter-plumage heads and half-tails of Baird (*above*) and white-rumped (*below*) sandpipers.

In the field (7 inches), this sandpiper is slightly larger than the other small "peeps" but has a breastband like the least sandpiper, has a face that is paler than its body, and has distinctive pale buffy or grayish edging on its upperpart feathers, producing a scaly (juvenile) or blotchy (adult) effect. The birds mix with other species but are usually relatively tame. When flushed they utter a guttural *quoit* or *kreep* note. They are often found in grassy marshes but sometimes frequent muddy shorelines, where they seem to pick up food rather than probe for it.

NATURAL HISTORY

Habitats and foods. The breeding habitat of this species consists of dry tundra, both in coastal situations and on inland mountains, though in Alaska it is most common near the coast. Habitats such as limestone flats, dry upland plateaus, arid hillsides, and rocky slopes are all utilized, especially where these over-look wet tundra, coastlines, or areas of fresh water. A five-year study in northern Alaska indicated summer densities of from 0.0 to 6.0 breeding birds per square kilometer (Bergman et al. 1977). Its wintering habitats are poorly known, but it occurs primarily from Peru to northern Tierra del Fuego and winters abundantly in the north Chilean deserts where water is present (Jehl 1979). In wintering areas of coastal Argentina the species occurs on interior ponds that lack vegetation or on saline ponds. It feeds above the waterline on drying mud and usually occurs in flocks of 20–30 birds. Some individuals defend territories at this time (Myers and Myers 1979). In the Barrow area of northern Alaska it competes with several other species of *Calidris,* and in early June it and *pusilla* both forage for midge larvae along the edges of streams and lakes. By mid-June *bairdii* begins to forage for cranefly larvae, and still later it feeds on beetles, spiders, and adult dipterans, which are fairly abundant on its upland tundra habitats. Unlike the other three *Calidris* species in the area, the young of *bairdii* remain on the upland tundra with the adults, feeding largely on the same food items, until both adults and young leave the area in early August (Holmes and Pitelka 1968). Foraging in this species is done primarily by picking items from dry surfaces, and little if any probing seems to be performed (Drury 1961).

Social behavior. At Cambridge Bay, Victoria Island, this sandpiper begins arriving early in June and immediately initiates flight display behavior. The birds soon begin to settle on drier slopes and ridges, which are among the first areas to become snow-free and thus provide early nesting opportunities. A peak in display activity occurs within a week after arrival, at which time the birds can be seen displaying all day long and late into the evening. Early territories are small and indefinable, but after dispersal the birds become distributed at about 20–25 pairs per 100 acres of suitable habitat. Displaying males rise at a steep angle, usually to 30–60 feet, and begin uttering a froglike trilled song near the peak of the ascent that continues during a downward glide back to earth with the wings held high above the level of the back. Sometimes the male will remain aloft for 2–3 minutes, circling and trilling. Shorter flight notes sounding like *cree* or *dree* are also sometimes uttered and closely resemble those of the stilt sandpiper (Par-

melee et al. 1967). Details of pair-forming behavior are not yet known, but it probably follows the pattern described for the western sandpiper and other *Calidris* species. Although the species is known to have a seasonally monogamous mating system, it is somewhat anomalous in that it shows opportunistic elements in its lack of site attachment and absence of mate-faithfulness between successive years.

Reproductive biology. Nesting at Victoria Island occurs on relatively dry ridges, including two types of terrain. One is stony, barren ground with little plant growth. The other, more common situation is dry slopes with from little to fairly dense plant growth, often near wet tundra or the bank of a pond or lake. Most nests are simply shallow depressions in bare ground or amid a mat of vegetation, with a lining of lichens and bits of other plants. Compared with those of most *Calidris* species, the nests are highly exposed, like those of plovers or turnstones. The egg-laying period lasts at least 2 weeks on Victoria Island, and eggs are apparently laid at daily intervals. Of 30 complete clutches seen at Cambridge Bay, all contained 4 eggs. Both sexes incubate, and there is a 19.5–21 day incubation period (Norton 1972; Drury 1961). The young birds leave the nest within 24 hours of hatching, and for the first 5–7 days they are accompanied by both parents. The parent-offspring bond sometimes dissolves before the chicks fledge at about 20 days of age, but some males apparently remain to attend fledged young. However, most adults apparently leave the area at about the time fledging occurs (Parmelee et al. 1967).

Status and relationships. The widespread breeding range of this species indicates a large total population, especially considering the moderately high densities reported for Victoria Island on the dry tundra habitats that are so widespread across the high arctic. The nearest relative of *bairdii* is probably *fuscicollis* (Larson 1957), the latter having a somewhat more subarctic breeding distribution and being more mesic in its ecological distribution. Drury (1961), however, suggested that *fuscicollis* should be separated (along with *melanotos* and *acuminatus*) in the genus *Hetero-pygia,* while *bairdii* should be retained in the genus *Calidris.* This was apparently based partly on the basis of an enlarged throat or pectoral sac in males of *fuscicollis,* which, though of behavioral interest, is probably not a character of generic significance. Holmes and Pitelka (1962) also criticized Drury's conclusion. Although the two species may not be sufficiently allopatric to be considered a superspecies, I believe that the Baird and white-rumped sandpipers possibly represent a species pair. It is quite possible that these two, especially the latter, provide a link between the pectoral and sharp-tailed sandpipers and the more typical "peeps."

Suggested reading. Drury 1961; Parmelee et al. 1967.

Temminck Stint

Calidris temminckii (Leisler) 1812
(*Erolia temminckii* of Peters, 1934)

Other vernacular names. None in general English use; bécasseau de Temminck (French); Temminck-strandläufer (German); correlimos de Temminck (Spanish).

Subspecies and range. No subspecies recognized. Breeds in Eurasia from northern Scandinavia eastward through the tundras of Russia and Siberia to the Chukotski Peninsula and Anadyrland. Also breeds on Kolguev and the Vaigach islands. Win-

ters from the Mediterranean Basin south to central Africa, Arabia, India, the Indo-Chinese countries, Malaya, and Borneo. See map 79.

Measurements and weights. Wing: males 91–101 mm; females 95–103 mm. Culmen: males 15–19 mm; females 15–19 mm. Weights: 31 males 22–26 g, average 24.3 g; 28 females 26–29.5 g, average 27.8 g (Hilden 1975). Eggs ca. 28 x 20 mm, estimated weight 5.8 g (Schönwetter 1963).

DESCRIPTION

Breeding adults of both sexes have a sepia crown and nape, a whitish stripe streaked with dusky from the base of the bill to the eye, the sides of the head light buff streaked with dusky, the chin and throat white, the breast with a pale ashy brown tinge, and the rest of the underparts white. The upperparts are generally blackish brown, the feathers having lighter edges, and the sides of the rump and lateral upper tail coverts are white. The central tail feathers are dark sepia, the outer ones becoming lighter, and the outer three pairs are almost entirely white. The wing co-

79. Breeding distribution of the Temminck stint.

28. Winter-plumage heads and half-tails of Temminck (*above*) and little (*below*) stints.

In the hand, the small size (wings under 105 mm) separates this from most "peeps," and the outer four to six tail feathers are nearly white and about equal in length, separating it from the little stint and the rufous-necked stint.

In the field (5¾ inches), the tiny size and yellow to greenish legs serve to identify this "peep," and in winter it is also more uniformly grayish above than the little stint, with which it often mixes. In flight, the white outer tail feathers may be visible, and the usual flight call is a distinctive short, high-pitched trill *tir-it-tit-tit,* sounding almost like a cricket. The flight is erratic and twisting, somewhat batlike.

NATURAL HISTORY

Habitats and foods. This is a subarctic species, with the center of its breeding distribution occurring along the northern edge of the boreal forest, only rarely penetrating into tundra. Nonetheless, it breeds in open habitats, preferring areas with a combination of grasses and sedges or *Empetrum* and scrub willow or birch thickets with gravelly or sandy stretches. Both dry and wet habitats are used, but habitats providing elevated locations such as boulders or even buildings are highly preferred because of their use as song perches. Foraging is done by running along the surface and picking up materials, rather than by probing, and during the breeding season the adults and young feed mostly on insects and their larvae, particularly craneflies and midges. Outside the breeding season they apparently concentrate on mollusks, small crustaceans such as sand fleas, and worms, but details of their foods are still unavailable. In the winter they tend to occupy relatively inland habitats, concentrating around freshwater pools and marshes but also occurring around tidal mud flats and on coastal lagoons. They are among the most abundant of small Eurasian sandpipers, occurring in favorable habitats in densities of from 1 to 2 breeding pairs per hectare (40–50 pairs per 100 acres) (Uspenski 1969).

Social behavior. Although it is known that females mature and initially nest when a year old, the average age of initial breeding in males is 1.8 years, with a range from 1 to 3 years. Quite probably some of these males are capable of breeding earlier, and they often do establish territories but fail to attract mates. The mean longevity of breeding adults is about 7 years, and a few individuals live to more than 10 years, so it is apparent that a good part of the breeding population is relatively experienced. Pair

verts are blackish brown to sepia, edged with white, especially on the greater coverts. The primaries are sepia, the inner ones margined with white, and the secondaries are sepia to olive brown, tipped with white. The iris is brown, the bill is brownish black, and the legs and feet are variably described as greenish, brownish, or yellowish, rarely blackish. *Adults in winter* are very uniformly dark grayish brown above, with white underparts and the breast strongly washed with dark grayish brown, sometimes forming a complete breastband. *Juveniles* are dark grayish brown above, with buffy fringes and dark subterminal bands and with the coverts fringed with buff or chestnut buff. The breast is washed with brownish. *First-winter* birds retain some inner median coverts with worn pale buff to whitish fringes (Prater et al. 1977).

bonds are rather loose, lasting only a week or so, and there is little fidelity to previous mates or nest sites. About a third of the females shift their breeding area from year to year, and apparently most females accept new mates to fertilize their second clutch of each breeding season (Hilden 1975, 1978a). Upon arriving on their breeding grounds, the males quickly establish territories, which they advertise by singing and display flights. Typically the displaying male flies in a wide circle, with the tail spread and the wings vibrating rapidly. He may also hover in the air or quickly gain and lose altitude. Throughout the aerial display he utters a dunlinlike tinkling song that may continue as long as 2 minutes. The display often ends with the bird's gliding back to earth with his wings held high above the back. This wing-raising display also sometimes occurs on the ground, especially while the male is perched on a boulder or other elevated site, and at times it becomes a wing-flicking, with the pale axillaries and undersides of the wings alternatively flashed open and hidden from view (Bannerman 1961). Vertical wing-raising also occurs in aggressive encounters, as is typical of many *Calidris* species. Besides forming definite, though short-lived, pair bonds, some males also engage in simultaneous polygyny, polyandry, or promiscuous mating. In general, the male's territory serves to repel other males and attract females, which are immediately courted. Breeding females are fairly sedentary, and they also are somewhat intolerant of other females close by; thus nesting females tend to be well spaced and most matings are monogamous. Not until the female has completed her first clutch and turned it over to the resident male for incubation does she leave her first mate and pair with a new one. The second nest is usually at least 200 meters away from the first one (rarely more than 900 meters and once as close as 4 meters), with about 25 percent of the second nestings between 100 and 200 meters away. Frequently males do not begin the incubation of the clutch immediately, but rather continue to display for some days, thereby probably increasing their probability of fertilizing another set of eggs (Hilden 1975).

Reproductive biology. Nests are often at the bases of small willows, junipers, or other shrubs, and their tiny size and excellent concealment make them hard to locate. They are usually lined with grasses or the dead leaves of birches or willows, and the full clutch is almost always 4 eggs. Of 182 clutches noted by Hilden (1978a), 86.8 percent had 4 eggs, and the range

was 2–5. Second clutches also typically have the full number of eggs, and a few known third clutches have likewise contained 4 eggs. However, these birds apparently only rarely lay replacement clutches after the incubated one is lost. Nests incubated by males are apparently somewhat less successful than those incubated by females, perhaps because of the delay in the start of incubation that is typical in that situation. The incubation period is from 19 to 22 days, averaging 20.8 days, the young begin to fly at 15–16 days, and by the time they are 17–18 days old they can fly more than 100 meters. The parents abandon their young when the latter are from 14 to 21 days old. Although the double clutch and associated bigamous mating systems were first discovered in Finland by Hilden, a similar report of double clutching has been made by Kokhanov (1973), based on studies in Kandalaksha Bay on the White Sea. However, Kokhanov believed that the pair bond persisted between successive nestings, and he found that most nests of such a pair are only 5–45 meters apart. He also noted that nest sites were frequently used for 2–4 years in a row and that successive eggs were laid every day, with the entire period of egg-laying lasting only about 11–15 days in most years.

Status and relationships. This is one of the most abundant shorebirds of Eurasia, and Uspenski (1969) has estimated that its population probably numbers more than 10 million individuals. Its nearest relatives are probably the North American species *pusilla* and *mauri* (Larson 1957; Lowe 1931b).

Suggested reading. Hilden 1975; Southern and Lewis 1938.

Semipalmated Sandpiper

Calidris pusilla (L.) 1766
(*Ereunetes pusillus* of Peters, 1934)

Other vernacular names. None in general English use; Le bécasseau semipalmé (French); Sandstrandläufer (German); correlimos semipalmeado (Spanish).

Subspecies and range. No subspecies recognized. Breeds in North America from the arctic coast of Alaska eastward through Mackenzie, on Victoria, Banks, King William, Baffin, and Southampton islands, in Keewatin, the coastal portions of Man-

80. Breeding distribution of the semipalmated sandpiper.

loral area that below is broken into streaks that extend back into the ear coverts as grayish brown markings, becoming whitish on the chin and throat. There is a chest band formed by blackish streaks on a gray background, which terminates abruptly where the white underparts begin. The upperparts are mostly brownish gray, with the central parts of the feathers blackish and the pale edges often tinged with buffy cinnamon, especially on the upper back. The rump, middle upper tail coverts, and central tail feathers are dusky, the lateral upper tail coverts are white, and the outer tail feathers are pale brownish gray. The wing coverts are brownish gray, darker centrally and with lighter margins, but white is limited to narrow margins of the greater coverts. The entire shaft of the outer primary is also white, as are parts of the others. The iris is brown, the bill is black, becoming yellowish at the base, and the legs and feet are dull greenish yellow to black. *Adults in winter* are grayish brown above, with a white superciliary stripe and underparts, and the chest is faintly streaked. *Juveniles* lack breast streaking, and the dorsal feathers are edged with white and pale buff, producing a scaly appearance. The legs are olive colored. *First-winter* birds closely resemble adults but may retain some buff-edged inner median coverts until spring (Prater et al. 1977).

In the hand, the presence of partial webbing between the bases of all of the front toes separates this species from other *Calidris* forms except *mauri,* and the slightly shorter and undrooped bill tip usually distinguishes it from that species. Additional information on separating semipalmated and western sandpipers has been summarized by Stevenson (1975), who found that the western sandpiper's greater bill depth at its base is a possible field mark, but that the best distinction between the two species is the bill-length/wing-length ratio (maximum of 0.183 for semipalmated, minimum of 0.188 for western). Phillips (1975) also provided some bill-length data for these two species.

In the field (6 inches), this species is somewhat grayer than the other "peeps," and both bill and legs appear black at any distance. The least sandpiper is smaller and browner, with yellow legs; the Baird is larger and is more scaly above; and the western is browner, with a longer and slightly drooped bill. In flight, this species often utters a hoarse, shrill *cherk* or an abrupt *kr-i-ip* when flushed. In general its calls are rather medium or low-pitched, and include "r" sounds (Wallace 1974). Its call is much lower-pitched

itoba, Ontario, northern Quebec, and coastal Labrador. Winters on the Atlantic and Pacific coasts from Guatemala and the West Indies south to southern Peru and central Argentina. See map 80.

Measurements and weights. Wing (chord): males 88–99 mm; females 92–102 mm. Culmen (to feathering): males 17–20 mm, females 18–22 mm. Weights (on breeding grounds): 27 males 20–29 g, average 24 g; 14 females 21–27 g, average 26 g (Irving 1960). The average of 102 fall migrants of both sexes was 28.1 g (Murray and Jehl 1964). Eggs ca. 30 x 21 mm, estimated weight 6.9 g (Schönwetter 1963).

DESCRIPTION

Breeding adults of both sexes have a brownish gray crown, a streaked white superciliary stripe, a dusky

29. Winter-plumage heads and feet of least (*top*), semipalmated (*middle*), and western (*bottom*) sandpipers.

than the call of the western and sounds like a "distant female great-tailed grackle" (Phillips 1975).

NATURAL HISTORY

Habitats and foods. The breeding habitat of this species consists of coastal or low inland tundra, especially the well-vegetated shorelines of rivers, pools, and lakes. It overlaps with the pectoral sandpiper in its breeding habitats, and, though abundant on areas of hummocky, wet tundra associated with marshy ponds, it sometimes also breeds on drier and more exposed peaty areas. When it overlaps with least sandpipers it is found on more coastal areas of

flat tundra, while the least sandpipers is most abundant in sedge meadows near timberline (Jehl and Smith 1970). Rocky ridges and barren sandy flats are also avoided (Parmelee et al. 1967). In the winter, the species is essentially limited to coastal areas, and winter foraging behavior patterns are very similar to those in summer. Like other *Calidris* species, the birds tend to be "generalist" foragers, and they typically forage by performing single, halting pecks (Baker and Baker 1973). In Alaska, where they breed with at least three other *Calidris* species (Baird, dunlin, and pectoral), there is generally a wide overlap in the diets of all these species, but the similar-sized *bairdii* breeds on higher and more barren habitats than does *pusilla*. It has an overlapping diet and apparently competes strongly for food with the larger *alpina* and *melanotos*, which may account for an early departure of adult *pusilla* from the breeding areas (Holmes and Pitelka 1968). In this area the summer density of breeding birds ranged from 3.2 to 47.0 per square kilometer over a five-year period (Bergman et al. 1977). On the wintering areas of coastal Peru it overlaps and certainly competes with *mauri* on coastal mud flats but probes less and forages mainly on wet and dry mud, while the longer-billed *mauri* forages mostly by probing in wet mud and in water (Ashmole 1970).

Social behavior. Little is known of the social behavior of these birds on their wintering grounds, or even of their patterns of migration, but it is now becoming clear that many birds use an elliptical migration pattern involving a spring northward migration through central North America followed by a southward fall migration via coastal areas (Harrington and Morrison 1979). Pair bonds are not formed until arrival at the breeding grounds, which may occur when about 90 percent of the area is still snow-covered. First arrivals are in small flocks of unknown sex, presumably males, since territorial advertisement begins almost immediately after the first spring arrivals appear. Many females probably return to their previous year's territories, especially if the previous breeding was successful, and males spend most of their time within such territories, gradually intensifying their display behavior and aggressive encounters (Ashkenazie and Safriel 1979). Display consists mostly of aerial flights often only 25–50 feet above the tundra and not covering much area. Often the bird hovers in midair, trilling during much of the flight. On one occasion 3 birds were seen displaying about 200 feet above the ground, flying this way and that for fully a minute, and monotonously uttering a single-syllable note (Parmelee et al. 1967). On Victoria Island, display continues into early July, or considerable later than that of other scolopacids in the area. Studies near Barrow, Alaska, indicate that pair-formation may occur 3–6 days after territorial establishment. It probably involves the females' being attracted to advertised territories and thereafter being accepted or chased off by the resident male. Another 4–6 days elapse between pair-formation and the laying of the first egg; the frequency of territorial display flights is reduced then, and the male typically follows the female, performing much the same activities as she does (Ashkenazie and Safriel 1979).

Reproductive biology. Nests of this species are usually placed in grassy hummocks, often over water on damp ground and sometimes hidden in willows. Eggs are laid one per day over a 4-day period; incubation behavior typically begins before the clutch is complete and usually become continuous a few hours before the laying of the last egg (Ashkenazie and Safriel 1979; Norton 1972). Of 36 nests with complete clutches found on Victoria and Jenny Lind islands, all but one (with 3 eggs) contained 4 eggs. The egg-laying period there lasts nearly a month (Parmelee et al. 1967). Both sexes incubate at about equal intensities throughout the incubation period, which lasts 20 days. The period between the hatching of the first and last eggs averages about 24 hours. Both sexes initially tend the young, though females may desert the family as early as 2–6 days after hatching. This is evidently brought about by the male's evicting her from the feeding area. After the female leaves the family there is a general exodus out of the original territory, sometimes resulting in a movement as distant as 2–3 kilometers. The fledging period is about 16 days, with the rate of brooding gradually diminishing and with the male sometimes deserting the brood a few days before they fledge. Soon after fledging, flocks begin to assemble that may consist of young birds or of mixtures of adults and young (Ashkenazie and Safriel 1979).

Status and relationships. The population status of the semipalmated sandpiper can at present only be guessed at, but at least along the Atlantic coast it is probably the commonest of the "peeps" during fall migration. The ease of confusing it with *mauri* to the west makes its status there difficult to judge. There can be little doubt that the nearest relative of *pusilla* is *mauri*, and in some Russian literature (e.g., Kuzy-

akin 1959) the two forms have at times been regarded as only subspecifically distinct. However, their breeding ranges clearly overlap in Alaska, and there seems no basis for considering them conspecific. Why only these two species of *Calidris* should have semipalmated toes remains an unsettled issue, as they are not known to swim more than the other *Calidris* forms. However, both species often feed on wet, sandy beaches and mud flats, and this may be the significant feature they have in common.

Suggested reading. Ashkenazie and Safriel 1979.

Western Sandpiper

Calidris mauri (Cabanis) 1856 (1857)
(*Ereunetes mauri* of Peters, 1934)

Other vernacular names. None in general English use; bécasseau d'Alaska (French); Bergstrandläufer (German); correlimos occidental (Spanish).

Subspecies and range. No subspecies recognized. Breeds in Alaska from Nunivak Island and the Kashunuk River north to the Seward Peninsula and less frequently to Point Barrow and Camden Bay. Winters from the coast of California and the Gulf of Mexico south to central South America and along the southern Atlantic and Caribbean coasts to northern South America. See map 81.

Measurements and weights. Wing (chord): males 91–99 mm; females 90–100 mm. Culmen (to feathering): males 20–24 mm; females 23–28 mm. Weights: 15 males 21–38 g, average 25.6 g; 11 females 24–42 g, average 31.6 g (various sources). Females average several grams heavier than males during the breeding season (Holmes 1972). Eggs ca. 31 x 22 mm, estimated weight 7.5 g (Schönwetter 1963).

DESCRIPTION

Breeding adults of both sexes have a superciliary stripe of white streaked with grayish, bordered above by a rusty cinnamon crown, and light rusty to cinnamon on the sides of the head. The rest of the head is whitish with narrow streaks of dusky. The foreneck and sides of the neck are more heavily streaked with the same, but the rest of the underparts are white, with V-shaped marks of dusky on the chest and sides. The upperparts are generally rusty

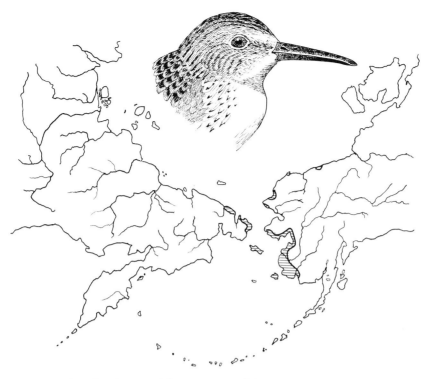

81. Breeding distribution of the western sandpiper.

cinnamon, the feathers with blackish centers, while the coverts are grayish brown or brownish gray with paler margins, and the greater coverts are tipped with white. The rump, central upper tail coverts, and central tail feathers are also grayish brown, with the lateral tail coverts white and the outer tail feathers pale brownish gray with narrow white margins. The flight feathers and primary coverts are dark grayish brown, the primaries with white shafts. The iris is brown, the bill is black, and the legs and toes are blackish (Ridgway 1919). *Adults in winter* are nearly plain grayish brown to brownish gray above, with a white superciliary region and underparts, the chest faintly streaked with dusky or grayish and almost inseparable from the corresponding plumage of *pusilla*, but slightly browner. *Females* are identical in plumage to males but have longer bills (exposed culmen usually 24.2 mm or less in males and 24.8 mm or more in females [Page and Fearis 1971]). *Juveniles* have prominent bright chestnut and whitish edges on the feathers of the upperparts and bright buff-edged wing coverts. The breast is washed with pale buffy and is finely streaked. *First-winter* birds resemble adults but retain some buff-tipped inner median coverts (Prater et al. 1977).

In the hand, the combination of partial webbing between all of the front toes and a longer (a total culmen of at least 22 mm in males and 24 mm in females) bill than that of *pusilla*, which bends downward toward the tip, should identify most individuals. Quellet et al. (1973) provide additional criteria for identifying these two species, as does Stevenson (1975).

In the field (6½ inches), this small "peep" closely resembles the least sandpiper, from which it can be told by its black legs, heavier bill, and lack of a dark wash on the breast. In breeding plumage it is rustier than the semipalmated sandpiper, but in winter the western sandpiper's longer bill, somewhat larger size, and at least some rusty back feathers help separate these two species. In flight, westerns utter a *cheet* or *cheep* note, more squeaky than that of the semipalmated and not so prolonged as the least's.

NATURAL HISTORY

Habitats and foods. Compared with its close relative the semipalmated sandpiper, this species is more subarctic in distribution and occupies drier and more upland tundra for breeding, where islands or ridges of heath tundra are close to marshy areas. The heath tundra is the primary nesting habitat of the western sandpiper, and both the marshy and heath areas are used for foraging. In the Hooper Bay area of Alaska the only other *Calidris* species using the same general habitat is the dunlin, but this species occurs only in the lower marshy habitats. There is a regular pattern of dispersal over suitable habitats, with the birds maintaining small territories that permit remarkably high population densities of from 3 to 7 pairs per hectare. Although the birds obtain some of their food within their small territories, much foraging is done in communal foraging areas near shorelines. Apparently the abundant food supply and the absence of foraging competition associated with the longer summer season in subarctic Alaska allow for these dense nesting concentrations and do not favor an early departure of males from the nesting grounds as is typical of several *Calidris* species farther north (Holmes 1971). During the breeding season, adults feed mostly on larval and pupal insects, especially flies of the families Chironomidae and Muscidae, as well as on craneflies, beetles, and various other invertebrates. Before fledging, young birds feed mostly on surface-active insects such as adult and larval flies and beetles. After fledging the birds begin a diet very similar to that of adults (Holmes 1972). On their winter quarters in Peru the birds forage coastally on tidal mud flats, primarily probing in relatively wet situations (Ashmole 1970). On wintering areas in California they coexist with dunlins, eating invertebrates such as amphipods and ostracods but competing relatively little with dunlins because of their shorter bills that are less suited to deep probing (Recher 1966).

Social behavior. By mid-May western sandpipers begin to arrive in small flocks on their breeding areas of western Alaska, and within a few days aggressive behavior among the males begins. This results in male dispersal to territories, and there is a strong tendency for males to return to the same territories they occupied the previous year. Females show a similar tendency to return to the same area, and they often nest in nearly the same sites they used the previous year. A strong pair bond is formed in this species, with no evidence of polygyny or polyandry. Additionally, the pair bond persists through the entire breeding season, with the male remaining to participate in rearing the young. This is apparently a result of the increased protection thus given to the young, in the face of rather intense predation pressure. Once the male has established its tiny territory (average size 0.2–0.3 hectare), it begins to advertise

actively by singing and calling. The advertising song is a series of ascending notes, followed by a buzzing trill. It is usually uttered during a display flight, which may be either a slow patrolling flight from 2 to 30 meters above the territory or a low, rapid flight followed by a quick stall and a steep gliding descent. Females begin arriving a few days after the males and tend to return to previously occupied areas, which may promote mating with the male of the previous year. After settling on a male's territory, the female becomes passive and feeds almost continuously, ignoring the male's threats. Soon the male begins to perform courtship displays toward her, including a stance in which the tail is vertically cocked, a nest-scraping display, and neck-preening. Copulation is typically preceded by the tail-up courtship stance and a courtship trill. If the female is receptive, copulation follows almost immediately, and afterward both birds perform neck-preening behavior (Holmes 1973).

Reproductive biology. Once pairing has occurred, it is only a matter of a few days before the nest site is chosen and the first egg is laid. Nests are generally placed in rather dense cover, usually under dwarf birches where they are very well concealed. Eggs are laid at approximately 24–25 hour intervals at various times of the day, usually beginning in late May. However, some females continue to lay until mid-June, suggesting that renesting after loss of the initial clutch may be fairly common in this subarctic environment. However, there is no indication that second clutches are produced after successful hatching of the first, or that the female turns over her first clutch to the male to incubate alone while she begins a second clutch. Of 215 clutches observed near Hooper Bay, all but 13 had 4 eggs, and the remainder had 2 or 3. Incubation begins with the laying of the last egg and is performed by both sexes. It lasts from 20.5 to 22 days, averaging 21 days. The young hatch over approximately a 24-hour period and leave the nest shortly thereafter. They are very mobile and soon leave the limits of the nesting territory. Growth is rapid, and fledging probably occurs about 19 days after hatching, by which time the adult weight is attained. Hatching success apparently is high and perhaps is related to the protection provided by the well-hidden nest sites these birds choose (Holmes 1972).

Status and relationships. Although this species has a very small breeding range, its population density is surprisingly high, and in winter it occurs in large numbers along the coastlines of California, Panama, and Colombia. On the Atlantic side its migrational status is extremely uncertain because of confusion with the semipalmated sandpiper (Phillips 1975; Quellet et al. 1977). The relationships of the western sandpiper are obviously closest to *pusilla*. Larson (1957) regards the two as constituting a species pair and as part of a larger species group including *C. temminckii*. Mayr and Short (1970) regard *mauri* and *pusilla* as a species group. I believe that the two might best be considered a superspecies and that *temminckii* is perhaps their nearest Old World relative.

Suggested reading. Holmes 1971, 1972, 1973.

Spoon-billed Sandpiper

Calidris pygmeus (L.) 1758
(*Eurynorhynchus pygmeus* of Peters, 1934)

Other vernacular names. None in general English use; bécasseau spatule (French); Löffelstrandläufer (German).

Subspecies and range. No subspecies recognized. Breeds on the coast of northeastern Siberia, west to Ukouge Lagoon on the Chukotski Peninsula, and south to Litke Strait (Karaginski Bay). Winters in southeastern China to Hainan and the Indo-Chinese countries. See map 82.

Measurements and weights. Wing: males 95–102 mm; females 95–104 mm. Culmen (from feathering): males 19–22 mm; females 21–24 mm. Males average about 29.5 g (Portenko 1957). One female weighed 34 g (*Auk* 96:189). Eggs ca. 30 x 22 mm, estimated weight 8.0 g.

DESCRIPTION

Breeding adults of both sexes have the entire head cinnamon rufous to hazel, with the crown broadly streaked with black, and dark streaking from the lores through the upper ear coverts, becoming whitish on the forehead and chin. The rufous fades to rufous whitish on the chest, which is also streaked with dusky, and the underparts are white with scattered dusky markings on the sides and flanks. The upperparts are mostly of feathers that are blackish centrally and edged with cinnamon rufous to dull buffy whitish. The wing coverts are brownish gray with paler margins, and the greater coverts are tipped with white. The primary coverts are dusky, with white tips proximally, the primaries are grayish brown becoming darker outwardly, and their shafts are mostly

82. Breeding distribution of the spoon-billed sandpiper.

white. The rump, middle upper tail coverts, and central tail feathers are brownish gray to dusky, with the lateral upper tail coverts white and the outer tail feathers pale brownish gray. The iris is brown, the bill is black, and the legs and toes are blackish. *Females* are slightly larger than males and sometimes are slightly less rufous around the head (Ridgway 1919). *Adults in winter* are brownish gray above, with a white forehead and superciliary streak, the mantle is margined with whitish, the wing coverts are grayish with white margins, and the underparts are white. *Juveniles* have brownish black upperparts with white and buff edges, brown wing coverts that are fringed with pale buff to orange buff, and white underparts, with a complete buffy breastband or the sides of the breast washed with buff. *First-winter* birds resemble adults but retain orange buff margined inner median coverts (Prater et al. 1977).

In the hand, the distinctive spoon-shaped bill is diagnostic at all ages.

In the field (6½ inches), the distinctive bill is often not obvious in side view, and the birds are otherwise remarkably similar to the rufous-necked sandpiper in breeding plumage and also resemble that species as well as the least stint in winter plumage. The call is a high-pitched, trilling twitter. The birds are said to feed in a typical "peep" manner by probing or dipping the bill rather than by sifting organisms from the surface in shoveler fashion.

NATURAL HISTORY

Habitats and foods. Little has been written of this species on its wintering areas, where it apparently occupies mud flats and other areas similar to those used by small *Calidris* species. In the breeding season the birds prefer feeding on grassy banks along bodies of fresh water, as well as sandy-banked lagoons. Stream-deposited gravel strips that alternate with marshy sections amid dry and gravelly tundra are apparently also favored. In spite of their highly specialized bill structure, there is still no good evidence that the bill is used primarily as a scoop or sieve for obtaining surface-dwelling organisms. Reports by

Dixon (1918) and Portenko (1972) both suggest that the birds feed primarily by dipping the bill in the water almost at a right angle. However, they do turn the head and neck from side to side while foraging in a manner apparently like avocets, "gulping" or "chattering," and thus the increased surface area of the bill is probably a distinct foraging advantage over a normal *Calidris* type of bill. The foods are known to include a variety of terrestrial insects (ground beetles, hymenopterans, dipterans), small seeds, larvae of beetles, and small aquatic amphipods. Bits of lemming bones have also been found, apparently the result of gleaning for calcium salts. Portenko has suggested that the enlarged bill provides an advantage in hunting insects near the ground, and additionally that the numerous nerve endings may help the bird find larvae in semiliquid soils.

Social behavior. These birds feed alone or in groups, sometimes in association with such species as rufous-necked sandpipers, with which they are very easily confused. Even the courtship displays of the two species seem to be very similar, judging from the descriptions of Dixon (1918) and Portenko (1972). Evidently both species rise to a height of 40–50 feet, perform a few dips, then hover in the air with rapid wingbeats and slowly drop back toward earth while singing a buzzing trill, *tsee-e-e, tsee-e-e, tsee-e-e.* This hovering may last 20–30 seconds. Portenko observed this display in July and saw two birds performing simultaneously. One, later collected, proved to be a male with enlarged gonads.

Reproductive biology. Portenko (1972) has reviewed the past records of breeding in this species. Two nests located by V. Leonovitch were in an area of mosses, sedges, and creeping osiers. One of the nests had a lining of dwarf willow leaves, and the clutches, both being incubated, consisted of 3 and 4 eggs. Dixon (1918) found a nest with 2 fresh eggs and another with 3 young that had just hatched. It seems that 3–4 eggs represent the normal clutch. Although it has been suggested that the male does all the incubation, this is very unlikely, and Portenko reported that both sexes were encountered at the nest or with hatched chicks, with the males undertaking the more active role in defense. The injury-feigning behavior is apparently like that of the other small *Calidris* forms. Dixon also observed defensive behavior on the part of both members of a pair that were tending downy young. He believed that incubation probably lasts

18–20 days. According to A. A. Kistchinski (in litt.), nesting occurs in a variety of dry-tundra habitats, and both sexes incubate. Sometimes both sexes also attend the young, but more typically it is only the male that broods them.

Status and relationships. Apparently this species has a sporadic but locally common distribution on the Chukotski Peninsula (Portenko 1972). The size of the population is uncertain, but has been estimated to perhaps number 2.0–2.8 thousand pairs (A. A. Kistchinski, in litt.). Portenko (1957, 1972) has reviewed the taxonomic status of this species, and noted that its summer plumage is remarkably similar to that of the rufous-necked sandpiper. Likewise, the immature plumage is almost identical to that of the rufous-necked sandpiper, as well as to those of the little stint and the semipalmated sandpiper. Further, Portenko noted that the downy plumages of these forms are very similar and that the eggs of the spoon-billed sandpiper are evidently nearly identical to those of the rufous-necked species. As I mentioned earlier, the male aerial displays are likewise almost identical. All these lines of evidence favor the idea that the spoon-billed sandpiper is a typical *Calidris,* and an extremely close relative of *ruficollis,* though Burton (1971) has urged its retention in *Eurynorhynchus.*

Suggested reading. Dixon 1918; Bent 1927.

Little Stint

Calidris minuta (Leisler) 1812
(*Erolia minuta* of Peters, 1934)

Other vernacular names. Lesser stint; bécasseau minute (French); Zwergstrandläufer (German); correlimas minuta (Spanish).

Subspecies and range. No subspecies recognized. Breeds in Eurasia from about 30° E longitude east to the Indigirka delta, on Kolguev and Vaigach islands on southern Novaya Zemlya, and locally on the New Siberian Archipelago. Breeds sporadically on the Chukotski Peninsula. Winters to Africa, Arabia, India, and the Maldives. See map 83.

Measurements and weights. Wing: males 90–100 mm; females 94–103 mm. Culmen (to feathering): males 16–19 mm; females 17–20 mm. Weights (on breeding grounds): 6 males 25–28 g, average 26.6 g; 4 females 22–40 g, average 30.6 g (Glutz et al. 1975). Wintering-ground weights average only about 22 g (Skead 1977). Eggs ca. 29 x 21 mm, esti-mated weight 6.3 g (Schönwetter 1963). Portenko gives an average weight of 5.75 g for half-incu-bated eggs.

DESCRIPTION

Breeding adults of both sexes have a white forehead, an orange cinnamon crown streaked with blackish and edged with white, the nape similar but paler, the ear coverts, sides of neck, and cheeks orange cinnamon with darker streaks, and the chin and throat white. The breast is spotted and shaded with dusky, especially at the sides, while the rest of the underparts are white. The mantle and scapulars are brownish black with cinnamon edging and ashy gray tips; the rump and upper tail coverts are blackish brown with cinnamon edging. The central tail feathers are sepia with cinnamon edging, while the others are pale ashy brown, except for the primary coverts, which are brownish black, and most are tipped with white, especially the greater coverts. The primaries are sepia, and the secondaries are the same but paler on the inner webs, with the bases and tips white. The iris

83. Breeding distribution of the little stint.

is brown, the bill is black, and the legs and toes are black. *Adults in winter* have the upperparts grayish brown with darker shaft streaks. The underparts are white with grayish smudges at the sides of the breast, sometimes forming a gorget. The upper coverts are uniformly pale grayish brown, with paler or whitish fringes. *Juveniles* are generally brownish black above, with bright chestnut and whitish edges, forming V-shaped marks on the back. The underparts are whitish, with a buff band on the upper breast, and the coverts are brown with chestnut fringes. *First-winter* birds resemble adults but retain some chestnut-fringed inner median coverts to late winter (Prater et al. 1977).

In the hand, the very small size (wing under 105 mm) and black legs separate this from most other "peeps"; the ashy gray tail and blackish legs separate it from the similar-sized Temminck stint. The long-toed stint has a longer middle toe (at least 22 mm, vs. under 20 mm), and the least sandpiper has yellow legs.

In the field (5½ inches), this tiny "peep" is among the smallest in Europe, comparable to the Temminck stint, with dark rather than yellowish legs and having grayish rather than white outer tail feathers. Its usual call is a short, clipped *chit* note, rather low in pitch, instead of the high-pitched trill typical of the Temminck stint. It is usually found around mud flats and marshes, mixing with larger species but usually showing more activity than they do. In all plumages this species is more reddish on the mantle than is the Temminck stint, which is distinctly grayish dorsally.

NATURAL HISTORY

Habitats and foods. The breeding habitat of this species consists of subartic tundra, usually where there are mosses and sedges interspersed with hummocks covered by *Empetrum*. In Norway, at the edge of its breeding range, the species is limited to areas around freshwater lakes, ponds, or river deltas where there is an abundant food supply and where there are flat, sparsely vegetated shorelines surrounded by heath tundra (Hilden 1978b). In the center of its Siberian breeding range, densities may reach about 1 to 3 breeding pairs per hectare (or about 40–120 pairs per 100 acres) (Uspenski 1969). During the breeding season the birds primarily eat insects such as dipterans, beetles and their larvae, and particularly the larvae of mosquitoes and craneflies. Some small plant seeds and bits of vegetation are also eaten. Studies in southern Sweden indicate that this species is pri-

marily insectivorous and obtains its food close to the water's edge by making rapid dashing movements, apparently foraging visually. However, it also walks slowly along the shoreline, taking materials from the surface. Where it occurs with dunlins, the two feed in quite different ways and exploit different areas of the shoreline, with dunlins concentrating on *Nereis* worms and the little stint feeding slightly, if at all, on this food resource (Bengston and Svensson 1968). Small mollusks, crustaceans, and earthworms are sometimes also eaten outside the breeding season, and while on migration the bird frequently forages along the shallow and muddy margins of lakes, reservoirs, and sewage farms as well as along tidal mud flats and beaches.

Social behavior. Although the species is similar to the Temminck stint in its general behavior and reproductive biology, some important differences do exist (Hilden 1978b). For example, this species is not as strongly territorial as the Temminck stint, and it defends no clearly definable areas. The pair bond is also stronger than in the Temminck stint, and unlike that species both sexes perform the song flight. The aerial display is performed 20–30 meters above the ground, with the bird singing while hanging almost motionless in the air and rapidly beating its wings (Gladkov 1957) or singing while descending in a spiral (Hilden 1978b). The song is a series of soft *svii-svii-svii* notes, quite different from that of the Temminck stint.

Reproductive biology. It is probable that, as in the Temminck stint, a double-clutch system of reproduction is the normal breeding strategy, and although this has now been reported from Norway as well as Siberia, it has not yet been proved with banded birds. Each clutch is apparently incubated by a single bird, with the male incubating the first clutch and the female the second one. In one area inhabited by only 4 or 5 pairs a total of 8 clutches was found. Nests are built in highly exposed positions on the tundra, and there is a layer of dead brown willow, birch, or other leaves in the nest cup. Of 19 completed clutches, 16 had 4 eggs and 3 had 3 eggs in one study (Hilden 1978b). The two nests of a pair may be as close as 7 meters apart (Flint and Kistchinski 1973). However, as many as 7 eggs have been reported in a single nest, which certainly were laid by two females. In most early reports, males have been found to be the incubating sex, and in at least one instance an observation of two parents tending a single brood has been made, suggesting that double clutching is not invariable. Incubating birds are extremely tame and frequently can nearly be touched before they leave the nest. At that time they typically perform a "rodent-run" display. The incubation period is probably 21 days (Hilden 1978*b*). The fledging period is still apparently unreported.

Status and relationships. This species is relatively common on the wintering grounds of Africa, and in some regions it is second only to the ruff in total numbers. Probably its total population numbers in the millions of individuals (Uspenski 1969). It thus does not seem to pose any special conservation problems at present. The species is clearly an extremely close relative of *ruficollis,* and sometimes the two have even been treated as subspecies, but this is clearly not an acceptable practice (Gladkov 1957). In addition to these two species, the North American *minutilla* is also obviously related, and Larson (1957) treats the three forms as a species group. I suggest that *minuta* and *ruficollis* be regarded as a superspecies and that these three forms and the Asian *subminuta* be considered a species group.

Suggested reading. Bannerman 1961; Uspenski 1969.

Rufous-necked Sandpiper

Calidris ruficollis (Pallas) 1776
(*Erolia ruficollis* of Peters, 1934)

Other vernacular names. Red-necked stint; eastern little stint; bécasseau à col roux (French); Rotkehlstrandläufer (German).

Subspecies and range. No subspecies recognized (this form is considered by some as probably conspecific with *C. minuta*). Breeds in Siberia on the east and south parts of the Chukotski Peninsula south to Koryakland, with a few isolated areas of known or probable breeding well to the west (Taimyr Peninsula and near the Lena delta). Also has bred in Alaska (Seward Peninsula near Wales, at Barrow, and perhaps elsewhere). Winters from China, Taiwan, and the Philippines south to the Indo-Chinese countries, Australia, and New Zealand. See map 84.

Measurements and weights. Wing: males 99–106 mm; females 97–107 mm. Culmen (to feathering): males 16–19 mm; females 18–19 mm. Weights: 62 males 21–47 g, average 32 g; 29 females 27–51 g, average 36 g (Shaw 1936). Thomas and Dartnall

(1971*a*) give a lower range (18–27.5 g) for wintering birds. Eggs ca. 32 x 23 mm, average weight 8.28 g (Gladkov 1957).

DESCRIPTION

Breeding adults of both sexes have the face, throat, foreneck, and upper breast rich chestnut red extending to the superciliary area, while the crown is a darker brown with chestnut edging. The nape is pale brown with chestnut fringes, and the back is dark brown with paler or chestnut edging. The wing coverts are gray brown with pale or whitish tips. The underparts are white, with some dark brown spotting on the breast and upper flanks. The tail feathers are similar to those of *minuta* but are whiter on the inner webs, and the central pair of feathers are more rounded, with narrower white edging. The iris is brown, the bill is black, and the legs and toes are blackish. *Adults in winter* have a white forehead and superciliary stripe, with the rest of the upperparts pale grayish brown, the inner ones grayer with whitish fringes. The underparts are white, with gray lateral breast patches and almost no breast streaking. *Juveniles* have a brown crown with pale chestnut edges, the grayish neck is streaked with brown, the back is brown with chestnut edging, and the scapulars are tipped with white and chestnut. The coverts are grayish brown, with paler fringes, and the underparts are white with a buffish gray breast and a few brown streaks. *First-winter birds* retain some buff-fringed inner median coverts through the winter (Prater et al. 1977).

In the hand, this species very closely resembles *minuta,* but in breeding plumage the rusty head color distinguishes *ruficollis* from this and other small "peeps." In winter or juvenile plumages the more rounded central tail feathers, somewhat larger wings (usually over 103 mm in males and 105 mm in females), and shorter tarsus (usually under 20 mm) of *ruficollis* help to separate these two species.

In the field (5½–5¾ inches), this small "peep" has a distinctive rufous head in spring (but in all plumages is very like the spoon-billed sandpiper except for bill shape), but in winter it is likely to be confused with the little stint in the Old World or the least sandpiper in the New World. However, it has a paler breast and a whiter forehead than either of these. When flushed, it is said to utter a *pit-pit-pit* that is similar to but coarser than that of the least sandpiper, sounding something like the squeak of saddle leather (Wallace 1974). During display flights the birds often

84. Breeding records and presumptive breeding distribution of the rufous-necked sandpiper.

fly 20–30 meters above the ground, uttering a repeated *wek-wek-wek* . . . call (A. A. Kistchinski, in litt.).

NATURAL HISTORY

Habitats and foods. On its wintering areas this species is associated with tidal and mud flats, sandy shores of seacoasts, bays, inlets, and estuaries, and with saltwater or freshwater swamps of coastal and inland locations (Frith 1976). The birds forage in dense flocks, picking up minute bits of food, presumably mostly small invertebrates. Thomas and Dartnall (1971a) reported that in Tasmania the foods consisted of items mostly obtained from the water surface and depths of no more than 20 mm, which they obtained by pecking, jabbing, and probing. These foods include polychaete worms, crustaceans, insects, mollusks, and seeds. The breeding habitats consist of sub-arctic tundra, but the birds nest only on rather dry and raised areas, rather than on flat and low-lying tundra. Nesting has been reported in the foothills and at the base of mountains, in mountain valleys, on low or high knolls, or where grassy or tussocky tundra borders alluvial deposits. Summer foods are diverse and include beetles (especially carabids and curculionids), insect larvae, hymenopterans, and tiny seeds (Portenko 1972).

Social behavior. The social behavior of this species is not yet well known, but the birds are highly gregarious during the nonbreeding season and often occur in dense flocks. Courtship flights have been described by Gladkov (1957) as being distinct from those of the little stint and the Temminck stint; they consist of several rapid towering ascents and descents while calling. However, Portenko (1972) described the flight as being only a few meters above the ground, with rapid wing-beating and gliding. Birds were sometimes also observed beating their wings slowly, like a bat that had just taken flight. Sometimes a *wav* or *wek* note was emitted.

Reproductive biology. Practically the only observations on nesting are those of Portenko (1972). He described several nests found near Providence Bay and near Uelen. In at least three cases, nests have been found in pairs, apparently each tended by one of the members of the pair. Apparently 4 eggs are the normal clutch, though some 3-egg clutches have been found as well. Recent observations by A. A. Kistchinski (in litt.) indicate that both sexes sometimes incubate, and that males usually tend the young after the first few days following hatching. He found no evidence of double-clutching.

Status and relationships. This species is sporadic but locally common on the Chukotski Peninsula, and judging from its abundance on the wintering areas it is probably not in any present danger from a conservation standpoint. Its taxonomic history is complex and has been summarized by Gladkov (1957), who argued that it should not be considered a subspecies of *minuta*. On the basis of behavioral differences, differences in the coloring of the eggs and chicks, and observed sympatry in one area (Tiksi-Bucht, near the Lena River), he believed that *ruficollis* should be regarded as a distinct species. However, Portenko (1972) failed to observe the distinctive courtship flights described by Gladkov, and he argued that such things as differences in degree of cautiousness, egg coloration, and chick patterning were quantitative rather than qualitative. He thus did not believe that proof of specific differences between *minuta* and *ruficollis* was yet at hand. Dementiev and Gladkov (1969) considered the two forms distinct, and I likewise believe that the preponderance of evidence favors this view.

Suggested reading. Dementiev and Gladkov, 1969; Uspenski 1969.

Least Sandpiper

Calidris minutilla (Vieillot) 1819
(*Erolia minutilla* of Peters, 1934)

Other vernacular names. American stint; least stint; bécasseau miniscule (French); Wiesenstrandläufer (German); correlimos menudillo (Spanish).

Subspecies and range. No subspecies recognized. Breeds in North America from central western Alaska east through the Yukon, northwestern British Columbia, Keewatin, Southampton Island, coastal Manitoba and Ontario, northern Quebec, Labrador, and Newfoundland, and on Anticosti, Sable, and Cape Sable islands. Winters from Oregon and North Carolina south to central South America. See map 85.

Measurements and weights. Wing (chord): males 86–94 mm; females 87–95 mm. Culmen: both sexes 17–20 mm. Weights: 6 fall migrants of both sexes weighed 18.4–32.3 g, average 23.9 g (Murray and

Jehl 1964). The averages of 16 males and 14 females in summer were 20 g and 22 g, respectively (Irving 1960). Eggs ca. 29 x 21 mm, estimated weight 6.4 g (Schönwetter 1963).

DESCRIPTION

Breeding adults of both sexes have a broad whitish superciliary stripe, becoming streaked with dusky behind the eye, the sides of the face pale grayish brown flecked with dusky, especially on the lores, and a brownish crown that is broadly streaked with blackish. The chest, foreneck, and sides of the neck and breast are pale brownish or grayish, with dusky streaks. The rest of the underparts are white. The mantle and scapulars are generally blackish, with pale fulvous or cinnamon margins and whitish tips. The rump and middle upper tail coverts are blackish, while the middle two tail feathers are grayish brown.

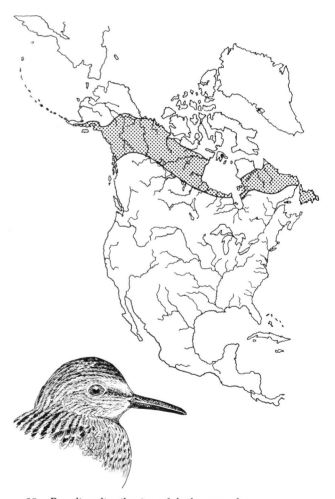

85. Breeding distribution of the least sandpiper.

The lateral upper tail coverts are white, and the outer tail feathers are light grayish brown. The wing coverts are deep grayish brown with paler margins, and the greater coverts are narrowly tipped with white. The primaries and their coverts are dusky, and the secondaries are similar but paler toward the base and are margined with white. The iris is brown, the bill is blackish, and the legs and feet are dull greenish yellow. *Adults in winter* are generally deep brownish gray or grayish brown on the upperparts, with the feathers darker centrally and edged with grayish. The coverts are similar but edged with whitish, and the upperparts show less streaking than in the breeding plumage. *Juveniles* have a scaly chestnut pattern on the upperparts, and the scapulars are fringed with whitish. The coverts are edged with bright buffy brown or chestnut. The breast is suffused with buff and lightly streaked. *First-winter* birds resemble adults but often retain some chestnut-fringed inner median coverts (Prater et al. 1977).

In the hand, the small size (wing no more than 95 mm) and short bill length (exposed culmen no more than 20 mm), together with a middle toe that is no more than 19 mm without the claw, serve to separate this form from other *Calidris* species.

In the field (5 inches), this is the smallest of all the "peeps" and is also the only small *Calidris* in North America with yellowish legs. However, yellowish legs also occur on *subminuta* and *temminckii,* and in areas where either of these others might also occur, field separation may be nearly impossible in winter. Least sandpipers often feed with other "peeps" and usually appear darker above and on the breast than

do semipalmated or western sandpipers. They lack the buffy scaling on the back typical of Baird sandpipers. When they fly the birds utter a prolonged or repeated *scree-ee-ee* or *threeep* note, and while foraging they often maintain an almost constant twittering. Compared with those of the similar-sized little stint, their calls are considerably higher-pitched and more prolonged (Wallace 1974).

NATURAL HISTORY

Habitats and foods. The breeding habitat of this species generally consists of fairly wet tundra where there is a mixture of tussock grasses and heath, boggy tundra sites having moss hummocks, or wet sedge meadows near the tree line. It generally avoids both very dry tundra and open tundra far from trees. In the Sheenjek Valley of Alaska, where it is the most common nesting shorebird, the species occurs on the tussock-heath tundra of the valley floor where the substrate is damp but not excessively wet (Kessel and Schaller 1960). In the area around Churchill, Manitoba, this is one of the most abundant breeding shorebirds, and it has foraging and microhabitat adaptations that overlap with those of both *alpina* and *pusilla,* which like it tend to be foraging "generalists." This species forages mainly by making pecking rather than probing movements, either singly or in multiple series (Baker and Baker 1973). Its foods are still not well documented but are known to include amphipods, the larvae and pupae of midges, small beetles, and small amounts of vegetation (Bent 1927).

Social behavior. In the Sable Island area of Nova Scotia this species is the only breeding calidridine shorebird, but there it nests abundantly in the lush vegetation surrounding freshwater and brackish ponds. Like other calidridines, sexually mature males occupy territories that are often adjacent and advertised by display flights. These flights are most frequently performed in early spring during morning hours, and they are least frequent during strong winds and periods of poor visibility. Each area is maintained for exclusive use by males and always includes the eventual nest site, but it only occasionally includes areas for foraging and brood-rearing. The apparent purpose of display flight activity is not nest spacing but rather mate attraction; thus sexual selection probably has played an important role in evolution of the behavior. Males typically fly upward at a

Distraction display

steep angle, leveling out at 15–20 meters and gradually replacing rapid wingbeats with an alternation of rapid fluttering movements and holding the wings outstretched but motionless. The flight is terminated by a rapid stooping or a slow parachuting. The associated vocalizations are initially a series of rather short calls, followed by a longer song that occurs at the flight's apex or sometimes is produced throughout a slow parachuting descent. The flights average 3.4 minutes in duration, and a male may cover an area as much as 200 meters in diameter during a single flight. Songs are also uttered during ground courtship, and various chattering calls are common during aggressive encounters. Males often fight with one another, but females fight only rarely. In the first few days after pairs are formed the paired birds spend nearly all their time together, but this incidence of shared time declines rather rapidly, and the first egg may be laid as early as 5 days after initial association (Miller 1979a).

Reproductive biology. Nests typically are placed on low moss cottongrass hummocks in the Sheenjek Valley of Alaska, and around Churchill they are frequently at the bases of small willows. Almost invariably the nest is lined with the small dead leaves of various shrubs (*Salix, Betula, Arctostaphylos, Dryas*), which seem to be placed there purposefully rather than accidentally, and which often are of approximately the same color and shape as the eggs themselves. Of 70 completed clutches found in the Churchill area, 65 had 4 eggs, 4 had 3 eggs, and a single nest had 2 eggs. The incubation period ranged from 19.5 to 23 days, averaging 20.5 days (Jehl and Smith 1970). The males play a major role in incubation, typically incubating during most of the daylight hours, and also are primarily responsible for raising the brood. Although double clutching has not been proved, several instances of renesting in the same season have been found. The fledging period has apparently not been established for this species, and neither the pair's foraging area nor the brood-rearing area are necessarily within the male's former display-flight area (Miller 1979a).

Status and relationships. This species is a common to abundant breeder over much of its range, and so it poses no special conservation problems at present. This species is certainly a close relative of the Old World complex made up of *minuta, subminuta,* and *ruficollis;* at times these have been considered to constitute two (*ruficollis-minuta* and, *subminuta-minutilla*) or even only one species. Larson (1957) believed that *subminuta* and *minutilla* represent a Pleistocene species pair, and I consider them members of the same superspecies.

Suggested reading. Miller 1977, 1979a.

Long-toed Stint

Calidris subminuta (Middendorff) 1853
(*Erolia subminuta* of Peters, 1934)

Other vernacular names. None in general English use; bécasseau à longs doigts (French); Langzehenstrandläufer (German).

Subspecies and range. No subspecies recognized. Breeding range not well established but includes Anadryland south to Kamchatka, Bering Island, and the northern Kurils, along the Sea of Okhotsk between Gizhiga and Magadan to the lower Amur River, and on northern Sakhalin. It has nested in the Baikal region, and it probably also breeds in the valleys of the Kolyma and Amur rivers and has been found in the breeding season as far west as western Siberia in the upper Ob drainage. Breeding also occurs on the Commander Islands and perhaps at times on the western Aleutians (*Condor* 80:312). Winters from China, Taiwan, and the Philippines south to India and the Sundas, sometimes reaching Australia. See map 86.

Measurements and weights. Wing (both sexes): 90–100 mm. Culmen (from feathering: both sexes 17–19 mm (Ali and Ripley 1969; Prater et al. 1977). Weights: 12 males 23–33 g (average 29 g; 8 females 28–37 g, average 32 g (Shaw 1936); 3 breeding specimens weighed 23.9–29.2 g. (Portenko 1972). Eggs ca. 31 x 23 mm, average weight 7.48 g (Leonovitch 1973a).

DESCRIPTION

Breeding adults of both sexes are generally brownish black on the upperparts with extensive bright chestnut and some grayish white fringes on the feathers. The head coloration is much like that of the little stint, but the feather edges are less rufous and more ochre-tinted, and the breast is less well developed, being finely streaked with brownish buff. The underparts are otherwise whitish, with some dark spotting on the sides of the neck and the upper flanks. The wing coverts are brown, edged with pale chestnut or fringed with whitish. The primaries are like those of the little stint, but only the outermost primary has a

86.   Breeding records and presumptive breeding distribution of the long-toed stint.

30. Winter-plumage heads and feet of rufous-necked sandpiper (*above*) and long-toed stint (*below*).

23 mm excluding claw—compared with 16–19 mm in *minutilla*) should separate this species from other *Calidris* types. The dorsal patterning is much like that of *minutilla,* but there is no pale nape, and the bill shape resembles *temminckii* but is slightly heavier.

In the field (5½ inches), this species' small size separates it from most other Asian species except the red-necked stint, which is very similar but has blackish legs. It is very difficult to separate from the least sandpiper and little stint in the field, but at least in breeding plumage the long-toed stint is distinctively black-backed, and the scapulars have wide orange fringes, producing a spangled appearance that is distinctive (*British Birds* 71:559). Its usual call note is a dry purring *prrp.* In flight it utters a rapid, high-pitched *che-che-che* note, apparently much like that of the least sandpiper. The birds are found on shorelines and estuaries, but they prefer grassy areas rather than muddy ones and tend to hide in depressions. They are also relatively solitary, and rarely occur in flocks of more than 6–8 birds.

NATURAL HISTORY

Habitat and foods. This little-known species winters in southern Asia, where its habitats are said to include marshy areas, paddy fields, and mud flats. In Australia it is said to prefer the edges of coastal and inland swamps and lakes to coastal habitats. The water can range from fresh to salty, but there must be soft mud present, with scattered grass and low herbage (Frith 1976). Its summer habitat is uniquely centered on the taiga zone, but actual nests are in tundralike or boggy openings. Mosses, low sedges, and dwarf, prostrate willows are favored vegetational cover for nesting (Leonovitch 1973*a*). No detailed information is available on foods, except that the remains of carabid beetles, small gastropod mollusks, and amphipods have been noted in the few stomachs so far examined (Dementiev and Gladkov 1969).

Social behavior. It is said that these birds may be seen in flocks of as many as 50 birds, but they are usually in groups of 3–7, or occasionally as many as 15. The little that is known of their social behavior comes from Leonovitch (1973*a*), who noted that in early June the males performed courtship flights. Flying at a height of about 100 meters, the males would circle over the taiga, alternately flapping their wings and gliding, all the while uttering a relatively soft

white shaft, while the others are entirely brown. The outer tail feathers are grayish brown, the iris is brown, the bill is black, and the legs and toes are grayish yellow to olive yellow, or sometimes brownish. *Adults in winter* are blackish brown above, with fringes of dark grayish brown, and the coverts are dark grayish brown with paler edges. The underparts are white, with the breast showing some fine streaking and dark grayish brown smudges. *Juveniles* are blackish on the crown and mantle, with bright chestnut fringes and spots and some whitish edging on the scapulars. The coverts are broadly tipped and spotted with whitish to buff, and the underparts are whitish, with the breast smudged with grayish brown and some dark streaking. *First-winter* birds resemble adults but retain chestnut tips on their inner median coverts until spring (Prater et al. 1977).

In the hand, the combination of small size (wing no more than 100 mm and fairly long middle toes—20–

trryui-trryui-trryui. These almost certainly represent territorial advertisement flights. Similar flights have been seen in the Kuril Islands (Taka-Tasukasa 1967).

Reproductive biology. All of 5 nests reported by Leonovitch (1973*b*) were found near pools in rather damp areas and in low tundralike vegetation. These nests were all found in mid-June, and nests have been found in Yenissey valley, Transbaikalia, and in the northern Kurils in late June. The birds mostly nest in open or boggy areas of tundra that are surrounded by boreal forest or taiga. Although both parents were seen near an incomplete clutch, complete clutches were tended by only a single bird. Four eggs are a complete clutch. Both males and females have been reported to incubate, but A. A. Kistchinski (in litt.) states that incubation by females is still unproven. Rather, all birds collected on nests or leading young have proven to be males. Broods have been found in the Koryak Highlands in mid-July.

Status and relationships. The scattered breeding distribution of this species suggests that it is a relict form that has been restricted to small tundralike openings in the boreal forest or to human-caused forest openings (Leonovitch 1973*a*). This suggests that the species may be declining in numbers, but there are no data on which to base such an opinion. It is quite probably an Asian representative of *minutilla,* and at times they have been considered conspecific. However, the two forms are quite distinct and differ considerably in toe length (presumably reflecting varying adaptations to soft-mud substrates) and in breeding plumage. It is also obviously a close relative of *minuta* and might represent a southern disjunct derivative from it.

Suggested reading. Dementiev and Gladkov 1969; Uspenski 1969.

Curlew Sandpiper

Calidris ferruginea (Pontoppidan) 1763
(*Erolia testacea* of Peters, 1934)*

Other vernacular names. None in general English use; bécasseau cocorli (French); Sichelstrandläufer (German); correlimos zarapitín (Spanish).

---

*The use of the specific epethet *testacea* for this species has been suppressed.

Subspecies and range. No subspecies recognized. Breeds in Siberia in the Taimyr Peninsula from Yenisei Bay to Katanga Bay, and locally farther east as well as on the New Siberian Archipelago and possibly the northern Gydan Peninsula. Breeds rarely in Alaska (Kessel and Gibson 1979). Winters from Africa, the Persian Gulf, India, and the Indo-Chinese countries south to the Sundas, Australia, and New Zealand. See map 87.

Measurements and weights. Wing: males 118–37 mm; females 119–37 mm. Culmen: males 31–42 mm; females 34–44 mm. Weights (in breeding season): males 44–91 g, average of 11 was 63.2 g; females 55–83 g, average of 16 was 63.3 g (Glutz et al. 1975). Eggs ca. 36 x 26 mm, estimated weight 12.0 g (Schönwetter 1963).

DESCRIPTION

Adult males in breeding plumage have the head, to hazel, with darker streaking on the crown, and the feathers of the underparts increasingly tipped with whitish and becoming white on the under tail coverts. The chin, forehead, and eye region are whitish, and the mantle area is strongly marked with blackish spotting. The wing coverts are grayish brown, with paler margins and the greater coverts tipped with white. The secondaries and inner primaries are slightly darker grayish brown, the secondaries are margined with white, and the outer primaries and their coverts are dusky. The rump is grayish brown to brownish gray, the upper tail coverts are white with dusky barring, and the tail is grayish brown with the outer feathers narrowly margined with whitish. The iris is brown, the bill is blackish becoming olive basally, and the legs and toes are blackish (Ridgway 1919). *Females* average slightly paler than males and have more white and distinct brown barring on the underparts. The bill is also longer (birds with a culmen of 36 mm or less are usually males, and of 40 mm or more usually are females). *Adults in winter* are mostly plain brownish gray or grayish brown on the head and upperparts, with a white superciliary stripe, white underparts, and upper tail coverts, and the chest indistinctly streaked with grayish. *Juveniles* have pale buff fringes on their dark brown upperparts and pale buff edging on their wing coverts. There are faint brown streaks on the breast, which is slightly buffy. *First-winter birds* retain inner median wing coverts with rich buff fringes through much of the winter (Prater et al. 1977).

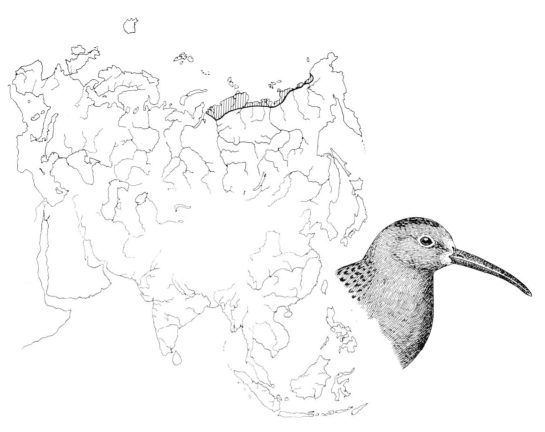

87. Breeding distribution of the curlew sandpiper.

In the hand, the moderate size (wing 118–40 mm) and distinctly lengthened and decurved bill (culmen at least 31 mm), separate this from other *Calidris* forms, as do the rather long legs (tarsus 27–32 mm) and white rump.

In the field (6½–7½ inches), this rather small sandpiper is notable for its gradually decurved bill (rather than terminally drooping as in dunlins) and white rump, which are good field marks even in winter plumage. In breeding plumage the rich chestnut body coloration is distinctive. The birds frequent various coastal habitats and sometimes are found on inland mud flats and sandbars as well. In flight the white rump shows up well, as does a white wing stripe, and the usual calls are loud *chirrup* or *chirreep* notes and a sharp whistle.

NATURAL HISTORY

Habitats and foods. The breeding habitats of this Eurasian species are areas of arctic tundra that include combinations of lowlands having abundant lakes and ponds, relatively well-drained polygonized ground, and low ridges. The birds prefer the somewhat elevated areas to the lowlands, especially south-facing slopes, which tend to become snow-free early in the season. In favored habitats they reach a breeding density of 1–2 pairs per hectare (Uspenski 1969). While on the breeding ground the primary foods are insects, especially the larvae and adults of midges and craneflies. Outside the breeding season such foods as polychaete worms, small mollusks such as snails, and crustaceans are all important items. A study in Tasmania (Thomas and Dartnall 1971*a*) on the winter food habits of this species and *C. ruficollis* indicated that the curlew sandpiper consumes a relatively larger proportion of polychaete worms, dipteran larvae and pupae, mollusks, and plant materials but smaller proportions of ostracod and amphipod crustaceans. This species has the longest and most decurved bill of any of the *Calidris* species. It often forages by "stitching," making a series of closely spaced probes, but in this species such foraging is often done while standing still (Burton 1974).

Social behavior. Males probably become territorial shortly after arriving on the breeding grounds, and in an Alaskan study they were found to defend areas of from 4 to 10 acres on any given day when actively courting. However, they are highly mobile, sometimes traveling a mile or more without regard to obvious boundaries, perhaps because of the low density of birds in the area. However, their mobility may also be a reflection of the short-term pair bond typical of the species, which results in a loose social structure. The males show a remarkably varied vocal repertoire, which has some surprising similarities to those of several other *Calidris* species as well as the stilt sandpiper. For example, the song is composed of several parts, sometimes lasting 10–15 seconds, but often with various parts omitted. In its complete form it includes an initial series of chattering notes, a number of trilled doublets, a complex four-part phrase, and finally a varying number of drawn-out whining notes. Thus it is quite unlike most *Calidris*, which typically have rather monotonous trilled songs. The song is often given near the end of the flight with full and rapid wingbeats, sometimes interrupted with slower wingbeats or glides. During the gliding phase the bird raises its head and utters the doublet or whine notes. In some cases it ascends to 12–15 feet, uttering the song and concluding with the four-part phrase, then gliding down to the ground. This is a display similar to that of *melanotos*, but it is appreciably lower and lacks the hovering characteristic of many of the other *Calidris* species. When landing, the bird often assumes a wing-up display posture, much like that found in other *Calidris* species. One apparently unique display occurs on the ground and consists of uttering whining calls while fluffing the plumage and periodically moving the wings outward and downward to expose the white rump. An important ground courtship display is the scraping or "nest-cup" display, in which a male settles over a depression, scratching and kicking and sometimes calling. If the female approaches, the male leaves the cup and displays beside it, while the female may enter it. In one observed case this was followed by a short flight and attempted copulation. An elaborate ground display precedes copulation, with the male raising both wings and fanning his tail, moving back and forth in an apparent effort to expose his tail and white rump to her view. It is probable that the pair bond is broken and the males leave the breeding grounds shortly after the eggs are laid (Holmes and Pitelka 1964). Not only is the pair bond of this species apparently short-lived, but some males maintain

31. Winter-plumage heads and feet of dunlin (*above*) and curlew sandpiper (*below*).

pair bonds with two or three females. In association with this trend toward polygyny, the males are noticeably brighter in color than are females, and the breeding distribution approaches a clumped pattern (Pitelka et al. 1974).

Reproductive biology. Nests are typically placed on low mounds among well-developed polygons. They are very exposed, and there is little vegetation surrounding the immediate nest site. Nests are sometimes on ridges that are slightly higher than the surrounding swampy tundra, in sites similar to those chosen by gray plovers. There are typically 4 eggs present, but there is one record of a nest with only 3 eggs. The incubation period is still unreported, and the fledging period is likewise still unknown. Although some early reports suggest that males might participate in incubation, this is now believed to be an erroneous observation, and apparently all the incubation is performed by the female (Holmes and Pitelka 1964).

Status and relationships. Although no specific numbers are available, this species is fairly common in many areas, and its total population probably numbers in the hundreds of thousands of individuals (Uspenski 1969). The species' display behavior, especially the male's vocalizations and aerial song flight, is very similar to that of *melanotos* (Holmes and Pitelka 1964). Yet both species have short pair bonds, and perhaps some of these similarities are convergently evolved. Further, *melanotos* apparently lacks a nest-cup display, a behavior typical of several of the *Calidris* forms. In its bill shape and foraging behavior the species approaches *alpina*, but these similarities may also be the result of convergence. There is thus no obvious close relative of *ferruginea*, although Larson (1957) suggested that *ferruginea* and *alpina* constitute a species group. In the absence of other information, I am inclined to accept that view.

Suggested reading. Holmes and Pitelka 1964; Portenko 1959.

Dunlin

Calidris alpina (L.) 1758
(*Erolia alpina* of Peters, 1934)

Other vernacular names. Red-backed sandpiper; bécasseau variable (French); Alpenstrandläufer (German); correlimos común (Spanish).

Subspecies and ranges. See map 88.

C. a. arctica: Greenland dunlin. Breeds in northeastern Greenland. Wintering area uncertain; see *schinzii*.

C. a. schinzii: Southern dunlin. Breeds in southeastern Greenland, Iceland, the British Isles, and in Europe from Holland to Finland, southern Sweden, and southwestern Norway. Winters (with *arctica* and *alpina*) from the British Isles and western Africa to India.

C. a. alpina: Northern dunlin. Breeds from Norway (interior and north) and Spitsbergen east through Russia, including Kolguev and Vaigach islands and southern Novaya Zemlya to the Taimyr Peninsula and the Kolyma, grading toward *sakhalina* on the Taimyr Peninsula. Winters as indicated for *schinzii*.

C. a. sakhalina: Siberian dunlin. Breeds in Siberia east of *alpina* to the Bering Sea, possibly also on the northern coast of the Sea of Okhotsk. Also breeds in northern Alaska (Maclean and Holmes 1971), a population sometimes separated as *arcticola* (Todd 1953). Winters from Japan and China to Taiwan and Hainan, rarely farther south.

C. a. pacifica: American dunlin. Breeds in North America from southwestern Alaska eastward across the northern Yukon, northern Mackenzie, Keewatin, northeastern Manitoba, northern Ontario, and on Southampton and Somerset islands. Includes *hudsonia* of northeastern Canada (Todd 1953), a race not recognized by the A.O.U., but probably valid (Browning 1977). Winters from British Columbia to western Mexico, the Gulf coast, and on the Atlantic coast north to Massachusetts.

Measurements and weights. Males: 105–25 mm; females 107–28 mm. Culmen (to feathering): males 24–40 mm; females 26–43 mm. Weights (fall migrants): 9 males 38–50 g, average 43.9 g; 13 females 35–61 g, average 50.0 g (Glutz et al. 1975). Eggs ca. 34 x 25 mm, estimated weight 9.6 (*arctica*) to 11.7 g (*sakhalina*) (Schönwetter 1963).

DESCRIPTION

Breeding adults of both sexes have a crown that is streaked with dark reddish brown and dusky, with the rest of the head mostly white with dusky streaks, the streaks gradually becoming larger and more definite on the chest and breast, below which there is a large area of dull black on the abdomen, while the sides, flanks, and under tail coverts are white with a few narrow streaks of dusky. The upperpart feathers are black centrally, with broad margins of reddish cinnamon, especially on the scapulars. The wing coverts are brownish gray with whitish margins, the greater coverts being tipped with white. The primary coverts and flight feathers are dusky, the inner primaries edged with white basally and the secondaries tipped with white. The tail coverts are dusky, while the rest of the tail feathers are pale brownish gray and edged with white. The iris is brown, the bill is black, and the legs and toes are blackish (Ridgway

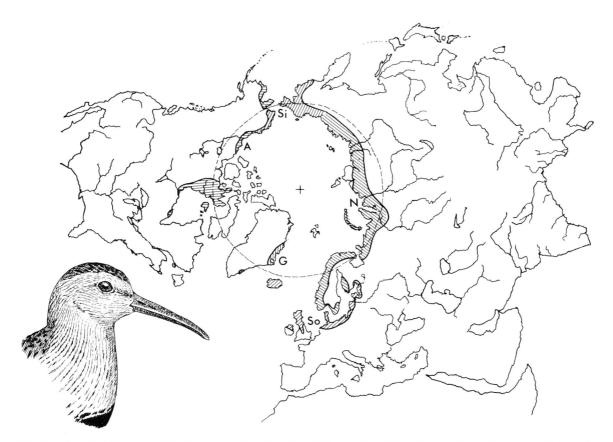

88. Breeding distributions of the American (A), Greenland (G), northern (N), Siberian (Si), and southern (So) dunlins.

1919). *Adults in winter* are mostly plain brownish gray above, with an indistinct superciliary streak, and the underparts are white except for faint grayish streaks on the neck and chest. The wing coverts are gray with whitish fringes. *Juveniles* are blackish brown above, with whitish buff and chestnut fringes, and the wing coverts have broad buff or chestnut fringes. The underparts are white, with brownish streaks on the breast and dark spots on the flanks. *First-winter* birds resemble winter adults but retain some chestnut to buff-fringed inner median wing coverts (Prater et al. 1977).

In the hand, this species differs from most other *Calidris* of comparable size (wings 105–28 mm) by its rather long bill (24–40 mm), which is decurved near the tip. In any plumage it can be separated from *ferruginea* by the dark upper tail coverts, and it also resembles *Limicola falcinellus* but has a narrower bill and lacks blackish lesser coverts.

In the field (7–7½ inches), this species is found on muddy flats or sandy beaches, often with sanderlings or other small "peeps," and in spring plumage the

reddish back and black abdominal patch are unmistakable. In winter plumage the drooping bill tip and rather uniformly grayish color pattern are fairly distinctive, but in some areas confusion with the curlew sandpiper is possible. There the absence of white upper tail coverts and the main bend of the bill near the tip rather than in the middle are useful characters. The birds utter a rapid, low grating trill in flight or an abrupt *chu* when flushed.

NATURAL HISTORY

Habitats and foods. In Alaska, the primary breeding habitat of this species consists of wet coastal plain tundra, but the birds also sometimes extend into areas of low foothills. In general they seem to prefer low, grassy, or sedge-covered tundra, and in Finland the race *schinzii* breeds in shore meadows covered by short grasses. During the winter the birds are usually found on estuarine mud flats, often in flocks numbering in the thousands. They move out onto the flats to forage at low tide, probing for invertebrates just

below the mud surface. Migrant birds often eat large quantities of nereid polychaete worms, whereas knots foraging in the same areas concentrate on bivalve mollusks, and sanderlings concentrate on insects (Ehlert 1964). Studies in Alaska indicate a fairly stable population density, of about 7 breeding pairs per 100 acres (Holmes 1966a), while various estimates of population density in Eurasia range from 0.001 to 1.0 breeding pairs per hectare (Uspenski 1969). A five-year study in northern Alaska indicated summer densities of from 0.0 to 21.2 birds per square kilometer (Bergman et al. 1977). In the Barrow area of Alaska the birds initially occupy territories in upland tundra, since these are the first habitats to become snow-free and also provide nest sites. Further, the upland tundra provides an abundant early summer food source in the form of cranefly larvae. After hatching has occurred, the adults and young move into the wet marshes, where insect food is then readily available (Holmes 1966a). While on their breeding ground, the birds feed almost exclusively on insects, primarily the adults and larvae of craneflies and midges, and hatching is apparently timed to coincide with the maximum abundance of available insect food (Holmes 1966b). In the Barrow area, the dunlin shares its breeding habitat with three other common species of *Calidris*, all of which feed on much the same foods, but there appears to be a considerable food competition only between dunlins and pectoral snadpipers. Adults of dunlins remain on the tundra longer than any of the other three species,

and during August they return from the marshes to forage again on upland tundra, eating cranefly larvae (Holmes and Pitelka 1968).

Social behavior. Although it is known that at least some females are able to reproduce at the end of the first year of life, the available data suggest that most dunlins are 2 years old or older when they first successfully breed. Frequently males establish territories in the summer before their initial breeding, and probably about two-thirds of the population (both sexes) initially breed when 2 years old. Almost without exception the males return to the same territory year after year, and females also often return to the same territory. In both sexes this is usually close to where the birds were hatched. Probably more than half the birds remate with their mates of the original year when available, and mate-changing is usually the result of the late return to the breeding grounds of one of the original pair members (Soikkeli 1967). Pairing takes place on the male's territory, and territoriality is well developed. Territories in Alaska range in size from 12 to 17 acres, with little year-to-year change associated with varying population densities. Territories are claimed as soon as the tundra begins to become snow-free and are mainly advertised by song flights. The male typically rises steeply from the ground until it reaches 10–150 feet, begins hovering on outstretched and rapidly quivering wings, and periodically interrupts the quivering to glide for a few seconds. The flight display may last from a few seconds to several minutes, and throughout it the bird utters a series of short trilled notes, repeated at the rate of about one per second. The trill continues as the bird terminates his display and glides to the ground on canted wings. As he lands the wings are held upward above the back for a few seconds before being lowered. Wing-raising displays are also performed while on the ground, sometimes being repeated rapidly to produce a wing-flashing display. Courtship activities include long chases above the tundra and nest-scraping ceremonies. In this display the male settles into a depression and performs nest-building movements. The female often then moves into the cup herself and probably eventually chooses one of the sites for her nest. Copulation is preceded by the male's closely following the female, raising one or both wings, and tilting his tail while uttering a trilled note. The male then takes off, hovers over the female, and lands on her back (Holmes 1966a).

Reproductive biology. Nests are usually at the base of a grass tussock and are well lined with grasses or

1. European oystercatcher, adult at nest. *Photo courtesy J. B. & S. Bottomley*

2. Ibis-bill, adult incubating. *Photo by Loke Wan Tho, courtesy Bombay Natural History Society*

3. American avocet, precopulatory display. *Photo courtesy Ed Bry*

4. American avocet, copulation. *Photo courtesy Ed Bry*

5. Eurasian lapwing, adult male. *Photo courtesy J. B. & S. Bottomley*

6. Lesser golden plover, adult in nuptial plumage. *Photo by author*

7. Greater golden plover, adult in nuptial plumage. *Photo courtesy J. B. & S. Bottomley*

8.　Dotterel, juvenile in fall. *Photo courtesy J. B. & S. Bottomley*

9.　Thick-billed plover, adult in fall. *Photo courtesy J. B. & S. Bottomley*

10. Semipalmated plover, downy young. *Photo by author*

11. Kentish sandplover, male incubating. *Photo courtesy J. B. & S. Bottomley*

12. Ruddy turnstone, female incubating. *Photo courtesy J. B. & S. Bottomley*

13. Ruddy turnstone, male in nuptial plumage. *Photo courtesy J. B. & S. Bottomley*

14.   Ruddy turnstone, male in nuptial plumage. *Photo by author*

15. Pectoral sandpiper, female incubating. *Photo courtesy D. F. Parmelee*

16. Pectoral sandpiper, female incubating. *Photo courtesy D. F. Parmelee*

17. Red knot, adult incubating. *Photo courtesy D. F. Parmelee*

18. Eastern knot, female near nest. *Photo courtesy A. A. Kistchinski*

19. White-rumped sandpiper, adult incubating. *Photo courtesy D. F. Parmelee*

20. White-rumped sandpiper, adult incubating. *Photo courtesy D. F. Parmelee*

21. Baird sandpiper, adult incubating. *Photo courtesy D. F. Parmelee*

22. Temminck stint, adult incubating. *Photo courtesy J. B. & S. Bottomley*

23. Semipalmated sandpiper, adult incubating. *Photo courtesy D. F. Parmelee*

24. Little stint, adult incubating. *Photo courtesy J. B. & S. Bottomley*

25. Least sandpiper, downy young. *Photo by author*

26. Least sandpiper, adult in nuptial plumage. *Photo by author*

27. Curlew sandpiper, juvenile in fall. Photo courtesy *J. B. & S. Bottomley*

28. Dunlin, adult incubating. *Photo courtesy J. B. & S. Bottomley*

29. Sanderling, adult in spring. *Photo courtesy Ed Bry*

30. Sanderling, adult incubating. *Photo courtesy D. F. Parmelee*

leaves. The eggs are laid at intervals that range from approximately 24 hours in northern Alaska to about 36 hours in southern Finland (Soikkeli 1967). Females laying for the first time begin later than do older birds, and those having strong pair bonds likewise tend to begin nesting earlier than others. More than 90 percent of 122 nests found in a Finland study had 4 eggs, and a 4-egg clutch is apparently typical throughout the dunlin's range. This species is the only arctic-breeding shorebird that regularly lays replacement clutches when the first is lost, and second clutches may be begun from 4 days to 2 weeks after the loss of the initial clutch. After the completion of the clutch, incubation begins. The male typically incubates by day and the female by night. Incubation lasts from 20.5 to nearly 24 days, with an average of 21.5–22 days typical. Although the birds are normally single brooded, there are a few records of females beginning a second nesting after the successful hatching of their first brood. In these cases the females mated with unpaired males that were breeding for the first time. By the time the young hatch territorial defense has ceased, and the birds gradually move to damper areas that often are well beyond their original territories. Both sexes tend the young, but the males show the stronger brooding tendencies, and after a few days the female often abandons the brood. Generally males leave the brood about the time they fledge, some 18–20 days after hatching. Probably about 20 percent of the hatched young survive to the end of their first year of life, and thereafter the annual adult mortality rate is about 27 percent (Soikkeli 1967).

Status and relationships. This species is often regarded as the most abundant wader in the northern hemisphere (Soikkeli 1967), and its Eurasian population probably numbers in the millions (Uspenski 1969). Its relationships are not immediately apparent, though Larson (1957) considers *alpina* and *ferruginea ("testacea")* to constitute a species group. Although it is quite possible that the bill and foraging similarities of these two species are simply the result of convergence, I am unable to suggest a more likely near relative than that species.

Suggested reading. Holmes 1966*a*, *b*; Soikkeli 1967.

Purple Sandpiper

Calidris maritima (Brunnich) 1764
(*Erolia maritima* of Peters, 1934)

Other vernacular names. None in general English sue; becasseau violet (French); Meerstrandlaufer (German); correlimos oscuro (Spanish).

Subspecies and range. No subspecies recognized. Breeds from Devon and Bylot islands to Southamton and Baffin islands in North America, on the coasts of Greenland, in Iceland, the Faeroes, Norway, Sweden, Spitsbergen, Franz Josef Land, the Kola Peninsula, Vaigash Island, Novaya Zemlya, Severna Zemlya, and the Taimyr Peninsula. Winters on the Atlantic coast of North America, the British Isles, and the Atlantic coast of France, sometimes reaching the Mediterranean. See map 89.

Measurements and weights. Wing: males 122–41 mm; females 123–43 mm. Culmen (to feathering): males 27–35 mm; females 30–36 mm. Weights (in summer): males 59–74 g, average of 6 was 63.3 g; females 72–80 g, average of 6 was 76.6 g (Glutz et al. 1975). Eggs ca. 37 x 26 mm, estimated weight 13.3 g (Schönwetter 1963).

DESCRIPTION

Breeding adults of both sexes have the entire head and neck narrowly and indistinctly streaked with light buffy and dusky, except for a grayish white chin and superciliary stripe, a dark stripe from the lores through the upper auriculars, and a crown heavily streaked with dusky. The foreneck and chest are more broadly streaked with dusky, and the rest of the underparts are mostly white, the sides and under tail coverts being streaked with light grayish. The upper feathers tend toward blackish with a faint purplish gloss and have edgings of dull buff and whitish. The rump, upper tail coverts, and central tail feathers are sooty blackish, also faintly glossed with purplish, and the outer tail feathers are light brownish gray with white shafts and white tips. The wing coverts are deep brownish gray with a purplish gloss and pale margins, the greater coverts being tipped with white, and the flight feathers are dark toward the tips and whitish basally. This white increases inwardly, so that the innermost secondaries are mostly white. The iris is brown, the bill is blackish at the tip and yellowish at the base, and the legs and toes are dull

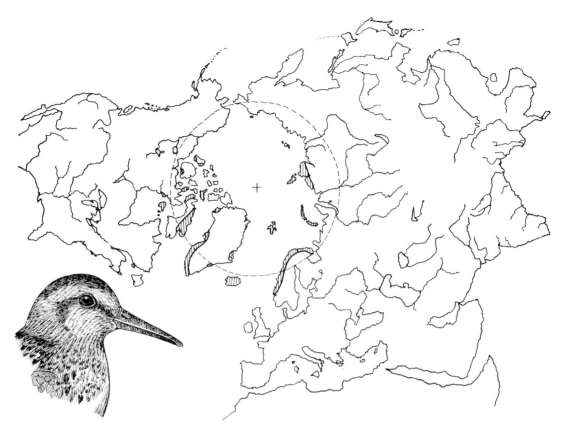

89. Breeding distribution of the purple sandpiper.

yellowish. *Adults in winter* are more uniformly gray above, with blackish back markings, a whitish area around the eyes, and a whitish spot on the upper lores, as well as a whitish throat. The middle fore-neck is pale gray, becoming brownish gray on the chest, these feathers having whitish tips that become broader on the lower chest and grade to a whitish abdomen. *Juveniles* resemble summer adults but have wing coverts with broader and more buffy margins, and the hindneck and cheeks are uniform brownish gray. *First-winter* birds retain some buff-tipped inner median coverts until spring (Prater et al. 1977).

In the hand, this medium-sized *Calidris* has rather short legs (20–23 mm tarsus) a slightly decurved bill (culmen 26–34 mm), and a purplish gloss on the feathers of the upperparts.

In the field (8–8½ inches), these birds are usually found on rocky coastlines, where their dark winter plumage makes them almost invisible. The yellow legs may be visible, and a white wing stripe and under wing lining flashes when they open their wings to fly. They usually fly only short distances, some-what in the manner of a spotted sandpiper, and they utter a weak *whit* or *twit* as well as some high-pitched twittery notes. They might easily be confused with surfbirds or rock sandpipers, but they are limited to the Atlantic coast, rarely reaching the center of the continent.

NATURAL HISTORY

Habitats and foods. Although this species is to be found in early spring along arctic seashores, it moves inland to breed. During the nesting season it is associated with upland lichen-covered tundra. In Spitsbergen, at least, the only other abundant nesting bird in this habitat is the snow bunting, and the nesting density of sandpipers is rather low, probably 2.5–3.5 breeding pairs per square kilometer. Although the birds do feed on the dry tundra, they prefer moister situations, such as the margins of ponds, edges of melting snowdrifts, and areas of thick mosses, that have an abundance of dipterans and springtails. Early in the season the birds feed mainly on the

32. Breeding-plumage heads of purple (*above*) and rock
(*below*) sandpipers.

Social behavior. In the high-arctic nesting areas of West Spitsbergen, these birds arrive in June, long before the tundra is snow-free. At that time they gather on the shorelines and evidently form pairs. As the tundra thaws, the birds abandon the shorelines and move into snow-free areas, especially those near small pools. Territorial and courtship activity is frequent during the last part of June, especially as the tundra becomes progressively snow-free. As the pairs occupy snow-free areas, the males claim territories that they advertise with song flights and by chasing intruding birds. The territories tend to be well spread out, owing to the distribution of snow-free patches, and thus boundary conflicts are apparently very rare. During display flights the male will ascend to 36 meters or more and fly in wide circles for varying lengths of time before descending quickly with his wings held high above the back. The flight song typically starts with a rapid and accelerating "announcement call," a repeated *prit-prit-prit,* during the ascent phase; during the gliding phase it shifts to a prolonged *bi-bi-bi-bi-bi* series of notes. A moaning call is uttered during the final descent to the ground. Courtship chases are usually performed close to the ground and may cover as much as 700 meters. When the male catches the female he usually glides out a short distance ahead of her, holding his wings above his back and uttering the moaning or announcement call. When a male chases a female on the ground he often performs a wing-lifting display, involving either one wing or both wings, and often uttering the moaning call. Copulations are preceded by ground chases and by a presentation display in which the tail is cocked and the male assumes a rigid posture. The same or a similar posture occurs in association with a scraping or nest-cup display, which is apparently similar to the corresponding display of the curlew sandpiper. Evidently the male prepares several nest cups within his territory, one of which the female might choose for depositing eggs (Bengtson 1970).

shores but display over the tundra. Later, as the young hatch, the males care for the broods while the females gather on the shorelines. The fledged young are the last to leave the uplands. As the birds leave the tundra and gather on the shores, their foods change to marine gastropod mollusks and amphipod crustaceans (Bengtson and Fjellberg 1975). Winter foods are diverse and include small fish, insects, crustaceans, and mollusks. Unlike most *Calidris* species, the bird's winter habitats are rocky rather than muddy or sandy substrates, and it is thus mainly limited to pecking rather than probing in soft substrates (Burton 1974). Studies from the coast of Yorkshire in Great Britain indicate the importance of mollusks, especially *Littorina* species, in the winter diet (Feare 1966). The birds feed at the very edge of the water, pecking or probing in rock crevices and resting during periods of high tide when their foraging areas are covered. When feeding, they sometimes swim across areas of water, but they do not typically feed while standing in water.

Reproductive biology. The nest is typically placed in upland tundra moss, often near clumps of *Dryas* or *Arctostaphylos,* and is extensively lined with plant materials. There are normally 4 eggs, but rarely 3 are found in a complete clutch. Although both sexes are known to assist in incubation, the male evidently plays a considerably larger role, and most observers have reported seeing only males on the nest. Perhaps the female broods at night, or possibly she abandons the clutch to the male several days before hatching. The incubation period is probably 21–22 days. In 20 observed cases of parental care of the brood, it was accompanied by the male, while in 2 cases a female was also seen in the same vicinity. The fledging period is still unreported. Recovery of some banded birds indicate that yearlings return to the place of hatching and that females may breed when a year old (Bengtson 1975).

Status and relationships. In many areas of Asia the purple sandpiper is fairly common, and its Eurasian population has been estimated by Uspenski (1969) to be several hundred thousand individuals. There are also breeding birds in arctic Canada, Iceland, and Greenland to add to this number. One of the interesting taxonomic problems associated with this species is its relationship to the rock sandpiper, which is discussed in that species' account. Beyond its obvious relationship to that species, its affinities are certainly not evident. It forages in a manner similar to that used by surfbirds, but nothing else suggests close relationships with *Aphriza*. Its downy pattern is much like that of *alpina* (Fjeldså 1976), and the two species are also very similar in bill configuration, though they forage in very different ways. Its general proportions and gait are also quite dunlinlike, and perhaps these two species are fairly closely related.

Suggested reading. Bengtson 1970.

Rock Sandpiper

Calidris ptilocnemis (Coues) 1873
(*Erolia ptilocnemis* of Peters, 1934)

Other vernacular names. Aleutian sandpiper; Pribilov sandpiper; bécasseau Aléoutien (French); Beringstrandläufer (German).

Subspecies and ranges. See map 90.
- *C. p. tschuktschorum:* Siberian rock sandpiper. Breeds in northeastern Siberia on the Chukotski Peninsula from Providence Bay to Kolyucha Bay, and on the west coast of Alaska south to the Alaska Peninsula. Winters from Alaska to Oregon and extreme northern California.
- *C. p. couesi:* Aleutian rock sandpiper. Resident on the Aleutian Islands from Attu to Unimak, and in the Shumagin Islands. Intergrades with the previous subspecies on Nunivak Island.

- *C. p. ptilocnemis:* Pribilof rock sandpiper. Breeds on the Pribilof Islands, on St. Matthew and Hall islands, and on St. Lawrence Island (Fay and Cade 1959). Probably winters along the Alaska Peninsula and the southern coast.
- *C. p. quarta:* Commander rock sandpiper. Breeds on the Commander Islands and probably the northern Kurils. Occurs south to the central Kurils in winter.

Measurements and weights. Wing: males 108–33 mm; females 112–35 mm. Culmen (to feathering): males 23–36 mm; females 26–37 mm. Weights: 4 males 80–132 g, average 112 g; 6 females 118–35 g, average 126 g (U.S.N.M. specimens). Eggs ca. 38 x 27 mm, estimated weight 13.3–15.0 g (Schönwetter 1963).

DESCRIPTION

Breeding adults of both sexes have a fulvous crown with heavy blackish streaks, and an otherwise whitish face except for a grayish loral streak and auricular patch, becoming pure white on the throat and underparts except for grayish dusky blotching or clouding on the breast. The upperparts are mostly ochre (the intensity varying with the race) to rufous with blackish spotting, and the tips of some feathers are margined with whitish. The rump and middle upper tail coverts are dusky slate, as are the central tail feathers, the outer tail feathers being much paler. The wing coverts are grayish slate, margined with white, especially on the greater coverts. The flight feathers are dark grayish with white shafts, the inner ones with increasing white basally, and the innermost secondaries are almost entirely white. The iris is brown, the bill is blackish at the tip and yellowish to greenish basally, and the legs and toes are dull yellow to greenish (Ridgway 1919). *Adults in winter* have a white breast that is irregularly streaked with dark gray, the sides of the head are pale gray, with an indistinct superciliary stripe and whitish around the eyes, and there is a grayish cast on the upperparts, with no rufous tones. The wing coverts are brownish gray fringed with white. *Juveniles* are blackish brown on the upperparts, with chestnut and whitish buff edging and with the wing coverts broadly edged with chestnut to pale buff. The underparts are white, with the breast buff and grayish. *First-winter* birds retain some buff-tipped inner median coverts until spring (Prater et al. 1977).

In the hand, the measurements of this species completely overlap those of *maritima*, but *ptilocnemis*

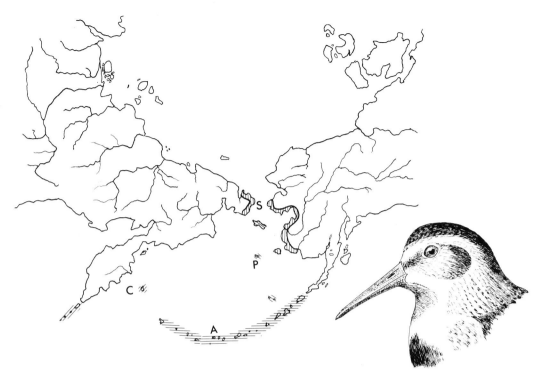

90. Breeding distributions of the Aleutian (A), Commander (C), Pribilof (P), and Siberian (S) rock sandpipers.

tends to have more greenish legs, is paler on the sides of the face and upperparts, and in spring plumage tends to be blotched rather than streaked on the breast. The white on the outer webs of the inner primaries is broader, and in *ptilocnemis* it extends to the shaft.

In the field (8–9 inches), these birds are found on rocky coastlines, often with surfbirds, which they closely resemble, though they lack white rumps and tend to be less uniformly grayish on the breast and back in winter plumage. The birds often stretch their wings when disturbed, raising first the near wing and then the farther one, exposing an extensive white wing stripe. In flight they also show a relatively dark tail. In winter plumage they are probably very difficult to separate from *maritima*, but they may have somewhat more greenish legs and are usually paler throughout. Except on its breeding grounds, the species is very quiet, but the mating call sounds like a droning trill of toads, varied by a repeated *per-derrr* (Murie 1959).

NATURAL HISTORY

Habitat and foods. Apparently the habitats used by this species are much like those of the purple sand-piper, though rather little has been written on the subject. During spring on St. Michaels Island they are said to occur on larger boulders and rocky shelves covered with seaweed, where they seek out slugs and marine worms, while during the same period at Hooper Bay they are said to feed on ice-bound sea lice. They have also been observed feeding on mud flats during low tide. Rocky shores and stony beaches seem to be their preferred habitats, and they have been observed feeding on sand fleas and gam-marids in such locations (Bent 1927). They often feed in water up to their breast and can swim readily. Preble and McAtee (1923) reported the contents of 192 stomachs as containing 33 percent mollusks, 29 percent crustaceans, 23 percent flies, 10 percent beetles, and small amounts of marine worms and vegetation, mostly algae.

Social behavior. In the fall, flocks of 30–40 birds are evidently common, but pair-formation must occur during winter, since the birds are said to be paired on arrival at their breeding grounds on St. Paul Island. On Amchitka, the birds form rather large flocks between January and March while foraging on tide-exposed rocks along the beach, but these numbers decline in April and by early May the birds are occupying territories well away from the beaches (Kenyon 1961). One of the few descriptions of the male's display flight is that of Bent (1927), who said that the birds perform song flights by "rising 30–40 feet in the air and fluttering down while pouring out a delightful twittering song." Murie (1959) compared the song to the droning trill of toads. They also utter a call similar to that of the upland sandpiper while in flight or while standing on an elevated hummock. One early description of what apparently was this song compared it to the bleating of a common snipe, during which the bird quivered with a tremulous motion as if highly excited. Eggs have been reported as early as May 6 in the Pribilovs but have also been found as late as July 24, suggesting a surprisingly long nesting season. Renesting has been found to occur (Hanna 1921).

Reproductive biology. These birds nest in upland tundra, as much as 500 feet above sea level, often in areas of light gray tundra where they are well concealed among an abundance of "reindeer moss" lichens. Some nests have also been found where there is a mixture of *Hypnum* mosses and where there are prostrate dwarf willows. However, nests have also been found in beach gravel, in debris just above tide line, and on a patch of snow on a mountainside that was at least 700 feet above the valley floor. The incubation period is believed to be about 20 days, and both parents are said to share in incubation. The normal clutch is 4 eggs, laid at periods ranging from 1 to 3 days apart. Collection of the first clutch has been found to result in renesting, but not in the same nest site. The young birds are taken from the nest soon after hatching, and for a time they remain in the upland tundra with the adults, eating flies and beetles. When they are able to fly they move to the beaches, where they are later joined by the adults and soon begin to form large flocks that may number more than 100 birds (Hanna 1921; Bent 1927).

Status and relationships. It is impossible to judge the status of individual subspecies or even the species, but obviously the total populations of some of the races must be very small and possibly subject to losses from foxes or other predators that occur on many of the Aleutian islands. At various times *ptilocnemis* has been considered conspecific with *maritima,* with which it is allopatric. Generally, Soviet investigators have listed the forms as conspecific, while in North America they have been considered separate species, since *ptilocnemus* was originally described by Robert Ridgway as a specifically

distinct form. Stepanjan and Flint (1973) have recently reviewed the evidence, pointing out the consistent differences in the two forms' breeding plumages and noting the absence of intermediate forms. They have also compared the strong differences in the patterning and coloration of the eggs, which they regard as additional evidence that the birds represent different species. Although I doubt that egg color and patterning has great taxonomic value, I do agree that the forms are best regarded as distinct species pending further study.

Suggested reading. Murie 1959; Bent 1927.

Sanderling

Calidris alba (Pallas) 1764
(*Crocethia alba* of Peters, 1934)

Other vernacular names. None in general English use; bécasseau sanderling (French); Sanderling (German); correlimos blanco, correlimos tridactilo (Spanish).

Subspecies and range. No subspecies recognized. Breeds in North America in northern Alaska (Point Barrow), northern Mackenzie, Banks, Victoria, Prince, Patrick, Melville, Devon, Bylot, and Ellesmere islands; also on the Melville Peninsula, Cape Fullerton, and Southampton Island. Also breeds on Greenland, Spitsbergen, the Taimyr Peninsula, Severnaya Zemlya, the mouth of the Lena, and on the New Siberian Archipelago. Winters in the western hemisphere from British Columbia and Massachusetts south to southern South America, and in the eastern hemisphere from the British Isles and China south to South Africa, India, and New Zealand. See map 91.

Measurements and weights. Wing (flattened): males 113–25 mm; females 113–27 mm. Culmen (to feathering): males 23–26 mm; females 24–28 mm. Weights (on breeding grounds): males 50–64 g, average of 5 was 53.7 g; 2 females 52 and 62 g (Glutz et al. 1975). Eggs ca. 35 x 25 mm, estimated weight 11.2 g (Schönwetter 1963).

DESCRIPTION

Breeding adults of both sexes are cinnamon rufous to rusty on the crown and hindneck, streaked with dusky and grayish white; the sides of the head and

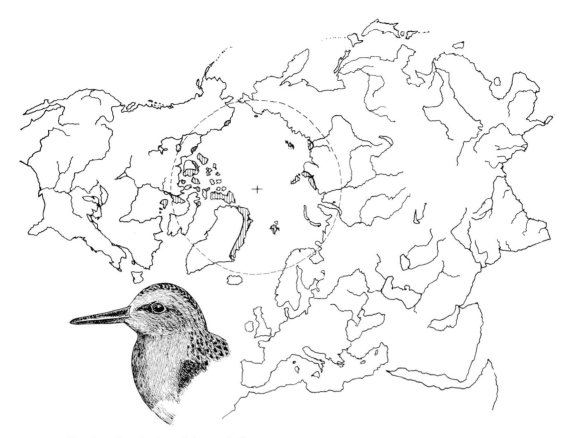

91. Breeding distribution of the sanderling.

neck are similar but paler, and the foreneck and chest are the same but barred. The chin, throat, and underparts are mostly pure white, with some spotting on the sides of the breast. The scapulars are black centrally, with pale whitish or brownish margins, and the rump and median tail coverts are brownish gray with white borders, the white increasing laterally. The tail feathers are mostly gray with white margins, the outermost pair being largely white. The wing coverts are mostly dusky with whitish margins, but the greater coverts are brownish gray with broad white tips. The secondaries are white on the basal half and edged with white, and the primaries are dusky, with the inner ones having considerable white on the basal parts of the outer webs. The iris is dark brown, the bill is black, and the legs and toes are black. *Adults in winter* are gray on the crown, back, and upper wing coverts (except for the anterior lesser coverts, which are dusky), with dusky shaft streaks. The underparts are completely white, and there is a white superciliary stripe. *Juveniles* have a blackish crown and

back, spotted with whitish buff. The wing coverts have wide buffy fringes, and the underparts are white with a buffy wash on the breast. *First-winter* birds resemble adults but have remnants of the buffy spots on the wing coverts (Prater et al. 1977).

In the hand, the absence of a hind toe, a straight bill about as long as the tarsus, and an extensive white wing stripe contrasting with gray upper coverts that become blackish at the leading edge and toward the wrist serve to identify this species.

In the field (8 inches), sanderlings are found on sandy beaches, where they remain very close to the water's edge, often following retreating waves. In winter plumage the birds appear very pale gray, except for a blackish wrist mark at the bend of the wing. In spring they are reddish brown on the head and neck but remain white underneath. In flight the birds show a striking white wing stripe, and they utter sharp *twit* notes when alarmed. They fly and forage in compact flocks and prefer to run rather than fly when approached.

❖ ❖ ❖   277

NATURAL HISTORY

Habitats and foods. This is a high arctic breeding species, primarily associated with rock desert and stony,
barren tundra supporting scant vegetation. The birds
often breed well away from the coast but usually are
not far from fresh water. Most territories and nests
are on stony, well-drained ridges, gentle slopes, or
level alluvial plains, all of which usually support
scattered willow, *Dryas*, and saxifrages and usually
are less than 200 feet above sea level (Parmelee 1970).
The habitats are similar to those used by red knots,
and, as in that species, breeding densities tend to be
very low. On Bathurst Island the breeding density
has been estimated as 3–4 pairs per square mile, but it
sometimes reaches 6–8 pairs in the best habitats and
is 1–2 pairs in some poor years. Higher densities of
perhaps as high as about 17 pairs per square mile
might be typical of a few habitats such as reported on
Prince of Wales Island (Parmelee 1970). This species
typically forages by quickly picking materials from
the surface of the substrate, but shallow probing is
also used to some extent. Like the red knot, the birds
are sometimes forced to eat vegetable matter when

they first arrive on their high arctic breeding
grounds, before invertebrate life becomes generally
available. The buds of saxifrages as well as moss and
algae may be eaten at such times. Later in the summer and autumn the birds eat many insects such as
midge, mosquito, and cranefly larvae. On autumn
migration they often feed in wet sand by running
quickly and producing a series of closely spaced
shallow probes or "stitches" in the sand, capturing
sand flies, small mollusks, ostracods, polychaete
worms, and the like. On wintering areas of coastal
Argentina the birds typically forage in small flocks
on sandy areas of tidal flats, usually in groups of up
to about 10 birds. Some winter territorial behavior
has been observed, but not all individuals establish
territories (Myers et al. 1979).

Social behavior. Sanderlings arrive on their high arctic nesting grounds near the end of May or in late
June, usually in small flocks, both sexes arriving together. Within a few days at least some of the birds
are apparently paired, suggesting a reestablishment
of old pair bonds. The remainder form pairs very
soon after the migrant flocks break up, and single

and unmated sanderlings seem to be a small minority. The territories encompass a portion of level or sloping ground above which flight displays are performed and include an area about 400 yards in diameter. The territories seem to be widely scattered, and thus there are few territorial conflicts over boundaries. However, intruders are quickly chased away if they are males or courted if they are females. The flight display thus seemingly serves mostly as a mate advertisement device, and it apparently ceases with the formation of a pair bond. The displaying bird ascends slowly or rapidly to a usual height of about 30 feet but sometimes reaches 60 feet. It then levels off and begins to fly with rapidly vibrating wings, alternated with brief glides, with the head somewhat drawn back into the body, producing a hunched appearance. The song is uttered in bursts, possibly synchronized with the wingbeats, and is loud and "froglike." The bird flies either in a straight line or in varying directions but usually lands within 200 yards of its point of takeoff. It descends either abruptly or in a long, gliding flight, sometimes ascending again into a new display sequence just before reaching the ground. Probably the female performs such flights occasionally, but she does not produce a flight song. Ground displays by the male include following a female with hunched head, drooping wings, erected feathers, and tail feathers that are spread and depressed. Females apparently initiate copulation by displaying in special nest scrapes that are used especially for this purpose and are not the sites of future nests. These are usually old nest sites, possibly even of other shorebird species. Copulation is initiated by the female's squatting in such a scrape and allowing the male to approach and ease her out of it. The two birds quickly run off side by side for about 10 feet, after which the male stops the female by pressing his bill against her breast. Copulation then follows and sometimes lasts several minutes. It is believed that this elaborate ceremony may be a device for strengthening a fairly weak pair bond (Parmelee 1970).

Reproductive biology. Nests are typically placed in open situations, fully exposed to the sky, often in a tuft of *Dryas*, willows, or saxifrages. They usually contain dried willow or saxifrage leaves and other dried leaves that seem to accumulate accidentally. Eggs are laid at about 26–29-hour intervals, and incubation probably does not begin until the entire clutch is completed. The clutch is normally 4 eggs, but in exceptional cases 3 have been recorded. The pattern of incubation is a matter of some interest and controversy. Parmelee (1970) reported that the pair bond terminates before incubation, and only a single adult, male or female, incubates the eggs. Parmelee and Payne (1973) later suggested that on Bathurst Island female sanderlings lay 2 clutches in rapid succession, each being independently raised by one of the two parents. However, in Greenland, recent observations by Pienkowski and Green (1976) indicate that both sexes attend the nest and help raise the brood. This controversy might be resolved by the possibility that different breeding strategies occur in different parts of the species' range, or possibly even in different years in the same area, depending on ecological conditions. In any event, incubation requires from as little as 24.2 days to as much as 31.6 days, the latter apparently being exceptional inasmuch as nearly 6 days elapsed between the completion of the clutch and the start of incubation. It is probable that in multiclutch species the male may regularly delay the onset of incubation until the female has completed her second clutch, thereby insuring that he will be available to fertilize the second set of eggs (Hilden 1975). Although shorter fledging periods have been estimated for the species, it is likely that about 17 days are required for fledging to occur (Parmelee 1970). Adults sometimes can be found with flying juveniles, but many adults leave the breeding grounds during July. By early August nearly all the birds have left the breeding grounds; thus the nesting areas are occupied by sanderlings for only about two months of the year.

Status and relationships. This is one of the most widespread of all the species of shorebirds and also has one of the longest migration routes, comparable to that of the golden plover. Its widely dispersed breeding and wintering grounds probably favor its prospects of long-term survival, and at present it certainly does not seem to merit any special conservation concern. The evolutionary relationships of the species are rather uncertain; it has often been placed in a monotypic genus (*Crocethia*) on the basis of its lack of a hallux, but few if any contemporary authorities support this position. Yet the relationships with *Calidris* are certainly obscure, and it seems most likely that the species has been derived from the general "peep" assemblage.

Suggested reading. Parmelee 1970; Pienkowski and Green 1976.

Genus *Limicola* Koch 1816 (Broad-billed Sandpiper)

The broad-billed sandpiper is a small Eurasian shorebird with a well-developed hind toe, a bill that is long but relatively wide, soft and flexible with a hard tip. The bill is strongly flattened in front of the nostrils and straight or slightly recurved, but with a decurved tip. The tail is nearly square, with the central pair of feathers longer and more pointed than the others. The downy young are *Calidris*-like. One species.

Broad-billed Sandpiper

Limicola falcinellus (Pontoppidan) 1763

Other vernacular names. None in general English use; bécasseau falcinelle (French); Sumpfláufer (German); correlimos falcinelo (Spanish).

Subspecies and ranges. See map 92.

L. f. falcinellus: Western broad-billed sandpiper. Breeds in Norway, Sweden, Finland, and on the Kola Peninsula of Russia. Winters in the Mediterranean and from the Persian Gulf to India.

L. f. sibirica: Eastern broad-billed sandpiper. Breeding area unknown, but probably in eastern Siberia east of the Yenisei River. Winters from China and Hainan south to the Indo-Chinese countries, eastern India, and Australia.

Measurements and weights (of falcenellus). Wing: males 99–109 mm; females 104–11 mm. Culmen (from feathering): males 26–34 mm; females 28–36 mm. Weights vary between 25 and 50 g, averaging about 40 g in summer (Glutz et al. 1975). A sample of 35 wintering *sibirica* ranged 32–56 g, averaging 40 g (D. Purchase, pers. comm.). Eggs ca. 32 x 23 mm, estimated weight 8.6 g (Schönwetter 1963).

DESCRIPTION

Breeding adults of both sexes have a brownish black crown, a white eye stripe spotted with dusky and

92.   Breeding distributions of the eastern (E) and western (W) broad-billed sandpipers.

becoming branched toward the nape, which is buffy with sepia streaks, a dusky line extending from the lores to the ear coverts, and the sides of the head and neck buffy with sepia streaking, the streaking heaviest on the upper breast and grading to pure white on the underparts. The mantle and scapular feathers are black with creamy, buffy, or tawny edging, the rump and middle upper tail coverts are blackish, and the tail is blackish brown centrally and pale ashy brown on the outer feathers, which are edged with white. The upper coverts are mostly ashy brown with white edging, especially on the greater coverts, while the flight feathers are sepia with white tips on the secondaries. The iris is brown, the bill is blackish, becoming yellowish toward the base, and the legs and toes are blackish to greenish or greenish brown. *Adults in winter* have the upperparts dark gray brown, with darker shaft streaks, white underparts, and the breast tinged and lightly streaked with gray. *Juveniles* have chestnut and whitish edges on the feathers of the upperparts, and the wing coverts are pale brown with broad edging of pale buff. The breast is faintly streaked with dull brown. *First-winter* birds are blotched with gray and brown on the upperparts, and the fringing of the wing coverts soon becomes

faded to resemble that of adults, except for the inner medians, which remain buffy (Prater et al. 1977).

In the hand, the relatively long (culmen 26–36 mm) bill, which, rather than being expanded toward the tip, is relatively wide from the base to near the tip, is unique to the group. Additionally, the forked superciliary stripe and rather short legs (tarsus 20–23 mm) are useful criteria.

In the field (6½ inches), this small shorebird closely resembles some *Calidris* forms, especially winter-plumaged dunlins. It is darker in general plumage tones than the *Calidris* species, almost snipelike in its back pattern and its strongly patterned head with the forked superciliary stripe. The birds are usually found singly, in intertidal flats or marshy areas, and on flushing very dark brown above, with no white wing stripe present.

NATURAL HISTORY

Habitats and foods. During the breeding season this species is associated with marshes having adjacent swampy meadows surrounded by coniferous forests. Secondarily it occurs in swampy wooded tundra habitats and montane moorlands. It overlaps more with the jack snipe in its ecological needs than it does with typical *Calidris* sandpipers. However, outside the breeding season it is often associated with muddy seashores, more like other *Calidris* species than like snipes. During migration in Europe the species is usually found on salt marshes, muddy creeks and inlets, or in muddy areas near the coast rather than on open shorelines, but in wintering areas it is reportedly often associated with open tidal flats, the mouths of rivers and creeks, and other areas having muddy rather than sandy substrates. Population densities are apparently very low even in favored habitats, ranging up to about one breeding pair per square kilometer in Finland (Hakala, cited in Glutz et al. 1975). Foods consist of insects, earthworms, snails, and other small invertebrates, most of which are probably obtained by probing rather than by visual foraging. Almost nothing is known of the foods during the breeding season, and a general analysis of available information on foods (Glutz et al. 1975) indicates the probable importance of univalve mollusks and insects in the diet, with little or no usage of bivalves, crustaceans, or other invertebrates. The heavy bill of this species is probably related to its adaptations for foraging on relatively large prey organisms (Burton 1971).

Social behavior. The social behavior of this elusive species is very poorly understood at present. In a study on the lower portion of the Indigirka River, USSR, where the tundra meets the forest, Flint (1973b) found these birds to be occupying an area where a large flooded lake basin dominated by sedges was near dry and hillocky tundra. Within that general area the birds established small "territories," which, however, they did not defend from other males. Instead they occurred in small colonies, and the males performed their aerial displays both singly and in groups above flat areas of the basin. Evidently the males do not fight among themselves, and during courtship flights they utter a soft buzzing call somewhat resembling that of a dunlin. The call is uttered as the birds circle over the marsh, presumably only by males, but participation in the flights by females has not been completely ruled out. It is believed that each male makes several nest scrapes, in one of which the female chooses to lay her eggs (Flint 1973a). The significance of colonial nesting, and of the apparent lack of territorial defense, is still unknown, and the relative role of the female in the later stages of breeding is still very uncertain.

Reproductive biology. Nests of this species are often on small sedge hillocks in the middle of flooded polygonized tundra, or less frequently are placed in shrubs surrounded by water. The nest is usually raised well above the water level and is often on a wet sedge or moss cushion, above which the nest cup is lined with the leaves of birch, willow, or dried grasses or sedges (Bannerman 1961). The completed clutch is usually 4 eggs, but there are records of nests containing 3 and 5 eggs as well. Since the birds are semicolonial, the nests are sometimes rather closely spaced. In one study the usual distance between nests was 80–100 meters, but some nests were as close as 9 meters apart. In this area, each colony consisted of from 2 to 4 "nesting areas" (Flint 1973a). Although most early accounts suggest that both sexes participate in incubation, Flint reported that only males were found near the nests, and there was no evidence that females participated in either incubation or rearing of the young. There is still no information on the lengths of the incubation and fledging periods. According to Flint, males that are left without clutches to incubate continue their courtship activities until the autumn migration begins.

Status and relationships. This species is apparently uncommon to rare throughout its range, though its ecological preferences place it out of view much of the time, and so its population is likely to be underestimated by the casual observer. It is obviously a specialized variant of the *Calidris* group of sandpipers, as is strongly indicated by its downy pattern (Fjeldså 1977). Strauch (1976) advocated merging *Limicola* with *Calidris,* and Voous (1973) also indicated that this procedure would meet his approval. Burton (1971) suggested that the large and heavy bill of this species evolved as a mechanism to allow the bird to more easily capture prey larger than that typical of other calidridine sandpipers of comparable size, and he questioned Larson's (1957) suggestion that *Limicola* and *Eurynorhynchus* might represent a species pair of rather ancient (Miocene) origin. He also questioned the desirability of transferring either *Limicola* or *Eurynorhynchus* to *Calidris.* Inasmuch as it is clear that the affinities of *Eurynorhynchus* are clearly with the "peeps" or stints, I believe that a merger of that species into *Calidris* is desirable, but since the possible affinities of *Limicola* to a particular subgroup of *Calidris* are still completely unknown, I favor retaining *Limicola* as a valid genus at this time.

Suggested reading. Nisbet 1961.

Genus *Micropalama* Baird 1858 (Stilt Sandpiper)

The stilt sandpiper is a small North American shorebird with very long, slender legs, a hind toe, and webbing between the bases of all the front toes. The bill is slender, straight, and compressed but is expanded toward the tip. The tail is nearly square and of 12 feathers, with the middle pair not projecting beyond the others. The downy young are *Calidris*-like. One species.

Stilt Sandpiper

Micropalama himantopus (Bonaparte) 1826

Other vernacular names. None in general English use; bécasseau échasse (French); Bindenstrandläufer (German); correlimos patilargo (Spanish).

Subspecies and range. No subspecies recognized. Breeds in northern Canada on Victoria Island, northern Mackenzie, Keewatin, northeastern Manitoba, and north coastal Ontario. Also breeds locally in northern Alaska, west possibly to Barrow. Winters in South America, from Bolivia and Brazil to northern Chile and northern Argentina. See map 93.

Measurements and weights. Wing (chord): males 116–35 mm; females 120–37 mm. Culmen (to feathering): males 35–41 mm; females 36–44 mm. Weights (on breeding grounds): 24 males 48–60.8 g, average 53.8 g; 15 females 52–68 g, average 60.9 g (Jehl 1973). Eggs ca. 36 x 25 mm, estimated weight 11.2 g (Schönwetter 1963).

DESCRIPTION

Breeding adults of both sexes have a dusky crown and a nearly white superciliary stripe, chin, and upper throat. A rusty brown stripe extends above the superciliary stripe around the back of the nape, and a diffuse rufous area covers the lores and the ear

93. Breeding distribution of the stilt sandpiper.

coverts, which are also streaked with dusky. The breast, sides, flanks, and upper abdomen are broadly barred with dusky, the under tail coverts are irregularly marked with the same, and the underwing surface is sparsely marked with grayish. The feathers of the upperparts are variegated with black and pale gray intermixed with buffy. The rump is brownish gray, the upper tail coverts are white with some black barring, and the tail is brownish gray with white on the inner webs. The wing coverts are deep brownish gray with paler margins, the secondaries are darker and edged with white, and the primaries and their coverts are dusky. The iris is brown, the bill is blackish, and the legs and toes are greenish to olive yellowish. *Adults in winter* are plain brownish gray on the upperparts, with a white superciliary stripe, white upper tail coverts, mostly white underparts with narrow grayish streaking, and no rufous on the face. *Juveniles* have blackish brown upperparts with pale whitish to buff fringes, and the coverts are grayish brown with broad whitish buff edging. There are a few dark streaks on the breast, which is washed with buff. *First-winter* birds retain some buff-fringed inner median coverts and gradually become more grayish on the upperparts (Prater et al. 1977).

In the hand, the long greenish to yellowish legs (tarsus 34–45 mm) and rather long and slightly decurved bill (culmen 34–44 mm) are useful marks, as are a white rump (slightly barred in spring) and a pale, unbarred tail.

In the field (8 inches), stilt sandpipers are usually found in small groups feeding in shallow prairie marshes, wading in breast-deep water. The rusty head markings and barred underparts are distinctive field marks in spring, but in winter plumage the birds might be mistaken for lesser yellowlegs, which have similar proportions and also have a whitish rump area. In flight, the birds utter a low *querp* or *whruu*, sometimes ending with a sort of chatter, quite different from the call of the yellowlegs.

NATURAL HISTORY

Habitats and foods. This subarctic to low arctic species breeds north of timberline in habitats ranging on Victoria Island from wet tundra areas having fairly high willow growth to higher and much drier slopes. However, in the Churchill area of Manitoba it is mostly limited to well-drained sedge meadows having scattered elevated areas that provide dry nesting sites early in the spring. Foraging in early summer is typically done by wading belly-deep in water and making thrusts at organisms that apparently are detected visually. Later in summer the birds tend to probe in soft mud. Foods of adults in summer include adult and larval beetles, especially dytiscids, adult dipterans, snails, and larval insects. Young birds often eat larval midges, but the stilt sandpiper does not appear to be as dependent on midges and craneflies as is typical of *Calidris* (Jehl 1973). Yet midge larvae have been found in stomachs of birds taken during late summer in Puerto Rico, along with a considerable quantity of seeds (Bent 1927). Winter habitats in South America include muddy tidal flats, shallow pools, marshy fields, and sometimes outer beaches (Blake 1977). In coastal Argentina the species uses coastal estuaries and flooded ponds and pastures, foraging in belly-deep water in groups of 20–100 birds. It usually shows considerable gregariousness and absence of territoriality at that season (Myers and Myers 1979).

Social behavior. In the Churchill area, birds begin arriving in late May, either singly or in small flocks. Males establish territories immediately upon arrival, and begin to advertise such territories with prolonged display flights lasting several minutes. Normally the male flies at a height of 20–60 meters, making shallow wingbeats and spreading the tail and sometimes also gliding and singing. Before landing he sings once more, then raises his wings and drops quickly back to earth. Territories average 15–20 acres in area, but in some areas may cover only 3–5 acres. There is a high degree of fidelity to previously held territories and mates, with remating rates of nearly 50 percent apparently typical. Jehl (1970, 1973) believed that initial pairing is based on size differences in the sexes, with relatively small males and large females being among the first to pair and begin nesting. Apparently the strong site and mate fidelity are adaptations that promote early nesting by experienced birds. Much of the courtship occurs in the air, with males chasing females while singing song after song, attempting to fly ahead of the females and then raising their wings almost vertically and singing as they fall while tilting from side to side. After dropping nearly back to earth they resume their chase and repeat the performance. Males also take the initiative in nest-scraping behavior, sometimes making five or more scrapes before obtaining a final nest location (Jehl 1973). Compared with those of typical *Calidris* species, this species' display tends to be fairly high and rangy, with the birds sometimes reaching heights of 200 feet or more, and the male sometimes alights a quarter-mile away from where he first took off. However, in its display calls the bird closely resembles the Baird sandpiper, and in its general aerial behavior it is similar to the red knot (Parmelee et al. 1967).

Reproductive biology. Nests are sometimes well hidden in dry vegetation near water, but often they are almost wholly exposed. Around Churchill the nests are usually atop small sedge hummocks or on low gravel ridges near sedge meadows, while on Victoria Island the birds sometimes nest in wet tundra habitats such as those used by pectoral and semipalmated sandpipers. The egg-laying interval averages about 36 hours, and the clutch size is almost invariably 4 eggs. Of 47 nests found near Churchill, 39 had 4 eggs, with a range of 2–5, and an average of 3.9 (Jehl and Smith 1970). Incubation is continuous after the last egg is laid, with males typically incubating during the day and females at night. The male maintains territorial defense for a time after the clutch is completed, but his role in incubation soon effectively terminates this behavior. Distraction displays by birds flushed from the nest include injury-flight and "rodent-run" behavior, though Jehl (1973) considers the latter descriptive name misleading and inappropriate. Incubation requires 19.5–21 days, and pipping may begin 3 to 4 days before hatching. Generally the entire clutch hatches within a span of 10–14 hours, and either parent or both may be present at that time. Departure from the nest occurs after all the chicks are dry, and the young are soon moved to relatively wet and marshy areas, sometimes moving

as much as 2 miles in the first 2 weeks after hatching. Fledging occurs when the young are 17–18 days of age, and by then both parents have abandoned them, with the females leaving first and the males going after about 2 weeks. Young birds begin to leave the nesting areas when they are still less than a month old (Jehl 1973).

Status and relationships. No information is available on population size in this species, but its wide breeding range over an area subject to little human disturbance seems to make its status quite secure. Its evolutionary relationships are clearly with *Calidris*, and indeed there is little morphological basis for retaining it in a monotypic genus (Jehl 1973). Merger of *Micropalama* with *Calidris* has been suggested by various workers such as Strauch (1976) and has also been adopted elsewhere (Glutz et al. 1975). This approach would certainly not be objectionable, especially if definite affinities could be established with a particular species or species group within that genus as here constituted. Jehl (1973) has suggested that its nearest relative may well be *Calidris ferruginea*, based on body proportions, juvenile and winter plumages, and vocalizations. Patterns of aerial display are also very similar in the two species, and both use a whining note in territorial defense. The two forms tend to occupy comparable ranges in Eurasia and North America, but they also exhibit pronounced differences in their social organization. This is probably not of real taxonomic significance, and thus I submit that this species might as well be included in the genus *Calidris*, probably somewhere near *ferruginea*.

Suggested reading. Jehl 1973; Parmelee et al. 1967.

Genus *Philomachus* Anon. (= Merrem) 1804 (Ruff)

The ruff is a small Eurasian sandpiper with moderately long legs, no hind toe, and a small but distinct web between the middle and outer toes. The bill is about as long as the head and is straight and tapering with a slightly expanded tip. The wing is pointed and the tail is rounded, with twelve rectrices, the middle pair not projecting beyond the others. The sexes are seasonally highly dimorphic in plumage. The downy young are *Calidris*-like. One species.

Ruff

Philomachus pugnax (L.) 1758

Other vernacular names. Reeve (applies to female); bécasseau combattant (French; Kampfläufer (German); combatiente (Spanish).

Subspecies and range. No subspecies recognized. Breeds in Europe and Asia from France north to Norway, and eastward across Germany to Russia and Siberia to Anadyrland. Also breeds on Kolguev and Vaigach islands and has bred in the New Siberian Archipelago, as well as in the Ukraine, Austria, Hungary, and England. Winters from the British Isles, the Mediterranean, and the Persian Gulf south to southern Africa, India, the Indo-Chinese countries, Malaya, and Borneo. See map 94.

Measurements and weights. Wing: males 173–210 mm; females 139–75 mm. Culmen (to feathering): males 29–39 mm; females 26–36 mm. Weights: males show marked seasonal variations in average weights ranging from 180 g in spring to about 230 g in fall, and females average from about 110 g in spring to about 135 g in fall (Glutz et al. 1975). Eggs ca. 44 x 31 mm, estimated weight 21.0 g (Schönwetter 1963).

DESCRIPTION

Adult males in breeding plumage are highly polymorphic; the head, neck, chest, and breast, together with paired erectile occipital tufts and a pectoral ruff, range from entirely white through sandy buff or chestnut to purplish black, often interspersed with white or brownish as streaks, bars, or freckles. The back and scapulars are variably variegated with black, gray, buffy, and whitish, the wing coverts are grayish brown with white edging, the primaries and their coverts are darker grayish brown with whitish shafts, and the secondaries are grayish brown with

94.   Breeding distribution of the ruff.

white edging. The upper tail coverts are blackish with variable barring medially and entirely white laterally, and the central tail feathers are light grayish brown with variable barring. The front of the face is bare and covered with red to yellowish papillae, the iris is brown, the bill is blackish distally and yellowish, pinkish, or orange red toward the base, and the legs and toes are usually orange red or pinkish. *Males in winter* are mostly grayish brown on the upperparts, the feathers darker centrally and with paler margins, the greater wing coverts are tipped with white, and sides of the head and neck and most of the underparts are light grayish brown, the feathers having whitish tips. The chin and upper throat are mostly white, the occipital tufts and pectoral ruff are absent, and the bill is more brownish, while the legs and toes are pinkish to orange red. *Adult females* resemble males in winter plumage but are smaller, and in summer plumage are much darker with more extensive barring on the tail feathers. *Juveniles* are mostly dark grayish brown above with warm buff feather edging, and with white underparts except for

a streaked breast. The legs are yellowish brown to greenish. *First-winter* birds resemble adults but retain bright buff-fringed inner median coverts. The legs may become brownish green to orange green. First-year males in spring have shorter pectoral ruffs than do older males (Prater et al. 1977).

In the hand, spring-plumaged males are unmistakable, but otherwise the birds resemble a short-billed *Tringa* or a long-legged *Calidris*. The combination of a long tarsus (38–54 mm), a rather short bill (culmen 28–39 mm), and a tail with a subterminal black band or irregular barring helps identify this species.

In the field (9–12 inches), breeding males have distinctively colored ruffs, but females and winter-plumaged males are more difficult, with pinkish to yellowish or greenish legs, a bill that is usually yellowish at the base, and a rather uniformly grayish brown breast, lacking definite streaking. The birds closely resemble redshanks or greenshanks in winter plumage, but they have a buffy scaly pattern on the upperparts that sets them apart from most *Tringa* forms. In flight they sometimes utter a *too-i* call

33. Heads of female (*above*) and breeding male (*below*) ruffs.

habitats the breeding density may reach 2.5 breeding females per 10 hectares, but often it is much less than this density (Glutz et al. 1975). Outside the breeding season it inhabits the muddy shorelines of lakes, pools, and wide estuaries having muddy banks or exposed mud flats (Voous 1960). Its bill is *Calidris*-like, and it evidently feeds on a rather large variety of terrestrial insects, as well as earthworms, snails, and small crustaceans. At times or locally it also consumes plant seeds, including cultivated grains such as rice, plus the seeds of weeds and grasses and even berries.

Social behavior. This species probably exhibits the most remarkable social behavior of any shorebird and is indeed perhaps unique among all birds in its degree of male plumage polymorphism associated with sexual behavior. All courtship and mating in this species occurs in communal display grounds, or leks, which are usually raised grassy hillocks on which varying numbers of males assemble. Each lek is subdivided into individual display territories or "residences." These are bare spots about 30 centimeters in diameter and about a meter apart. Displays consist of ritualized aggression, including raising of the head tufts and neck ruffs, wing-raising, and also overt threats and flights. The great variation in the coloration of the head tufts and neck ruffs results in strikingly diverse visual aspects of display among the males. The first person to prove the reproductive significance of such plumage polymorphism was Hogan-Warburg (1968), who found that two groups of males occur within leks—independent males and satellite males. Independent males are mostly dark-colored, and their behavior includes much fighting and other related behavior associated with territoriality. Within this category are two types; resident males that defend territories that they occupy almost continuously, and marginal males that do not defend territories but stay at the edge of the lek, sometimes acquiring resident status by establishing peripheral territories. Satellite males, which are mostly white, do not defend territories but rather visit the territories of resident males, avoiding aggressive behavior toward these males. Satellite males also include two types, central and peripheral satellites. The central satellite males spend more time on the lek and are less easily driven away by resident males. Transformations between independent and satellite status do not occur, but the status type within each of the two major categories is influenced by age. Females visit the lek for short periods, and on their arrival the resident

but are generally quite silent. They are often associated with damp meadows and marshes, and less frequently occur on coastal shorelines.

NATURAL HISTORY

Habitats and foods. This species breeds over a rather wide range of temperate to subarctic climatic zones, mainly in low-lying grassy marshes in slightly wooded or treeless portions of the northern birch zone and the shrub and moss tundra habitats. It also breeds in the marginal vegetation of freshwater lake shorelines, in wide river meadows, in extensive coastal salt marshes, and locally in meadows or hayfields of cultivated areas (Voous 1960). In favorable males adopt a motionless squatting posture. Satellite males also squat with the resident males. Females are evidently stimulated to step onto the territory, and they usually crouch during periods when the resident male is attacking or threatening nearby satellite males. Copulations often occur at this time. On large leks, the resident males that are alone on their territories are almost exclusively the successful copulating partners, while on small leks the presence of one or more satellite males increases the chances for copulation by both resident and satellite males. Thus the species shows balanced behavioral polymorphism, in which the reproductive value of satellite males lies in the fact that their presence increases the chances for copulation on small leks and promotes the establishment of new leks or maintenance of several leks in an area (Hogan-Warburg 1968). The status category of males (independent or satellite) depends on both genetic and environmental factors, with some plumage types occurring exclusively in one status category or the other, while males with intermediate plumage types may belong to either category. The maintenance of this plumage and behavioral polymorphism may thus be the result of a reproductive superiority of these intermediate or heterozygotic types, or the mutual dependence of independent and satellite males (van Rhijn 1973). The success of an individual resident male in attracting females for copulation appears to be related to his ability to defend a centrally located display territory, his rate of display, and the presence of satellite males on display territories when receptive females are also present (Shepard 1976). The reproductive value of the two major status categories probably varies locally and with time, with independent males being favored during periods when there are low numbers of males present and satellites having the reproductive advantage when there are many males available (van Rhijn 1973).

Reproductive biology. After they have been fertilized, the females leave the lek and begin to construct their nests. Nests are typically in rather damp situations among deep grass and are usually at least 100 meters away from the nearest lek site. There are almost invariably 4 eggs in the clutch, though 3-egg clutches have sometimes been reported. The eggs are normally laid at 24-hour intervals, and all the incubation is performed by the female, since the male plays no role in nesting or brooding phases of reproduction. The incubation period is 20–23 days, usually 21 days. Although the young become relatively independent in 7–10 days, the fledging period is probably 25–27 days (Glutz et al. 1975).

Status and relationships. Throughout much of its European range the ruff has decreased considerably in numbers, probably because of increasing cultiva-

tion of its favored breeding grounds. Though once abundant in Great Britain, it was nearly extirpated by about the end of the last century, probably because of drainage and hunting, and it last bred in England in 1922. However, it returned to breed again in 1963, and a small population again exists on the Ouse Washes. It also returned after a thirty-year absence to the Dovrefjell, in Norway (Gooders 1969). Based on anatomical evidence and the plumage of the downy young, this species is clearly close to the *Calidris* assemblage (Jehl 1968a; Lowe 1915). Strauch (1976) went so far as to merge *Philomachus* with *Calidris,* and it certainly must be admitted that at least some of the form's "generic" traits (cf. Ridgway 1919) are male features associated with the

highly competitive and promiscuous mating system. Yet, since it is not clear that it is more closely related to any single species or group within *Calidris* than any other, I see no real disadvantage in retaining the genus. It has been suggested that this species may be most closely related to *Calidris melanotos* (Pitelka 1959; Voous 1973), but I suspect that this is a reflection of their similar breeding systems. I also doubt that it is very close to *Tryngites;* instead, I suspect that behavioral similarities between the genera are likewise the result of covergences associated with similar mating systems.

Suggested reading. Shepard 1976; Hogan-Warburg 1968.

Genus *Tryngites* Cabanis 1856 (Buff-breasted Sandpiper)

The buff-breasted sandpiper is a small North American shorebird with moderately long legs, a hind toe, and no webbing between the front toes. The bill is shorter than the head, slender and tapering. The wing is long and pointed, and the flight feathers are marked with black marbling on the inner vanes. The tail is rounded, of twelve rectrices, with the middle pair not projecting beyond the others. The downy young are *Calidris*-like. One species.

Buff-breasted Sandpiper

Tryngites subruficollis (Vieillot) 1819

Other vernacular names. None in general English use; bécasseau rousset (French); Grasläufer (German); correlimos canelo, correlimos ocraceo (Spanish).

Subspecies and range. No subspecies recognized. Breeds locally in northern Alaska (Prudhoe Bay, Point Barrow), perhaps in the northern Yukon, and in northwestern Mackenzie, Banks Island, Victoria Island, Bathurst Island, Melville Island, and King William Island. Winters in Paraguay, Uruguay, and Argentina. See map 95.

Measurements and weights. Wing (chord): males 129–36 mm; females 122–32 mm. Culmen (to feathering): males 19–21 mm; females 17–20 mm. Weights (summer): 4 males 64–80.5 g, average 71 g; 6 females 50–58 g, average 53 g (Irving 1960). Eggs ca. 37 x 26 mm, estimated weight 13.0 g (Schönwetter 1963).

DESCRIPTION

Adults of both sexes have the sides of the head and neck as well as the underparts pale pinkish cinnamon to pinkish buff, the crown streaked with blackish, with a pale buffy eye ring and a pale throat. The breast feathers are tipped with white, with the lower abdomen and under tail coverts becoming buffy white. The under wing surface is mostly white except for the primary coverts, which are buffy with blackish markings, and the inner webs of the primaries are marbled with black. The upperparts are blackish and grayish buff, with the coverts edged with buffy white. The primary coverts are tipped with black and white, the secondaries are marbled with black and buffy and tipped with black and white, and the primaries are blackish toward the tips and narrowly tipped with white. The rump and upper tail coverts are buffy and black, and the tail feathers are grayish brown with a dusky subterminal area, becoming paler outwardly and narrowly tipped with buffy white. The iris is brown, the bill is black toward the tip and brownish basally, and the legs and toes are yellow. *Females* are smaller than males and have a maximum wing length of 132 mm and a tarsus of 30 mm or less, while most males have minimum measurements of 134 mm and 32 mm. *Juveniles* have whitish buff rather than orange buff edgings on the feathers of their upperparts, pale buffy fringes on their wing coverts, less black on the undersides of the outer primaries, and a narrower black band on their wing coverts (Prater et al. 1977).

In the hand, the distinctive blackish markings on the undersides of the primaries, plus the scaly buff pattern on the upperparts and short bill (18–21 mm), are unique.

In the field (8 inches), this species is associated with shortgrass habitats often well away from water. The birds are tame and gregarious, flying only reluctantly, and then remaining close to the ground. The distinctive buffy plumage and yellow legs are distinctive; the calls include harsh *crik* and sharp *tik* notes, the latter often uttered in a series.

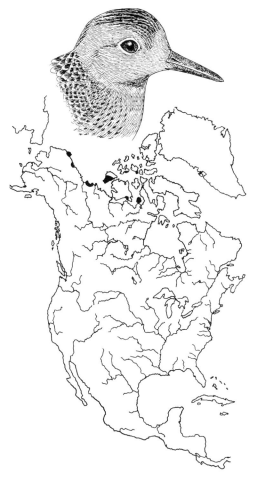

95. Breeding distribution of the buff-breasted sandpiper.

Habitats and foods. The typical breeding habitat of this species consists of grassy arctic tundra, especially the relatively large stretches of low and nonrocky areas that are well drained and sandy, supporting rather scant vegetation. The tops of ridges having fairly well developed grassy cover are also used, and in particular the well-vegetated hummocky ground around marshy ponds that are surrounded by extensive marsh tundra are favored sites for territories, though not for nests. However, marshy areas themselves are avoided by the birds. On the wintering grounds of Argentina, the birds forage in uplands where the grass is extremely short, and only in midday do they move to local water for bathing and drinking. There the birds roost in flocks of as many as 200 birds on dry grassy areas, and during the day they forage in much smaller groups that usually number only 5–6 birds. Individual birds often show winter territoriality, defending small areas of about 0.04 hectare (Myers and Myers 1979). On migration the birds typically frequent shortgrass plains, stubble fields, and dry uplands. Their spring foods consist of terrestrial invertebrates, especially the adults and larvae of beetles and the larvae and pupae of dipterans. In fall they are also known to eat copepods, craneflies, and gammarid crustaceans (King 1972), but so far nothing specific is known of their foods on the breeding and wintering grounds.

Social behavior. This species is of special interest, since it is the only North American shorebird that exhibits true lek behavior, and thus it provides some insight into the evolution of the complex lek behavior found in the ruff. Males typically gather in groups of from 2 to 10 males, each defending areas 10–50 meters in diameter, or considerably larger than the tiny lek territories of ruffs (Pitelka et al. 1974). These display sites are usually well-drained grassy areas with closely spaced sedge tussocks approximately 20 centimeters high, and they have no specific spatial relationship to nesting areas. Males actively chase other males from their tiny territories, and they apparently concentrate their display activities during the hours of low illumination. Males typically display before any moving object, including humans. In both defense and courtship display, wing-flashing or wing-waving is of prime importance, and in low light the exposure of the wing undersides is highly conspicuous. Display birds also often stand on tiptoe and sometimes also jump up and down in a "flutter jump" display. At times two competing birds will rise

side by side 20–40 feet in the air, their wings fluttering and their legs dangling (Parmelee et al. 1967; Prevett and Barr 1976). Besides displays on breeding grounds, communal displays sometimes are also performed while the birds are on migration, and in these circumstances the displays appear to be of the same general form, though flock sizes are larger and probably no territorial defense as such exists. At least some of the male displays are also performed by females (Oring 1964), though the functions of nearly all the displays are still unstudied. Almost all involve exposing the brilliant white undersurfaces of the wing. Displays usually begin by the male's raising one wing vertically in a single-wing-up display, occasionally tilting over to direct the underwing surface toward the sky or sometimes cocking the tail. Additionally, double-wing-raising and flutter-jumping are performed, as well as a display in which both wings are held forward in a parabolalike shape. All these displays make the male more conspicuous to other males and also to females, which are attracted to particular territories. After a female arrives, the male tries to lead her to a relatively secluded copulation area, simultaneously running toward the mating site and displaying by raising one wing or erecting the dorsal feathers. Copulation is preceded by the male's performing a double-wing embrace, with the attracted female or females coming within the arc of his wings. Although such attempts are often disrupted by territorial patrolling, the female typically then spreads her wings, which is the last step before copulation. Other males frequently attack the pair at this point, chasing the female away or attracting her to a different territory (Myers 1979).

Reproductive biology. There is no clear relationship between this species' display areas and its nesting areas, the latter always being on drier slopes while the former occur in both wet and dry habitats. Probably the female alone chooses the site and constructs the nest, which is a shallow scrape on dry grassy or "black" lichen tundra, usually sheltered to some degree by grassy cover. It is lined with mosses, lichens, sedges, willow leaves, or other vegetation that blends with the spotted eggs. There are typically 4 eggs present, though there is at least one record of a nest with 5 eggs. Practically the only observations of incubating birds are by Sutton (1967); they suggest that only the female incubates and that the male plays no role in nest defense. In Sutton's observations, even the female did not perform injury-feigning behavior when flushed from a well-incubated clutch. There is no information on the length of the incubation of fledging periods. In the vicinity of Victoria and Jenny Lind islands the young hatch from mid-July to late July. They are attended by the female on dry slopes and sometimes also on wetter areas. Many males flock during July, and unattended juveniles occur as late as the third week of August on the inland tundra (Parmelee et al. 1967).

Status and relationships. In some arctic areas this is an abundant breeder; Parmelee et al. (1967) stated that on Jenny Lind Island it was the second most abundant sandpiper in 1962. In northern Alaska, Bergman et al. (1977) reported that its density varied from 0.0 to 10.0 birds per square kilometer during five breeding seasons. Yet its status and biology are still extremely poorly known, and so far as is known its total breeding range is actually very restricted and it may have once been more numerous that it is at present. Its relationships are of special interest and are quite clearly close to *Calidris*. Strauch (1976) advocated merging *Tryngites* into *Calidris,* and the downy pattern is certainly *Calidris*-like. Yet the species has unique juvenile and adult plumages, and its displays are also distinctive in their lack of song flights and other typical *Calidris* features. In some respects the genus seems somewhat intermediate between *Calidris* and *Philomachus,* though the sharing of behavioral features with the latter genus may only be a convergent result of the two forms' evolving similar social systems. I thus do not accept Larson's (1957) suggestion that *Tryngites* and *Philomachus* may be a Miocene species pair, and I believe the latter is considerably closer to *Calidris* than is *Tryngites.* I believe the genus *Tryngites* should be retained as a separate taxon close to *Calidris* but having no obvious close relatives within that genus.

Suggested reading. Oring 1964; Sutton 1967; Prevett and Barr 1976.

Subfamily Tringinae (Tattlers and Allies)

Wilson Phalarope

This subfamily consists of small to medium-sized tropical to arctic-breeding sandpipers with moderately long bills, moderately to extremely long legs, a hind toe, and webbing or scalloping often present between the front toes. The nasal groove usually does not extend more than two-thirds the length of the upper mandible. The tarsus is scutellated in front and usually also behind but sometimes is reticulated behind. The downy young are usually strongly striped dorsally with brown or blackish bands, but they lack powder-puff feather tips. The sexes are alike in plumage as adults, but seasonal variations are common. Five tribes containing 37 species are recognized here.

Tribe Phalaropini (Phalaropes)

Phalaropes are small temperate to arctic-nesting shorebirds with a short but strongly compressed tarsus, scutellated in front and behind, a hind toe, and anterior toes that are webbed basally and scalloped laterally. The bill is straight and relatively weak, usually shorter or only slightly longer than the head, and tapering or slightly expanded toward the tip. The downy young are yellowish, with median and lateral blackish dorsal spots or stripes. Females are larger and more brightly colored than males, and polyandrous mating occurs in at least 2 species. One genus and 3 species are recognized here.

Genus *Phalaropus* Brisson 1760

This genus of 3 species has the characteristics of the tribe Phalaropini.

KEY TO THE SPECIES OF *Phalaropus*

A Bill about as long as head, slightly expanded toward the tip . . . . *fulicarius*
A′ Bill longer than head, slender and tapering toward tip
  B Bill no more than 25 mm, tarsus under 24 mm. . . . . *lobatus*
  B′ Bill at least 28 mm, tarsus over 28 mm. . . . . *tricolor*

Wilson Phalarope

Phalaropus tricolor (Vieillot) 1819
(*Steganopus tricolor* of Peters, 1934)

Other vernacular names. None in general English use; phalarope de Wilson (French); Amerikanische Odinshühnchen (German); faláropo tricolor (Spanish).

Subspecies and range. No subspecies recognized. Breeds in North America from British Columbia, Alberta, Saskatchewan and Manitoba south to California, Nevada, Utah, Colorado, Nebraska, and Minnesota. Also breeds locally to southern Ontario. Winters in southern South America and the Falkland Islands. See map 96.

Measurements and weights. Wing: males 115–33 mm; females 124–43 mm. Culmen (from feathering): males 28–34 mm; females 30–38 mm. Weights (breeding grounds): males 30–64 g, average of 100 was 50.1 g; females 55–85 g, average of 53 was 68.0 g (Höhn 1967). Eggs ca. 33 x 23 mm, estimated weight 9.4 g (Schönwetter 1963).

DESCRIPTION

Adult females in breeding plumage have the forehead and crown a pale bluish gray, the forehead with a blackish line along each side, the back of the head white, becoming bluish gray or slate gray on the back and scapulars. The rest of the head is white, except for a black line extending from the lares through the eyes, across the ear coverts, and down the side of the neck, where it becomes a dark chestnut and continues backward along the inner scapulars. The outer scapulars also have a streak of chestnut, and the foreneck and chest are a soft buffy cinnamon, fading to creamy buff on the breast. The rest of the underparts are white, as are the upper tail coverts. The wings are brownish gray, the coverts and tertials having paler margins, the rump is brownish gray, and the tail is mouse gray, the feathers narrowly tipped with white. The iris is brown, the bill is black, and the legs and toes are black. *Males in breeding plumage* are smaller and duller in color, with the female's pattern only faintly indicated (Ridgway 1919). *Adults in winter* are plain gray on the upperparts except for the white upper tail coverts and white superciliary stripe, and the underparts are white. The wing coverts are uniform gray, and the legs are yellow. *Juveniles* are brownish on the upperparts and wing coverts, and the feathers have extensive buff fringes. *First-winter* birds retain buff fringes on the inner median coverts until spring (Prater et al. 1977).

In the hand, the scalloped toe webbing identifies this as a phalarope, and the bill length (at least 28 mm) distinguishes it as *tricolor*.

In the field (7½ inches), the long and extremely narrow black bill separates this species from the

other phalaropes, even in winter plumage. The northern phalarope also has a very thin bill, but it is shorter, and the upperparts of that species are much darker. Like all phalaropes, the birds forage while swimming in circles, and in flight they show no white on the upper wing surfaces, though the tail coverts and tail are white or very pale gray. The usual call is low-pitched and honking, or like the barking of a distant dog.

NATURAL HISTORY

Habitats and foods. The breeding habitats of this grassland adapted species consist of ponds and lakes containing expanses of shallow water that are close to wet-meadow vegetation, as well as swales along intermittent streams. The wetlands used range from fresh to strongly saline, and in North Dakota nearly half of 438 pairs were found on semipermanent ponds ranging from fresh to subsaline, about 40 percent were found on seasonal ponds and lakes, and the remainder occurred on alkali ponds or lakes, fen ponds, temporary ponds, or other wetlands (Stewart 1975). Population densities of as much as 24 pairs per square mile occur in favorable North Dakota breeding habitats (Stewart and Kantrud 1972). Lakes, reservoirs, and ponds are all used by migrating birds, and in winter the birds remain relatively inland compared with other phalaropes, using freshwater marshes, shallow lakes, mud flats, and seashores, and to some extent salt marshes and coastal waters. In coastal Argentina the birds typically forage in small groups on inland vernal ponds, marshes, sloughs, and coastal estuaries (Myers and Myers 1979). During the breeding season they forage almost exclusively on insects, including the adults and larvae of beetles, flies, and hemipterans, with adult dytiscid and hydrophilid beetles and midge larvae being par-

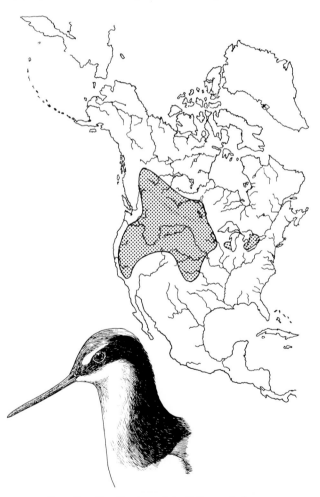

96. Breeding distribution of the Wilson phalarope.

ticularly prevalant. Small crustaceans, spiders, and the seeds of aquatic plants constitute a very small part of the diet (Wetmore 1925; Höhn and Barron 1963). Like the other phalaropes, this species typically forages while actively swimming, wheeling in tight circles, scarcely slowing when striking at prey, and moving either clockwise or counterclockwise (Höhn 1971). The birds also often forage while wading in shallow water or while walking about on muddy shorelines, constantly sweeping the bill sideways. At times they walk about in deeper water with the head entirely submerged, apparently performing their usual side-sweeping movements (Bent 1927).

Social behavior. Unlike most shorebirds, females tend to precede the males on the spring migration, though mixed sex groups and courtship displays are common among spring flocks. Typically a female selects a male and defends his position against other females, either by threats toward them or by overt chases. Several vocalizations are present, but most are of rather low amplitude and apparently serve for short-range communication. One call used exclusively by females and directed toward males is a *chug* note, uttered while expanding the neck feathers and extending the bill forward. This call seems to be associated with pair-formation and also occurs before

copulation. Aerial chases are frequent and are often initiated by the male's taking flight, followed closely by the female. She often utters a series of loud, hollow *wa* calls, followed by a hunchback posture or "loon flight" as she moves ahead of the male. These aerial chases may serve to synchronize the sexual cycles of the birds and do not seem to result in pair spacing (Howe 1975a). Although females are territorial in the sense of defending males, they apparently do not defend specific areas, nor do males significantly defend their nest sites. Copulation mainly occurs among well-established pairs while they are standing in shallow water, or sometimes while swimming. The most frequent precopulatory display is an upright posture of the male, which is held for several seconds and sometimes is the only obvious precopulatory signal. However, the *chug* call and posture is sometimes exhibited by the female. The male quickly hovers up onto the back of the female and copulates while holding the wings lifted but not quivering them. No specific postcopulatory displays are present (Howe 1975b).

Reproductive biology. Nest-site searches are initiated by the female, and during such activities either sex may perform scraping behavior at various locations. A day or two before the first egg is laid, the male returns to some of these scraping sites and begins to denude them of most vegetation. Typically the male thus readies two or three scrapes for eggs, with the female playing little or no role in nest preparation. The female evidently chooses one of these sites, and begins her clutches in it, laying eggs on consecutive days. Nests are often closely spaced; Kagarise (1979)

reported an average distance between nests of only 24 meters for 20 nests, and a surprisingly high breeding density of 1 nest per 0.17 hectare, whereas Howe (1975b) found only 1 per 14.3 hectares. Incubation probably begins on the day the third egg is laid (Howe 1975b). Normally the clutch is 4 eggs; 38 nests in one study all had this number (Stewart 1975), and in another study 18 of 19 clutches were of 4 eggs (Kagarise 1979). There is one case of a nest found with 8 eggs, suggesting simultaneous polygyny, nest-sharing, or nest parasitism (Kagarise 1979). Whether sequential polyandry occurs in this species is still somewhat speculative. Howe (1975b) found no definite evidence for it on the bases of ovarian examinations but found that the pair bond breaks immediately after the clutch is completed, and one of three marked females was found in the company of a new male only a few hours later. There is also no good evidence yet on the incidence of renesting after the loss of a clutch, though there is suggestive evidence of this (Kagarise 1979). The incubation period is reported to range from 16 to 21 days (Johns 1969), but no information is available on the fledging period.

Status and relationships. This species is relatively common over much of its range, and unless drainage of its prairie marsh habitat becomes much more serious it does not seem to be in any immediate danger. The phalarope relationships are still unsettled. Although the phalaropes are frequently separated as a distinct family and sometimes regarded as relatives of the Recurvirostridae, Jehl (1968a) disputed this view. He included them as a subfamily of the Scolopacidae, as probable tringine derivatives. Strauch (1976) likewise believed the tattlers are probably the nearest relatives of the phalaropes, but he considered the evidence for this rather weak. Both included all of them in the single genus *Phalaropus*, and Jehl argued that the sequence of species *tricolor, lobatus, fulicarius* best reflects their progressive development of aquatic adaptations.

Suggested reading. Howe 1975b; Kagarise 1979.

Northern Phalarope

Phalaropus lobatus (L.) 1758
(*Lobipes lobatus* of Peters, 1934)

Other vernacular names. Red-necked phalarope; phalarope à bec étroit (French); Odinshühnchen (German); faláropo picofino (Spanish).

Subspecies and range. No subspecies recognized. Breeds Greenland, Iceland, Spitsbergen, Scandinavia, northern Russia, and northern Siberia to the Chukotski Peninsula, and in North America from the Aleutians and the Alaska Peninsula east through northern Canada to James Bay, southern Labrador, and the adjoining arctic islands (Victoria, Southampton, Baffin). Winters off the west coast of South America, in the eastern Pacific, and off the west coast of Africa. See map 97.

Measurements and weights. Wing: males 102–12 mm; females 107–18 mm. Culmen (from feathering): males 19–22 mm; females 18–23 mm. Weights (on breeding grounds): males 29–35 g, average of 14 was 32 g; females 29–43 g, average of 7 was 35 g (Irving 1960). Eggs ca. 30 x 21 mm, estimated weight 6.3 g (Schönwetter 1963).

DESCRIPTION

Adult females in breeding plumage have the upperparts including the upper half of the head dark slate gray to blackish, which is sharply separated from a white chin and throat, with a darker spot in front of the eye and small white spots above and below the eye. There is a bright cinnamon area on the side of the neck that extends to the upper chest, and the underparts are otherwise white except for a slate gray on the sides of the breast and body that become streaks on the flanks. The back is mostly grayish except for pale cinnamon streaks on each side of the body that converge toward the rump to form a V, the rump and tail are dusky, the outer tail feathers becoming largely white. The upper wing surface is dusky brown, with the greater coverts broadly tipped with white, forming a wing stripe that extends out to the bases of the primaries. The iris is brown, the bill is black, and the legs and toes are bluish gray. *Adult males in breeding plumage* are smaller and duller, with the cinnamon mostly confined to the sides of the neck, and the chest is mixed with white and grayish. *Adults in winter* have a pale bluish gray mantle and wing coverts, a blackish band on the ear coverts, and otherwise are nearly white on the sides of the head and underparts. *Juveniles* have a brownish cap and a dark brown mantle with golden buff fringes on the feathers. The underparts are whitish, with a pinkish buff wash on the throat. The legs are yellowish to bluish gray. *First-winter* birds gradually become grayer on the upperparts, but some yellowish fringes on the scapulars and tertials remain until spring (Prater et al. 1977).

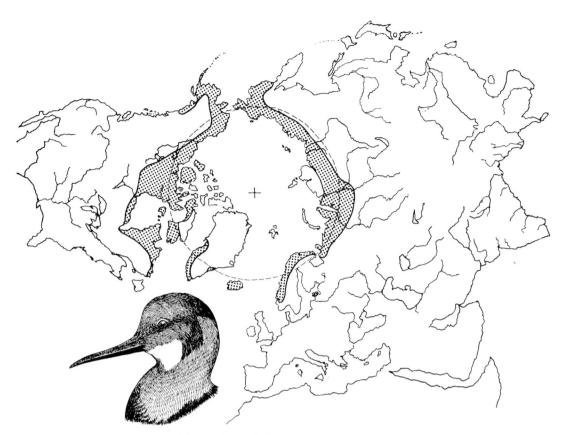

97. Breeding distribution of the northern phalarope.

In the hand, the combination of scalloped webbing on the toes and a fairly short black bill (culmen 18–23 mm) serves to identify this species.

In the field (6½ inches), this species of phalarope is smaller and darker dorsally than the Wilson phalarope and has a noticeably shorter bill. In winter plumage it has a more definite blackish eye stripe as well, and at any season shows a strong white wing stripe when flushed. It utters various low-pitched scratchy notes, usually sounding like *whit* and sometimes given in series.

NATURAL HISTORY

Habitats and foods. Summer habitats of this low arctic species consist of ponds, lagoons, and streams having adjacent grassy or sedge vegetation. This species is more subarctic in its distribution than the red phalarope and more often tends to breed near lakeshores or other fairly permanent water bodies, according to Kistchinski (1975). Yet in most parts of its range it inhabits marshes having small ponds, where moss or sedge cover near the waterline provides suitable nest sites. On typical wet arctic tundra the red phalarope usually outnumbers the northern phal-

arope, but in more southerly habitats, where river channels are bordered by willow thickets, the two species sometimes nest in equal densities (Kistchinski 1975). Apparently the highest nesting density so far reported for the species is an estimate of 6.6–8.8 nests per hectare on Victoria Island, but nesting densities seem to vary considerably from year to year and in different habitats of the same general area. Studies in Finland indicate that in some habitats the predominant food taken during the breeding season was midges, including larval forms as well as hatching and swarming adults, while in other nearby areas the major foods were trichopteran larvae, water fleas, tadpoles, water spiders, and springtails. Evidently the availability of such food items regulates the choice of feeding areas and thus influences nest-site selection. The start of the egg-laying season is apparently timed to coincide with the major midge hatching period, while later in the summer a much greater diversity of invertebrate foods is available (Hilden and Vuolanto 1972). Like the other phalaropes, the birds do much foraging while swimming and spinning in a rotary manner, which tends to be performed faster than in red phalaropes but slower than in Wilson phalaropes. Although spinning is usually

◆ ◆ ◆ 299

34. Winter-plumage heads and feet of red (*top*), northern (*middle*), and Wilson (*bottom*) phalaropes.

believed to help bring subsurface items within reach, Höhn (1971) has suggested that it may simply be the most efficient manner of gathering a highly concentrated food source. Compared with the other phalaropes, this species probably eats a larger proportion of crustaceans than does *tricolor* but less than is typical of *fulicarius,* and in all three species insects seem to be the major dietary component (Wetmore 1925).

Social behavior. Studies in Finland and Iceland indicate that a substantial proportion of birds are already paired when they arrive on their breeding grounds, and those that are not become paired within a few days. Apparently the female takes the initiative in courtship, and a frequent display is the "imposing posture," with the neck enlarged, the breast puffed

out, and the tail depressed. A female typically selects and follows a particular male either on the water or, if he takes flight, on the wing. Should another female approach too close to an incipient pair she is threatened or chased away, whereas males evidently rarely threaten other males when they approach a pair. Unlike the other phalaropes, copulation in this species apparently always occurs in water of swimming depth. It can be initiated by either sex, but frequently females initiate it by a wing-whirring display in which they rise in the water and beat the wings rapidly for a short time. This is followed by the female's lying motionless on the water with her tail toward the male. He may respond with wing-whirring, then he rises in the air and copulates, with the female being almost entirely submerged. Although females do not defend a territory as such, they do have an advertisement display in which they fly 10–20 yards in a distinctive manner while uttering a characteristic note. Before egg-laying begins, the female may perform such ceremonial flights, calling the male to follow her and then performing nest-scraping displays at potential nest sites. The nest-scraping behavior sometimes immediately follows copulation, with the male usually following the female but sometimes going on land alone and making scrapes of his own. In one of the several scrapes thus made the female begins to deposit her clutch, and thereafter the male becomes progressively attached to the nest and less attached to his mate (Tinbergen 1935). Nest-tending may begin as early as the laying of the first egg, and thus the female becomes free to begin a second clutch, sometimes within 7 days of completing her first clutch. Evidently successive polyandry is a regular feature in this species, provided excess males are available. According to Hilden and Vuolanto (1972), females may lay two clutches in a single season if an early nest is destroyed and the male is able to remate either with his original mate (one known case) or a new one (two known cases), or if there are excess males in the population available for incubating a second clutch (five known cases).

Reproductive biology. In most areas, nests are very close to water, usually on hummocks surrounded by water, but in one study area in Finland most nests were 5–50 meters from the nearest water (Hilden and Vuolanto 1972). In that study most nests were among or near colonies of arctic terns, but this tendency does not seem to be a general one for the species. Eggs are laid 24–30 hours apart, averaging about 26.5 hours, and in the Finnish study all but one of 71 completed clutches contained 4 eggs, the exception being a clutch of 3. Males begin to incubate during the egg-laying period, spending increasing periods of time on the nest from the laying of the first egg and spending little time off the nest after the clutch is completed. The incubation period ranges from 16.8 to 20.7 days, averaging about 18 days. As in the other phalaropes, this is an extremely short period and may be related to the relatively small egg sizes of phalaropes. In spite of the apparent onset of incubation early in the clutch, hatching of the entire brood usually occurs within about 8 hours, ranging from 4 to 12, and the brood often leaves the nest 3–6 hours later. If hatching occurs late in the day, however, the chicks typically remain on the nest through the night. Young birds often travel several hundred meters a day, usually remaining on dry soil or on mud along the waterline. However, they are able to swim like ducklings, and sometimes broody males will adopt young that are not from their own brood. The fledging period is probably about 20 days, and about a week before the young are able to fly the male stops giving alarm calls toward them. Although females have rarely been seen near broods, they have not then been concerned with the young, but rather were exhibiting sexual behavior toward the male (Hilden and Vuolanto 1972).

Status and relationships. Like the red phalarope, this species has an extremely broad breeding range that is well beyond the range of most human activity, and its wintering areas are primarily oceanic. It thus no doubt has a very large, though essentially inestimable, total population. It is the most aquatic-adapted of all the phalaropes and thus may be regarded as constituting one extreme of the genus *Phalaropus*, with *tricolor* being the other extreme.

Suggested reading. Hilden and Vuolanto 1972; Tinbergen 1935.

Red Phalarope

Phalaropus fulicarius (L.) 1758

Other vernacular names. Gray phalarope; phalarope à bec large (French); Thorshühnchen (German); falaropo gris, falaropo picogrueso (Spanish).

Subspecies and range. No subspecies recognized. Breeds in North America from St. Lawrence Island and northwestern Alaska eastward across northern Canada to Hudson Bay, northern Quebec, and the arctic islands of Canada north to Ellesmere. Also breeds on Greenland, Iceland, Spitsbergen, southern Novaya Zemlya, and arctic Siberia from the Taimyr Peninsula east to the Chukotski Peninsula and the coast of Anadyrland. Winters at sea off the west coast of Africa, along the coast of Chile, and probably elsewhere in the Pacific. See map 98.

Measurements and weights. Wing: males 123–35 mm; females 128–41 mm. Culmen (from feathering): males 21–24 mm; females 21–25 mm. Weights (breeding season): males 41–60 g, average of 69 was 50.8 g; females 49–73 g, average of 51 was 61 g (Kistchinski 1975). Eggs ca. 31 x 22 mm, estimated weight 7.5 g (Schönwetter 1963).

DESCRIPTION

Adult females in breeding plumage have the upper and anterior parts of the head black, including the lores and chin, while the sides of the head are entirely white, and the throat, sides of neck, and entire underparts are a deep cinnamon to reddish brown. The hindneck is a mixture of cinnamon and slate, and the mantle feathers are blackish with broad buffy margins. The wing coverts are slate gray, the greater coverts broadly tipped with white, and the flight feathers are dusky slate. The central tail feathers are sepia, and the outer ones are paler gray, sometimes with pinkish buff toward the tip. The iris is brown, the bill is black at the tip and yellow basally, and the legs and toes are brownish with yellowish lobes on the toes. *Males in breeding plumage* resemble females but are smaller and duller, with the white on the sides of the head more restricted and less well defined and the underparts broken up by varying amounts of white. *Adults in winter* have gray upperparts with darker shaft streaks and have white-tipped gray median coverts. The legs are also more grayish at this period. *Juveniles* have rich, dark brown upperparts with buffy feather edges and white underparts, with pinkish buff on the face, neck, and upper breast. The bill is blackish, becoming brown at the base. *First-winter* birds show a mixture of gray and brown upperpart feathers, and buff fringes on the wing coverts may be present until spring (Prater et al. 1977).

In the hand, the combination of scalloped webbing on the toes and a bill between 21 and 25 mm long that is rather stout for much of its length, instead of uniformly tapering, provides certain identification.

In the field (7 inches), this phalarope is most often seen in marine situations, and in winter plumage it

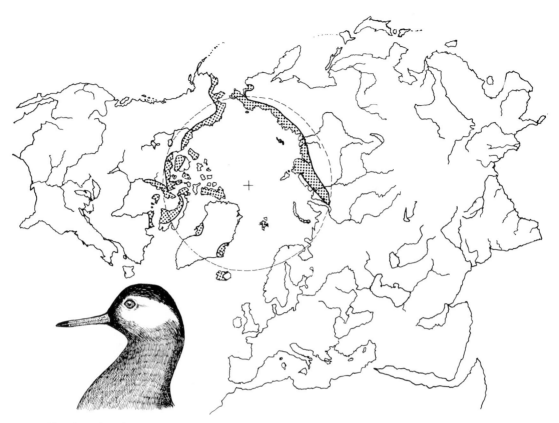

98. Breeding distribution of the red phalarope.

closely resembles the northern phalarope, but it has a much stouter bill and a rather uniformly gray mantle with white underparts and a dark gray eye stripe. The red body coloration in breeding plumage is unmistakable. In flight, a white wing stripe is visible. The usual calls are a series of whistled *twit* notes. Like all phalaropes, it swims well and often turns in tight circles while foraging.

NATURAL HISTORY

Habitats and foods. This is the most arctic-adapted of the three phalaropes, and it breeds on flat, low tundra with numerous ponds. In Canada the birds favor wet meadows or marshy tundra with grass or moss cover and networks of shallow ponds (Parmelee et al. 1967; Snyder 1957), and likewise in Siberia they are associated with polygonal and tussocky moss-sedge tundra rich in ponds, lakes, and marshes (Kistchinski 1975). Breeding densities on these habitats are quite variable, such as a pair per 0.05–1.1 hectares in the USSR (Kistchinski 1975), a pair per 0.10–0.13 hectares on Victoria Island (Parmelee et al. 1967), a nesting male per 0.63 hectares on Bathurst

Island (Mayfield 1979), and a pair per 2.2–4.2 hectares in Alaska (Schamel and Tracy 1977). Outside the breeding season the birds are almost entirely pelagic, but they also occur on bays and coastal estuaries. A sample of foods taken from 36 birds during the spring, summer, and fall in North America suggests that crustaceans constitute about a third of the diet, beetles and flies about a quarter each, and the remainder is mostly small fishes (Wetmore 1925). Samples from Greenland during the breeding season had a high incidence of adults and larvae of dipterans, such as gnats and their larvae, small beetles, and other insects. During their period at sea these birds eat very small fishes, small jellyfish, crustaceans, and sometimes even parasites plucked from the backs of whales. Like the other phalaropes they often forage by swimming in tight circles, but they also hunt by stalking insects on land or by standing on floating masses of seaweed, probing or pecking for invertebrates (Bent 1927).

Social behavior. On West Spitsbergen, these birds arrive on their breeding ground in mid-June, and at least some of them appear to be paired on arrival, though other observers have suggested that females arrive in advance of the males, with pair bonds being formed almost immediately after the males appear (Mayfield 1979). Egg-laying may begin only a week or so after arrival, so any late pair-formation must occur very rapidly at this latitude. The two sexes show considerable hostility toward one another, though actual attacks are quite rare. When approached by a potential mate or intruder, birds typically assume an alert posture, accompanied by a series of alarm notes. An important part of courtship is a "ceremonial flight" or "circle flight," in which the female flies in wide circles over an area, sometimes alone but often chasing males. In this case the female attempts to get below and slightly ahead of the male, apparently exhibiting her white upper wing markings to his view. A display called "wing-whirring" or "rattling," which involves hovering while making a rattling sound with the wings, often terminates such a circle flight. This display is also performed by males just before copulation. In courting on the water the female swims while she lowers her head and presses it down between her shoulders with the bill pointing forward, while uttering chattering notes. A variant of this is to withdraw the head into the shoulders and point the bill downward in a "pushing" display. This latter posture often occurs in the precopulatory situation. Copulation occurs on land, but in one case the female initiated it by performing pushing and rattling displays. The male then hovered over her back for a few seconds, and copulation followed. Evidently females normally take the initiative in copulatory behavior, and if the males are unwilling they simply chase the females away. No obvious postcopulatory displays are present (Bengtson 1968). In rare cases the male may initiate copulation, and copulation sometimes occurs without any preceding display behavior (Kistchinski 1975). Territorial defense as such does not exist in this species, and home ranges are large. According to Kistchinski, pair-formation can occur either before or after arrival on the breeding ground, and promiscuous mating seems fairly common, with the only biological purpose of pair-bonding being to obtain a mate to incubate the eggs. Although successive polyandry has been speculated for this phalarope, it has only recently been proved to occur. In an Alaska study, 4 definite instances of successive polyandry were found in a color-banded population of 11 pairs. All 4 of these females formed their second pair bonds during a brief period, and the original pair bond either ended before the initial clutch was completed

or persisted for nearly 2 weeks after incubation had begun. Replacement laying was also found to occur and to involve either the original mate or a new one (Schamel and Tracy 1977). Mayfield (1979) recently observed 3 suspected cases of polyandry among a local population of about 100 birds on Bathurst Island. However, Ridley (1980) considers the mating system basically monogamous and derived from double-clutching.

Reproductive biology. Nests are among sedges or grasses in wet areas of moss-sedge tundra, often at the edge of a temporary pond or on a tussock in flooded areas. Less frequently they are on islands or along the shore of a permanent lake, or on dry tussocks or ridges in polygonal tundra. Frequently the nests are clustered, as when they are placed in colonies of terns or gulls, and at times they may only be a few meters apart. Evidently several nest scrapes are made before one is selected and lined for receiving the eggs, and the others remain unlined. Both sexes have been observed nest-searching and making nest scrapes (Kistchinski 1975; Mayfield 1979). Males increase in their attentiveness as the clutch increases in size and may begin incubating after the second or third egg is laid. Eggs are laid at approximate 24-hour intervals, and in most cases the clutch numbers 4 eggs (Bergman et al. 1977). Extremes of as few as 2 and as many as 7 eggs have been reported, and lower clutch sizes may be typical in years when nesting is delayed (Kistchinski 1975). The incubation period is usually about 18.5–19 days, with reported extremes of 15 and 24 days. The young birds remain with the male parent throughout the 18–21-day fledging period. There are no records of females tending broods, and together with nonbreeders they soon begin to flock and leave the breeding grounds (Kistchinski 1975; Mayfield 1979).

Status and relationships. Although no specific estimates are at hand, there is no indication that this is a rare species, and in many high-arctic tundra areas it is one of the commonest of breeding shorebirds (Bergman et al. 1977; Kistchinski 1975). Its relationships have been discussed in the section on *tricolor*; it is the most aquatic of the three phalaropes and apparently is more closely related to *lobatus* than to *tricolor*.

Suggested reading. Kistchinski 1975; Schamel and Tracy 1977; Mayfield 1979.

Tribe Tringini (Tattlers)

Lesser Yellowlegs

Tattlers are small to medium-sized temperate to subarctic nesting shorebirds with a relatively long tarsus that is not compressed and is scutellated in front and usually also behind, and with anterior toes that usually are not webbed and are never scalloped. The bill is moderately slender and is straight or slightly recurved, and it often is distinctly longer than the head. The sexes are alike as adults, but seasonal variations in plumage are frequent. The downy young are rather grayish, with black dorsal markings that usually include one major medial stripe and two smaller lateral stripes. Three genera and 16 species are recognized here.

KEY TO GENERA OF TRINGINI

A Secondaries and bases of primaries uniformly white, wing at least 180 mm, culmen over 50 mm . . . . *Catoptrophorus* (1 sp.)

A' Secondaries and bases of primaries not extensively white, wing and culmen measurements rarely so great
  B Bill distinctly recurved . . . . *Xenus* (1 sp.)
  B' Bill never distinctly recurved; either slightly decurved, straight or almost imperceptibly recurved . . . . *Tringa* (14 spp.)

Genus *Catoptrophorus* Bonaparte 1827 (Willet)

The willet is a medium-sized North American sandpiper with a stout but compressed bill that is nearly as long as the tarsus. The tarsus is long and scutellated both in front and behind, a hind toe is present, and all the front toes are joined by webbing. The downy young are *Tringa*-like. The sexes are alike as adults and show some seasonal plumage variation. One species.

Willet

Catoptrophorus semipalmatus (Gmelin) 1789

Other vernacular names. None in general English use; chevalier semipalmé (French); Nordamerikanischer Schlammtreter (German); playero aliblanco, archibebe aliblanco (Spanish).

Subspecies and ranges. See map 99.
  C. s. semipalmatus: Eastern willet. Breeds along coastal Nova Scotia and on Cape Breton Island, and from New Jersey southward along the Atlantic coast to Florida, also along the Gulf coast from Texas to Louisiana, and locally in the West Indies. Winters locally in the Atlantic coastal states, the Gulf of Mexico, in the West Indies, and along the Atlantic coast of Mexico and Central America to northern South America.
  C. s. inornatus: Western willet. Breeds from eastern Oregon, Idaho, and central Alberta eastward to Colorado, Nebraska, and the Dakotas. Formerly bred to Iowa and Minnesota. Winters from California and South Carolina south to northern South America.

Measurements and weights. Wing (chord): males 180–218 mm; females 175–220 mm. Culmen (to feathering): males 53–64 mm; females 52–65 mm. Weights (in winter): males 173–90 g; females 202–36 g (Haverschmidt 1968). Males (spring to fall): 221–308 g, average of 10 was 273.0 g; females 214–375 g, average of 5 was 301.4 g (M.V.Z. specimens). Eggs ca. 54 x 38 mm, estimated weight 39.5 g (Schönwetter 1963).

DESCRIPTION

Breeding adults of both sexes have a crown that is brownish gray with dusky streaking, a whitish supraloral stripe that passes above and behind the eye, dusky lores, and the sides of the head finely streaked with dusky. Dusky streaking or irregular spotting also occurs on the foreneck and chest, becoming barring on the sides, while the rest of the underparts is white. The axillaries and most of the under wing coverts are plain sooty blackish, contrasting with the white bases of the flight feathers. The back and scapulars are brownish gray, irregularly spotted with

99. Breeding distributions of the eastern (*hatched*) and western (*shaded*) willets.

In the hand, this *Tringa*-like bird is readily identified by its large size (wing at least 175 mm) and the black-and-white primaries.

In the field (13 inches), this large shorebird is godwit-sized but is much grayer, with the distinctive black-and-white wing pattern not evident until the bird flies. Its bill is heavier than that of the greater yellow legs, but not so long as a godwit's, and its legs and bill are gray. In flight, a distinctive *pill-will-willet* call is loud and frequent.

NATURAL HISTORY

Habitats and foods. The two subspecies of willets have rather different breeding habitats. The eastern subspecies is primarily coastal-breeding, nesting in salt marshes that are dominated by salt hay (*Spartina patens*), particularly those that are periodically burned or mowed or where the grasses are cropped by wild geese. The western subspecies is associated with prairie marshes, and in North Dakota nearly half of 219 pairs were distributed around semipermanent ponds and lakes. Another 43 percent were found near seasonal ponds and lakes, and the rest were associated with permanent ponds and lakes, alkali ponds and lakes, or intermittent streams. Highest densities were found on brackish to subsaline semipermanent ponds and lakes (Stewart 1975). In the winter the birds preferentially occupy such coastal habitats as open mud flats associated with estuaries, bayous, or coastal marshes, but at times they also occur around ponds in coastal prairies or pine woods or on sandy shorelines. Foraging methods include probing, pecking at surface items, and "mowing," during which the bird walks over the mud, rapidly opening and shutting the bill and moving the head up and down very quickly. They sometimes also do this while wading in water, and they frequently stalk fiddler crabs or small fish in belly-deep water. Additionally, they eat small mollusks, fish fry, marine worms, and aquatic insects. Willets spend more time probing for food than do *Tringa* species, but they are not so highly specialized for this method of foraging as are snipes and woodcocks (Burton 1974; Bent 1929).

Social behavior. Willets are highly aggressive birds and are probably monogamous, with pairing usually occurring before territorial behavior has begun. Males appear to defend their mates before establishing territories, and when the territory is defended by both sexes. Early stages of pair-formation are still

blackish, the upper tail coverts are white, and the tail is light brownish gray, more or less mottled with darker gray. The wing coverts are nearly plain brownish gray with narrow white fringes, and the basal halves of the primaries as well as most of the secondaries except for their tips are white. The tips of the primaries, outer secondaries, and primary coverts are dusky. The iris is brown, the bill is black, becoming grayish basally, and the legs and toes are blue gray. *Adults in winter* are plain brownish gray above and almost uniformly dull white below except for pale gray shading on the foreneck (Ridgway 1919). *Juveniles* have grayish brown upperparts with broad buff fringes and subterminal dark bars. The underparts are white, except for grayish on the breast. *First-winter* birds resemble adults but retain buffish fringes and subterminal dark bars on the wing covers until winter (Prater et al. 1977).

unstudied, but the distinctive *pill-will-willit* call of the male is a conspicuous part of territorial proclamation. It is often uttered during a display flight called wing-waving by Tomkins (1965) and "spottying" by Vogt (1938), in reference to the similar flight behavior of the spotted sandpiper; it consists of flying in circles with the wings arched downward and moving in short and rapid strokes. At times during this display the birds rise so high that they are nearly out of sight, and the display often has a contagious effect on other males in the vicinity. Courtship is usually initiated by the male, as he walks slowly toward the female uttering deliberate *dik-dik* notes, depressing the closed tail, and raising the wings and waving them above the back in a narrow arc as the speed of calling increases. If the female is responsive, she crouches slightly, and the male soon flutters up to her back and copulates. The treading is terminated by the female's "tossing" the male lightly over her head and is normally followed by quiet feeding. Wing-vibrating or wing-waving displays appear to be associated only with these two specific display patterns, and the white markings of the wings are certainly emphasized at that time. Females at times also perform the territorial "spottying" display, but do not exhibit the vertical wing-waving (Vogt 1938). Although the birds are strongly territorial, they are also semicolonial, and they frequently gather in groups to display. However, part of this seeming gregariousness is a reflection of the high degree of contagious response the birds show to disturbance, and, as with stilts or avocets, an intruder may draw the attention of all the breeding birds of a marsh. Tompkins found that willets either feed within their nesting territories or defend nearby foraging territories where small crabs are abundant. In the coastal populations of the southeast most of the eggs are laid between April and June, whereas farther north on the Atlantic coast and in the northern Great Plains most records fall between mid-May and mid-June.

Reproductive behavior. Willets breed initially when they are 2 years old. The nest-spacing pattern has been studied by Burger and Shisler (1978), who found that the birds "spaced out in a clump," or probably responded to the conflicting tendencies toward gregariousness and territorial spacing advantages. Nest sites are usually at least 200 feet apart but may be as close as about 40 feet apart on rare occasions. The site is apparently chosen by the female, since during the prenesting period the male spends much time following the female about. The nests are often well concealed in heavy grass, but at times they are placed on open sand and almost completely exposed. The clutch size is normally 4 eggs, which may be laid at intervals of from 1 to 4 days. Incubation is performed primarily by the female, with the male possibly taking over at night. The incubation period is rather variable, lasting from 21 or 22 to 29 days, and at times it apparently begins before the clutch has been completed. This results in a staggered hatching, and the possible abandonment of some unhatched eggs after the first young have hatched. This behavior certainly affects the nesting success, although its frequency is undetermined. After hatching the young birds become extremely furtive, and the adults leave the breeding area before the young are fledged. There is an early observation of an adult bird carrying each of 4 newly hatched birds between its thighs across several creeks and over marshland for a distance of about a quarter of a mile, in a manner similar to that frequently described for woodcocks. The fledging period is still unreported (Bent 1929; Tompkins 1965).

Status and relationships. Both subspecies of the willet are seemingly still moderately abundant over much of their original range, perhaps in part because of their considerable tolerance for human activities such as mowing and burning of marshes. The evolutionary relationships of this species are clearly with the "shank" group of tattlers, and at times it has been included in the genus *Tringa* (Stout 1967, Strauch 1976). Yet its foraging adaptations differ somewhat from those of typical *Tringa* forms (Burton 1974), and none of the typical "shanks" have such distinctive wing patterning and conspicuous associated aerial displays. Thus there seems no strong reason for merging this rather well-marked form into *Tringa*.

Suggested reading. Tompkins 1965; Sordahl 1979.

Genus *Tringa* Linné 1758 (Typical Tattlers)

The tattlers and allies are small to medium-sized temperate and subarctic-nesting shorebirds with a long, slender bill that is straight or very slightly recurved and has a hard and slightly decurved tip. The tarsus is long, with a well-developed hind toe, and webbing is usually limited to the base of the middle and outer toes. The downy young are variably striped with dark markings along the dorsal midline and usually also laterally. The sexes are alike as adults, but seasonal plumage variations are common. Fourteen species are recognized here.

KEY TO SPECIES OF *Tringa*

A Inner webs of tail feathers crossed by a broad white band ("*Actitis*")
  B Adults in summer spotted below with blackish, tail usually under 50 mm . . . . *macularia*
  B' Adults in summer with white to buff underparts, tail usually over 50 mm . . . . *hypoleucus*
A' Inner webs of tail feathers not crossed by a broad white band
  B Web present between base of inner and middle toes . . . . *guttifer*
  B' No web between inner and middle toes
    C Axillaries and tail plain brownish gray ("*Heteroscelus*")
      D Upper tail coverts uniform gray or slightly edged with white . . . . *incana*
      D' Upper tail coverts distinctly barred with white . . . . *brevipes*
    C' Axillaries white, sometimes barred with gray, tail barred
      D Tarsus less than 1½ times as long as middle toe without claw
        E' Outermost primaries with white shafts . . . . *glareola*
        E' Outermost primaries with dark brown shafts
          F Upper tail coverts white, middle tail feathers barred with white . . . . *ochropus*
          F' Upper tail coverts dusky with white barring, middle tail feathers dusky, spotty along edges with white . . . . *solitaria*
      D' Tarsus more than 1½ times as long as middle toe without claw
        E Bill slender, tarsus more than a third as long as wing
          F Bill blackish, legs greenish . . . . *stagnatalis*
          F' Bill and legs reddish . . . . *erythropus*
        E' Bill stouter, tarsus less than a third as long as wing
          F Rump white or mostly white, legs and feet greenish or reddish
            G Larger (wing over 175 mm), greenish legs . . . . *nebularia*
            G' Smaller (wing under 165 mm), reddish legs . . . . *totanus*
          F' Rump not white or mostly white, feet and legs yellow
            G Larger (wing over 175 mm) . . . . *melanoleuca*
            G' Smaller (wing under 165 mm) . . . . *flavipes*

Spotted Redshank

Tringa erythropus (Pallas) 1764

Other vernacular names. None in general English use; chevalier arlequin (French); Dunkler Wasserläufer (German); archibebe oscuro (Spanish).

Subspecies and range. No subspecies recognized. Breeds from northern Scandinavia west across the Kola Peninsula and brushy tundras of northern Russia and northern Siberia to the Anadyr Basin, and south to the northern limits of dense forest. Winters in the Mediterranean Basin, Persian Gulf, India, the Indo-Chinese countries, and eastern China. See map 100.

Measurements and weights. Wing: males 158–71 mm; females 164–76 mm. Culmen: males 53–59 mm; females 57–62 mm. Weights: males 106–83 g; females 97–205 g. In fall, males average about 148 g and females about 157 g (Glutz et al. 1977). Eggs ca. 47 x 32 mm, estimated weight 24.5 g (Schönwetter 1963).

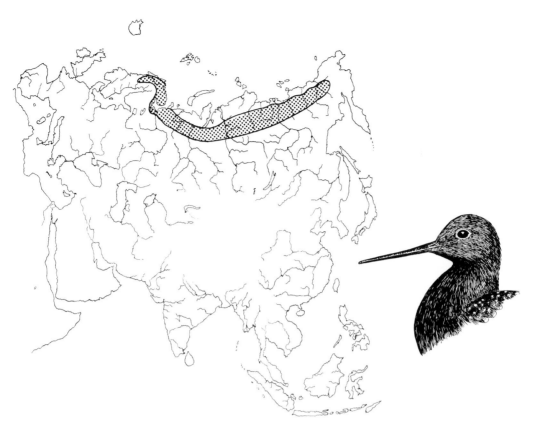

100. Breeding distribution of the spotted redshank.

DESCRIPTION

Adult males in breeding plumage have a sooty black crown, and the sides of the head, neck, throat, upper breast, abdomen, and flanks are slate black, the feathers narrowly edged with white. The under tail coverts are barred with slate black and white and are edged with white. The mantle and scapulars are sooty black, the feathers narrowly tipped with white and notched with white or pale pinkish buff. The wing coverts are grayish brown with large white fringes and subterminal brown bars. The primaries are sepia, the inner ones edged with white, and the secondaries are pale sepia with edges and notches of white forming irregular bars. The back and rump are white, and the upper tail coverts are white with blackish brown barring. The tail feathers are ashy brown with whitish edging and barring or notching. The iris is brown, the bill is dark brown, becoming dusky red at the base, and the legs and toes are very dark red (Witherby et al. 1941). *Females* resemble males but are larger and have white rather than

blackish central under tail coverts, and the crown feathers have pale whitish edges. *Adults in winter* are gray to ashy brown on the upper parts, with some white spotting, and the underparts are generally white with dusky streaking, especially on the breast. The legs and toes are orange. *Juveniles* are dark brown on the upperparts, with white spotting. The wing coverts have large white spots at the sides and toward the tips the latter with a dark central area. The underparts are whitish, heavily washed and barred with brownish gray. *First-winter* birds resemble adults, with underpart barring being lost by early fall and the spotting on the coverts lost by winter (Prater et al. 1977).

In the hand, this *Tringa* can be identified by its long (tarsus 52–61 mm) red legs, its straight reddish bill (culmen 54–62 mm), and the extensive whitish notching and tipping on the secondaries and inner primaries.

In the field (12 inches), the blackish plumage of birds in breeding plumage is unique, but in fall or winter the long reddish legs and the barred secon-

❖ ❖ ❖ 309

daries, which appear to be pale gray rather than white as in the redshank, are important field marks. Winter habitats include various brackish and coastal habitats as well as inland lakes and reservoirs. Swampy forested areas are the breeding habitats, and during that season the birds often perch on trees and bushes. In flight, a two-syllable *tchuet* call is frequent, which is less musical than the redshank's.

NATURAL HISTORY

Habitats and foods. The breeding habitat of this species consists mainly of marshy and swampy places and heathlands in lightly wooded regions near arctic timberline, ranging from open pine and birch forests northward well into the shrub tundra. Fairly dry areas of forest are often used for nesting, but these usually have a peat bog in the vicinity for foraging. The species' foods are almost entirely of animal origin. Although still not well studied, they are known to primarily include aquatic insects and their larvae, especially swimming hemipterans and beetles, terrestrial flying insects such as craneflies, small crustaceans, mollusks, polychaete worms, fish, amphibians, and crustaceans. Outside the breeding season, the species occurs in a rather wide diversity of habitats, including coastal mud flats, salt marshes, freshwater marshes, sewage farms, and the borders of lakes, reservoirs, and other water areas. Its modes of foraging are not well described, but the long, delicate bill suggests that both pecking and probing are used. Yet these birds are sometimes able to cope with fairly large prey such as frogs, newts, and fish up to 6–7 centimeters in length. At least at times the birds forage while wading in fairly deep water, or even while swimming with the neck and head submerged. Probably no other *Tringa* species swims as frequently while foraging as does this one, and virtually all of its foraging is done while standing or swimming in water rather than on muddy or sandy substrates. Sometimes the birds forage in an avocetlike manner, and occasionally several individuals perform coordinated foraging in this fashion (Glutz et al. 1977).

Social behavior. Spotted redshanks are territorial during the breeding season, but otherwise the birds are moderately gregarious. Their sexual behavior is only very slightly known, but the song flight is similar to that of the redshank, consisting of an undulating display flight and a somewhat grating repeated vocalization *tjuitt-tjuee-tjuee-tjuee*. Pair-formation probably occurs upon arrival at the breeding grounds, though pigeonlike strutting displays and song flights have been reported among spring migrants. Aggressive displays include a strong wing-lifting, exposing the white wing undersides, which contrast strongly with the surprisingly blackish body. The function of this uniquely dark-colored breeding plumage is somewhat speculative, but it has been suggested that it may be related to the reduced value of countershading in arctic environments, where the sun's rays are frequently almost parallel to the ground (Gooders 1969). However, it may also play an advertisement role, since the loss of countershading is certainly not general in most shorebirds of even higher arctic latitudes. Although it is likely that monogamy is the normal system of breeding, there have been a few observations of apparent polyandrous matings (Raner 1972). In general the birds seem to be well dispersed on their breeding areas, with pairs separated by 200–50 meters, and breeding occurs at a density of only about 2–3 nests per square kilometer (Glutz et al. 1977).

Reproductive biology. Nests of this species are usually in grass tussocks or mosses, with a scanty lining of vegetation in the scrape. The birds typically select a site near a dead tree or other suitable lookout perch, from which one member of the pair can alert the other to intruders. The nest is usually in rather low herbage, such as dwarf willows, and the eggs closely match their surroundings. The normal clutch is 4 eggs; there are possibly occasional clutches of only 3. It has been reported that both sexes have incubation patches, but most observers have noted only the male on the nest. Unless the female incubated at night, as is typical of several arctic-breeding shorebirds, the evidence is that the male undertakes most of the incubation. The suggestion that polyandry, presumably successive, might sometimes occur in this species would also be favored by the adoption of male incubation and brood care. Neither the incubation nor the fledging period has been established for this species, but they are likely to be very similar to those of the redshank. The female is said to help take care of the young when they are quite small, but some females apparently begin migration as soon as the eggs hatch (Bannerman 1961).

Status and relationships. At least in Scandinavia this species is fairly abundant; the combined estimated population for Sweden and Finland is about 48,000 birds, compared with 22,000 for the redshank and 80,000 for the greenshank (Glutz et al. 1977). Farther east the population is much more uncertain, but the species seems to be in fairly comfortable numbers over much of its Eurasian range. In spite of its distinctive nuptial plumage, this species is a typical member of the "shank" assemblage and clearly has affinities with both *flavipes* and *totanus*. Larson (1957) suggests that *erythropus* and *totanus* are a species pair, with *erythropus* being derived from an arctic invasion of early *totanus* stock, probably in Miocene or early Pliocene times. I believe it is just as likely that both *totanus* and *erythropus* resulted from a dual invasion of early *flavipes* stock, with the differences in plumage and ecology of these two Eurasian species resulting from needs for reduced competition and for promoting species recognition.

Suggested reading. Bannerman 1961.

Redshank

Tringa totanus (L.) 1758

Other vernacular names. Common redshank; chevalier gambette (French); Rotschenkel (German); archibebe común (Spanish).

Subspecies and ranges (after Hale 1971). See map 101.

T. t. robusta: Icelandic redshank. Breeds in Iceland and the Faeroes. Winters in the British Isles and along the coast of the North Sea. The resident British population ("*britannica*") is considered part of *robusta* by Hale (1971).

T. t. totanus: European redshank. Breeds in northern Norway, northern Sweden, northern Finland, the Kola Peninsula, Iberia, southern France, northern Italy, Turkey, and from Belgium and Greece east to western Siberia. Hale (1971) considers much of this area to constitute a hybrid zone.

T. t. eurhinus: Himalayan redshank. Breeds in northern India, Ladakh, Sikkim, in the Himalayas to 16,000 feet, and in central and southern Tibet. This and the other Asian races winter in southeastern China, the Philippines, southern Asia south to India and the Indo-Chinese countries, and the East Indian islands to the Sundas, while *totanus* winters from the Sundas west to India, the Red Sea, tropical Africa, and the Mediterranean Basin.

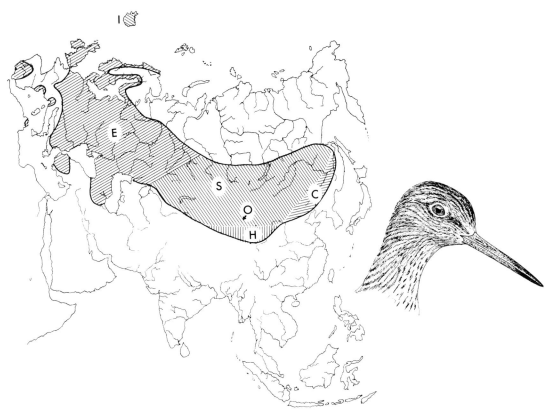

101. Breeding distributions of the Chinese (C), European (E), Himalayan (H), Islandic (I), and Siberian (S) redshanks, and the type locality of the Oriental redshank (O).

T. t. terrignotae: Chinese redshank. Breeds in eastern China and as far north as Vladisvostok, USSR.

T. t. ussuriensis: Siberian redshank. Breeds in eastern USSR, Manchuria, Mongolia, northern Tibet, Sinkiang, and central USSR north of latitude 45° N to the Urals.

T. t. craggi: Oriental redshank. Range unknown; type specimen from the oasis of Tcha Tcheu, northwestern China (Sinkiang).

Measurements and weights. Wing: males 149–76 mm; females 154–75 mm. Culmen (from feathering): males 34–46 mm; females 38–50 mm (Hale 1971). Weights (of breeding *totanus*): 100 males 107–42 g, average 123.3 g; 100 females 121–52 g, average 134.9 g (Glutz et al. 1977). Eggs ca. 45 x 32 mm, estimated weight 22.5–23.4 g (Schönwetter 1963).

DESCRIPTION

Breeding adults of both sexes have the crown and hindneck streaked with grayish brown and blackish, the sides of the head and neck pale brownish gray with narrow dusky streaking, with darker lores and a whitish eye ring. The chin, throat, foreneck, and underparts are white, variably flecked, streaked or spotted with dusky, except on the lower abdomen, which is usually entirely white. The axillaries and under wing coverts are also usually barred with dusky. The upper parts are generally grayish brown, the back and scapulars streaked or marbled with blackish, becoming bars posteriorly. The wing coverts are grayish brown, margined with whitish, especially on the greater coverts, and the secondaries are white distally and barred basally. The primaries and their coverts are dusky black, with the innermost primaries extensively barred with white and dusky. The iris is brown, the bill is dusky toward the tip and reddish basally, and the legs and toes are orange red. *Adults in winter* are grayish brown above, with the blackish summer markings only slightly indicated. The underparts are white, with fewer dusky markings, the breast being washed with gray and finely streaked with brown or dusky (Ridgway 1919). *Juveniles* are dark brown above, with white spotting. The wing coverts have distinct large white spots at the sides and toward the tip. The underparts are whitish, heavily washed and barred with brownish gray. *First-winter* birds closely resemble adults, with most spotting on the median coverts lost by winter (Prater et al. 1977).

In the hand, this is a moderately large (wing 150–75 mm) *Tringa* with reddish bill and legs (tarsus 41–55 mm), extensive white on the inner primaries and secondaries, and a white rump.

In the field (11 inches), this is the only *Tringa* with white secondaries and inner primaries, which together with the red legs provides for easy identification. The birds are found in wet grassy areas, and they sometimes bob the head when alarmed. They fly with quick, clipped wingstrokes and utter a variety of loud calls, mostly variations of *tu* or *teuk* notes.

NATURAL HISTORY

Habitats and foods. The breeding habitats of this species include grassy marshes, natural and cultivated wet meadows, shoreline meadows, coastal salt marshes, and swampy heathlands or moors as high as 1,500 feet. In Asia it also occupies high grass steppes and subalpine areas up to 15,000 feet. Unlike the lesser yellowlegs, this species does not occur in continuous forests. In highly favored habitats the species may reach densities as great as 80–90 pairs per 100 hectares (Glutz et al. 1977). Outside the breeding

35. Winter-plumage heads and secondaries of spotted redshank (*above*) and redshank (*below*).

season the birds are usually associated with muddy flats, tidal estuaries, or sometimes sandy or rocky shorelines. Their foods are nearly all of animal origin, and insects predominate in most samples, with mollusks, crustaceans, and annelid worms of generally lesser significance. The birds typically feed at a brisk walk, with occasional runs. Foraging is done by pecking, jabbing, probing, and "mowing" movements, the last being done by moving forward with the bill held nearly vertical and its tip in contact with the mud while the head is usually moved from side to side detecting the food by touch. Pecking is perhaps the most common of these movements, and thus most prey is probably obtained at or near the surface by visual means. *Hydrobia* gastropods are one of the items thus obtained in large quantities, while the amphipod *Corophium* is probably obtained primarily by probing (Burton 1974). Many of the species' foods are very small invertebrates, and during winter the birds often feed well into the night to fulfill their energy requirements, sometimes eating as many at 40,000 individual prey items per day (Goss-Custard 1969). Breeding birds also feed mostly on invertebrates, including insects and earthworms, but there are considerable seasonal differences in the specific items they eat (Goss-Custard and Jones 1976).

Social behavior. Since these birds breed over a broad latitudinal range, their time of arrival on the breeding grounds may vary as much as two months between March and May. Although some displaying may occur before arrival, it is likely that pair-formation occurs on the breeding areas. The display flight is normally performed by a single male on territory, but it is occasionally observed in flocked birds, with as many as 40 individuals taking part. In the normal situation a bird takes flight and begins a series of switchbacks. At the bottom of each switchback the male begins to flap his wings very quickly, causing him to rise almost vertically in the air in a distinctive "shivering" manner. Toward the end of the ascent phase he sets his wings, spreads his tail, and throws the head upward, calling with a series of shrill and piping notes. In this posture the bird glides forward and downward with the wings curved downward until the "shivering" phase is started again. This undulating flight may go on for several minutes. Before copulation the female runs away from the male in an erratic course, while the pursuing male spreads his tail and stretches his neck. When the female stops, the male advances toward her, initially raising his wings vertically and then fluttering them, while pro-

ducing a rolling or rattling call. As the male approaches he vibrates his wings more rapidly and performs high-stepping movements, finally fluttering off the ground and landing on the female's back; copulation then occurs (Huxley 1912; Grosskopf 1958, 1959).

Reproductive biology. Although initial breeding probably normally occurs in the second year of life, there are some records of both sexes successfully breeding when a year old. The nest is typically at the base of a tall clump of grass, with a side entrance and the grass overhead twined together to conceal the eggs from above. The normal clutch size is 4 eggs, with this number found in 94 percent of 379 nests observed in one study, and clutches of 3 are next most frequent. There is only one brood per season, but replacement clutches have been reported. Both sexes participate in incubation, which usually requires 23–24 days, though there is considerable variation in this, with extremes of 22 and 29 days being reported. Likewise, there seems to be substantial variation in the time of initial fledging, with records of wild birds fledging when from 23–35 days old, while hand-reared birds have fledged when 25–28 days old (Glutz et al. 1977). Probably about half the chicks that are hatched survive to fledging, and likewise about half of the young that fledge survive until the end of their first year (Boyd 1962).

Status and relationships. Although certainly an abundant bird over much of Eurasia, this species has declined greatly in Europe as a breeding species, probably as a result of land reclamation (Voous 1960). Nevertheless, it is one of the more common of the tringine sandpipers and is in no apparent danger from a collective population standpoint. The two most obvious close relatives of this species are *erythropus* and *flavipes,* and collectively these forms probably constitute a species group. The juvenile plumages of the three species are very similar, as are their downy young and winter plumages. Larson (1957) suggested that *totanus* and *erythropus* might represent an incomplete Pliocene species pair, with the latter the result of an early invasion of the arctic region by ancestral populations of *totanus.* The two Eurasian species might also have been derived from a double invasion of ancestral *flavipes* stock.

Suggested reading. Huxley 1912; Grosskopf 1958, 1959.

Lesser Yellowlegs

Tringa flavipes (Gmelin) 1789

Other vernacular names. Yellowshank; la petit chevalier à pattes jaunes (French); Gelbschenkel (German); archibebe patigualdo chico, archibebe pati-amarillo menor (Spanish).

Subspecies and range. No subspecies recognized. Breeds from north-central Alaska eastward through the Yukon, Mackenzie, Keewatin, northern Manitoba, northern Ontario, and western Quebec, south to about 54° N latitude. Winters in the Gulf states, eastern Mexico, Central America, and South America. See map 102.

Measurements and weights. Wing (chord): males 149–63 mm; females 149–57 mm. Culmen (to feathering): males 35–38 mm; females 30–39 mm. Weights (in winter): males 68–88 g; females 70–87

g (Haverschmidt 1968). Males on the breeding grounds range 69–94 g, average of 14 was 81 g; females range 77–100 g, average of 6 was 81 g (Irving 1960). Eggs ca. 42 x 29 mm, estimated weight 17.5 g (Schönwetter 1963).

DESCRIPTION

Breeding adults of both sexes have a streaked grayish white and dusky crown and have white on the sides of the head with dusky streaking, especially on the lores, with a clear whitish area around the eyes and a whitish superciliary stripe. The lower neck, breast, and other underparts have dusky streaking or spotting on the foreneck, chest, and breast, becoming barring on the sides and under tail coverts. The axillaries and under wing coverts are also barred with dusky. The back and scapulars are brownish gray with black and whitish spotting, the rump and upper tail coverts are gray to white with some dusky barring, and the tail feathers are white with broad dusky barring, the bars becoming narrower and more broken on the outer feathers. The secondaries are brownish gray with narrow white margins, the primaries are blackish outwardly and more grayish inwardly, the latter narrowly margined with white. The coverts are light brown, with pale whitish tips and darker subterminal spotting. The iris is brown, the bill is black, becoming brownish toward the base, and the legs and toes are yellow. *Adults in winter* have grayish brown upperparts, sometimes nearly uniformly so but usually slightly broken with dusky and whitish spotting. The breast, neck, and head are nearly uniform light gray, lighter on the throat and above the lores. *Juveniles* are brownish above, with buff spots. The breast is indistinctly streaked, with a grayish wash. *First-winter* birds show extensive notching of buff on their scapulars and tertials (Prater et al. 1977).

In the hand, this medium-sized *Tringa* can be identified by its long tarsus (45–58 mm), bright yellow legs, and straight blackish bill that does not exceed 40 mm.

In the field (10 inches), this sandpiper is larger than the "peeps", with which it often associates, and it has bright yellow legs as well as a straight and gradually tapering black bill. It is appreciably smaller than the greater yellowlegs, but unless the two are together or with other sandpipers of known size such as dowitchers, the size differences are not so apparent. The greater yellowlegs, however, often has a slightly upturned bill, which is both longer and heavier, and

102. Breeding distribution of the lesser yellowlegs.

when alarmed the lesser typically utters a single or double *wheo* note, rather than the three or four louder calls of the greater yellowlegs. In flight, both species show whitish rumps and tails, compared with uniformly dark upper wing surfaces.

NATURAL HISTORY

Habitats and foods. This species breeds in habitats that provide a combination of rather open and tall woodlands, having sparse and low undergrowth, and that are fairly close to marshy or grassy ponds. In Alberta the birds have been found nesting among broken hills that are covered with burned and fallen timber and have a second growth of low poplars, while in the Churchill area of Manitoba they frequent dry, lichen-covered, and often burned-over ridges in forested areas, where there are muskegs nearby for foraging. Outside the breeding season they are usually found along mud flats and shallow pools, in areas similar to those used by greater yellowlegs but they generally seek out somewhat more sheltered locations. In coastal Argentina the birds commonly occur in grassy ponds, marshes with emergent vegetation, and flooded fields, but they sometimes also occupy coastal estuaries. At that time of year flock

sizes are small, and many individual birds defend feeding territories of from 0.1 to 0.5 hectare in sloughs, or linear areas of up to 100 meters (Myers and Myers 1979). While foraging the birds usually wade in shallow water, often up to the breast feathers, and they obtain most of their food by pecking at the water surface or mud, rather seldom probing for it. Unlike the greater yellowlegs, the birds do not regularly skim the bill from side to side over the water surface. Foods on the breeding grounds are only poorly studied, but foraging methods and the foraging niche of this species overlap considerably with several species of calidridine sandpipers in the Churchill area (Baker and Baker 1973). The density of breeding population in some areas of concentrated habitat may reach 3–4 pairs per 100 acres, but in general the density is much lower (Bannerman 1961). Outside of the breeding season the foods include a diverse array of insects, crustaceans, worms, small fish, and other animal life. The adults and larvae of terrestrial and aquatic insects are mentioned prominently in most reports, while fish and crustaceans seem to play a smaller role in the diet than is true for the greater yellowlegs.

Social behavior. These birds are territorial during the breeding season and appear on the breeding grounds of western Canada in late April. The birds arrive in groups ranging from 2–3 birds to as many as 20 and quickly disperse over the available habitat. The song flight is performed with undulating movements over a prolonged and switchback course, accompanied by a nearly continuous yodeling *pill-e-wee, pill-e-wee, pill-e-wee*. Evidently both sexes perform the song flight, as pairs have been seen flying only a few yards apart during the display. The same call is sometimes uttered from a stump or treetop, and occasionally the undulating flight is performed without the accompanying vocalizations. Copulation often occurs on a dead tree stump (Rowan 1929).

Reproductive biology. The usual nest site of this species is in an area where there is an abundance of tree stumps, charred tree trunks, and similar debris of a forest fire. Often nests are on a ridge, where the new growth among the stumps is no more than 8–10 feet high. Nests are usually placed at the base of such a stump, under a small bush such as Labrador tea, under a brush pile, or rarely in the open. The plumage coloration of the bird is such that it blends remarkably well with the color of dead wood, and a bird on the nest is almost impossible to see. All but one of 6 nests found in the Churchill area by Jehl and

Smith (1970) had 4 eggs, and the other had 3. At least one clutch of 5 eggs has also been noted, and one nest with 6 eggs was found being simultaneously incubated by two birds. Incubation begins when the last egg is laid. Although it is probable that both sexes incubate, the relative role of the sexes is still undetermined. The incubation period is probably 22–23 days, and the young almost immediately leave the nest, with both parents in attendance. They are strongly protected by the parents, which persistently dive at human intruders but never actually strike them. The fledging period is also still unestablished but is probably similar to that of the redshank (23–25 days), which the young of this species closely resemble.

Status and relationships. This is one of the most common medium-sized shorebirds in North America, and its population certainly does not appear to be in any difficulty at present. As to the species' relationships, the close affinities of several *Tringa* species allow for various speculative models of speciation, but it seems probable that *totanus* is the nearest relative of *flavipes*. The two species have complementary ranges in the New and Old Worlds, and in plumage and behavior the species seem to be extremely similar. Mayr and Short (1970) suggested that either this species or *erythropus* is likely to be the nearest relative of *flavipes*, and certainly *erythropus* is the ecological counterpart of *flavipes* in its breeding ecology.

Suggested reading. Bannerman 1961; Bent 1927.

Greater Yellowlegs

Tringa melanoleuca (Gmelin) 1789

Other vernacular names. Greater yellowshank; chevalier criard à pattes jaunes (French); grosser Gelbschenkel (German); tigüi-tigüi grande, archibebe patiamorillo mayor (Spanish).

Subspecies and range. No subspecies recognized. Breeds from southern Alaska and south-central British Columbia eastward through the central and northern parts of all the major provinces to Labrador and Newfoundland, as well as on Anticosti Island and northeastern Nova Scotia. Winters from British Columbia and South Carolina south-ward through Mexico, Central America, and South America to Tierra del Fuego. See map 103.

Measurements and weights. Wing (chord): males 180–98 g; females 180–97 g. Culmen (to feathering): males 52–61 mm; females 53–58 mm. Weights: 5 males 155–212 g, average 165 g; 3 females 132–205 g, average 174 g (various sources). Eggs ca. 49 x 33 mm, estimated weight, 27.5 g (Schönwetter 1963).

DESCRIPTION

Breeding adults of both sexes have a crown that is streaked with blackish and grayish white, the sides of the head whitish with dusky streaking, heaviest on the lores and lightest above the eyes, and the underparts white with dusky streaking except on the chin, throat, abdomen, and middle under tail coverts, and heaviest on the sides of the neck, breast, and sides, becoming barring on the sides and flanks. The axillaries and under wing coverts are barred. The back and scapulars are marked with black and edge spotting of grayish white, the lower back and rump are dusky gray with dusky and whitish barring, and the upper tail coverts are white, becoming barred posteriorly. The tail coverts are brownish gray centrally, becoming increasingly spotted and barred with dusky and whitish. The wing coverts are brownish gray with paler margins, the secondaries are brownish gray with white edging or spotting, and the primaries and their coverts are blackish, with the innermost primaries paler and spotted or edged with whitish. The iris is brown, the bill is blackish at the tip, becoming more brownish basally, and the legs and toes are bright yellow (Ridgway 1919). *Adults in winter* are grayish above, with white spotting but no black markings; the foreneck and breast are narrowly streaked, and the sides and flanks are irregularly marked with grayish. *Juveniles* are brown on the upperparts, with extensive buff spotting, and the wing coverts have a series of light spots along the edges, leaving a broad darker central bar at the tip. *First-winter* birds resemble adults but are browner, with some spotted inner median coverts persisting until spring (Prater et al. 1977).

In the hand, the long yellow legs (tarsus 55–68 mm) and the long wing (180–200 mm) set this apart from other *Tringa* species.

In the field (14 inches), this is the largest of the *Tringa* species, and its large size and bright yellow legs provide excellent field marks. Compared with

103. Breeding distributions of the greater yellowlegs (*hatched*) and the greenshank (*shaded*).

the lesser yellowlegs, it is generally more robust, with a heavier and often slightly upturned bill, and it sometimes has more orange yellow legs as well as more strongly marked upperparts. In flight, it usually utters a series of three or more-loud *wheu* notes and exhibits a whitish rump and tail, while the upper wing surface is quite dark. It inhabits a variety of coastal and freshwater habitats, often feeding by wading while chasing small invertebrates in breast-deep water.

NATURAL HISTORY

Habitats and foods. This species generally is associated with muskeg forests, with mixed trees, clearings, and usually also ponds, though in many areas the birds breed on hills or ridges near swamps that are mostly burned over and grass-covered, or which are still covered with dense forests of poplars, birches, and spruces. The breeding habitats also extend into subarctic tundra and into subalpine scrub, but the optimum combination seems to be rather open and high woodlands with only low and sparse undergrowth, and within reasonable distances of marshy or grassy ponds. When feeding, the birds usually stand in shallow water, skimming the surface for foods and sometimes also probing in mud. Unlike the lesser yellowlegs, which sometimes swims over the deeper parts of a pond to reach the other side, this species typically will wade only until it reaches belly-deep water, when it takes flight for a shallower areas. The birds will sometimes also forage on the shore, quickly running over the grasses and snatching terrestrial insects such as small grasshoppers. Foods on the breeding grounds are unstudied, but while on migration the birds evidently feed on insects such as ants, flies, and grasshoppers and on small crustaceans, fishes, and worms. Adult aquatic insects such as water boatmen, and aquatic larvae of swimming insects such as dytiscids and hydrophilids have also been found in the stomachs of migrating birds (Bent 1927; Bannerman 1961). On the wintering grounds of Argentina the birds occupy various wetlands, especially ponds with emergent vegetation, but they also use littoral areas on tidal flats with vegetation nearby (Myers and Myers 1979).

Social behavior. While on migration and on wintering grounds this species is only moderately social. In

36. Breeding-plumage heads of lesser (*above*) and greater (*below*) yellowlegs.

coastal Argentina it occurs in flocks of up to about 10 birds, but more commonly individuals forage alone, with many birds defending feeding territories. During migration there is a limited amount of social display, but it is likely that pair bonds are formed after arrival on the breeding grounds. The birds become highly dispersed after arrival, with no more than a dozen pairs spread out over an area of several square miles. Like the lesser yellowlegs, this species also performs a switchback type of display flight, but it lasts appreciably longer, up to 10 or 15 minutes. During that time the undulating flight is continuous, but there are intervals during which the yodeling call ceases for as long as 30 seconds. This call is a rolling *toowhee, toowhee,* scarcely distinguishable from the call of the lesser yellowlegs. It has also been described as sounding like *tweda* or *wull yer?* and is said to have a flickerlike quality. In southern Alaska the birds have been observed running in circles on sandbars around courted individuals, uttering shrill whistles and posing with upraised and quivering

wings. Nothing else seems to have been written on the specific aspects of pair formation and courtship behavior (Bent 1927; Bannerman 1961).

Reproductive biology. The nests of this species are extremely hard to locate, and thus little has been written on this phase of the species' biology. The nests are usually placed rather close to water but rarely may be as far as a mile from the nearest pond. They are often placed beside a branch or log lying half-hidden in the mosses, and they may be sheltered by the stems of dwarf birch growing through the mosses. Sometimes they are on higher ground, where the soil is relatively dry, and one observer reported that in Newfoundland the species never nests on ground that is too soft to scratch in, preferring areas with dry, hard peat or actual soil for a substrate. It has been stated that if the first clutch is lost a second or even a third may be started, usually no more than 20–30 yards from the original site. Nearly all of nests described have had 4 eggs. The incubation period is reportedly 23 days, and though it is likely that both sexes participate this is still unsubstantiated. It is known that both sexes attend the young, taking the chicks from the vicinity of the nest to a source of water only a day or two after hatching (Bannerman 1961; Bent 1927).

Status and relationships. This species is certainly far less common than is the lesser yellowlegs, but no quantitative information is available for either species. The relationships of these two remarkably similar species have not received sufficient attention. Nichols (1923) suggested that the two are actually not as closely related as they seem and that *melanoleuca* may actually be a mimic of *flavipes,* possibly to elude predation by peregrine falcons. Jehl (1968a) doubted this interpretation and placed the two side by side in linear sequence within the "shank" group of species. The downy young of both closely resemble the young of *erythropus* and *totanus,* but pattern variation is great in this group, between and even within species. Larson (1957) suggested that *melanoleuca* is part of a species group that includes *nebularia, stagnatalis,* and *guttifer,* while *flavipes* is part of a group that also includes *totanus* and *erythropus,* plus the extinct form *numenoides.* Mayr & Short (1970) believed that *nebularia* is a geographical replacement of *melanoleuca* and considered the two a superspecies. I favor the view Larson proposed, though a close relationship between *melanoleuca* and *flavipes* is a distinct possibility.

Suggested reading. Bannerman 1961.

Greenshank

Tringa nebularia (Gunnerus) 1767

Other vernacular names. None in general English use; chevalier aboyeur (French); Grünschenkel (German); archibebe claro (Spanish).

Subspecies and range. No subspecies recognized. Breeds in Scotland, and from Norway east across Scandinavia, Estonia, Russia, and Siberia to the northern limits of the taiga and the Bering coast, south through Kamchatka and southern Siberia to the mouth of the Amur. Winters from the Mediterranean Basin, Iraq, the Persian Gulf, eastern China, Taiwan, and the Philippines south to southern Africa, India, the Indo-Chinese countries, the Sundas, New Guinea, and Australia. See map 103.

Measurements and weights. Wing: males 178–88 mm; females 182–203 mm. Culmen: males 46–58 mm; females 49–60 mm. Weights: immatures and adults on fall migration range 128–270 g; males in spring average about 163 g and females probably average slightly heavier (Glutz et al. 1977). Eggs ca. 51 x 33 mm, estimated weight 30.5 g (Schönwetter 1963).

DESCRIPTION

Breeding adults of both sexes have a crown that is grayish white broadly streaked with dusky, a broad white superciliary stripe, a white eye ring and chin, and the rest of the head finely streaked with dusky, especially on the lores. The sides of the neck, breast, and anterior portions of the sides of the body are coarsely and irregularly streaked with dusky, with these markings becoming somewhat arrowlike or V-shaped posteriorly. The rest of the underparts are entirely white, while the axillaries are sometimes marked with grayish and the under wing coverts are irregularly marked with the same. The upperparts including the scapulars are blackish and pale gray, the wing coverts are nearly uniform deep brownish gray, the inner coverts are fringed with white, and the flight feathers are nearly uniformly dusky. The rump and upper tail coverts are white, the larger coverts having dusky markings, and the tail is white with grayish dusky barring. The iris is brown, the bill is blackish, becoming grayish green basally, and the legs and toes are pale olive green or very rarely yellowish. *Adults in winter* have grayish upperparts with the wing coverts gray, the inner ones fringed with white. The underparts are white, with very fine streaking on the breast. *Juveniles* are dark brown above, the feathers broadly edged with buffy brown. The wing coverts are grayish brown with buffy fringes that are interrupted centrally by brown bars, and the central tail feathers are fringed with whitish and edged with small brown bars. The lower breast and flanks have many narrow crescentic bars; otherwise the breast is only sparsely streaked and mottled. *First-winter* birds retain some buff-fringed inner median wing coverts to spring (Prater et al. 1977; Ridgway 1919).

In the hand, this rather large *Tringa* (wing 178–203 mm) is distinguishable on the basis of its rather long and greenish legs and its long and straight or very slightly upturned bill (culmen 46–60 mm). It is closely similar to *guttifer* but has longer legs (tarsus at least 50 mm, vs. maximum of 45 mm) and also has slight spotting on the axillaries and under wing surface, whereas *guttifer* is immaculate white in these areas.

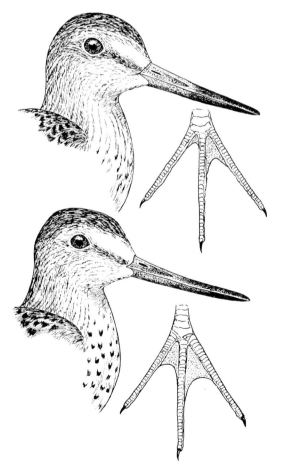

37. Breeding-plumage heads of greenshank (*above*) and spotted greenshank (*below*).

In the field (12 inches), the greenish legs, white rump that extends well up the back, and a tapering, grayish bill that is very slightly upturned help identify this species. It is larger than the marsh sandpiper and lighter colored above than the green sandpiper. It is very close in appearance to the spotted greenshank but is grayish rather than yellowish at the base of the bill, has longer and more grayish legs, and has gray markings on the under wing lining and axillaries. In flight, this species shows no white on the upper wing surface but has extensive white on the lower back and rump, and it often utters a whistled *tew* or *chew* that may be repeated up to six times, though it usually consists of three notes.

NATURAL HISTORY

Habitats and foods. This species breeds primarily in the boreal climatic zone, in habitats that include open marshes, bogs, eutrophic lakes with wide margins of dead and decaying vegetation, swampy clearings in coniferous forest, and the birch zone where mosses and lichens are abundant (Voous 1960). The birds breed from altitudes close to sea level to as high as about 3,600 feet in the mountains of eastern Norway. Although their nests are found in wooded areas, they sometimes also nest in treeless areas having boulders, heather, and lichens for cover. Breeding densities are low, and in the most favored forest habitats of Norway they range from a maximum of about a pair per 100 acres to a pair per 640 acres. Likewise in Scotland the maximum density is about a pair per 100 acres (4 pairs in a 400-acre area), while often it is only a pair per several thousand acres (Bannerman 1961). The birds prefer to forage in sheltered situations at the edge of the water, feeding especially on insects and their larvae. They often rush about through the shallows chasing prey, but at other times they walk slowly about, picking up food in a leisurely manner. Besides insects, they also are known to eat crustaceans, annelid worms, mollusks, amphibians, and small fish up to 3 inches in length. On their wintering grounds the birds are associated with estuaries, sandy or muddy coastal flats, salt, fresh, or brackish swamps, and lakes. At times they also occur along quiet stretches of rivers well away from coastlines.

Social behavior. Male greenshanks typically arrive on their breeding grounds in advance of the females, and before establishing territories they often visit a large loch, where mating occurs. Females are able to recognize their mates of the previous year and often

remate with former partners. The average breeding life for females is 3–4 years, and in one case a female bred in the same territory for 6 years. During early stages of pairing the male often chases the female in dramatic flights that may cover several miles, often ending at about the same place they started. Sometimes during these flights both sexes sing in duet, and the female at times will also accompany the male on his song flight, gliding below him while he performs switchbacks above her. In his territorial song flight the male alternates climbing and gliding phases, usually singing a rich *too-hoo-hoo* during the gliding phase. A special "sex call" is sometimes also mixed with this song when the male is trying to attract a mate or to attract a second female after his first mate begins incubating. At times when the female is nest-prospecting, or sitting on the eventual nest site, the male glides over her and swings in an inverted arch to a second perch on which he shakes his wing or holds it over his back. Before copulation the male may raise both his wings and wave them violently overhead, as he stalks toward the female and stands behind her, beating his wings before fluttering up onto her back. Before he mounts her the male sometimes also lifts a wing above the hen and calls in short, guttural bursts. Or he may bow forward and fan his tail, or chase the female in erratic patterns. The hen squats or tilts forward when ready for copulation, and the male beats his wings during treading to maintain his balance. Although normally performed on the ground, copulation sometimes also occurs on fenceposts, rocks, stumps, or even in trees. Copulation may occur as early as 18 days before the first egg is laid or as late as during the incubation period. In bigamous matings the two females may actually lay their eggs in the same nest, or they may have separate nests from 4 to 200 yards apart. Both sexes search for nest sites, and the male will often lead his mate to a succession of possible nest locations. The female makes the final choice, lining it several days later and perhaps not laying the first eggs until as late as 10 days afterward (Bannerman 1961; Nethersole-Thompson 1951).

Reproductive biology. Nests are typically placed close to fragments of dead wood, or less often are beside a granite block, a small or large tree, or some similar landscape feature. New mates of an experienced male will sometimes nest in the previous year's site, indicating the role of the male in the choice of a nest site. Females will also sometimes nest in the same site for as many as three consecutive years. The eggs are not laid at fixed intervals, but usually are produced at intervals of 36–48 hours, occasionally as much as 72 hours apart. The clutch is almost invariably 4 eggs; of 111 initial Scottish clutches, 105 were of 4 eggs and the remainder were of 3. A very few 5-egg clutches have been reported. Incubation usually begins immediately after the last egg is laid. The role of the male varies considerably. In some monogamous matings the male incubates by night and the female by day, but some males seeking out second mates do little incubation. The incubation period ranges from 22–25 days and averages nearly 24 days. Both parents tend the young birds, but frequently one of the adults leaves the brood before the young are able to fly. In some cases it is the male that leaves, in others it is the female. Fledging occurs between 26 and 31 days after hatching, and soon thereafter the fall migration begins (Bannerman 1961; Nethersole-Thompson 1951).

Status and relationships. The broad breeding range of this species, which centers on the relatively unmodified boreal forest of Eurasia, suggests that the population is secure. Together with the very similar Asian form *guttatus* and the North American *melanoleuca*, it is part of a species group within the "shank" assemblage of *Tringa*. Mayr and Short (1970) supported the suggestion of Voous (1960) that *melanoleuca* and *nebularia* are geographic replacement forms and constitute a superspecies. I suspect that *guttifer* may be more closely related to *nebularia* than is *melanoleuca*, and I consider them a probable superspecies.

Suggested reading. Nethersole-Thompson 1951; Bannerman 1961.

Spotted Greenshank

Tringa guttifer (Nordmann) 1835
(*Pseudototanus guttifer* of Peters, 1934)

Other vernacular names. Armstrong's sandpiper; Nordmann's greenshank; chevalier tachete (French); Fleckengrünschenkel (German).

Subspecies and range. No subspecies recognized. Breeds on Sakhalin, and possibly also on southern Kamchatka and along the coast of the Sea of Okhotsk. Winters in northeastern India, Burma, the Malay Peninsula, and on Hainan. See map 106.

Measurements and weights. Wing: males 174–81 mm; females 169–83 mm. Culmen (from feathering): both sexes 48–58 mm. Weights: 1 female 158 g; 3 males 136–40.7 g (Nechaev 1978). Eggs ca. 49 x 34 mm, weight about 28 g (Nechaev 1978).

DESCRIPTION

Breeding adults of both sexes closely resemble those of the greenshank, but with the dark breast spotting well out onto the flanks and abdomen and with the neck area tending toward spotting rather than streaking as in *nebularia*. Additionally, the under wing surface and axillaries are entirely white, rather than flecked and barred with brownish. The upperparts are blackish brown with white spotting while the underparts are white with blackish brown spots on the breast, neck, sides, and upper flanks. The iris is brown, the bill is blackish toward the tip and yellowish basally, and the legs are pale green or yellowish. *Adults in winter* are grayer on the upperparts, the wing coverts are gray with white edging, and the tertials are gray. *Juveniles* are brown on the upperparts, with extensive buff streaking and spotting. The coverts are brownish, with narrow edging of pale whitish buff. The tertials are brownish with buffy spotting, and the underparts are white with a brown wash on the neck and breast. *First-winter* birds resemble adults but retain brownish inner median coverts and tertials (Prater et al. 1977).

In the hand, the semipalmated conditions of the inner front toes is unique in *Tringa*, the webbing being 5–6 mm long. Additionally, the tarsus is relatively short (49–45 mm, or shorter than the culmen), while in *nebularia* the tarsus is longer (usually over 55 mm) than the culmen. Last, the entirely white under wing surface distinguishes this species from *nebularia*.

In the field (12 inches), this species is most likely to be seen in winter plumage, when it is very much like the common greenshank, but it is paler in winter, with paler barring on the tail and white under wing linings and axillaries; the bill is slightly stouter, with yellowish rather than grayish at the base. The usual call is a loud, piercing *keyew*, sharper and less musical than that of the greenshank.

NATURAL HISTORY

Habitats and foods. On its wintering grounds this species is associated with coastal mud flats and deltas. More generally, it occurs along the shorelines of shallow lagoons, in swamps, willow beds, and along streamside mossy meadows. The nesting habitat consists of marshy coastlines having lagoons and adjacent sparse larch forests (Nechaev 1978). Its foods are poorly studied, but at least on Sakhalin Island it includes small fish such as sticklebacks. Crustaceans and mollusks are apparently also eaten. An early observer in the 1840s reported that the birds foraged in large flocks in shallow water, at least half a mile from shore, seemingly swimming with their wings raised. He later judged that the birds were actually wading, fluttering their wings to keep their balance against the tide. The semipalmated condition of this species' toes suggests that it probably is a good swimmer, and it may well regularly forage while wading or swimming.

Social behavior. Not much is known of this species' social behavior, since it has been seen on its breeding grounds by so few people. It is too uncommon everywhere to judge normal flock size, but a group of 22 adults has been seen on Sakhalin Island during the breeding season, suggesting considerable sociality. After leaving their coastal shallows in late May, males perform nuptial flights and produce associated loud calls over the adjacent larch forests during the month of June (Nechaev 1978).

Reproductive biology. Eggs are laid during the first half of June, and in all of 5 cases the nest site was the crotch of a larch branch from 2.3 to 4.5 meters aboveground, supported by "old-man's-beard" lichens. Two full clutches were found to have 4 eggs each, and incubation was by both sexes. Hatching begins in late June, and within a day or two after hatching the adults begin to lead their young toward the coastal lagoons (Nechaev 1978).

Status and relationships. This species is rare and apparently is disappearing from its highly restricted range (Nechaev 1978). It is almost certainly a very close relative of *nebularia*, differing mainly in its toe palmation and its bill shape; its downy young are said also to be closely similar to those of *totanus*. In my opinion it is a coastline-adapted derivative of ancestral *nebularia* stock, and its minor structural differences certainly do not warrant generic (*Pseudototanus*) distinction.

Suggested reading. Kuroda 1936; Nechaev 1978.

Marsh Sandpiper

Tringa stagnatalis (Bechstein) 1803

Other vernacular names. Little greenshank; chevalier stagnatile (French); Teichwasserläufer (German); archibebe fino (Spanish).

Subspecies and range. No subspecies recognized. Breeds in southeastern Europe from Bulgaria and Romania east to the Ukraine, Crimea, and eastern Russia, and in western Siberia east to the Kuznetsk and Kirgiz Steppes. Also breeds in Transbaikalia, north perhaps to the Okekma River and possibly east to Lake Khanka in Ussuriland. It has also bred in Hungary, Austria, and elsewhere in Europe. Winters from the Mediterranean, Persian Gulf, Indo-Chinese countries and Hainan south to southern Africa, India, the Sundas, and Australia. See map 104.

Measurements and weights. Wing: males 133–45 mm; females 135–46 mm. Culmen: males 37–45 mm; females 36–45 mm. Weights (spring birds): 4 males 64–73 g, average 68 g; 5 females 70–85 g, average 77.8 g (Glutz et al. 1977). A sample of 77 wintering birds averaged 74.3 g and ranged 55–94 g (Summers and Waltner 1979). Eggs ca. 38 x 27 mm, estimated weight 14.0 g (Schönwetter 1963).

DESCRIPTION

Breeding adults of both sexes have a forehead and crown that are whitish, heavily spotted with sepia, a whitish superciliary stripe, and the sides of the head and neck white with heavy sepia streaking, especially on the lores and the sides of the neck. The sepia marks become spots on the breast and grade into irregular bars on the flanks and under tail coverts. The mantle and scapulars are light drab, the feathers faintly edged with buff and with large blackish brown centers. The back, rump, and upper tail coverts are white, the longer coverts being barred or marked with ashy brown. The tail feathers are white with sepia and ashy brown markings, except for the central pair, which are usually more grayish, barred with ashy brown. The wing coverts are ashy brown, with sepia streaks and narrowly tipped with white, the primaries are blackish brown, and the secondaries are ashy brown with mottled white on the inner webs (Witherby et al. 1941). The iris is brown, the bill is blackish with a greenish tinge at the base, and the legs and toes are olive green. *Adults in winter* are grayish brown on the upperparts, the wing

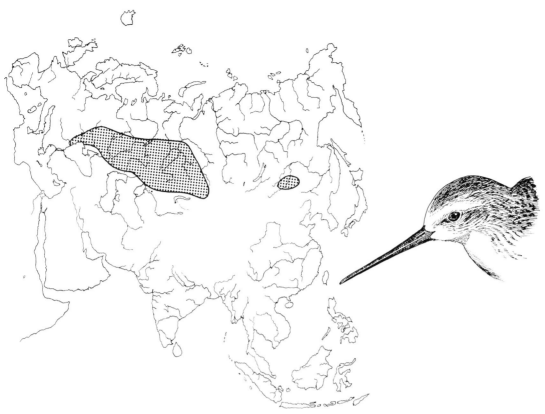

104. Breeding distribution of the marsh sandpiper.

coverts being grayish with narrow white fringes. The sides of the breast have irregular ashy brown markings, but the rest of the underparts are white. *Juveniles* have dark brown upperparts and wing coverts, heavily spotted or edged with buff, and white underparts. *First-winter* birds resemble winter adults but retain some brown on the back for a time, and some well-worn juvenile inner median coverts are held until winter (Prater et al. 1977).

In the hand, this *Tringa* has an extremely fine-tipped, straight bill (culmen 36–45 mm) that is almost as long as the legs (tarsus 47–57 mm), as well as a white rump and lower back contrasting with a plain brownish upper wing surface. It is most like *nebularia*, but with a much more slender bill, and the wing length is appreciably less (maximum 148 mm, vs. 184 mm in *nebularia*).

In the field (9 inches), this bird is usually found around the margins of freshwater habitats and resembles a greenshank, but it is smaller, has a much weaker bill, is more obviously spotted on the upperparts in breeding plumage, and has relatively longer legs. In flight the birds utter various mellow whistles, such as *teu*, *keeooo*, and *ke-weep*, and the white

rump contrasts strongly with the fairly dark wings and mantle.

NATURAL HISTORY

Habitats and foods. This is a relatively southern and steppe-adapted species, breeding in the grassy and marshy shores of predominantly freshwater pools in warm steppe areas, and to a lesser extent around salt lakes. Brackish marshes, where the water is quite shallow and where there are patches of low and scanty vegetation, are favored both during the breeding season and outside it. Although this species is said to sometimes nest in a semicolonial manner, population densities are highly localized. At one time in Hungary a population of 50–60 pairs bred in an area 5–6 kilometers in circumference, and in the USSR populations of up to 3–4 pairs per hectare have been reported (Glutz et al. 1977; Hudson 1974). Freshwater marshes, the marshy borders of lakes or pools, flooded areas of cultivated lands, and paddy fields are used by migrating and wintering birds, but they are only rarely found on tidal shorelines. Most of the foraging is done in shallow water, either singly or in small and often tightly packed groups, and the birds

38. Breeding-plumage heads of greenshank (*above*) and marsh sandpiper (*below*).

generally resemble redshanks in their foraging methods. They probe in mud and sand and sometimes seize prey in their long bills, or at times they wave their bills back and forth in the water. Their long legs allow them to forage in water up to 6 inches deep. Their foods are not yet extensively studied, but they include small crustaceans such as *Gammarus*, gastropods and other small mollusks, and a variety of aquatic insects. A sample from 12 birds taken in the USSR between June and August consisted almost entirely of insects, including *Corixa* bugs, the larvae of tendiped and dytiscid beetles, and some gastropod mollusks. Terrestrial insects are eaten less frequently but are sometimes plucked from aquatic plants or ground surfaces. Marsh sandpipers and greenshanks often forage in mixed groups while on migration, presumably using much the same foods.

Social behavior. The birds arrive on their breeding grounds during late April and early May and soon establish territories. Song flights by males have been seen as late as June, but they probably peak in May, shortly after arrival. The flight is accompanied by a trilling song, uttered as the male alternately rises on rapidly vibrating wings and descends in a steep glide with the wings held curving downward. After such a flight the male sometimes approaches the female while singing and fluttering his wings over her, moves aside, then returns toward her with out-

stretched wings, and finally rises into the air again, still singing. Territories must be small or poorly defended, since there are records of nests being placed only 5–10 yards apart. Copulatory behavior has not yet been described, but it probably resembles that of the redshank.

Reproductive biology. Nests are constructed on mounds at the marshy edges of lagoons, lakes, or pools and sometimes are placed on stacks of old hay that protrude from the water. The nest is usually close to water and is lined with dried grasses. The clutch is typically 4 eggs, but sometimes is 3 or 5 eggs, and exceptional cases have been found with as many as 7 eggs, certainly laid by two females. As I noted earlier, the nests are often rather closely spaced, and it has been reported that as many as several dozen pairs may sometimes nest on a marsh only 1–2 hectares in area. It is believed that both sexes incubate, but neither the incubation period nor the fledging period is known. Only one brood is reared, apparently tended by both parents, and shortly after fledging the birds begin to leave their nesting areas. Fledging occurs from late June to late July, and the young gradually move from damp meadows to open riverbanks and finally depart in groups of 10–15 individuals.

Status and relationships. This species is included in the "threatened" birds of Europe by Hudson (1974), inasmuch as the total population breeding in Europe outside of Russia is perhaps only 100 pairs. Most of these are in northeastern Bulgaria, and probably some still occur in the Danube delta of Romania. It once nested in Hungary and has very rarely nested Yugoslavia, but these are all marginal and insignificant in terms of the present total population. In the USSR the species is said to be abundant in eastern Kazakhstan (Dementiev and Gladkov 1969), and the center of the wintering range is probably east Africa, where the birds are often abundant in the rift valley lakes of Ethiopia, Kenya, and Tanzania. The relationships of the species are certainly with *nebularia*, and it might be regarded as a specialized offshoot of early greenshank stock that adapted to a more steppelike habitat and smaller food items. Larson (1957) considers it part of a species group that includes *nebularia*, *guttifer*, and *melanoleuca*, while Fjeldså (1977) noted that the downy young closely resemble those of *totanus*. I agree with Larson (1957) that it is a part of the greenshank assemblage, though a somewhat peripheral member.

Suggested reading. Hudson 1974; Tolchin 1976.

Green Sandpiper

Tringa ochropus L. 1758

Other vernacular names. None in general English use; chevalier culblanc (French); Waldwasserläufer (German); andarríos grande (Spanish).

Subspecies and range. No subspecies recognized. Breeds from Scandinavia and Denmark east through eastern Europe, Russia, and Siberia, perhaps to the Sea of Ohkotsk, where it occurs in summer but is not yet proved to breed. Sporadic or local breeding has also occurred to the west and south of these limits. Winters from southern Europe, the Mediterranean Basin, Iraq and the Persian Gulf, eastern China, Taiwan, and the Philippines south to tropical Africa, India, the Indo-Chinese countries, and Borneo. See map 105.

Measurements and weights. Wing: males 138–46 mm; females 140–52 mm. Culmen (from feathering): males 30–42 mm; females 32–38 mm. Weights (fall migrants): males 73–104 g; females 67–119 g. Males in spring average about 75 g, and females

average about 85 g (Glutz et al. 1977). Eggs ca. 39 x 28 mm, estimated weight 15.5 g (Schönwetter 1963).

DESCRIPTION

Breeding adults of both sexes have a crown that is deep grayish brown streaked with grayish white, a white stripe extending from above the lores to behind the eyes, a darker streak through the lores, and fainter streaking on the sides of the head with the chin and throat white. The lower throat is heavily streaked with dusky, these markings becoming larger on the chest and approaching transverse barring on the sides and upper flanks. The rest of the underparts are white, while the axillaries and under wing coverts are deep grayish brown. The scapulars and back are blackish with dull white spotting, the wing coverts have smaller whitish spots, and the flight feathers are blackish. The lower rump and upper tail coverts are white, and the tail is mostly white, the inner feathers becoming progressively more barred with dusky. The iris is brown, the bill is black with a greenish tinge toward the base, and the legs and toes are dark

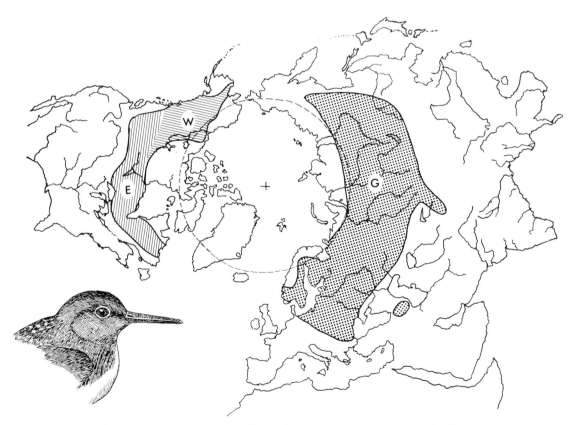

105. Breeding distributions of the green sandpiper (G) and the eastern (E) and western (W) solitary sandpipers.

39. Breeding-plumage heads and outer tail feathers of green (*above*) and wood (*below*) sandpipers.

olive green. *Adults in winter* have a plain brownish gray crown, the upperparts have fewer and smaller white spots, and the chest and sides of the neck are nearly uniform brownish gray, with less distinct streaking than during summer. The coverts are olive brownish, with indistinct pale fringes. *Juveniles* are heavily spotted with buff on the olive brown upperparts. *First-winter* birds retain some buff-spotted inner median coverts until late winter (Prater et al. 1977; Ridgway 1919).

In the hand, this rather small *Tringa* is notable for its rather short greenish legs (tarsus 31–37 mm), its tapering, straight bill (culmen 30–42 mm), and its very dark upper and lower wing surfaces. The similar *glareola* is smaller and has a shorter bill (culmen 27–31 mm).

In the field (9 inches), this species is found in wooded areas during the breeding season and is usually in freshwater areas at other times. Its very dark wings and back contrast strongly with the white

rump, and the heavily striped breast contrasts with the white abdomen. The birds are usually found alone or in small groups, and when flushed they tower upward while uttering a shrill *weet-a-weet* or *tit-looet* call. In size and pattern the species most closely resembles the wood sandpiper, but it is darker throughout, especially on the underwings.

NATURAL HISTORY

Habitats and foods. During the breeding season this species is closely associated with the boreal climatic zone, but it does extend into montane regions in some areas. It breeds in old, swampy forests rich in pines and spruces and with many fallen and rotten tree trunks. It also breeds in alder swamps. It prefers pine woods, but in any case the woods must be marshy, with a heavy carpet of lichens and mosses and with ponds, lakes, or streams nearby. Outside the breeding season it occurs on sewage farms, marshes, narrow ditches, riverbanks, and streamsides and occasionally also on small ponds. It is rarely found on tidal flats or the open seashore, but it does occur on the channels of salt marshes. The birds are rather nongregarious and are only infrequently seen in flocks larger than a few dozen birds. During the breeding season the birds are probably mostly insectivorous, feeding on both adult insects and their larvae, including beetles, dipterans, and tricopterans. They also consume annelids, small, thin-shelled crustaceans, spiders, small mollusks, and some vegetable matter (Witherby et al. 1941). Feeding is done in sheltered locations, in much the same manner and habitats as are used by the wood sandpiper, but while the green sandpiper prefers to feed at the water's edge the wood sandpiper spends more time in open water and regularly probes for food (Burton 1974). Yet in India the birds have been observed wading and sometimes even swimming, usually capturing prey by momentarily submerging the head and neck and even the shoulders. Swimming and even diving have also been observed in Great Britain, where a bird fed in a swiftly flowing stream with water running over its back, almost in the manner of a dipper. In one interesting observation, a group of these birds were seen swimming, diving, and rushing about in a shallow pond in Kent, apparently to bring mudworms out of the mud or to the edge of the pond, where they were quickly snatched up (Bannerman 1961).

Social behavior. Green sandpipers return to their breeding grounds fairly early, and they sing from the time of arrival until about the time the young are fledged. There are two major song types, the first of which is done from the ground, from elevated singing perches, or in the air. When on the ground, singing birds often raise and spread their tails, and flight songs may be uttered while taking off or landing or during prolonged aerial displays. In a typical aerial display the bird undulates above the trees, alternately climbing with rapid wingbeats and sharply gliding with the wings bowed downward. This is usually done in a circular fashion, but at times the bird will speed swiftly and erratically above the forest. During the typical aerial display the bird sings a second song type, having five musical subunits that are repeated several times to form a complete song. Song-flight displays occur not only over feeding and nesting territories, but also over intervening areas when these are not adjacent. Additionally, a similar aerial display of small amplitude is sometimes directed toward the female during pair-formation and at the time of copulation (Oring 1968). The song elements of the two song types in this species have many characteristics in common with those of the solitary sandpipers, but tend to be lower in frequency and thus carry farther. This may be related to the fact that the birds tend to be more dispersed than are solitary sandpipers. On Oring's study area of about 50 square kilometers, some 20–25 breeding pairs were present, and the most closely associated pair of nests were 400 meters apart. Copulatory behavior is still only very poorly described, but according to an early description the courting bird's head is lowered, the tail is partly spread and raised, and the wings are drooped. In another account the two performing birds alternately raised their wings and fanned their tails, then one fluttered up and over the bird in front of it, which in turn did the same (Witherby et al. 1941). This description approaches that of typical precopulatory behavior in the other *Tringa* species. Wing-raising, which exposes the dark undersides of the wings, is a common threat display in this species, as it is in many other scolopacids.

Reproductive biology. Like the solitary sandpiper, and otherwise almost unique among the shorebirds, this species regularly uses the old nests of passerine birds. The number of species whose nests are used is quite large, including several species of thrushes, crows, jays, shrikes, and even squirrel dens. It sometimes also nests on accumulated pine needles among branches, on stumps, among fallen trees, and occasionally on the ground itself. The nest is usually be-

tween 3 and 60 feet aboveground, and it is usually little if at all modified, with at most a lining of moss or lichens being added. The nests used are generally those of the previous year, but at times a freshly made nest is occupied. Yet the birds nest very early, and by that time very few nests of the current year are available for their use. The clutch is almost invariably 4 eggs; clutches of 3 or 2 are apparently very unusual. Replacement clutches have been found, and in at least one reported case a new nest site was chosen for the second nesting attempt. Both sexes incubate, with the female reportedly taking on the major part. The exact incubation period is still unknown, but it probably is about 20 days. The chicks are rapidly led away from the nest after they jump down to the ground; in one case day-old chicks were found some 300 meters from the nest (Glutz et al. 1977). The fledging period is 28 days (Fjeldså 1977), and probably after the young are about 9 days old they are tended only by the male.

Status and relationships. This is a rather inconspicuous and nongregarious species, and thus its population status is impossible to judge with any certainty. The species is nearly the exact ecological counterpart of the North American *solitaria*, and additionally these two forms have greatly overlapping vocabularies (Oring 1968). Mayr and Short (1970) tentatively considered the two conspecific, with *glareola* as part of the same species group, while Larson (1957) considered *ochropus* and *solitaria* as "closely related" and as part of the same species group as *glareola*. I believe all three should be given species rank and am uncertain of the relationships among them; I suspect that both the Eurasian forms may have evolved from a dual invasion by pre-*solitaria* stock. I have listed *solitaria* and *ochropus* as a superspecies and consider *glareola* part of the same species group.

Suggested reading. Oring 1968; Bannerman 1961.

Solitary Sandpiper

Tringa solitaria Wilson 1813

Other vernacular names. None in general English use; chevalier solitaire (French); Amerikanischer Waldwasserläufer (German); andarríos solitario, archibebe solitario (Spanish).

Subspecies and ranges. See map 105.

    T. s. solitaria: Eastern solitary sandpiper. Breeds from Labrador and Quebec west across Ontario to the southern parts of the Prairie Provinces and eastern British Columbia. Winters from the southern United States south through Central and South America to Argentina.

    T. s. cinnamomea: Western solitary sandpiper. Breeds from central Alaska and Mackenzie south to northern British Columbia, the southern Yukon, southern Mackenzie, and northeastern Manitoba. Winters from northern South America to Argentina.

Measurements and weights. Wing (chord): males 121–37 mm; females 126–42 mm. Culmen (to feathering): males 27–32 mm; females 28–32 mm. Weights: 10 males 38–62 g, average 54.1 g; 5 females 37–75.5 g, average 55 g (various sources). Eggs ca. 36 x 26 mm, estimated weight 12.0 g (Schönwetter 1963).

DESCRIPTION

Breeding adults of both sexes have a deep grayish crown that is indistinctly streaked with whitish, a whitish stripe extending from above the lore to the eye, where it meets a whitish eye ring, and the rest of the head streaked dusky brown, except on the chin and throat. Dusky brown streaking extends down the side and front on the neck to the breast and sides, where it grades into paler barring. The axillaries and under wing coverts are slate gray with white barring. The upperparts are generally grayish brown, with sparse speckling of whitish on the back, scapulars, and wing coverts, while the upper tail coverts are barred with white. The tail is dusky grayish brown centrally, with whitish spotting, and the rest of the tail is broadly barred with white and dusky. The primaries and their coverts are dull blackish, and the secondaries are grayish brown, becoming spotted inwardly. The iris is brown, the bill is blackish at the tip and greenish basally, and the legs and toes are olive greenish. *Adults in winter* are more grayish on the upperparts and less distinctly speckled with white, while the anterior underparts are very indistinctly streaked or washed with grayish. *Juveniles* have brown median wing coverts, with lateral buffy spots, and the upper breast has a reddish brown tinge. *First-winter* birds rapidly lose their buff spotting, and this may be visible only on the inner median coverts after early winter (Prater et al. 1977; Ridgway 1919).

In the hand, this rather small *Tringa* has relatively short legs (tarsus 29–34 mm) and a fairly short (culmen 27–32 mm), straight bill; the strong tail barring and lack of a white patch on the upper tail coverts distinguish this from other species in the same size range.

In the field (8 inches), this species is usually found in freshwater habitats, where it may be readily identified by its barred tail and lack of white on the upper tail coverts. It is smaller and darker above than the lesser yellowlegs and has darker legs, and its white eye ring is also distinctive. In flight it utters a series of high-pitched *weeep* notes, similar to those of a spotted sandpiper.

NATURAL HISTORY

Habitats and foods. When breeding, this species seeks out wet muskeg country, typically wet, open terrain with scattered trees or clumps of trees. The birds are closely associated with water and frequently nest in the same areas as does the rusty blackbird, whose nests they sometimes utilize. Besides breeding in muskeg forests, they also nest in montane forests, fairly heavy northern coniferous forests, and along the tundra/forest edge. Since the birds invariably utilize the nests of some passerine species such as robins or blackbirds, there is no close dependence upon specific vegetation or tree cover. Foraging is done in fairly wet habitats, especially in small, stagnant pools that are fringed with vegetation, tidewater ditches intersecting marshes, wet meadows, and the interiors of moist woods having matted layers of decaying vegetation. A favorite method of foraging is to wade in shallow water, slowly advancing and vibrating the leading foot, thus stirring up the bottom sufficiently to disturb insects but not so greatly as to muddy the water and obscure the view. In this way it captures aquatic insects and their larvae, including dragonfly nymphs and water boatmen, and small crustaceans. It also captures winged insects in the air, as well as spiders, grasshoppers, and even small frogs (Bent 1929). In the winter the species avoids coastal areas and beaches, though it sometimes occurs at the upper edges of tidal inlets. Generally it occurs in South America along small pools and quiet streams, using much the same habitats as it does during migration in North America.

Social behavior. As its vernacular and Latin names imply, this is a nongregarious species and the birds are rarely seen in groups of more than 2 or 3 individuals. On both spring and fall migration the birds seem to move as individuals, apparently migrating to a large degree at night, and thus large-scale migrations are never evident. Singing probably begins at the time of arrival on the breeding grounds and continues at least until the clutch is completed and possibly until the fall migration. Both sexes sing, and members of a pair frequently maintain contact by this means. Singing occurs on the ground, from elevated sites, and in the air, either during direct flight or during a song-flight display. This is a weakly undulating flight accompanied by a high-pitched, repetitive song. The bird may fly 100 meters or more before beginning his arcs, which are only 1–2 meters in height (Oring 1968). As it arcs upward, the wingbeats become shallower and more rapid, and the accompanying call is something like that produced by an American kestrel. At least early in the season, song flights have been noted as being circuitous, and they probably are associated with territorial establishment. The birds' territories are relatively large, and in one instance a territory consisted of approximately half a square kilometer. This pair's home range was considerably larger than that, since they often foraged outside the limits of the defended area.

Copulation is apparently largely limited to the period of egg-laying and about 5 days before. The male typically utters a series of "epigamic calls" while approaching the female from the rear or else from the side and then the rear. The female typically stands still or walks very slowly, while her mate performs a wing-raising display. If she remains still, the male begins wing-fluttering and finally mounts her. During copulation he pecks at the female's head, and afterward he usually performs a wing-raising display involving both wings. Essentially the same "wing-up" display occurs in hostile situations, especially in individuals having equal tendencies toward attack and escape, while during maximum threat intensity the white of the plumage is highly exposed. On the other hand, during appeasement the bird sits on its tarsi, holding its head low and its tail down, concealing the white areas (Oring 1973).

Reproductive biology. A large number of passerine bird species' nests have been reported used by this species for nesting. They include the rusty blackbird, American robin, common grackle, cedar waxwing, Bohemian waxwing, eastern kingbird, and gray jay.

Although the birds normally use nests of the previous year, they are also known to sometimes exploit recently made nests. Nests ranging in heights from 1.2 to 12 meters aboveground have been found to be used, and from shoreline to as much as 200 meters from water. Normally the nests are in coniferous trees, but sometimes those in deciduous trees are also used (Oring 1973). Both members of a pair seek out prospective nests, sometimes visiting several potential sites, and it is probable that the male is the primary nest prospector while the female undertakes the necessary adjustments to the nest and arranges the lining. In one observed case, 4 eggs were laid during a 5-day period (Oring 1973). The relative role of the two sexes during incubation is still unestablished, but it is probable that both sexes participate. The incubation and fledging periods are likewise still undetermined.

Status and relationships. It is impossible to judge this species' status, inasmuch as it is so elusive and nongregarious throughout the year. It is clearly a member of the "wood sandpiper" group, which sometimes has been generically separated from the "shanks" and which includes *solitaria, glareola,* and *ochropus.* All three are closely related, and, though Mayr and Short (1970) tentatively considered *solitaria* and *ochropus* conspecific, Oring regarded *solitaria* as being nearer to *glareola* than to *ochropus.* I tend to favor the view that *solitaria* and *ochropus* are a superspecies, with *glareola* part of the same species group.

Suggested reading. Oring 1968, 1973.

Wood Sandpiper

Tringa glareola L. 1758

Other vernacular names. None in general English use; chevalier sylvain (French); Bruchwasserläufer (German); andarrios bastardo (Spanish).

Subspecies and range. No subspecies recognized. Breeds from Scandinavia eastward through Russia and Siberia to Anadyrland, Kamchatka, and the Commander Islands, and south to Denmark, the northern Ukraine, the Kirgiz Steppes, Russian Altai, northern Mongolia, and northern Manchuria. Has bred rarely in the Aleutian Islands (At-tu and Amchitka) (*Auk* 91:175). Winters from the eastern Mediterranean, the Persian Gulf, China, Taiwan, and the Philippines south to southern Africa, India, the Indo-Chinese countries, the Sundas, New Guinea, and Australia. See map 106.

Measurements and weights. Wing: males 117–29 mm; females 123–34 mm. Culmen (from feathering): males 26–31 mm; females 27–31 mm. Weights (breeding grounds): 12 males 52–77 g, average 59.4 g; 10 females 54–67 g, average 62.2 g (Glutz et al. 1977). Wintering-ground weights average 56 g and range 34–89 g (Skead 1977). Another sample of 65 wintering birds averaged 64.1 g (Summers and Waltner 1979). Eggs ca. 38 x 26 mm, estimated weight 13.5 g (Schönwetter 1963).

DESCRIPTION

Breeding adults of both sexes have a crown that is dusky with whitish streaks, a broad white stripe extending from above the lores to behind the eyes, dusky lores, the sides of the head white with fine dusky streaking, a white chin and throat, and the neck and chest white to pale gray with dusky streaking and barring. The rest of the underparts are pure white, with irregular barring of dusky brownish gray on the sides, flanks, and under tail coverts. The axillaries and under wing coverts are white, with brownish gray barring or other markings. The back and scapulars are brownish gray, irregularly spotted with blackish and dull whitish. The rump has darker brownish gray, the upper tail coverts are white, with darker markings on the longer feathers, and the tail feathers are barred with white and dusky. The wing coverts are mostly dull brownish gray with irregular whitish spots, the primaries and their coverts are dusky, and the secondaries and greater coverts are dusky brownish gray, narrowly edged with whitish. The iris is brown, the bill is dark brown, becoming yellowish at the base, and the legs and toes are yellowish brown to greenish. *Adults in winter* are nearly uniform brownish gray on the upperparts, with few if any blackish spots, and the wing coverts are fringed with white. The breast is grayish brown with poorly defined darker streaking. *Juveniles* are dark brown on their upperparts and coverts, with many distinct warm buff spots, and the breast is grayish brown with pale mottling. *First-winter* birds gradually lose their buff spotting through abrasion, but spots on the inner medians are apparent until spring (Prater et al. 1977; Ridgway 1919).

106. Breeding distributions of the spotted greenshank (*hatched*) and wood sandpiper (*shaded*).

In the hand, this small *Tringa* has a relatively short bill (culmen 26–31 mm), shorter wings than *ochropus* (maximum 131 mm, vs. 137 mm in *ochropus*), and differs from *solitaria* in having white upper tail coverts.

In the field (8 inches), this species is smaller and less dark dorsally than the green sandpiper and also has a much lighter gray under wing surface. Both of these species have white upper tail coverts that contrast with the darker back plumage, and they lack any white on the upper wing surface. The leg coloration is more yellowish than in the green sandpiper, and the bird thus sometimes resembles the lesser yellowlegs, but the legs are shorter and are duller in color. In flight the usual call is a triple-noted *wee-wee-wee* or *wit-wit-wit*. The breeding habitats are bogs and scrubby woodlands, and in the winter it is mostly found along marshes and shorelines.

NATURAL HISTORY

Habitats and foods. The breeding range of this species centers on the boreal climatic zone, but it locally extends to the temperate and arctic zones and is somewhat more northerly in distribution than is the green sandpiper. Its primary habitats are bogs, swampy heathlands with or without scattered conifers, open marshes in forests, and flooded coniferous forests. Highest population densities occur at the northern edge of the taiga and in the birch and willow scrublands between coniferous forest and tundra (Voous 1960). In the southern part of its range it is largely limited to wet heaths, bushy peat moors, and similar habitats. Outside the breeding season it favors muddy areas of marshes, the muddy or marshy edges of lakes or reservoirs, sewage farms, and boggy places on heaths or moorlands. It seldom frequents open coastlines, but it sometimes occurs along the channels of salt marshes. In Africa it is found on stagnant ponds in semiarid regions, or where small pools or stream remnants persist, and is not associated with woodlands for the most part. In the Ethiopian uplands the birds occur in high moorland country to elevations of 8,000–10,000 feet, mainly in marshes and along grassy streambanks (Bannerman 1961). Foods during the breeding season mostly consist of small insects up to about 20 mm long, especially the aquatic forms such as dytiscid and hydrophilid beetles, various hemipterans, and the larvae of flies such as midges. Migrating birds also eat insects in considerable quantity, as well as small crustaceans, gastropod mollusks, and occasionally even small fish. Foraging occurs along muddy streambanks or in shallow water, with individual items obtained both by pecking and by probing. Relative to the larger sandpipers, much of the food is taken in more sheltered locations and on less exposed mud banks (Voous 1960; Glutz et al. 1977).

Social behavior. Very shortly after arrival on the breeding grounds the birds become territorial, and they soon become well dispersed. Figures presented by Kirchner (1978) suggest that breeding densities are less than a pair per square kilometer. The advertising flight song is performed by both sexes, although it is probably not performed as intensively or as often by females as by males. The flight is marked by a steep ascent, with rapid wingbeats, followed by a gliding descent with the wings downcurved. As the downward glide ends, the bird spreads its tail wide and breaks into a musical series of flutelike notes, sound-

ing something like a repeated *liro* (the Finnish name for the species), or *di-le* or *tu-le,* according to various descriptions. Most songs last 2-6 seconds, and a single song flight may last 5–10 minutes. During the ground display the male follows the female with lifted wings while uttering *bibibibi* notes. He then flutters up onto the female from behind, while still calling loudly, and settles on her back. The copulation lasts only a few seconds, and when completed the male flutters down to land beside his partner (Kirchner 1978). Like other *Tringa* species, vertical wing-raising is also used during threat displays, sometimes even while the birds are swimming.

Reproductive biology. Unlike the green sandpiper or the solitary sandpiper, this species uses ready-made nests only infrequently, but a few cases of nesting in old thrush, magpie, or dove nests have been found in Norway, Sweden, and the USSR. Although it has been suggested that this type of nesting is typical of flooded forests or other sites subject to inundation, such does not seem to be the case and may only reflect the relative availability of suitable old nests in the area. More typically the nest is in a mossy depression, often close to a dwarf birch or on a slight hillock among heather. At times the bird will even lay her eggs on dry ground carpeted with reindeer moss, but more often the eggs are very well concealed in grass or shrubbery. There are normally 4 eggs, with clutches of 3 being infrequent. The clutch is completed in 5–6 days, and both sexes participate in incubation. The incubation period is probably 22–23 days, and although both sexes initially tend the chicks the female deserts and leaves the brood to the male's care after about 7–10 days. The fledging period is 28–30 days.

Status and relationships. Although this species has disappeared from many of its breeding places in western Europe as cultivation has encroached on its breeding grounds (Voous 1960; Kirchner 1978), it is still widespread over much of the forested portions of Asia and is very common in much of the USSR (Dementiev and Gladkov 1969). Its probable relationships with *solitaria* and *ochropus* have been mentioned in the accounts of those species, and it is certainly part of the same species group as those two forms. My impression of its plumage and general appearance is that it is less like *solitaria* than is *ochropus,* though the relative relationship of these three forms to one another is still rather uncertain (Mayr and Short 1970).

Suggested reading. Bannerman 1961; Kirchner 1978.

Siberian Tattler

Tringa brevipes (Vieillot) 1816
(*Heteroscelus brevipes* of Peters, 1934)

Other vernacular names. Gray-rumped sandpiper; gray-tailed sandpiper; Polynesian tattler; chevalier cendre (French); Graubürzelwasserläufer, Ostsibirischer Wanderwasserläufer (German); archibebe paticorto (Spanish).

Subspecies and range. No subspecies recognized. Breeds in the mountains of eastern Siberia west to the Putorana Mountains and south to Lake Baikal. The Sayan Mountains near Lake Koso Gol. Overlaps with the range of *incana* in the Koryak Highlands and also possibly in the Anadyr Range. Winters from China, Taiwan, and the Philippines south to the Indo-Chinese countries, Malaya, and the Sundas, and east to New Guinea, Australia, and New Zealand. See map 107.

Measurements and weights. Wing: males 152–68 mm; females 156–79 mm. Culmen (from feathering): males 35–40 mm; females 36–42 mm. Weight: 6 males 92–118 g, average 101 g; 10 females 81–116 g, average 87 g (U.S.N.M. specimens). Both sexes of wintering birds 80–162 g, average of 161, 108 g (D. Purchase, pers. comm.). Eggs ca. 43 x 30 mm, average weight 19.0 g (Leonovitch and Kretzschmar 1966).

DESCRIPTION

Breeding adults of both sexes very closely resemble those of the wandering tattler, being generally lighter throughout, with flank barring fading ventrally, so that the lower abdomen and under tail coverts are entirely white, and the upper tail coverts are barred with white. The iris is brown, the bill is blackish to dark olive gray with a yellowish base, and the legs and toes are yellow. *Adults in winter* are virtually identical to the wandering tattler, and *juveniles* likewise appear to be indistinguishable.

In the hand, this species resembles a typical *Tringa* but is relatively long-tailed and short-legged (tarsus 29–34 mm), with a straight and tapering bill (34–42 mm) that has the front end of the nasal groove ending at least 17 mm from the bill tip, compared with 15 mm or less in *incana*. Additionally, the wing length is usually less than 170 mm (averaging 164 mm), com-

107.   Probable breeding distribution of the Siberian tattler.

40. Winter-plumage heads of wandering (*above*) and Siberian (*below*) tattlers.

pared with usually more than 170 mm in *incana,* and the rear edge of the tarsus is reticulated rather than scutellated.

In the field (10 inches), this species and the wandering tattler have almost uniformly slate gray upperparts, a pale superciliary stripe, barred flanks and breast patterning, and yellow legs. In breeding plumage this species averages paler and slightly smaller than the other, but in fall or winter the two cannot be distinguished by plumage alone. In this species the forehead is usually whitish, however, and the breast band is paler. Both inhabit rocky Pacific coast shorelines as well as reefs, mangrove swamps, and tidal flats. When disturbed they utter loud double whistled notes, *troo-eet* or *ter-wee,* or a soft trill of three or four notes. By comparison, the wandering tattler usually utters a quick series of about eight notes, with the second the loudest and the rest diminishing in volume and given more rapidly.

NATURAL HISTORY

Habitats and foods. On their wintering grounds in Australia, the favored habitats of these birds are coastal areas where there are large expanses of exposed sand and mud flats. They also are found on small coral islands, and at high tide they often can be found roosting in mangroves. Their foods there and elsewhere are still unstudied but probably are much like those of *incana.* On their breeding grounds they have been found to eat the aquatic larvae of blackflies, caddis flies, and stone flies and other insects such as beetles (Frith 1976; Dementiev and Gladkov 1969). Adults have also been seen eating mosquitoes and springtails.

Social behavior. Although they are still not well described, it has been said that the displays and call of this species are very similar to those of the greenshank, including the tendency to perch and call from dead tree branches. Its breeding-season calls are also apparently very similar to those of the wood sandpiper and green sandpiper (Leonovitch and Kretzschmar 1966; Neufeldt et al. 1961).

Breeding biology. It is only in the last few years that any information on the breeding biology of this species has become available. The first described nest, containing 4 eggs, was found in July 1959. The nest, a hollow among stones, was near the edge of a river in stony forest tundra 320 meters above sea level, close to the headwaters of the Manus River and about 80 kilometers WSW of Norilsk. At about the same time several broods were found, sometimes accompanied by the male, sometimes by the female, and sometimes by both parents. Young capable of flight were found in early August. Observations by K. Vorobiev on the upper Indigirka River indicate that the birds occupy mountain streams where there are islands covered with willow scrub, dwarf birches, shrubs, and open spaces with mosses and lichens. In 1960 another nest was located, this time in an old thrush (*Turdus naumanni*) nest situated about 2.5 meters aboveground in a larch. This nest also had 4 eggs present (Leonovitch and Kretzschmar 1966).

Status and relationships. This is regarded as one of the most common waders wintering in Australia (Frith 1976), so it apparently is in no danger at present. It seems to behave as a distinct species on its wintering grounds with respect to *incana,* having somewhat different calls and wintering in rather different areas. Differences in the bill structure are small but apparently consistent. There is now good evidence that both species breed sympatrically in eastern Siberia, since A. A. Kistchinski (in litt.) has found them breeding together in the central Koryak Highlands, and collected one apparent hybrid. He thus considers them good species. There is, however, considerable doubt whether the genus *Heteroscelus* can be accepted. Vaurie (1965) merged it with *Tringa*, as did Dementiev and Gladkov (1969) and Strauch (1976), while Jehl (1968*a*) maintained both it and *Actitis* as distinct genera, though citing similarities between them. I see no real reason for maintaining either *Actitis* or *Heteroscelus* distinct from *Tringa*, and I believe the nearest relatives of *brevipes* and *incana* are the group composed of *glareola, solitaria,* and *ochropus.*

Suggested reading. Leonovitch and Kretzschmar 1966.

Wandering Tattler

Tringa incana (Gmelin) 1789
(*Heteroscelus incanus* of Peters, 1934)

Other vernacular names. American gray-rumped sandpiper; chevalier errant (French); Amerikanischer Wanderwasserläufer (German); archibebe andante (Spanish).

Subspecies and range. No subspecies recognized. Probably breeds in Siberia (Anadyrland and the Chukotski Peninsula) and in North America in the mountains and highlands of south coastal, central, and western Alaska east to the mountains of the Yukon and northwestern British Columbia. Winters from southern California to Ecuador, the Galápagos Islands, the Hawaiian Islands, and the Central and South Pacific islands to Australia and New Zealand. See map 108.

Measurements and weights. Wing: males 161–74 mm; females 169–80 mm. Culmen (from feathering): males 34–41 mm; females 38–42 mm. Weights: wintering males 72–180 g, average of 28 was 111 g; females 80–205 g, average of 17 was 129 g (Lacan and Mougin 1974). Summer birds average slightly lighter (Irving 1960). Eggs ca. 44 x 32 mm, estimated weight 23.0 g (Schönwetter 1963).

DESCRIPTION

Breeding adults of both sexes are plain, deep slate gray on the upperparts, with a grayish white superciliary stripe, a dusky loral stripe, a grayish white to grayish suborbital and auricular region, and whitish underparts, with streaking on the foreneck and with

the rest of the underparts other than the chin and throat barred with dusky slate. Barring also extends to the axillaries and the under wing coverts, as well as the under tail coverts. The upper tail coverts are sometimes narrowly margined with whitish, and the primaries and their coverts are blackish, the primaries with partially white shafts. The wing coverts are gray with some narrow white fringes. The iris is brown, the bill is blackish at the tip and yellowish at the base, and the legs and toes are yellow. *Adults in winter* are slightly lighter dorsally than in summer, and the underparts lack bars or streaking and instead are shaded with dark gray on the foreneck, chest, and sides. The wing coverts are gray with paler edges. *Juveniles* have their upperparts and wing coverts with extensive buffy whitish fringes and subterminal bars, and the inner median coverts are spotted with buffish. The underparts are white, with mottled grayish on the breast. *First-winter* birds retain some buff spotting or fringing on the wing coverts until winter (Prater et al. 1977).

108. Probable breeding distribution of the wandering tattler.

In the hand, this is a rather large *Tringa*-like bird with relatively short yellow legs (tarsus 30–35 mm), a straight bill that has a long nasal groove (two-thirds the length of the bill, or reaching to within 15 mm of the tip), long wings (at least 163 mm), and slate gray upperparts.

In the field (11 inches), the dark slate gray upperparts, without white on the wings or tail, and the barred underparts (at least in the breeding season), combined with yellow legs and a whitish superciliary stripe, identify this as a tattler. Where there is a chance of confusion with the Siberian tattler, this species' extended series of calls—usually of about six to ten notes, increasing in rapidity while declining in volume—help to separate it from the Siberian form. Both species are often found along rocky shorelines, but they also inhabit reefs and pebbly beaches.

NATURAL HISTORY

Habitats and foods. The breeding habitat of this elusive species is apparently mostly restricted to the alpine zone, along mountain streams where gravel bars are available for nesting and where foraging opportunities are available in shallow but swift waters or in quiet pools. Tailings from placer-mining have also been found to serve as nesting habitat in Alaska. The habitats used outside the breeding season are quite different, and on their wintering grounds in the southern Pacific Ocean the birds are found on reefs, shorelines, and sometimes on ponds well away from salt water. On migration along the Pacific coast of North America they are typically seen on rocky shorelines, where, together with black turnstones and surfbirds, they feed among the breakers along kelp-covered rocks. However, they also utilize sandy or gravelly beaches, feeding largely on decapod crustaceans, marine worms, and minute mollusks. On freshwater streams they seem to specialize in capturing the larvae of caddis flies and aquatic dipterans, and they adeptly capture flying insects such as adult caddis flies (Bent 1929; Stout 1967). Like the spotted sandpipers, they often walk or run with a bobbing action while foraging, and when frightened they simply "freeze" so they blend in well with the rocky background. When foraging they thrust the bill vertically downward into the water, often immersing the head and part of the neck for as long as 10 seconds. They will occasionally even swim across narrow streams (Dixon 1933).

Social behavior. These are not particularly gregarious birds, and, though at times are seen in groups or in loose association with turnstones, this is a reflection of attraction to common feeding areas rather than indicating strong gregarious tendencies. Rarely are groups of more than 10 or a dozen birds seen in the same area. They are also highly dispersed on the breeding areas. Weeden (1965) reported that two pairs occupied territories about a mile apart on a creek, and the birds would sometimes fly down the center of the valley for nearly a mile, staying 500 to

1,000 feet above the creek and at times uttering a high, rapid whistle similar to that of a courting lesser yellowlegs. Territories must be established immediately after arrival on the breeding ground, since egg-laying begins only a week to a week and a half after arrival. These flights occurred from the time of arrival on the breeding grounds in latter May until mid-June, with usually only a single bird engaging in them, though occasionally two were aloft simultaneously. Clutches were complete by mid-June, when the flights terminated. Weeden noted that what was perhaps the same pair occupied the same territory and nested in the same site for 6 of 7 years between 1957 and 1963, with the only exception occurring one spring when snow and ice covered the old nest site, and a new nest was built about a quarter-mile downstream.

Reproductive biology. The nest of this species is sometimes a fairly substantial structure of roots and twigs; Murie (1924) reported that the nest he saw (containing the first complete clutch ever found) was so compact that he could readily pick it up and carry it. Nests found more recently have not been so well constructed, however, such as that found by Weeden (1959) or a second nest found by Murie (1946). Dixon (1933) noted that he had found several previous years' nests that had remained intact through the winter. Weeden's (1965) observations suggest that successive eggs may be laid as frequently as a day or so apart (one clutch increased from 2 to 4 eggs in 3 days), and that incubation is 23 to 25 days. The usual clutch is 4 eggs, judging from the few nests found so far. Both sexes incubate; the smaller male has a well-developed incubation patch and seems to show the greater anxiety when danger threatens. Pipping of the eggs requires about 3 days, and the young leave the nest within a day after hatching. They are brooded by both parents, at least for the first week, but later it is unusual to see both parents with the brood. Even when very young, the chicks are able to swim extremely well; Weeden saw 5-day-old chicks swimming across a swift 8-foot-wide stream. When about 12 days old the young are already well grown, with primaries over an inch long, and are readily able to capture flying insects along sandy stream banks (Dixon 1933).

Status and relationships. There is no specific information on the status of this elusive species, and its close similarity to the Siberian tattler makes judging its distribution and abundance even more difficult. The question of specific distinction of these two species has been dealt with in the account of the Si-

berian tattler, and perhaps only the question of the generic validity of *Heteroscelus* needs further attention. Various recent writers (Strauch 1976; Stout 1967; Condon 1975) have merged this genus with *Tringa,* while several others (Morony et al. 1975; Jehl 1968a; Voous 1973) have retained it. Jehl (1968a) notes that the downy plumages of *Heteroscelus* and "*Actitis*" (*Tringa*) have similar patterns and that there are other morphological and behavioral similarities between these types. Although they may be peripheral members of the tattler group, I think it is preferable to keep then united generically rather than to separate them from the rest of the *Tringa*-like forms.

Suggested reading. Dixon 1933; Weeden 1965.

Eurasian Sandpiper

Tringa hypoleucos L. 1758
(*Actitis hypoleucos* of Peters, 1934)

Other vernacular names. Common sandpiper; chevalier guignette (French); Flussuferläufer (German); andarrios chico (Spanish).

Subspecies and range. No subspecies recognized. Breeds in the British Isles and on the mainland of Europe from Spain and Scandinavia eastward across Russia and Siberia to Kamchatka, as well as on Sakhalin, Japan, and adjoining islands. Winters from southern Europe and the Mediterranean Basin, the Persian Gulf, China, and Japan southward to southern Africa, Saudi Arabia, India, the Indo-Chinese countries, the Sundas, New Guinea, and Australia. See map 109.

Measurements and weights. Wing: males 105–15 mm; Females 107–19 mm. Culmen (from feathering): males 22–27 mm; females 22–26 mm. Weights (fall migrants): males 50–70 g; females 55–67 g. Breeding males and females average 49.0 and 54.6 g respectively (Glutz et al. 1977). Eggs ca. 36 × 26 mm, estimated weight 12.5 g (Schönwetter 1963).

DESCRIPTION

Breeding adults of both sexes are very similar to those of the spotted sandpiper, but the underparts lack black spotting and instead are entirely white, except for strong sepia streaking on the sides of the head, neck, throat, and upper breast. The feathers of the upperparts are streaked, mottled, or otherwise

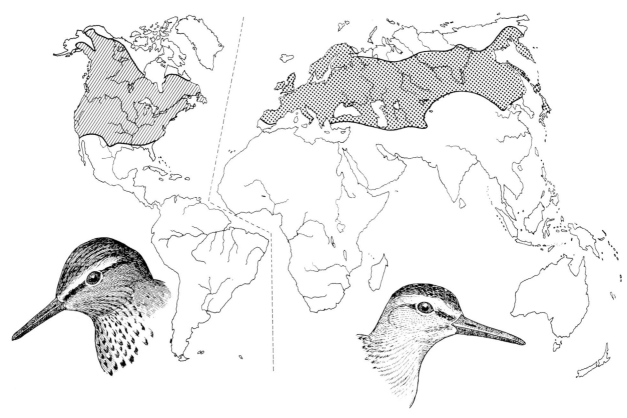

109. Breeding distributions of the spotted sandpiper *(hatched)* and Eurasian sandpiper *(shaded).*

marked with sepia, and the feathers are faintly tipped with light buff. The wing coverts are fairly uniform brown, with darker streaking. The central tail feathers are olive brown, the central ones with a subterminal sepia bar and notched with sepia, while the rest have broad white tips and sepia notching. The two outer parts are more or less white, with sepia barring or shading. The primaries are sepia, with a white patch on the middle of the inner webs, usually starting with the third primary. The secondaries are sepia, tipped with white and with a broad white band across the middle of both webs, usually extending to the eighth or ninth. The iris is brown, and the bill is black at the tip and bright straw-colored basally, with a reddish tint. The legs and toes are gray with darker joints (Witherby et al. 1941). *Adults in winter* are very similar but have barred rufous buff and brown tips on the wing coverts. *Juveniles* have brown upperparts with extensive buffish tips and dark subterminal bands. The coverts are brown, barred at their tips with dark brown and bright rufous buff (Prater et al. 1977).

In the hand, the white markings on the middle of the secondaries, along with the short legs (tarsus 22–25 mm) and short bill (culmen 22–27 mm) separate this species from all the *Tringa* forms except *macularia.* In juveniles or winter-plumaged birds of this species, white extends to the second or third pair of outer tail feathers, the lower throat is distinctly streaked with sepia rather than white with faint darker shaft streaks, and the middle portions of the eighth and ninth secondaries (counting inward) are usually white rather than sepia.

In the field (8 inches), this species bobs and teeters exactly like the spotted sandpiper, and it flies in the same distinctive way. It lacks spotting on the underparts in spring and instead has a distinct brownish breastband. Its call is a series of shrill piping notes, usually sounding like *twee-wee-wee,* the first note the loudest. On the rare occasions when *macularia* and *hypoleucos* might be seen together in winter plumage, *hypoleucos* can usually be recognized by its distinctly longer tail and by the fact that in flight the white wing bar of *hypoleucos* extends to the secondaries, while in *macularia* it is limited to the primaries (*British Birds* 70:346–48).

NATURAL HISTORY

Habitats and foods. Like its North American counterpart, this species occupies an extremely broad breeding range, from steppe and desert climates through temperate zones to boreal and mountain forests almost to the snow line of alpine or arctic tundra. It is associated both with flat, open country and with dense coniferous forests, breeding along sandy or pebbly shorelines of lakes or ponds, beside slowly or rapidly flowing streams, and even along mountain torrents (Voous 1960). Outside the breeding season it occurs on almost every aquatic habitat, from the smallest muddy and marshy temperate areas to tropical marshes, where it sometimes perches on the backs of floating hippopotamuses or picks ectoparasites from crocodiles. At all seasons the species feeds primarily by pecking rather than by probing and generally confines its activities to wet ground rather than wading as is typical of many *Tringa* species. It feeds almost exclusively on materials of animal origin, particularly adult insects of many families of beetles and dipteran flies. Mollusks, crustaceans, and annelid worms constitute most of the rest of the species' diet, but it sometimes eats frogs and tadpoles and, rarely, small fish. These foods are all secondary to insects, and, like their North American relative, the birds deftly capture small insects from the surface or pull them out from rocks or mud. Sometimes they stalk them with head held low and nearly parallel to the substrate while the hindquarters are well raised, thus somewhat hiding the approaching bird's head (Bannerman 1961; Glutz et al. 1977).

Social behavior. These birds arrive on their breeding grounds in small flocks or individually and begin to display very soon after their arrival. Territories are advertised by song flights that may last as long as 15 minutes. Typically the displaying bird flies in a circling flight, ascending higher and higher while uttering series of trilled whistles. It has been assumed that this display is done by the male, but based on studies of the North American species it seems probable that the female establishes the territory and is mainly responsible for its advertisement. Various chaselike flights have been described, and a batlike flight of a bird rising 50–100 feet in the air has also been witnessed. Ground display involves one or both sexes standing close together and raising one or both

41. Winter-plumage heads and half-tails of spotted (*above*) and Eurasian (*below*) sandpipers.

wings, holding them aloft as they utter trilled calls and either walk about or break into circular flights. The song uttered during display is a simple variant of the usual call, a repeated *kitt-weewit*. The same vocalization is used during ground display. In one probable precopulatory sequence the displaying bird stood bold upright before another, holding its wings perpendicular so that their points almost touched and revolving them with a slow fanning action. The other birds showed no response to this, other than occasionally flying off for a short time and then returning to the same place (Bannerman 1961). When the female is ready for copulation she remains still and allows the male to flutter up onto her back. In contrast to the North American species, there is little difference in the weights and measurements of the sexes, and no breeding plumage differences exist in the two sexes, so it is possible that the strong female dominance in sexual behavior that has been found in *macularia* does not apply to *hypoleucos*. Yet Oring and Knudson (1972) state that in both of these species the female is larger than the male, the relative egg size is small, the female is dominant in courtship, and the male does most of the incubating, and thus they suggest that polyandry may be as well developed in *hypoleucos* as it is in *macularia*. Yet definite pair

❖ ❖ ❖ 339

bonds are formed in *hypoleucos,* and thereafter a male will strongly defend his mate against the attentions of other males, sometimes resulting in intense fights between the males.

Reproductive biology. Nests of this species are typically built in sheltered depressions, usually fairly close to water. They may be placed on stream banks, on wooded slopes, in cultivated fields, or even in such unlikely locations as rabbit holes or the old nest of the ring ouzel *(Turdus torquatus).* The normal clutch is 4 eggs, but there sometimes are only 3, and rarely as many as 7 or 8 have been noted, the latter certainly laid by two females. A sample of 98 clutches from Finland included 94 of 4 eggs and the remainder of 3. Incubation is performed by both sexes, but probably the male undertakes the major part of the responsibility. It lasts 21–22 days. Both sexes brood, and, though some writers suggest that the male takes on the larger share of this, others have suggested that the female participates as actively as the male. In general the female probably loses interest in the brood first, leaving them to the male during the last part of their fledging period, which has been reported at 16–19 days (Bannerman 1961) and also as 26–27 days (Glutz et al. 1977). Although renesting is known to be common, it is believed that there is normally only one brood per season, and thus sequential polyandry seems unlikely.

Status and relationships. This species has been described as being perhaps the most numerous of any European *Tringa* (Voous 1960), and thus its status seems secure. As noted in the account of *macularia,* these two species are sometimes regarded as conspecific, but the apparent differences in degree of sexual dimorphism of size, plumage, and probable associated differences in mating systems make it rather difficult to consider them only racially different. The validity of the genus *Actitis* is another matter; Voous (1960, but not 1973), Stout (1967), Vaurie (1965), Mayr and Short (1970), and others have merged this genus with *Tringa,* while Jehl (1968a) maintained it and *Xenus* as separate genera close to *Tringa.* More surprisingly, Strauch (1976) not only recognized the genus but shifted it as well as *Xenus* to the calidridine group. I fail to find his arguments convincing, and I believe that merger with *Tringa* is reasonable inasmuch as there seem to be some real similarities between the two *Actitis* species and the two typical tattlers *(brevipes* and *incana).*

Suggested reading. Bannerman 1961.

Spotted Sandpiper

Tringa macularia L. 1766
(Actitis macularia of Peters, 1934)

Other vernacular names. None in general English use; chevalier grivelé (French); Drosseluferläufer, Amerikanischer Uferläufer (German); andarríos masculado, andarríos manchado (Spanish).

Subspecies and range. No subspecies recognized here *(rava* was found invalid by Monroe 1968). Breeds widely in North America from Alaska to southern California, and eastward to Newfoundland and the Atlantic coast south to South Carolina. Winters from British Columbia and South Carolina south through Mexico and the West Indies to central South America. See map 109.

Measurements and weights. Wing: males 89–105 mm; females 100–109 mm. Culmen (to feathering): males 21–26 mm; females 21–25 mm. Weights (in winter): males 31–46 g; females 40–48 g (Haverschmidt 1968). Breeding males and females average 47.2 and 33.2 g, respectively (Glutz et al. 1977). Eggs ca. 32 x 23 mm, estimated weight 9.0 g (Schönwetter 1963).

DESCRIPTION

Breeding adults of both sexes have a grayish brown crown that is streaked with dusky, a white superciliary stripe extending from above the lores to the ear coverts, a white lower eyelid, and the sides of the face streaked with brownish. The chin, throat, and underparts are white, with rounded spots of blackish throughout except sometimes on the chin. The upper-

340 ◈ ◈ ◈

parts are rather uniformly grayish brown, with blackish markings as transverse bars, arrowheads, or linear streaks. The tail is bronzy grayish brown, with most of the feathers barred and tipped with white. The flight feathers are brownish, the inner primaries having increasing amounts of white on their inner webs, and the secondaries are white at their tips as well as basally. The iris is brown, the bill is black at the tip and yellow to pinkish basally, and the legs and toes are pale grayish olive to pale yellowish pink. *Adults in winter* are plain grayish olive above, with a faint metallic gloss and with dark barring limited to the wing coverts. The underparts are white, with a brownish gray cast across the chest. *Juveniles* are plain gray on the mantle, which contrasts with the wing coverts, which are barred with brown and buff. The underparts are white. *First-winter* birds resemble adults but retain buff-barred inner median juvenile coverts (Prater et al. 1977).

In the hand, this bird resembles a small (wing 89–109 mm) *Tringa*, but it has white at the base and trailing edge of the secondaries and has a relatively short (culmen 21–26 mm) bill. Separation from *hypoleucos* in winter plumage is possible by the shorter tail (52 mm or less in males, 53 mm or less in females) and by white present on the outer webs of only the first one or two parts of outer tail feathers, instead of the first two or three pairs.

In the field (8 inches), this small and widely distributed shorebird is easily recognized by its "teetering" behavior and by its distinctive mode of flight, with vibrating and stiffly downcurved wing-strokes. When flushed, the birds do not fly far, and they usually utter a series of *weet* or *peet-weet* notes. By comparison, the Eurasian sandpiper utters a series of louder, more strident whistling notes, *twee-wee-wee*, which provides the best means of recognizing birds not in breeding plumage (*British Birds* 63:168–73). There are other minor plumage traits that help separate these two species, as mentioned in the account of *hypoleucos*.

NATURAL HISTORY

Habitats and foods. During the breeding season this species occurs over an extremely broad array of habitats, ranging from sea level to alpine timberline and north to arctic timberline. It is rarely found near the ocean but seems to require only open terrain with temporary pools, lakes, streams, rivers, marshes, or impoundments, and it sometimes breeds even in the absence of nearby water. During the nonbreeding season the species is typically found in shaded or sheltered watercourses, such as along forests or woodlands, but again its environmental tolerances are very broad. It also commonly winters along the seashores, where it forages both on beaches and along the muddy edges of creeks and inlets, typically working the water margins rather than wading and sometimes catching flying insects with remarkable agility. At times it will stalk a winged insect by stretching its entire body parallel to the ground and slowly moving up on it. It thus eats a vast diversity of aquatic and terrestrial insects, including grasshoppers, locusts, caterpillars, crickets, beetles and their larvae, and occasionally even small fish such as trout fry Bent 1920). Small crabs and gastropod mollusks have been noted as foods of wintering birds in Surinam. In Yellowstone Park the birds have been found to specialize on the larvae of brine flies living in thermal springs. There the birds tend to remain in small areas averaging about 3 acres per individual. They feed by using rapid pecking movements as well as by a slow approach followed by a quick forward thrust of the head (Kuenzel and Wiegart 1973).

Social behavior. Although it has long been believed that, like other tattlers, this species is monogamous, recent studies have indicated that at least a certain amount of polyandrous mating is regularly performed. One of these studies was done on Gull Island, New York, an island of 17 acres. This area supported 5–6 females and 11–12 males in the two years it was studied, and most of the females paired successively with 2 or more males. One female mated with 4 males in succession, while only 2 remained monogamous. Whereas the males became attached to particular nest sites, the females defended areas that included a number of sites or moved freely from one site to another after completing their clutches (Hays 1972). At the same time, a study was being performed at Itasca Park in Minnesota, where a open area of 6.8 hectares formed by a sewage lagoon provided an "island" of habitat surrounded by coniferous forest. A second study area was a small 1.6 hectare island on Leech Lake, Minnesota. At the lagoon site, from 3 to 5 females laid from 4 to 8 clutches between 1970 and 1972, or collectively 12 females laid 19 clutches for 14 males over the three-year period. On the island, females bred with 10 males in 1972. Sometimes females returned to the territories of the prior year, and in several cases both members of a pair returned and reestablished their pair bonds. Pair bonds are formed in minutes, and once established there is no indication of promiscuity (Oring and Knudson 1972). However, pair bonds apparently break as rapidly as they are formed, and more recently some cases of simultaneous polyandry have been noted (Oring and Maxon 1978). In 1975, one of 16 females on the Leech Lake island mated with a second male while still completing her first clutch with her original mate. During the following year she laid 5 clutches for 3 males, having simultaneous temporary pair bonds with 2 males on three separate occasions. The displays of the species are still rather poorly described, but the female apparently is the dominant sex. She sometimes flies upward, then glides down to land near another bird and display in a turkeylike manner with the tail spread, the head held high, and the wings drooped. A repeated *tweet* call is uttered during the display. During pair-formation both sexes sing repeatedly, sometimes about one song per minute (Oring and Knudson 1972). Copulation is not preceded by specific displays, though a wing-fluttering courtship has been described (Miller and Miller 1948).

Reproductive biology. Nests are almost anywhere—in grass, among rocks, under decaying logs, or among mosses, to mention only a few sites. They are often at considerable distance from water, and sometimes the birds nest in colonial fashion, especially on small islands. One island on the Detroit River having 13.5 acres of suitable habitat supported 43 pairs of birds, giving a density of 2.7 pairs per acre (Miller and Miller 1948). Nest prospecting is done by both members of a pair, sometimes the same day a pair bond is formed, and the first egg may be laid as early as 3 days after the male arrives on the female's territory. Likewise, replacement clutches are initiated an average of 5.3 days after predation or desertion, and eggs are laid daily. The normal clutch size is 4 eggs, with 5 rarely present, but in areas of high disturbance clutches of 3 or even 2 eggs are fairly frequent. The average clutch of 37 nests in Michigan consisted of 3.95 eggs (Miller and Miller 1948). Female aggressiveness and sexual activity increase after the laying of the third egg, and increased singing as well as aerial advertisement at this time may promote the changing of mates. In the New York study the females typically shared incubation duties with their mates during the final clutch, but in Minnesota the intensity of female incubation varied greatly among individuals and according to the stage of the reproductive cycle or even the weather. Although the incubation periods reported for this species vary greatly, they probably normally range from 20 to 24 days (Hays 1972). In general the male is believed to brood the young, but at least in some instances females may participate in this (Oring and Maxon 1978). Fledging is attained 17–18 days after hatching, and the brood scatters thereafter.

Status and relationships. This is certainly the most widespread and probably the most abundant North American scolopacid, and it is thus not of concern from a conservation standpoint. It is obviously a very close relative of *hypoleucos;* Mayr and Short (1970) followed Voous (1960) in considering them conspecific. I believe they constitute a superspecies, but I see no pressing need to consider them conspecific at this time.

Suggested reading. Nelson 1939; Oring and Knudson 1972.

Genus *Xenus* Kaup 1829 (Terek Sandpiper)

The Terek sandpiper is a small Eurasian sandpiper with a long, recurved bill that is much wider at the base than the tip and is at least 15 mm longer than the tarsus. The legs and toes are short, with a well-developed hind toe, and all the front toes are connected with webs. The downy young are *Tringa*-like, with reduced dorsal striping. One species.

Terek Sandpiper

Xenus cinereus (Güldenstaedt) 1774

Other vernacular names. Avocet-sandpiper; chavalier de Térek (French); Terekwasserläufer (German); andarrios de Terek (Spanish).

Subspecies and range. No subspecies recognized. Breeds in Russia and Siberia from the Severnaya Dvina River east to the Anadyr Basin, north locally to about 70° N. latitude, and south to about 55° N latitude in the west and about the mouth of the Amur in the east. Winters from the Persian Gulf, southern Red Sea, Indo-Chinese countries, and Hainan south to southern Africa, India, the Sundas, and Australia. See map 110.

Measurements and weights. Wing: males 121–34 mm; females 123–36 mm. Culmen: males 37–52 mm; females 37–52 mm. Weights (breeding grounds): 7 males 60–78 g, average 69 g; 6 females 65–77 g, average 74 g (Glutz et al. 1977). Wintering ground weights 54–119 g, average of 272 was 75 g (D. Purchase, pers. comm.). A sample of 105 wintering birds averaged 76.5 g (Summers and Waltner 1979). Eggs ca. 38 x 27 mm, estimated weight 13.5 g (Schönwetter 1963).

DESCRIPTION

Breeding adults of both sexes have a crown that is streaked with blackish and have darkish brown lores with dusky streaking extending to the upper ear co-

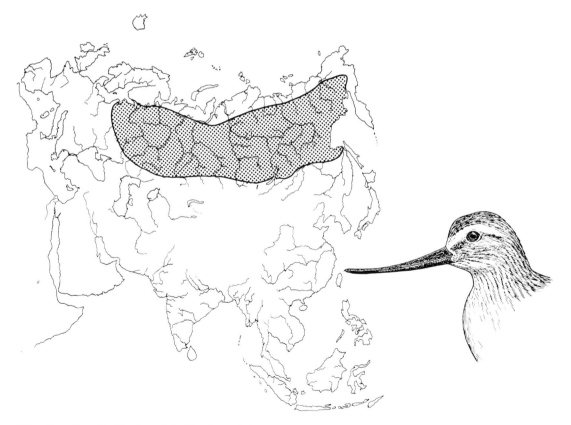

110. Breeding distribution of the Terek sandpiper.

verts, separated from the crown by a whitish superciliary stripe. The sides of the head, neck, upper breast, and sides of the breast are white with brown streaking, while the rest of the underparts are white. The upperpart feathers are black centrally with grayish brown margins, the upper tail coverts are grayish brown with sepia barring, and the tail is brownish, becoming whitish outwardly. The upper wing coverts are dull gray, the greater coverts are tipped with white, and the shaft of the first primary is white, while the others are brownish. The secondaries are also tipped with whitish. The iris is brown, the bill is brownish black, becoming orange yellow toward the base, and the legs and toes are orange yellow. *Adults in winter* are grayish brown on the upperparts, the wing coverts having narrow white terminal fringes. There is a white superciliary stripe, and the underparts are white except for some gray on the sides of the breast. *Juveniles* are brownish gray above, with darker shaft streaks, and the scapulars have large black V markings. The wing coverts are gray, with cinnamon buff fringes, and the upper tail coverts and tail are tipped with brown and cinnamon buff. *First-winter* birds usually retain some juvenile coverts and tertials until midwinter (Prater et al. 1977).

In the hand, the long and recurved bill (culmen 37–52 mm) compared with short legs (tarsus 26–32 mm) readily identifies this species.

In the field (9 inches), the long, slender, and up-turned bill together with short, orange legs on this small sandpiper provide easy identification, even in the dull winter plumage. In summer plumage a black stripe extends along the scapulars. In flight, a narrow white wing stripe is evident, and there is a prominent white trailing edge on the secondaries. The birds are active when feeding, often bobbing the head and jerking the tail, and sometimes swimming. They utter a *teeoo* alarm note, a repeated high-pitched *tee,* and a rapid *titter-tee.*

NATURAL HISTORY

Habitats and foods. The breeding habitats of this little-known species consist mainly of narrow, marshy openings in the Eurasian coniferous forest, especially boulder-strewn valleys where there are marshy grasslands and meadows, with an alternation of tall grasses and scrub willows. The species also extends northward into the scrub tundra a short distance and occurs locally in flooded districts with trees. Outside the breeding season it is primarily found on coastal lagoons and narrow creeks passing through salt

marshes, as well as on sandy and pebbly sea beaches. In Australia it favors muddy beaches near mangroves, but it is also seen around coral reefs and sometimes extends up to 10 kilometers inland around brackish pools. It resembles the Eurasian sandpiper in its short legs and body conformation, but it apparently uses open mud banks less often and is probably more specialized in its foraging, as is suggested by the bill shape (Voous 1960). On the breeding grounds it feeds largely on the adults and larvae of midges, while in the migratory and wintering areas it eats a variety of insects (beetles, ants, dipterans, etc.), small gastropods, crustaceans, and other invertebrates. In general the species is probably more adept at probing than are typical *Tringa* species, and the anterior portion of the upper mandible is very flexible. The birds are highly mobile when foraging, sometimes taking food from the water surface and chasing swimming prey, but usually they feed in shallow water or where there is an exposed mud substrate. In some respects the species is convergent toward the godwits in its bill shape and mode of foraging, although its evolutionary affinities are certainly with the typical tattlers.

Social behavior. Outside the breeding season this species is moderately gregarious, being found in groups of 5–25 birds, associating little if at all with other species. During the breeding season the birds are territorial, and like various *Tringa* species they frequently call from elevated sites. The display flight consists of rising to a considerable height, hovering for a few moments, then slanting back to earth on outstretched wings. The display call is a loud, clear note with the last syllable rising higher and being more drawn out, variously rendered by different observers as *koo-vedruh, koovitruu,* or *qu-widrruu.* This call is said to be uttered on the ground, but doubtless it also is performed while in flight. The male also sings while crouching before the female, with his tail raised and his wings fluttering. Remarkably little has been written on the social behavior of this species, considering that at least in some areas of Siberia it is abundant, reaching densities of up to about 10 pairs per square kilometer. (Glutz et al. 1975).

Reproductive biology. Nesting apparently occurs in dense growths of scrub willow, often in semisocial

fashion; as many as 10 nests have been found within a square kilometer. The nest is a shallow depression lined with various bits of vegetation, either in fairly open view or sheltered by grasses. It is usually rather close to water but is placed so as to avoid flooding, and it normally has 4 eggs, with occasional records of 3 or 5. The eggs are laid in late May and June, and incubation is said to begin after the laying of the third egg. At least the initial stages of incubation are performed by both sexes, but in the later stages the male alone may incubate. Incubation requires 20–21 days, and apparently there is a very short fledging period of only about 15 days (Glutz et al. 1977).

Status and relationships. Although this species is included in a list of "threatened" birds of Europe (Hudson 1974), this provides a rather misleading idea of its worldwide status. In fact, its range has recently expanded to include a few areas of coastal Finland, where breeding was first proved to occur in 1957. Perhaps 10–15 pairs now breed there, mainly along the Gulf of Bothnia (Hudson 1974). In the USSR as a whole the species is regarded as common to abundant in its taiga nesting grounds. In its behavior and its downy plumage it closely resembles the *"Actitus"* group of tringine sandpipers, and most current workers favor placing *Xenus* between *Tringa* and *Actitis* or even merging them all into *Tringa.* I am not sure it is justifiable to recognize *Xenus* and not *Actitis,* but the former is more specialized in its bill structure and foraging adaptations. Rather surprisingly, Strauch (1976) took the position that both *Actitis* and *Xenus* belong with the calidridine sandpipers rather than with the tringine forms, and he indeed questioned whether these two genera may actually be closely related to one another, suggesting that they are perhaps only convergent in their behavior. Strauch suggested a possible affinity between *Xenus* and *Limosa,* which I believe is more likely to be a convergent adaptation than is the suggested convergence of *Xenus* and *Actitis.* I thus favor retaining *Xenus* as a genus but keeping it adjacent to *Tringa,* probably most closely related to *hypoleucos* and *macularia,* which might be regarded as a subgenus *Actitis.*

Suggested reading. Hudson 1974; Hosking and Ferguson-Lees 1959.

Tribe Prosobonini (Polynesian Sandpipers)

Tuamotu Sandpiper

Polynesian sandpipers are small tropical Pacific Ocean shorebirds with a bill that is shorter than the tarsus, slender, straight, and not noticeably expanded toward the tip. The tarsus is scutellated in front and reticulated behind, and there is slight webbing between the middle and outer toes. The tail is relatively long and is strongly barred. The wings are relatively short (about twice as long as the tail) and are rounded. The downy plumages are undescribed. The sexes are alike as adults. One genus and 2 species are recognized here, of which one is extinct.

Genus *Prosobonia* Bonaparte 1850

This genus of two species has the characteristics of the tribe.

KEY TO SPECIES OF *Prosobonia*

A Underparts unbarred, lesser wing coverts with white edges . . . . *leucoptera*
A' Underparts barred, lesser wing coverts lacking white edges . . . . *cancellatus*

Tuamotu Sandpiper

Prosobonia cancellatus (Gmelin) 1789
(*Aechmorhynchus cancellatus* of Peters, 1934)

Other vernacular names. Barred phalarope; chavalier des Tuamotou (French); Südseeläufer (German).

Subspecies and range. No subspecies recognized. Now limited to a few isolated islands on the fringing reefs of atolls in the Tuamotu Archipelago (this population sometimes separated as *parvirostris*) of French Polynesia, including the Gambier and Actaeon Islands. Formerly also Christmas Island, where the type specimen was collected. See map 111.

Measurements and weights. Wing (chord): males 97–109 mm; females 97–112 mm. Culmen: both sexes 15–18 mm (Zusi and Jehl 1970). Weights: both sexes 32–44 g, average of 7 was 36 g (Lacan and Mougin 1974). Eggs: 2 in the American Museum of Natural History measure 35 x 25.5 mm and 36 x 25 mm, estimated weight 12.5–13.4 g.

DESCRIPTION

Adults of both sexes are rather variable but typically have an ashy white stripe above and behind the eye, an umber brown crown, a dark stripe from the base of the bill through the eye and along the upper ear coverts, and the rest of the head buffy to ashy col-ored. The upperparts are generally umber brown, with extensive white and reddish fulvous edging and tipping on the feathers, including the wing coverts. The flight feathers and under wing coverts are uniform umber, except for narrow white edging on the inner primaries and secondaries. The tail feathers are umber, with irregular and incomplete narrow bands of ashy and pale reddish white, and tipped with the same. The underparts are generally white, with an ashy tinge, the throat and abdomen unspotted, and the breast, sides, and under tail coverts are spotted or barred with brown, with barring especially evident on the sides and flanks. The iris is dark brown, the bill is blackish, and the legs and toes are gray, grayish brown, or yellowish, apparently varying individually. *Females* are said to be slightly paler than males (Sharpe 1896), but considerable variation in plumage darkness occurs and evidently is not related to sex. *Immature* plumages are undescribed.

In the hand, this species is identifiable by its unusually rounded wings (the first three primaries nearly equal in length and no longer than the fourth), the slender and ploverlike bill, and the rather long, barred tail.

In the field (6.5 inches), these birds are limited to some of the remote atolls of the Tuamotu group, where they are usually found on stretches of bare gravel. They evidently look somewhat like small upland sandpipers in life, but virtually nothing is known of their behavior and vocalizations. The birds are reported to call almost constantly in a soft, high-

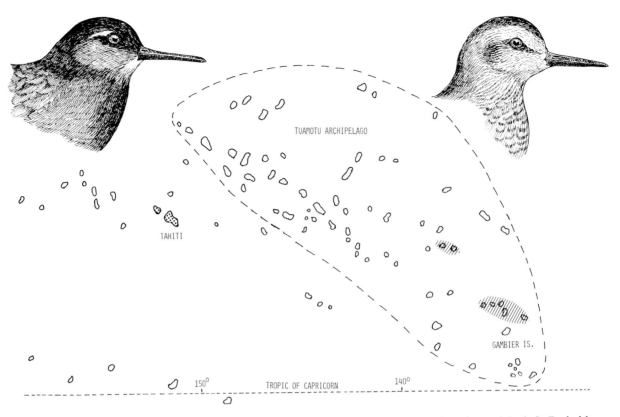

111. Original distributions of the Tuamotu sandpiper (*broken line*) and White-winged sandpiper (*shaded*). Probable current distribution of the Tuamotu sandpiper is indicated by hatching.

pitched voice, generally reminiscent of upland sandpipers.

NATURAL HISTORY

Habitats and foods. Evidently these birds occupy essentially all the habitats of the tiny atolls of the Tuamotu Archipelago, but they occur most commonly where there are stretches of open shingle or gravel (Greenway 1958). The birds' foods and foraging behavior are not well known, but Lowe (1927) noted that one specimen contained the remains of at least four species of ants, a few small leafhoppers, and a portion of what was probably a chalcid wasp. Lacan and Mougin (1974) noted that 6 stomachs they examined contained nothing but insect remains and some vegetational debris.

Social behavior. Nothing specific is known of the species' social behavior. The breeding season is evidently quite prolonged, beginning in some individuals in April or even earlier and persisting until early June or later (Greenway 1958). However, it is clear

that not all birds on any single island breed at the same time, since the majority of birds taken during this period have not been in breeding condition. It is possible that different atolls have somewhat different periods of breeding (Lacan and Mougin 1974), or perhaps only a small proportion of the birds are able to obtain suitable breeding territories and actually breed, while the remainder do not come into full breeding condition.

Reproductive biology. There seems to be only two reported cases of actual breeding in this species. One nesting pair was collected on Kauehi Island in late May, and a nest with 2 eggs was found in early May on Tunake Island. The nest was on the shoreline of the lagoon and consisted of small, dry grass stems in a slight hollow on a shingle substrate (Greenway 1958). The 2 eggs were rather heavily blotched, rather like those of an upland sandpiper, and this blotching suggests that a pebbly rather than sandy substrate is the usual one.

Status and relationships. The best current discussion of this species' status is that of Temple (1979), who

348 ✦ ✦ ✦

noted that recent sightings have been made from Maturei-Vaveo of the Actaeon group, Ragiroa (a single bird, probably only a visitor), Pinake, Maurtea, and Nakutayake. It is now evidently gone from Christmas Island, Makaroa, Kamaka, and Manui and perhaps has also vanished from several others, presumably because of the introduction of predatory mammals. The relationships of this interesting bird have been reviewed by Zusi and Jehl (1970). They have concluded that only a single species should be recognized and that, since *cancellatus* is the earliest available name, *parvirostris* should thus be regarded as a synonym of *cancellatus*. They have further concluded that *Aechmorhynchus* should be considered congeneric with *Prosobonia*, and that these two species have their nearest affinities with the typical tattlers rather than with the curlew group. Strauch (1976) has criticized these conclusions and suggested that the group's closest affinities are actually with *Bartramia* and the curlews. I believe all of these are quite closely related and regard the three tribes as only weakly separable.

Suggested reading. Greenway 1958; Zusi and Jehl 1970.

Subspecies and range. No subspecies recognized. Extinct; previously occurred on Tahiti and Eimeo (Moorea), Society Islands. Only the type specimen is extant, but others once existed. See map 111.

Measurements and weights. Wing (chord): 111 and 113 mm (right and left). Culmen (from feathering): 20 mm (Zusi and Jehl 1970). Eggs undescribed.

DESCRIPTION

Adults (based on the single specimen known) are generally plain-colored, except for barring on the tail and pale edges on the under wing coverts. The crown is blackish, the hindneck browner, with sides of the face also brownish. The chin and gular region are whitish with some buff. The malar region and underparts are russet, and there is a ringlet of paler russet around the eye. There is a broken superciliary line of pale russet from the bill to the eye, reappearing above the eye and changing to white above the auricular region. The back and wings are sooty brown, and the rump is russet. There is a crescent-shaped patch of white on the lesser wing coverts, continuous across the leading edge of the wing, with a similar but smaller patch on the underside. The wing lining is

White-winged Sandpiper

Prosobonia leucoptera (Gmelin) 1789

Other vernacular names. Tahitian sandpiper; chavalier à ailes blanches (French); Gesellschaftsläufer (German).

dusty brown and the axillaries are sooty brown. The tail feathers are sooty brown with russet tips and become progressively more heavily barred with russet laterally. The iris is dusky blackish, the bill is black, and the legs are greenish (Sharpe 1896; Zusi and Jehl 1970). Other plumages are undescribed.

In the hand, this extinct species may be easily separated from *cancellatus* by its white wing markings on the lesser coverts and by its unbarred underparts.

NATURAL HISTORY

Habitat and foods. Nothing is known, other than that the birds were said to occur near small brooks.

Social behavior. No information is available.

Reproductive biology. No information is available.

Status and relationships. Although at least 3 specimens of this species were collected during Captain James Cook's voyages to the South Pacific in 1773 and 1777, only a single specimen remains in Leiden, Holland. It has been suggested that pigs or goats released by Cook were the cause of this species' extinction, but more probably rats were responsible (Greenway 1958). Zusi and Jehl (1970) have reviewed the rather complex taxonomic history of this form and concluded that only one species should be recognized, since the specimen described from Eimeo Island (or Moorea Island) 10 miles to the west of Tahiti did not differ enough from the Tahitian form to warrant specific recognition. Zusi and Jehl compared measurements, skull features, and plumage characteristics of *leucoptera* and *cancellatus* and concluded that the two forms deserved only species separation rather than generic status. Thus *Prosobonia* is the appropriate name for both forms. Reviewing the evidence of Lowe (1927), Zusi and Jehl concluded that *Prosobonia* is probably not a close relative of the curlew group (Numenini), but instead is more closely related to the tattlers (Tringini). I have followed their recommendations that it be placed in a special tribe close to the tattlers.

Suggested reading. Zusi and Jehl 1970.

Tribe Numenini (Curlews and Godwits)

Eskimo Curlew

Members of this tribe are small to very large temperate to arctic-nesting shorebirds having a bill that is usually appreciably longer than the tarsus, a relatively long tail, and a tarsus that is scutellated in front and reticulated or scutellated behind. The hind toe is well developed, and the front toes usually lack webbing between the middle and inner toes. The wings are relatively long and pointed, and the rectrices and flight feathers are often heavily barred. The downy young are buffy to light cinnamon, usually with blackish spots or broken bands on the upperparts. The sexes are alike as adults. Three genera and 13 species are recognized here.

KEY TO GENERA OF NUMENINI

A Bill decurved and at least 40 mm long . . . . *Numenius* (8 spp.)
A′ Bill not decurved
   B Bill long (over 65 mm) and straight or slightly recurved, tail relatively short . . . . *Limosa* (4 spp.)
   B′ Bill short (under 35 mm) and straight, tail at least half as long as wings . . . . *Bartramia* (1 sp.)

Genus *Bartramia* Lesson 1831 (Upland Sandpiper)

The upland sandpiper is a small North American shorebird with a bill that is about half as long as the tarsus and a tail that is longer than the tarsus. The bill is relatively short and slender and is slightly decurved toward the tip. The tarsus is scutellated both in front and behind, the hind toe is well developed, and there is a small web between the outer and middle toes. The downy young are brownish dorsally, with buffy feather tips that form no definite pattern. The sexes are alike as adults. One species.

Upland Sandpiper

Bartramia longicauda (Bechstein) 1812

Other vernacular names. Bartram's sandpiper; upland plover; Bartramie des champs (French); Prärieläufer (German); batitú, correlimos de Bartram (Spanish).

Subspecies and range. No subspecies recognized. Breeds from northwestern Alaska (Brooks Range), the Yukon, and British Columbia south to Oregon and southeast through the Great Plains and Lake States to West Virginia and Maryland. Winters in southern South America. See map 112.

Measurements and weights. Wing: males 163–80 mm; females 164–91 mm. Culmen (to feathering): males 26–31 mm; females 27–32 mm. Weights: males 132–66 g, average of 8 was 137.3 g; females 132–75 g, average of 6 was 163.9 g (various sources). Eggs ca. 45 x 33 mm, estimated weight 25.2 g (Schönwetter 1963).

DESCRIPTION

Adults of both sexes have a blackish crown, divided by an indistinct median line of buffy, and a whitish line above the lores that merges with a white eye ring and postocular stripe; the sides of the head are white with dusky streaking. The sides of the neck, foreneck, and chest are pale buff with black streaking, and the chin, throat, and the rest of the underparts are white, with V-shaped markings of black on the breast, becoming vertical bars on the sides and flanks. The axillaries and under wing coverts are white, with blackish bars and grayish brown and buffy margins, and the rump and upper tail coverts are blackish, with the longer upper tail coverts becoming spotted with grayish brown and buffy. The tail grades from grayish brown on the middle feathers to pinkish buff on the outermost pair, with the four outermost pairs broadly tipped with white and having black subterminal bars, while all the feathers are variously spotted or barred with black. The wing coverts are mostly grayish brown, margined with buffy and sometimes barred with dusky. The primaries and their coverts are dusky, the inner primaries being spotted with white, and the secondaries are extensively barred and margined with grayish brown and buffy white. The iris is brown, the bill is blackish above, becoming brownish or yellowish below and basally, and the legs and toes are yellowish gray. *Juveniles* have dark blackish brown scapulars that

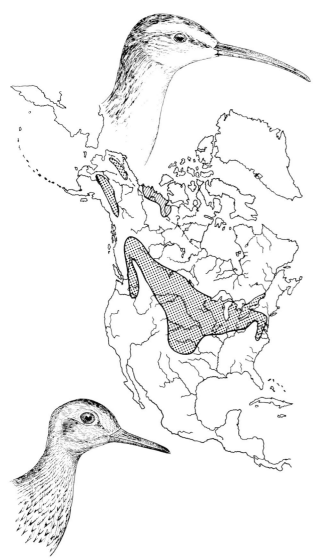

eyes, white eye ring, and curlewlike plumage pattern are distinctive, as is its long tail. It is highly vocal and utters a variety of calls, including a rapid *quip-ip-ip-ip* alarm call on the ground and a richer rolling trill in flight. It frequently flies with shallow, rapid, fluttering wingbeats.

NATURAL HISTORY

Habitats and foods. In the center of its breeding range in North Dakota, this species occurs equally commonly on mixed-grass prairies and locally extensive tracts of wet meadows. It also regularly breeds on grazed tallgrass prairie, domestic hayfields, fields of retired croplands, and mowed or burned railroad or highway right-of-ways. Rarely, breeding occurs on cultivated fields, stubble fields, and fields with growing small-grain crops (Stewart 1975). Maximum breeding densities of as many as 20 pairs per square mile occur in favorable habitats in North Dakota (Stewart and Kantrud 1972). Outside the breeding season the species is also associated with old fields, pastures, and natural grasslands, and in South America it occurs in fields and grasslands from sea level through the subtropical zone. Its diet is thus primarily composed of terrestrial insects; grasshoppers,

112. Breeding distribution of the upland sandpiper (*shaded*) and probable original breeding distribution of the Eskimo curlew (*hatched*)

lack distinct barring and dark brown tertials that have pale buffy edges and notches; the wing coverts are pale brown, with pale buff fringes and subterminal crescent-shaped brown spots (Prater et al. 1977; Ridgway 1919).

In the hand, this small species (wing 160–91 mm) has a plumage strongly resembling a curlew, but with a short (26–32 mm) *Tringa*-like bill and an unusually long (80–90 mm) tail.

In the field (11 inches), this species is sometimes found along wet shorelines, but more often it inhabits grassy uplands, and it frequently may be seen perched on fenceposts in prairie habitats. Its large

crickets, and weevils constitute almost half its diet, based on an analysis of 163 stomachs. Many larvae of beetles and lepidopterans are eaten, as well as adult beetles, moths, ants, flies, bugs, centipedes, millipedes, spiders, snails, and earthworms. Most of the seeds eaten are of weeds, but sometimes during the fall the birds gather in grainfields and eat waste wheat (Bent 1929).

Social behavior. In Wisconsin, upland sandpipers return to their nesting area in mid-April, with most of the birds already paired when they arrive, and they move onto their home ranges shortly after arrival. Nesting territories are generally grouped, and some aggressive behavior apparently associated with territories is typical early in the season but is soon terminated. Even after pairing, the birds sometimes regroup into flocks, such as during stormy weather. Two call notes are distinctive and diagnostic of the breeding season and are named the short and long whistles. The short whistle is a series of 6–10 rapid notes on the same pitch. This is uttered by both sexes, especially when they are flushed. The long whistle is also uttered by both sexes, usually when alighting or when circling in the sky. It is preceded by a throaty or whistled chatter, has an ascending slurred note, a silent pause, and a final ascending and descending wailing whistle. The song flight of breeding birds is spectacular, often performed at such a height that the bird is barely visible, and while it slowly circles about it utters the long whistle, a drawn-out *whip-whee-ee-you*. Sometimes after such a circling flight the bird will close its wings and plummet to the earth like a falling stone (Bent 1929). There is also another type of flight behavior typical of breeding birds. This is a rapid "flutter stroke" performed in low, horizontal flight and followed by a long glide, performed by both sexes until after the nesting season. At that time there is a gradual transition to a normal wing-stroking characterized by deeper and slower wingbeats than used during the flutter stroke phase (Buss and Hawkins 1939). Behavior associated with copulation is still only poorly understood, but Ailes (1976) reported that the male courts by raising his tail like a dancing prairie chicken, then runs toward his mate uttering a short guttural whistle. If the female allows the male to approach, copulation follows; otherwise she runs away. Both sexes participate in nest-scraping behavior, with the male being more persistent and initiating scrapes more often than the female.

Reproductive biology. Nesting in Wisconsin and North Dakota begins 15–20 days after initial arrival. In North Dakota, 183 of 199 nests were in grassland, most of which was native prairie (Stewart 1975). The nests are usually well hidden and frequently are in

grass tufts that hang over the nest, hiding it from above. Nests are sometimes fairly close to one another, but even in areas of dense nesting a single field will support no more than about one nest per 1.5 acres, and usually there are from 2.5 to 15 acres per nest (Buss and Hawkins 1939). Eggs are laid at intervals averaging about 26 hours between successive eggs, and all of 189 clutches in a North Dakota study had 4 eggs (Higgins and Kirsch 1975). There is apparently some renesting activity, resulting in a tailing out of hatching dates after an initial clustering of hatched nests in early to mid-June in Wisconsin and North Dakota. Both sexes incubate, and the incubation period averages 24 days, with extremes of 21–28 days reported by various observers. The young are tended by a single adult, often if not always the male (Ailes 1976). Other studies suggest that both adults remain with the young until they are nearly fledged, which occurs at about 32–34 days of age (Buss 1951).

Status and relationships. Although this species has certainly been lost from much of its original range, it is still fairly abundant in some parts of the Great Plains and has regained some of its lost range in Minnesota and Wisconsin. It is also known to breed much more extensively in Alaska than was previously believed. Its relationships are of particular interest. Always regarded as a monotypic genus, *Bartramia* has usually been placed near the tringine sandpipers. However, its affinities are almost certainly with the curlews (Lowe 1931b; Jehl 1968a), although its downy young are distinct from both of these groups of putative relatives. Larson (1957) believed that *Bartramia* resulted from a Miocene reinvasion of boreal areas by arctic ancestral *Numenius* populations. Strauch (1976) supported the idea that the curlews are the nearest relatives of *Bartramia* and went so far as to suggest that the genus be merged with *Numenius*. I do not accept such an extreme position, but I believe that a close relationship to the curlews is very probable.

Suggested reading. Higgins and Kirsch 1975; Ailes 1976.

Genus *Numenius* Brisson 1760 (Curlews)

Curlews are medium to very large temperate to arctic-breeding shorebirds with a bill that is distinctly decurved and is longer than the tarsus. The tarsus is scutellated in front and usually reticulated behind, the hind toe is well developed, and there is some webbing between the anterior toes, especially between the middle and outer toes. The downy young are tan or grayish, with brown or black dorsal spotting or striping, sometimes forming a diamondlike pattern on the back. The sexes are alike as adults. Eight species are recognized here.

KEY TO SPECIES OF *Numenius*

A Bill very long (culmen over 100 mm), longer than tail
  B Rump and lower back white . . . . *arquata*
  B′ Rump and lower back like rest of upperparts
    C Axillaries pale chestnut and lacking barring . . . . *americana*
    C′ Axillaries with brown barring present . . . . *madagascariensis*
A′ Bill shorter (culmen under 100 mm), not as long as tail
  B Bill slender, more than twice as long as middle toe and claw, head not distinctly striped . . . . . *tenuirostris*
  B′ Bill stouter, less than twice as long as middle toe and claw, head distinctly striped
    C Bill only slightly decurved, culmen no more than 60 mm
      D Exposed culmen under 47 mm, tarsus scutellated behind . . . . *minutus*
      D′ Exposed culmen at least 47 mm, tarsus reticulated behind . . . . *borealis*
    C′ Bill stouter and distinctly decurved, culmen over 65 mm
      D Thigh feathers with bristle tips, tail pinkish cinnamon with heavy darker barring . . . . *tahitiensis*
      D′ Thigh feathers normal, tail grayish brown with narrow barring . . . . *phaeopus*

Little Curlew

Numenius minutus Gould 1840 (1841)

Other vernacular names. Little whimbrel; courlis nain (French); Zwergbrachvogel (German).

Subspecies and range. No subspecies recognized. Breeding range not well known, but breeds in northern Yakutia between the drainages of the Moiero and Indigirka rivers. Also breeds in the Verkhoyansk and Tcherski ranges, and near the junction of the Yana and Adycha rivers. Winters in eastern Indonesia from the Moluccas east to New Guinea and Australia. See map 113.

Measurements and weights. Wing: males 172–94 mm; females 175–98 mm. Culmen: males 40–43 mm; females 40–46 mm. Weights: both sexes 300–440 g, average of 8 was 351 g (D. Purchase, pers. comm.). Eggs ca. 51 x 35 mm, weight 29 g (Vorobiev 1963).

DESCRIPTION

Adults of both sexes are almost identical in plumage to the Eskimo curlew but are slightly more buffy and less cinnamon in color, with a more whitish belly, less streaking on the breast and flanks, and the under wing coverts and axillaries pale buff rather than cinnamon buff. The iris is brown, the bill is dark brown becoming yellowish toward the base, and the legs and toes are gray or bluish gray. *Juveniles* have narrow whitish buff spots on the edges of the scapulars, and their coverts are broadly edged and indistinctly spotted with buff (Prater et al. 1977).

In the hand, the curlewlike but extremely short (40–46 mm) bill, the short wings (under 200 mm), and the moderately long legs (tarsus 47–51 mm) separate this species from all but some small specimens of *borealis.* The color differences mentioned above, as well as the scutellated condition of the rear tarsus surface, provide for distinction from that species.

In the field (10 inches), the extremely small but typically curlewlike configuration of this species

113. Probable breeding distribution of the slender-billed curlew (*broken line*) and known (*hatched*) breeding distribution of the little curlew.

42. Adult heads of Eskimo (*above*) and little (*below*) curlews.

should readily distinguish it from the other Asian curlews. Besides the bird's being smaller than the whimbrel, the striping above the eyes and on top of the crown is buffy and subdued rather than whitish, and the rump area is no paler than the back. The calls include a soft, musical *te-te-te* when feeding and a rather harsh *tchew-tchew-tchew* when alarmed.

NATURAL HISTORY

Habitats and foods. On its wintering grounds, this tiny curlew is found in grasslands and mud flats generally, and in northern Australia it occurs on coastal black soil plains and open, dry country, sometimes including suburban lawns. Few wintering birds have been examined for foods, but wild rice seeds and caterpillars are listed as known foods (Frith

1976). On their breeding grounds on the upper Yana watershed the birds were found in burned clearings of larches, osier, and drawf birches, as well as in river valleys covered with sparse stands of dwarf birch and osiers, in a valley covered with thick stands of dwarf birch, and on a ridge with a light cover of larches and a shrub undergrowth (Labutin 1959). This suggests that they are certainly not typical tundra birds, but rather are timberline species. Their foods in that area were found to be mostly terrestrial insects such as beetles, weevils, crickets, and ants, with a few berries and plant seeds also reported.

Social behavior. In winter these birds are often in rather larger flocks; groups of 50 to several thousand have been seen in northern Australia (Frith 1976). On the breeding grounds of northern Siberia they are also somewhat gregarious and throughout the nesting season groups of 6–12 birds can be seen, presumably representing nonbreeders or perhaps nonincubating males. The males apparently play no part in defending the nest during incubation, and when the female is frightened from the nest she usually flies away directly and calls from 20–30 meters away (Labutin 1959).

Reproductive biology. There are rather few observations of birds on the breeding grounds; those of Vorobiev (1963) and Labutin (1959) are among the few available. Labutin reported that in the Verkhoyansk region the birds begin egg-laying in late May, and that the nests found were 50–70 meters from a river. One was in a burned clearing with a sparse cover of charred larches and other small trees, while another was in a denser growth of birches of about 50 centimeters high. One nest contained 3, and the other 4 eggs. Evidently the usual clutch is 4 eggs (Kozlova 1962; Piechocki 1967). Apparently only the female incubates until near the end of the period, which lasts 22–23 days. Hatching occurs in late June, and flocking begins in late July, with departure from the Verkhoyansk region starting in early August. The fledging period is thus not much more than a month to 5 weeks in length.

Status and relationships. Although this species is known to breed in only a few localities, it is seemingly not endangered, judging from Frith's (1976) comments on the numbers sometimes seen in northern Australia. Dementiev and Gladkov (1969) considered it "rare" but noted that it was common in various localities of China during migration. If we can learn from the lesson of the Eskimo curlew, it will be important to watch closely the population trends of this seemingly common species. The relationship of these two forms is interesting, and at times they have been regarded as conspecific, as for example by Dementiev and Gladkov (1969). However, there are a number of distinct morphological differences between them (Farrand 1977), and I see no reason for believing them conspecific.

Suggested reading. Piechocki 1967.

Eskimo Curlew

Numenius borealis (J. R. Forster) 1772

Other vernacular names. None in general English use; courlis esquimau (French); Eskimobrachvogel (German); zarapito esquimal, zatapito polar (Spanish).

Subspecies and ranges. No subspecies recognized, but this form is sometimes regarded as conspecific with *N. minutus*. Nearly extinct; formerly bred in northern Mackenzie, probably extending west to northern Alaska, and perhaps east to Hudson Bay, wintering in South America from southern Brazil south to Chiloé Island and Chubut, Argentina. See map 112.

Measurements and weights. Wing: males 200–24 mm; females 189–215 mm. Culmen (to feathering): males 48–53 mm; females 47–60 mm. Eggs ca. 51 x 35 mm, estimated weight 33.0 g (Schönwetter 1963).

DESCRIPTION

Adults of both sexes have a sooty black crown that is streaked with pale buffy, especially along the median line, a broad superciliary stripe of pale buffy, and a blackish patch on the lares. There is blackish streaking on the sides of the head and neck, foreneck, chest, and upper breast, becoming V-shaped on the sides and forming broad bars on the flanks; otherwise the underparts are dull buff. The axillaries and under wing linings are cinnamon buff, with darker bars, spots, or streaks. The back and scapulars are sooty black with buffy feather edging and spotting.

The secondary coverts are mostly grayish brown to blackish with buffy margins, while the primary coverts are dusky grayish brown with narrow whitish margins. The primaries are uniformly brown and lack barring. The secondaries and tertials are lighter, with margins or notching of pale buffy. The rump and upper tail coverts are fuscous with buffy spotting, and the tail is grayish brown, barred with dark fuscous and narrowly tipped with buff. The bill is blackish, with the basal portion brownish, the iris is brown, and the legs and toes are dark gray (not greenish as often depicted).

Juveniles have narrow pale buff edging and spotting on the tertials and scapulars, which are unbarred, and the coverts have more extensive buffy edges than in adults (Prater et al. 1977; Ridgway 1919).

In the hand, the relatively short (48–60 mm) gradually decurved bill and very small size (wings under 230 mm) distinguish this from all other curlews except *minutus*, which differs in having a scutellate pattern on the rear of the tarsus, has buffy rather than bright cinnamon under wing linings and axillaries, and usually has a longer tarsus than culmen.

In the field (11 inches), this extremely rare bird is most likely to be confused with the whimbrel, but is smaller (size of greater yellowlegs), is more generally buffy, has a shorter and thinner bill, and is less distinctly striped on the crown. Additionally, the underwing is darker and more cinnamon, rather than pinkish buff. In flight, a series of soft, melodious, and tremulous whistling notes are uttered. The unbarred condition of the primaries may also be visible in flight.

NATURAL HISTORY

Habitat and foods. The winter habitats of this species were the pampas of southern South America, but no observations on them there are available. In the summer they were apparently associated with barren ground tundra rather than with wooded areas, according to the early reports of Roderick MacFarlane. If this is the case, the species differs rather strongly from the little curlew, which is associated with arctic timberline rather than true tundra. During spring migration the birds apparently fed largely on grasshoppers in the Great Plains, but also on grasshopper eggs, ants, and berries. Fall foods on the Labrador coast included crowberries (*Empetrum*), snails, and doubtless other foods as well (Bent 1929).

Social behavior. This was a highly gregarious species, with terms such as "immense flocks," "dense flocks," and the like used to describe it during the fall. They were said to occur in flocks of from 3 to many thousand birds and were sometimes associated with golden plovers.

Reproductive biology. Regrettably little is known of this. All the eggs that are known were collected by Roderick MacFarlane in northwestern Canada in the latter part of June, between the Mackenzie and Coppermine rivers. Other than that the birds nested in open tundra, and that the normal clutch is 4 eggs with occasional instances of 3-egg clutches, almost nothing has been written (Bent 1929). In the 1860s these birds were apparently breeding abundantly from the tundra area east of Fort Anderson up to the arctic coast. Since there are records of large flocks of birds arriving in very late July or early August in the Gulf of St. Lawrence, there must have been a short fledging period and an immediate departure from the breeding grounds thereafter.

Status and relationships. There are only a few records of sightings of this species in recent years. The best-documented cases were on Galveston Island, of single birds in 1959, 1960, 1961, and 1963 and of 2 or more birds in 1962 and 1964 (*Auk* 82:493; *Audubon Field Notes* 18:469). There was also a Martha's Vineyard sighting of 2 birds in early August of 1972 (*American Birds* 26:907). Lately, and most encouraging, there was a fairly convincing 1976 sight record of 2 birds in mid-August on a salt meadow marsh west of Hudson Bay (17 miles northeast of Moosonee), in an area surrounded by muskeg forest (*American Birds* 31:135). If breeding birds still survive, it seems possible that they must be nesting on the west coast of Hudson Bay, perhaps not far from this sighting.* The evolutionary relationships of this species are obviously with the little curlew. As I noted in that species' account, I believe that for morphological and ecological reasons they should not be considered conspecific. More distantly, the larger curlews and the upland sandpiper are self-evident relatives.

Suggested reading. Bannerman 1960; Greenway 1958, Johnsgard, 1980.

---

* See *American Birds* 34:788 for a 1980 Manitoba sighting, and 34:849 for a possible Guatemala record.

Whimbrel

Numenius phaeopus (L.) 1758

Other vernacular names. Hudsonian curlew (*hudsonicus*); courlis corlieu (French); Regenbrachvogel (German); Zarapito trinador (Spanish).

Subspecies and ranges. See map 114.

N. p. phaeopus: Eurasian whimbrel. Breeds in Iceland, the Faeroes, Scotland, and continental Eurasia from Sweden to the Yenisei. Winters in Africa, southern Iraq, and the coast of western India to Sri Lanka and the Andaman and Nicobar islands.

N. p. variegatus: Siberian whimbrel. Breeds in northeastern Siberia from the Verkhoyansk Range to the Kolyma and Anadyr basins, and probably at the northern end of the Sea of Okhotsk. Winters from Taiwan, China, and the Philippines south to eastern India, the Indo-Chinese countries, and Australia and New Zealand.

N. p. hudsonicus: American whimbrel. Breeds in North America from western and northern Alaska east to the western Yukon, northwestern Mackenzie, and the western side of Hudson Bay to James Bay. Winters from central California to the Galapagos Islands and southern South America.

Measurements and weights. Wing: males 230–55 mm; females 224–65 mm. Culmen: males 61–92 mm; females 68–99 mm. Weights (adults in fall): males 360–475 g; females 343–453 g (Glutz et al. 1977). Two males in summer averaged 355 g; 4 females averaged 456 g (Irving 1960). Eggs ca. 57 x 40 mm (*hudsonicus*) to 63 x 43 mm (*variegatus*), estimated weight 46 g (*hudsonicus*) to 59.5 g (*variegatus*) (Schönwetter 1963).

DESCRIPTION

Adults of both sexes have a dark grayish brown crown, divided by a median pale buffy stripe and bounded by a broader pale buffy superciliary stripe. There is also a broad dusky loral stripe and a less distinct postocular stripe, while the rest of the head is streaked with grayish brown. The neck and underparts are white to dull buff, with grayish brown streaking on the neck, chest, and upper breast, and the sides and flanks are cinnamon, barred or spotted with grayish brown, as are the under tail coverts.

114. Breeding distributions of the American (A), Eurasian (E), and Siberian (S) whimbrels.

The axillaries and under wing coverts are rich buffy *(variegatus* and *hudsonicus)* or white *(phaeopus)*, with brownish markings. The back and scapulars are grayish brown with paler marginal spotting, and the lower back, rump, and upper tail coverts are white *(phaeopus)* to grayish brown *(variegatus* and *hudsonicus)* with darker barring. The tail is grayish brown with dusky barring and tipped with whitish. The wing coverts and secondaries are similar to the scapulars, and the primaries and their coverts are dusky, with the inner primaries spotted on their outer webs. The iris is brown, the bill is dark brown, becoming pale brown or pinkish at the base, and the legs and toes are bluish gray. *Juveniles* have a more mottled upper surface, with extensive spotting of whitish or buffy on the feathers. *First-winter* birds retain their buff spots for some time, but the coverts become increasingly notched owing to wear of the spotting (Prater et al. 1977; Ridgway 1919).

In the hand, this medium-sized curlew is the largest of the curlews (wing 233–65 mm) that have strong crown striping, except for *tahitiensis,* which is considerably more cinnamon-toned on its rump, tail,

and underwing surface. Except for *tenuirostris*, it is smaller than the curlews that lack crown striping, and unlike that species it has no white on the secondaries and inner primaries.

In the field (16 inches), this curlew has a strongly striped head, lacks cinnamon on the under wing coverts, and (in *phaeopus*) has a white or whitish patch on the lower back and rump. Its calls include a series of loud, whistled *pip* notes, uttered rapidly six to seven times, and a harsh screech.

NATURAL HISTORY

Habitats and foods. This species breeds in subarctic and subalpine tundra, usually where stunted bushes are present amid a moss and lichen tundra environment, and where marshes are nearby. The birds seem most abundant near the tree line, dropping off rapidly in the open tundra and being absent from boreal forest, though they occur in willow scrub in the forest zone of Scandinavia. In the Churchill area of Manitoba they breed in hummock bogs, sedge

meadows and dry heath tundras (Skeel 1978). They are generally well dispersed in such habitats, and breeding densities typically range from about 0.4 to 2.2 pairs per square kilometer, with higher densities in some exceptional situations (Glutz et al. 1977). Outside the breeding season they are largely associated with muddy or sandy seashores, mangroves, tidal swamps, and estuaries. Sometimes they also occur on rocky shorelines but rather infrequently are found in interior locations except while on migration. Their foods are similar to those of the larger curlews but probably include more items obtained by pecking and fewer obtained by probing. Like the larger species, whimbrels often feed on crabs, but at least in some cases these are killed and partially dismembered before being swallowed, rather than being eaten alive and whole (Burton 1974). Spring and summer foods are especially high in insects such as adult and larval beetles and orthopterans, although as the summer passes the birds progressively begin to feed on ericaceous berries such as cowberries and cloudberries. These vegetarian food sources apparently play a larger role in the late summer diet of whimbrels than is true of the larger and more temperate-breeding species of curlews (Glutz et al. 1977).

Social behavior. Whimbrels arrive on their breeding grounds singly and in small parties. Apparently many birds return to their territories of previous years (Jehl and Smith 1970), and as the nesting areas become snow-free the territorial males begin their advertisement displays and calls. During the breeding season adults utter a variety of calls, including low whistles, low trills, and aerial display songs. The aerial display song consists of the low whistle repeated up to about twenty times, followed by a three-phrase low trill call. During the display the bird ascends quickly until it is 150–300 meters high, then begins making circles about 200 meters in diameter, alternately gliding and climbing again. During the gliding phase the bird usually utters the low whistle followed by the low trill call, and the entire display may last 1–10 minutes. Toward the end of the display the bird loses altitude and descends in a final glide at about a 45° angle until finally leveling off and landing with the wings raised vertically. The aerial display song may begin as early as two days after arrival on the nesting grounds, and it continues with decreasing frequency for more than a month. The low trill call is probably the counterpart of the *curloo* call of the larger species, and when used during the aerial display it probably serves as self-advertisement, while at other times it probably functions in maintaining contact between pair members. A total of ten adult calls have been recently distinguished, though vocalizations associated with copulation were not included in the list (Skeel 1978). Copulation is said to occur on the ground in the vicinity of the nest, but details of it are still unknown.

Reproductive biology. Nests are placed in exposed locations, often in short heather or grass, or sometimes on a hummock in rather dry surroundings well away from the nearest water. Typically the location provides an unobstructed view in all directions. There are normally 4 eggs in the clutch; this was the case with 26 of 35 clutches in the Churchill area (Jehl and Smith 1970) and with 47 of 52 clutches in Finland, with the remaining clutches including 2 or 3 eggs. Incubation is shared by the two sexes and probably begins with the completion of the clutch. The incubation period is 27–28 days, and the young hatch over a period of 1 or 2 days. They are tended by both parents, and fledging occurs in 5–6 weeks (Glutz et al. 1977). Almost as soon as the chicks fledge, the adults leave the nesting areas and begin their long fall

migration, leaving the juveniles to follow a few weeks later.

Status and relationships. There is no good information on the population status of this species as a result of its greatly scattered breeding and wintering areas, but it is perhaps the most abundant of all the curlews in view of its extremely extensive breeding range. Larson (1957) considered *phaeopus* and *tahitiensis* a closely related species pair, in turn fairly closely related to *minutus* and *borealis*. Mayr and Short (1970) suggested that *phaeopus, tahitiensis,* and *tenuirostris* constitute a species group. I believe that *phaeopus* and *tahitiensis* are very closely related and might be considered a superspecies, while *tenuirostris* is more probably derived from early *arquatus* stock.

Suggested reading. Skeel 1976, 1978; Bannerman 1961.

Bristle-thighed Curlew

Numenius tahitiensis (Gmelin) 1789

Other vernacular names. None in general English use; courlis d'Alaska (French); Borstenbrachvogel (German).

Subspecies and range. No subspecies recognized. Breeding has been reported only from near the mouth of the Yukon River, in Alaska. Winters from the Marshall and Hawaiian Islands south to the Santa Cruz, Fiji, Tongan, Samoan, Marquesa, and Tuamotu islands. See map 117.

Measurements and weights. Wing: males 222–30 mm; females 227–52 mm. Culmen (to feathering): males 69–88 mm; females 83–96 mm. Weights: 10 males 254–400 g, average 378 g; 10 females 372–618 g, average 489 g (U.S.N.M. specimens). Lacan and Mougin (1974) provide weights on wintering birds that average somewhat greater than these, as do Johnson and Morton (1976) for summer birds. Eggs ca. 60 x 42 mm, estimated weight 54.8 g.

DESCRIPTION

Breeding adults of both sexes have two broad stripes of dark sooty brown on the crown, separated by a narrower stripe of light buff, a broad superciliary stripe of light buff, a fuscous streak through the

lores, and the sides of the head buff with dusky streaking. The neck and entire underparts are light pinkish buff to very pale buff, with dusky streaking on the neck, lower throat, and chest, and with the sides and anterior flanks barred or spotted with dusky. The axillaries and under wing coverts are cinnamon buff, barred or spotted with fuscous, except for the under primary coverts, which are uniform grayish brown. The back, scapulars, and wing coverts are dark brown with buff spotting, the lower back and rump are nearly uniform dark brown, and the upper tail coverts are light cinnamon buff, paler at the tip, and crossed by about six fuscous bars. The primaries and their coverts are dusky, with some of the inner primaries having large buffy spots on their outer webs, and the secondaries are also extensively spotted with buffy. Compared with related species, there is a strong cast of pinkish or cinnamon throughout. The iris is brown, the bill is blackish toward the tip and dark pinkish at the basal half, and the legs

and feet are bluish gray. *Adults in winter* are similar, but they are more deeply colored, with more cinnamon spotting. Adult females have longer bills than do males, averaging 90 mm in females and 79 mm in males, but there is overlap between 83 and 88 mm. *Juveniles* and *first-winter* birds have coverts that are broadly edged with pale buff and that have large pale cinnamon buff spots. The primary coverts also have large white spots (Prater et al. 1977; Ridgway 1919).

In the hand, birds in breeding plumage can be identified by the bristlelike extensions of the shafts of some of the flank and thigh feathers, and in other plumages by the combination of a fairly long (69–96 mm) and rather heavy bill and a pinkish cinnamon cast throughout, but especially on the tail and upper tail coverts.

In the field (15 inches), this species closely resembles a whimbrel, but it has a cinnamon-tinted tail and under wing lining, compared with a more grayish brown tail and a whitish or buffy under wing lining in the whimbrel. Its calls include a long, drawn-out *aweu-wit,* similar to the call of a gray plover.

NATURAL HISTORY

Habitats and foods. These birds are known only to breed on the flat, dry, exposed tundra ridges east of the coastal range of low mountains in westernmost Alaska and north of the Yukon River. The only nests so far found have been on tundra that, in addition to the usual reindeer moss, also has numerous clumps of black, matted, and hairlike lichens that give the tundra a blackish speckled appearance (Allen and Kyllingstad 1949). The population density even in this limited area is apparently not very great, since within about 25 square miles a maximum of 20 curlews were found during the breeding season. During the winter this species occupies a much wider diversity of habitats, including beaches, sandy areas, grasslands, and areas of lava. It is relatively widespread in Polynesia, including the Tuamotu Islands and the Gambier Islands, where it occurs in small parties over many types of island habitats. An analysis of stomach contents of 14 individuals collected in Polynesia indicated that vegetation was present in 7 stomachs, crustaceans were present in 5, insects in 4, and gastropods and scorpions one stomach each (Lacan and Mougin 1974). One of the most interesting aspects of the foraging of wintering birds is their tendency to steal and eat the eggs of various seabirds, including terns, frigate birds, boobies, and even albatrosses. In the case of the larger eggs, the birds sometimes pick them up and drop them on a hard surface to help crack them, while the thin-shelled tern eggs are simply speared. Sometimes they consume addled eggs that are several months old, but at other times they actually steal eggs from incubating frigate birds. Much less is known of the foods taken on the breeding grounds, but at least some Alaskan birds have been found feeding on blueberries. One individual was also reported to have eaten crowberries (*Empetrum*) (Bent 1929; Stout 1967).

Social behavior. Essentially nothing is known of the social life or pairing behavior of this species, but it is presumably much like the patterns found in the other larger curlews. Apparently the species is not highly social at any time; the maximum number of birds ever seen at any one time seems to have been several hundred seen in early August in the Hooper Bay region. These early fall flocks consist of mixed adults and immatures, and thus the long and precise fall mi-

gration from the breeding grounds to a few scattered islands in the South Pacific is apparently done by the juveniles in the company of their parents or other adult birds.

Reproductive biology. Only two nests of this species have thus far been discovered. The first was found on June 12, 1948, on barren tundra about 20 miles north of Mountain Village. A second nest was located three days later about a mile southwest of the first, on an adjacent ridge. The nests are simple depressions in mossy tundra, with little or no lining added. One of the nests contained 4 pipping eggs; the other had only 2, but presumably the remaining eggs of the clutch had been lost earlier to jaegers. The first nest was beside a clump of black lichen and a mat of alpine azalea, and the eggs blended extremely well with the colors of the surrounding tundra. Both sexes presumably incubate and care for the young, but these details have not been determined, nor are the fledging and incubation periods known (Kyllingstad 1948).

Status and relationships. This must certainly be one of the rarest of the North American shorebirds, and its known nesting range is also the smallest. It is quite probable that other nesting areas exist in the mountain ranges of the lower Kuskokwim and Yukon rivers, or possibly in the mountains near Norton Sound and the Seward Peninsula, since the breeding density reported in the Mountain Village area is certainly too small to support a viable population. As to the species' relationships, it is similar in appearance to *phaeopus,* and Larson (1957) regarded the two as a species pair. Mayr and Short (1970) included it with *phaeopus* and *tenuirostris* in a species group, and I am inclined to favor this interpretation. The downy young appear virtually identical to those of *phaeopus,* and the adults are also extremely similar in plumage.

Suggested reading. Kyllingstad 1948; Allen and Kyllingstad 1949.

Slender-billed Curlew

Numenius tenuirostris Vieillot 1817

Other vernacular names. None in general English use; courlis à bec grêle (French); Dunnschnabelbrachvogel (German); zarapito fino (Spanish).

Subspecies and range. No subspecies recognized. Breeds on the steppes of western Siberia and northern Kazakhstan from the region of Uralsk northeast to the drainage of the upper Ob, and east to the steppes north of Lake Balkhash. Winters in Iraq and the eastern Mediterranean, and in smaller numbers as far as northwestern Africa. See map 113.

Measurements and weights. Wing: males 240–67 mm; females 250–66 mm. Culmen: males 69–89 mm; females 74–96 mm. Weights: 2 first-year females in fall, 255 and 360 g (Glutz et al. 1977). Eggs ca. 65 x 46 mm, estimated weight 69.0 g (Schönwetter 1963).

DESCRIPTION

Adults of both sexes are generally similar to *arquata* in plumage, differing from it primarily in having rather rounded or heart-shaped dark markings on the sides and lower breast and in having pure white axillaries and under wing linings. Like it, the head lacks a dark crown and definite superciliary stripe, and the

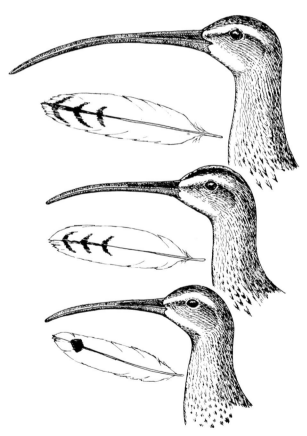

43. Adult heads and flank feathers of curlew (*top*), whimbrel (*middle*), and slender-billed curlew (*bottom*). After Etchécopar and Hüe 1967.

lower back and rump are white, while the upper tail coverts are white with sparse brown spotting and streaking. The tail is white with darker barring. The greater coverts, secondaries, and inner primaries are strongly spotted with white, and the underparts are also virtually white. The iris is brown, the bill is dark brown, becoming pinkish on the lower mandible toward its base, and the legs and toes are bluish gray. *Females* average larger in size and have bill lengths of 76 mm or more, while the males usually have bill lengths of 75 mm or less. *Juveniles* have brown flank streaks rather than heart-shaped spots and have their inner eight or nine primaries tipped with white, rather than only the first six or seven. *First-winter* birds begin to attain some heart-shaped spots by midwinter (Prater et al. 1977).

In the hand, this species is identified by its fairly long (69–95 mm) and unusually slender bill and its moderately long (240–67 mm) wings, which are almost entirely white underneath. The similar *arquata* also has white underwings, but the wings are longer (at least 270 mm).

In the field (16 inches), this species is the same size as a whimbrel, but it lacks the strong head striping pattern of it and the other smaller curlews. It is smaller and has a notably shorter bill than *arquata* and shows more white on the tail and upper wing surface when in flight than does this species, and its call is higher-pitched and shorter.

NATURAL HISTORY

Habitats and foods. The breeding habitats of this little-studied species consist of bogs or swampy areas on the steppe and marshy tracts on the edge of forests. Breeding apparently occurs in small colonies in such habitats, but densities are apparently very low. Since the first nests of this species were found in 1916, nothing more of substance has been learned of the breeding biology or habitats. The winter range includes the Mediterranean Basin as well as adjoining portions of southwestern Asia. It has most frequently been reported during winter in southern Spain and in Algeria, Tunisia, and Morocco, and its range probably extends east to Egypt. At that time of the year the birds are said to inhabit marshes and mud flats, more often in the interior than along the coast, and they perhaps do more wading than is typical of the larger species of curlews. The species' foods are poorly studied, but they apparently include a variety of worms, small insects, and crustaceans (Glutz et al. 1977).

Social behavior. Nothing specific has been written on the displays or sexual behavior of this species, but they are presumably much like those of the other large curlews.

Reproductive biology. The first nest of this species was found in late May, near Tara, USSR, in dense sedge growth and moss on a comparatively dry location. There were 4 highly incubated eggs present. No more nests have been found since the initial one in 1916. Breeding has also been reported near Smeinogorsk, on Lake Chany, and in Chalov district, and it is also believed to nest on the steppes between the lower Volga and Ural rivers (Bannerman 1961). However, no information is available on nesting behavior, incubation periods, or other aspects of the breeding biology.

Status and relationships. This is certainly one of the rarest of the Eurasian shorebirds, and it may well be a threatened species. It is apparently uncommon everywhere, and the only recent record of even moderate numbers of birds seems to be an estimate of 600–900 birds between Merja Zerga and Puerto Cansado, southwestern Morocco, during the winter of 1964 (Glutz et al. 1977). The relationship of the species is probably closest to *arquata,* though it is smaller and considerably paler. However, the eastern race (*orientalis*) of *arquata* shows reduced pigmentation and rather closely approximates the plumage condition found in *tenuirostris.* It thus seems likely that *tenuirostris* is a steppe-breeding derivative of early *arquata* stock. Presumed hybrids between this species and *arquata,* as well as *phaeopus,* have been reported.

Suggested reading. Bannerman 1961.

Curlew

Numenius arquata (L.) 1758

Other vernacular names. Common curlew; European curlew; Eurasian curlew; courlis cendré (French); grosser Brachvogel (German); zarapito real (Spanish).

Subspecies and ranges. See map 115.

  N. a. arquata: European curlew. Breeds from Great Britain and France across western Europe north to about the Arctic Circle and south to the Caspian Sea, eastward through the Ukraine, Crimea, and Sea of Azov to about the Volga, where it intergrades with *orientalis.* Winters from Iceland, the British Isles, and southern Europe through the Mediterranean Basin, Iraq, the Persian Gulf, eastern Africa, and western India.

  N. a. orientalis: Siberian curlew. Breeds in southeastern Russia east of the lower Volga and Ural steppes eastward to Transbaikalia and central Manchuria. Winters from the southern Caspian and Persian Gulf, China, and Japan southward to tropical Africa, the islands of the Indian Ocean, India, the Indo-Chinese countries, and the Greater Sundas.

Measurements and weights. Wing: males 273–305 mm; females 297–324 mm. Culmen: males 95–170 mm; females 116–81 mm. Weights (*arquata,* in breeding season): males 572–779 g, average 662.4 g; 62 females 680–919 g, average 787.7 g (Glutz et al. 1977). Eggs ca. 68 x 47 mm, estimated weight 76 g (Schönwetter 1963).

DESCRIPTION

Adults of both sexes are generally light grayish brown to grayish buffy above, with a more dusky crown, a dull white stripe above the eyes, and a darker loral stripe, and the sides of the head are streaked with dusky. The chin and throat are white, while the rest of the head, neck, and chest are very pale grayish buffy with dusky streaks. The breast, sides, and flanks are more whitish, and the flanks are spotted or barred with grayish brown, while the abdomen and under tail coverts are white, the latter with narrow dusky streaks. The axillaries and under wing coverts are white, sometimes with scattered brownish streaks or bars. The anterior upperparts are heavily streaked or spotted with dusky, while the lower back, rump, and upper tail coverts are mostly white, with dusky streaking limited to the tail coverts. The tail is barred with dull white and pale buffy grayish. The greater coverts and secondaries are deep blackish brown to grayish brown with pale grayish buffy spotting, while the primary coverts and primaries are dusky, becoming more broadly spotted and tipped with white inwardly. The iris is brown, the bill is blackish at the tip, becoming pinkish underneath basally, and the legs and toes are bluish gray. *Females* are larger and have bills that average about 30 mm longer than those of males. *Juveniles* have very extensive buffy edges on the upperparts,

31. Broad-billed sandpiper, adult incubating. *Photo courtesy J. B. & S. Bottomley*

32. Ruff, female incubating. *Photo courtesy J. B. & S. Bottomley*

33. Ruff, female in nuptial plumage. *Photo courtesy J. B. & S. Bottomley*

34. Ruff, male (white phase) in nuptial plumage. *Photo by author*

35. Ruff, male (dark phase) displaying toward female. *Photo by author*

36. Ruff, two males in aggressive display, female in foreground. *Photo by author*

37. Buff-breasted sandpiper, female incubating. *Photo courtesy D. F. Parmelee*

38. Buff-breasted sandpiper, juvenile in fall. *Photo courtesy J. B. & S. Bottomley*

39. Northern phalarope, male incubating. Photo courtesy *J. B. & S. Bottomley*

40. Northern phalarope, female in nuptial plumage. *Photo courtesy J. B. & S. Bottomley*

41. Wilson phalarope, female in nuptial plumage. *Photo courtesy Ed Bry*

42. Wilson phalarope, copulation. *Photo courtesy Ed Bry*

43. Spotted redshank, adult in nuptial plumage. *Photo courtesy J. B. & S. Bottomley*

44. Spotted redshank, juvenile in late fall. *Photo courtesy J. B. & S. Bottomley*

45. Redshank, adult incubating. Photo courtesy *J. B. & S. Bottomley*

46. Lesser yellowlegs, downy young. *Photo by author*

47. Greenshank, adult incubating. *Photo courtesy J. B. & S. Bottomley*

48. Wood sandpiper, adult incubating in song thrush nest. *Photo courtesy J. B. & S. Bottomley*

49. Green sandpiper, juvenile. *Photo courtesy J. B. & S. Bottomley*

50. Terek sandpiper, adult incubating. *Photo courtesy J. B. & S. Bottomley*

51. Black-tailed godwit, adult female in nuptial plumage. *Photo courtesy J. B. & S. Bottomley*

52. Black-tailed godwit, adult male in nuptial plumage. *Photo by author*

53. Marbled godwit, adult in nuptial plumage. *Photo courtesy Ed Bry*

54. Hudsonian godwit, downy young. *Photo by author*

55. Eurasian curlew, adult incubating. *Photo courtesy J. B. & S. Bottomley*

56. Whimbrel, downy young. *Photo by author*

57. Short-billed dowitcher, downy young. *Photo by author*

58. Long-billed dowitcher, juvenile in fall. *Photo courtesy J. B. & S. Bottomley*

59. Eurasian snipe, adult incubating. *Photo courtesy J. B. & S. Bottomley*

60. American woodcock, adult. *Photo courtesy Larry Stevens*

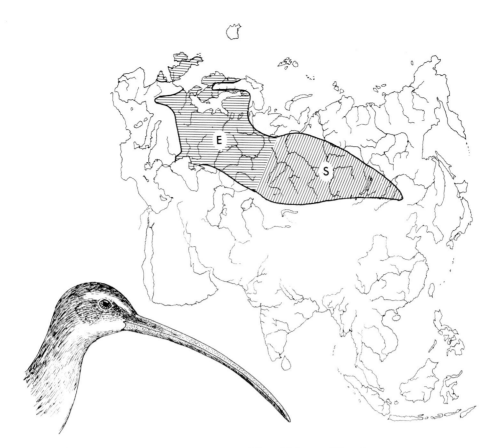

115. Breeding distributions of the Eurasian (E) and Siberian (S) curlews.

and the coverts have bright buffy edges and a distinct brown central mark, compared with grayish brown centers and paler edges in adults. *First-winter* birds are best distinguished by their scapulars, which are heavily notched with buff, while adults have brownish scapulars with distinct grayish brown barring (Prater et al. 1977; Ridgway 1919).

In the hand, this large curlew (wing at least 273 mm) is the only species of such size having a white lower back and rump.

In the field (22 inches), this is the largest of the European curlews, and the only large curlew with a white rump patch. It also lacks heavy striping on the head, and the underwing and axillaries are nearly pure white. Its usual calls are loud *quee* and *cooorwee* notes.

NATURAL HISTORY

Habitats and foods. The breeding habitats of this species include open marshy regions, moors, swampy and dry heathlands, dune valleys, natural and cul-

tivated meadows or grasslands that are not excessively damp, and open arable land. Compared with the black-tailed godwit, it occupies drier, less grassy areas having more heather and scattered trees (Voous 1960). In most habitats the breeding density is less than one pair per square kilometer, but under extremely favorable conditions the density may be as great as 10.4 pairs per square kilometer (Glutz et al. 1977). Outside the breeding season the birds are mostly found on open or muddy shorelines of rivers, lakes, coastal lagoons, and sandy beaches and coastal meadows, but only rarely on rocky shorelines (Voous 1960). The foods taken throughout the year are highly diversified and include earthworms, insects, spiders, crustaceans, mollusks, and amphibians as well as some berries and seeds. Samples taken during the spring, summer, and fall show a high incidence of insects, especially terrestrial forms such as beetles and orthopterans (Glutz et al. 1977). The birds feed almost entirely by day and use pecking, jabbing, and probing movements according to the substrate conditions and the kind of prey being

44. Adult heads and rump feathers of eastern curlew (*above*) and curlew (*below*).

hunted. Crabs of fairly large size (20–25 mm) are effectively captured and eaten, apparently by swallowing them whole and alive, in spite of possible hazards (Burton 1974).

Social behavior. Curlews arrive relatively early on their British nesting areas, with males typically preceding females by a few days. As soon as weather permits, these males begin their song flight displays, consisting of long, undulating flights with alternate ascending and gliding phases and with a short hovering at the top of the ascent, during which trilled, fluty calls are usually uttered. The trill is often preceded by a series of slow *coooo* calls, which increase in speed and pitch, finally becoming a rich descending trill. Although the *coooo* calls are used only during aerial display, the trill is used in other situations, as a greeting signal between the sexes, even outside the breeding season. A soft, repeated *gri-gri-gri* is used by the male during copulation. Near Munich, however, the birds arrive mostly already paired, and apparently they show constancy to their mates and territories of the previous year. In the view of von Frisch (1956), the aerial display and associated calling are a part of territorial advertisement and general excitement and have nothing to do with actual courtship. Territories are defended passively by these advertisement flights and actively by ritualized fighting behavior that rarely leads to actual attack. Paired females will sometimes participate in territorial defense. An important aspect of pair behavior before egg-laying is the scraping ceremony, during which the male makes several nest cups, usually in the presence of the female. Copulation is preceded by the male's approaching the female with withdrawn head and lowered breast. He performs trembling movements with his wings while uttering repeated *quoi* notes and pecking at the female's head and neck. Treading lasts for a surprisingly long period, 4 minutes or more, and during this time the male utters *gri* notes and waves his wings above his back to maintain his balance. The female is passive throughout copulation, and there are no specific postcopulatory displays (von Frisch 1956).

Reproductive biology. Nests are typically placed in quite open situations, often in grass or sedge cover, and sometimes far from water. The eggs are laid at intervals of 1–2 days. There are normally 4 eggs; a sample of 127 Bavarian clutches included 104 of 4 eggs, and 204 of 239 Finnish clutches were also of 4 eggs. Clutches of 3 and 2 are progressively less common, and those of 5 eggs are extremely rare. Incubation is performed by both sexes, which typically exchange places on the nest twice a day. The incubation period is 27–29 days. The fledging period for hand-reared birds may be as short as 28–36 days, whereas wild-reared young are fully fledged between 40 and 50 days (Glutz et al. 1977; von Frisch 1956).

Status and relationships. In much of Europe this species has suffered greatly as a result of hunting and habitat disturbance, but in a few areas such as Finland it has benefited by forest removal and a corresponding increase in open habitat. It is still surprisingly common; wintering ground counts in the late 1960s and early 1970s indicated more than 150,000 wintering birds, with more than a third of them in Great Britain (Prater 1974*b*). The species' relationships are clearly with such forms as *americana*, *tenuirostris*, and *madagascariensis*, which collectively seem to constitute a species group. I suspect that its nearest relative is *madagascariensis*, while the other two forms represent somewhat earlier divergences.

Suggested reading. Bannerman 1961; von Frisch 1956.

Eastern Curlew

Numenius madagascariensis (L.) 1766

Other vernacular names. Australian curlew; long-billed curlew; red-rumped curlew; courlis de Sibérie (French); Sibirischer Brachvogel (German).

Subspecies and range. No subspecies recognized. Breeding range uncertain but probably includes eastern Siberia from Kamchatka and the Sea of Okhotsk west to the upper Nizhnyaya Tunguska and the Vilyui River, and south to northeastern Mongolia, northern Manchuria, and Ussuriland. Records of definite breeding exist only for the Karaga River in northeastern Kamchatka, Manchuria, and the Maritime Territory. Winters from Taiwan and the Philippines south to the Sundas, New Guinea, and Australia. See map 116.

Measurements and weights. Wing: males 290–338 mm; females 302–30 mm. Culmen: males 128–70 mm; females 154–201 mm (Dementiev and Gladkov 1969; Prater et al. 1977). Weights: males 820–1,040 g; females 690–950 g (Shaw 1936); 12 wintering birds ranged 565–1,150 g and averaged 792 g (D. Purchase, pers. comm.). Eggs ca. 70 x 48 mm, estimated weight 80.0 g (Schönwetter 1963).

DESCRIPTION

Adults of both sexes are very similar to those of *arquata*, but the lower back and rump are brown, or the same color as the upper back. Additionally, the underparts are generally tinged with dull ochre and are extensively mottled. The axillaries and under wing coverts are white or buffy and are heavily marked with brownish black. The secondaries and inner primaries are extensively spotted with brownish white, as in *arquata*. The iris is brown, the bill is dark brown, becoming pinkish on the underside toward the base, and the legs and toes are bluish gray. *Females* have somewhat longer bills than do males, but there is much overlap. *Juveniles* have extensive buffy white edgings on their upperparts, especially the wing coverts. The underparts also tend to be more finely streaked than in adults. Usually the inner eight primaries as well as the outer primary coverts are tipped with white (Prater et al. 1977).

116.   Breeding distribution of the eastern curlew, including presumptive range (*broken line*).

In the hand, this species is notable for its extremely long decurved bill and long wings (at least 290 mm), which separate it from all other Old World curlew species except *arquata*. The plumage differences noted above provide for separation from that species.

In the field (23 inches), the extremely large size and strongly decurved bill, together with an unstriped head pattern, eliminate all curlews except *arquata*. The absence of a white rump patch provides for distinction from that species. The calls include a mournful *carr-er*, a more high-pitched *kerlee*, and a repeated whistling *tyui* during courtship.

NATURAL HISTORY

Habitats and foods. On its wintering grounds, this species occurs on tidal estuaries, mud flats, mud under mangroves, and soft sandy beaches. The birds apparently live largely on worms and small crabs (Frith 1976). However, on the breeding grounds they are associated with mossy peat marshes, reedgrass thickets on slightly boggy soil where there are burned-out or bare areas, and grassy marshes. Moorlands where the ground is mossy or slightly swampy, and where there are only scattered bushes, seem to represent the primary habitat. The foods there consist of the larvae of beetles and soldier flies, amphipods, and even frogs and crabs. During fall migration the birds sometimes also eat berries (Dementiev and Gladkov 1969).

Social behavior. On migration and in their wintering areas this species occurs in flocks of up to several hundred birds, especially on the east coast of Australia. At these times the birds associate with whimbrels, but the differences in size and bill length probably preclude much direct competition or social interaction between them. The displays and pair-forming behavior of this species are not well studied, but apparently courtship occurs among spring migrants in mid-April. At that time the male performs aerial displays in which he rises 10–15 meters, beating his wings slowly, and utters a repeated *tyui, tyui, tyui,* which gradually becomes more rapid as he slowly descends on outspread wings. Territoriality is apparently not well developed, since the birds are said to breed in small colonies of 2–3 pairs (Dementiev and Gladkov 1969).

Reproductive biology. The breeding ecology of this species is apparently much like that of *aquata* and *americana*, though the birds nest in more moist or boggy environments. The nests are usually on small mounds and are lined with dry grass. The clutch is normally 4 eggs. Maritime Territory nests with eggs have been found in early May, and young birds have been seen later that month, while on Kamchatka nests have been found as late as June 21. Fledged young have been noted around Lake Khanka, Maritime Territory, in late June, with flocking occurring by early July (Dementiev and Gladkov 1969).

Status and relationships. Apparently this species is not numerous, and it is irregularly distributed. It apparently suffers from the marsh burning that occurs in the Maritime Territory in spring, but it evidently is protected from hunting along much of its migratory route and on its Australian wintering grounds. Like the Eurasian curlew, it is apparently hunted for food in China (Chen Tso-hsin 1973), but there is no information on the size of the harvest there. The species is obviously a very close relative of *arquata* as well as *americana*, being somewhat closer to *arquata* in its under wing coloration but more heavily barred with brown on the back, rump, and axillaries. Dementiev and Gladkov (1969) indicate that specimens of *arquata* from the eastern part of its range begin to show some of these characteristics, suggesting intergradation or hybridization, but they urge that the two be retained as separate species. In any case it is evident that *arquata* and *madagascariensis* are very closely related to one another and to a lesser degree to *americana*.

Suggested reading. Dementiev and Gladkov 1969.

Long-billed Curlew

Numenius americanus Bechstein 1812

Other vernacular names. Sicklebill; courlis à long bec (French); Amerikanischer Brachvogel (German).

Subspecies and ranges. See map 117.

N. a. americanus: Greater long-billed curlew. Breeds from Nevada, Idaho, Wyoming, and South Dakota south to Utah, New Mexico, Texas, and northwestern Oklahoma, formerly to Kansas, Iowa, and Wisconsin. Winters from California, Texas, and South Carolina south to Mexico and Guatemala.

N. a. parvus: Lesser long-billed curlew. Breeds from British Columbia, Alberta, Saskatchewan, and perhaps Manitoba south to California, Nevada, Wyoming, and South Dakota. Formerly bred widely in North Dakota; now rare and limited to the southwestern corner. Winters from California east to Louisiana and south to Mexico. (Also known as *occidentalis*, as in Ridgway 1919.)

Measurements and weights. Wing: males 253–88 mm; females 251–308 mm. Culmen (to feathering): males 105–55 mm; females 118–219 mm. Weights: 12 males 445–792 g, average 640.1 g; 14 females 630–951 g, average 758.6 g (M.V.Z. specimens). Eggs ca. 65 x 46 mm, estimated weight 73 g (Schön-wetter 1963).

117. Breeding distributions of the bristle-thighed curlew (*shaded*) and the greater (G) and lesser (L) long-billed curlews.

DESCRIPTION

Adults of both sexes are generally light pinkish cinnamon, with a dusky black crown, streaked with buff and with buff sometimes forming a median streak posteriorly, an ill-defined pale line above the eyes and a dusky stripe across the lores, and the rest of the head streaked with dusky. The sides of the neck and the lower throat and breast are finely streaked with black, while the sides and the flanks are narrowly barred with dusky and the rest of the underparts are pale cinnamon. The axillaries and under wing coverts are deeper pinkish cinnamon, with sparse dusky markings. The back and scapulars are striped and barred with dusky grayish brown, buffy white, and pale cinnamon, the barring heaviest on the upper tail coverts. The tail is similarly barred with dusky brown and pale cinnamon. The wing coverts are margined, spotted, barred, or streaked with grayish brown, buffy gray, and pale cinnamon, while the exposed parts of the primaries and their coverts are dusky black. The inner webs of the two outer primaries are barred with cinnamon, while the inner primaries and secondaries are barred with pale cinnamon. The iris is brown, the bill is dusky toward the tip and pinkish to grayish pink basally on the underside, and the legs and toes are grayish. *Females* are identical in plumage to males, but the length of the bill is appreciably greater in females than in males. This difference is readily evident in the field (*Inland Bird Banding News* 50:15), and the average is about 40 mm in both subspecies (Ridgway 1919). *Juveniles* have tertials and scapulars with broad brown central marks and cinnamon buff notches (rather than narrow brown central marks and dull grayish buff pale areas), and most of their primaries have noticeable pale tips (Prater et al. 1977).

In the hand, this is the only large curlew (wing at least 250 mm) that has deep cinnamon under wing linings and axillaries.

In the field (20 inches), this very large curlew is the largest North American shorebird, and the only North American curlew with cinnamon-colored under wing coverts. The marbled godwit has very similar underpart coloration, but its bill is essentially straight rather than decurved. Unlike the smaller North American curlews, the head lacks a strongly streaked pattern. The usual flight call is a plaintive *curlew,* sometimes prolonged into a roll or rattle.

NATURAL HISTORY

Habitats and foods. In North Dakota, this species occupies a variety of grassland habitats including short-

grass, grazed mixed-grass prairies, and shortgrass prairies that have cactus and an open scrub layer of sages. Habitats on gravelly soils and gently rolling terrain seem to be favored over others (Stewart 1975). Maximum densities in favorable habitats may reach as high as 1.5 pairs per square kilometer but are usually under 1.0 pair (Fitzner 1978). Outside the breeding season the birds concentrate in much wetter environments, especially the edges of shallow inland and coastal waters, the open parts of marshes, the intertidal zone, and sandbars. Both on their breeding grounds and while on migration the birds tend to be upland foragers, but on migration they also concentrate on ocean beaches or the edges of large lakes, sometimes feeding in company with marbled godwits and wading out considerable distances. They often forage on rather large, burrow-dwelling crustaceans such as mud crabs and shrimps (Stenzel et al. 1976). Their foods are extremely diverse and include crustaceans, mollusks, worms, toads, the adults and larvae of insects, and sometimes even berries (Bent 1929). They are also effective predators on nesting birds (Sadler and Maher 1976). The birds forage both by probing and by pecking, but pecking seems to be the more common method on the breeding ground. Probing is done directly in front of the bird or toward one side, and a good many head movements are used at such times. When probing in sandy soil the birds sometimes violently jab into holes or depressions in the sand or pause before a burrow and then suddenly thrust the bill into it. At times they will also probe the bases of shrubs or poke among the branches with their bills. Evidently they need little or no direct water on the breeding ground, getting almost all their water from their diet (Fitzner 1978).

Social behavior. Breeding territories in this species are well defined and tend to be occupied from year to year by the same birds. The birds arrive on their territories either already paired or as unpaired males

that soon begin advertising for mates. Territory size varies with topography and cover and ranges from about 20 hectares in flat and open habitats to 6–8 hectares in more diversified habitats. Territorial defense and advertisement are performed by the male, and one of the major associated displays is the "bounding-SKK flight." In this display the male ascends almost perpendicularly, then glides with the wings curved downward, sometimes almost touching the ground before rising again. It utters a series of soft *kerr, kerr* (SKK) notes that are repeated in series and sound like the noise made by blowing through a pipe, beginning at the top of the ascent phase and continuing until the end of the gliding phase. This flight apparently serves to advertise a male's unpaired status. Ground calling is also performed and consists of drawn-out *whee* notes that vary in form. It is sometimes used by a member of a pair, but it usually serves to attract males to unpaired males. Pairs are apparently formed by females' visiting a variety of territories until each finally apparently chooses one territory and remains within it. Birds that arrive already paired begin their courtship activities with scraping and precopulatory displays. Paired birds perform a scraping ceremony during which the usual nest-scraping behavior is performed on a variety of potential nest sites throughout the pair's territory. Copulatory behavior typically follows the scraping ceremony. Precopulatory behavior begins with the male's running toward the female with his neck retracted and his wings held slightly away from the body. He then takes a position behind the female with the tail slightly cocked and the wings increasingly raised, as he begins a series of shaking movements of his head, causing his bill to rub over the female's shoulder feathers, ruffling them on each side of the neck. These shaking movements gradually increase as the wings are raised above the back, where they are fluttered in a jerky manner, and the male gradually straddles the female's tail and makes climbing movements with his feet. Throughout this phase the mate utters a call that sounds like a chorus of tree frogs, and finally he flutters up the female's back. He does not grasp her head, but balances by holding his wings above his back. Calling continues after mounting, which may last about a minute. There is no specific postcopulatory display (Fitzner 1978).

Reproductive biology. Nest sites are selected that offer a combination of relatively flat areas, fairly short grass cover averaging about 100–200 mm in height, with well-spaced grass clumps, and close proximity to a conspicuous object such as a shrub, rock, or dirt mound. Nests are usually well separated, though pairs may nest within sight of one another and in dense populations nests may be as close to each other as 250 meters. Eggs are laid at intervals of 1–2 days, and clutches are almost invariably 4 eggs, though a few 5-egg clutches have been reported. Apparently the male begins incubation after the laying of the last egg, and within a few days an incubation schedule has been established, with the male incubating at night and the female during the day. Incubation requires 27–28 days, and hatching is usually synchronous, with all the chicks emerging within about 5 hours. If hatching occurs late in the afternoon, the brood is likely to remain on the nest through the night, leaving the next morning. They are brooded by both sexes and soon become highly mobile. Females usually abandon their broods when they are 2–3 weeks old, but the male continues to guard them until they fledge, at 41–45 days of age (Fitzner 1978).

Status and relationships. The breeding range of this species has retracted considerably in this century,

and the bird is now absent from several midwestern states (Kansas, Iowa, Minnesota, Wisconsin, Illinois) where breeding once occurred. It is apparently also extirpated from Manitoba and southeastern Saskatchewan. The breeding range in North Dakota is also greatly restricted, and current populations are widely scattered pairs or small groups (Stewart 1975). It has apparently also decreased considerably in South Dakota in the past fifty years, but it is still a locally common breeder in the Nebraska Sandhills (Johnsgard 1979). The western populations have apparently not suffered as greatly as the more easterly ones. As to its evolutionary relationships, it is part of a group of similar and allopatric species that includes *arquata* and *madagascariensis*, which probably constitute a species group or superspecies. Mayr and Short (1970) regarded *americana* and *arquata* as a superspecies but were uncertain as to the affinities of *madagascariensis*. I suspect that both *madagascariensis* and *tenuirostris* are derivatives of early *arquata* stock, and that *americana* resulted from an earlier separation.

Suggested reading. Fitzner 1978; Bicak 1977.

Genus *Limosa* Brisson 1760 (Godwits)

Godwits are medium to large temperate or arctic-breeding shorebirds with a bill that is straight to slightly recurved and is longer than the tarsus. The tarsus is scutellated in front and behind, the hind toe is well developed, and there is a distinct web between the outer and middle toes, but webbing between the middle and inner toes is lacking or greatly reduced. In some species the inner edge of the claw of the middle toe is pectinated. The downy young are similar to those of *Numenius* but have a solid dark crown patch and an unbranched middorsal stripe that is continuous with the diamondlike marking on the back. The sexes are nearly alike as adults. Four species are recognized here.

KEY TO SPECIES OF *Limosa*

A   Bill straight and only slightly flattened toward tip . . . . *limosa*
A′ Bill recurved and distinctly flattened toward tip
  B   Tail black, with white base and tip . . . . *haemastica*
  B′ Tail barred
    C   Upper tail coverts pale cinnamon with black barring . . . . *fedoa*
    C′ Upper tail coverts white with dusky spotting . . . . *lapponica*

Black-tailed Godwit

Limosa limosa (L.) 1758

Other vernacular names. None in general English use; Barge à queue noire (French); Uferschnepfe (German); aguja colinegra (Spanish).

Subspecies and ranges. See map 118.
  L. l. islandica: Islandic black-tailed godwit. Breeds on Iceland; sometimes also on the Faeroes and perhaps northern Norway. Winters in Ireland and perhaps elsewhere.
  L. l. limosa: Eurasian black-tailed godwit. Breeds in Europe and Asia from Belgium north to southern Sweden, south to the southern Urals and the Kirgiz steppes, and east in western Siberia to the upper Chulym River. Winters from the British Islands south to tropical Africa, Iraq, and the Persian Gulf east to western India.
  L. l. melanuroides: Oriental black-tailed godwit. Probably breeds from Mongolia and perhaps the upper Yenisei east to the Sea of Okhotsk and Sakhalin, and from Anadyrland south to Kamchatka. Winters from Taiwan and the Philippines south to India, the Indo-Chinese countries, Malaya, New Guinea, and Australia.

Measurements and weights. Wing: males 158–228 mm; females 180–240 mm. Culmen: males 67–127 mm; females 73–122 mm. Weights (*limosa*, late spring): males 280–440 g; females 350–500 g (Glutz et al. 1977). Eggs ca. 53 x 37 mm, estimated weight 34–39 g (Schönwetter 1963).

DESCRIPTION

Breeding adult males have a cinnamon head, neck, and chest, with dusky streaking on the sides of the head and neck and with a paler chin, throat, and supraloral stripe. The rest of the underparts are white, with dusky barring on the breast, sides, and sometimes also the abdomen and under tail coverts. The axillaries and under wing coverts are white. The back and scapulars are mixed with black, cinnamon, and grayish, and the wing coverts are brownish gray, the greater coverts broadly tipped with white, forming with the white bases of the secondaries a broad white wing stripe. The primaries are dusky, and some have white bases. The rump, longer upper tail coverts and base of the tail are white, with the white increasing on the outer tail feathers. The tail is otherwise black, tipped with white. The iris is brown, the bill is blackish at the tip and orange to pinkish on the basal half, and the legs and toes are dark grayish. *Females* have longer bills and usually have little chestnut and fewer blackish bars on the underparts. *Adults in winter* have the head, neck, back, and scapulars nearly plain brownish gray, the head and neck more grayish, the chest unstreaked gray, and

118. Breeding distributions of the Eurasian (E), Islandic (I), and Oriental (O) black-tailed godwits and the Hudsonian godwit (G).

the wing coverts grayish brown. *Juveniles* are dark brown on the upperparts, with pale chestnut and whitish feather edging, and the wing coverts are brown with reddish buff fringes. The neck and breast are reddish buff, and the abdomen is white. *First-winter* birds resemble adults, but the buff-fringed inner median coverts are retained for varying periods (Prater et al. 1977).

In the hand, the bill of this species is of similar length but less recurved than that of *lapponica* or *haemastica.* The strong black tail band and white wing stripe separate it from *lapponica,* while the white under wing lining distinguishes it from *haemastica.*

In the field (16 inches), this species closely resembles the bar-tailed godwit in size and appearance, especially in winter, but it has a grayer head and breast, longer legs and bill, and in particular a black-and-white banded tail. In flight a conspicuous white wing stripe is evident, which is lacking in the bar-tailed godwit. The usual flight call is a loud *wicka-wicka-wicka,* and, unlike the otherwise similar Hudsonian godwit, the under wing lining is white.

NATURAL HISTORY

Habitats and foods. This species breeds well to the south of the bar-tailed godwit, mostly in temperate climatic zones but extending into steppe and boreal climates. Its preferred habitats are wet or dry meadows with high grass and soft soil, including cattle pastures, hayfields, swampy heathlands, grassy shores of lakes and ponds, and meadows in river valleys or forest clearings. Generally the birds are associated with a cover of turf and soft mud, and less often they are found on sandy ground. In favorable habitats of the Netherlands the breeding density is often 0.4–0.6 pairs per hectare, while on certain preserves it may reach 2–3 pairs per hectare (Bannerman 1961). Outside the breeding season it is also associated with muddy substrates, including the

45. Winter-plumage heads and half-tails of black-tailed (*above*) and bar-tailed (*below*) godwits.

flights" by males become frequent. These flights begin with a steep ascent while the bird utters a loud trisyllabic call. On reaching 150–200 feet, the bird levels out and alters the call to a low disyllabic note, while the wingbeats are slower and flight becomes somewhat undulating as the bird pitches from side to side, spreading his tail. This phase is often performed in a circuitous route 150–300 yards in diameter, but its direction and duration are quite variable. Then the bird stops its calling and rolling flight, and suddenly nose-dives toward the ground with wings and tail almost closed. About 50 feet above the ground the wings are opened and the bird sideslips downward, finally spreading the tail and raising the wings vertically just before landing. A second type of display flight is the "joint flight," performed by both sexes. In this flight the female typically takes the lead, with both sexes calling, though the male calls far more than the female. This flight may take the form of a pursuit, especially when other males join in. Additionally, pursuit flights are performed toward other godwits that fly over a male's territory, but fighting is relatively infrequent. There are several threat display postures performed toward rivals on the ground, including tail-spreading, ruffling of the back feathers, and bill-crossing. Displays toward females include several of the same elements, including ruffling of the back feathers and tail-spreading, sometimes accompanied by wing-raising. A nest-scrape ceremony is also present, during which a male runs to a depression, crouches in it, and tilts his tail high in the air while rubbing his breast against the ground. Sometimes the female will scrape alternately with the male in the same location, and such scrapes are usually close to eventual actual nest sites. Copulation is preceded by the female's standing fairly motionless or slowly walking forward while the male approaches from behind. He spreads his tail, utters a disyllabic note, and begins vibrating his wings, eventually fluttering above the female with legs dangling and wings still vibrating. He lands on the female, sometimes still calling, and copulation follows. Postcopulatory behavior lacks specific displays and often is marked by a nervous picking at the ground (Huxley and Montague 1926; Lind 1961; Kirchner 1969).

Reproductive biology. These godwits tend to be somewhat colonial in their nesting, and nests frequently may be found only 10–30 yards apart. Nests are usually in rather luxuriant grass cover, often 20–30 centimeters high, but in sandy areas the cover

shorelines of lakes, rivers, and estuaries, in both fresh and saltwater environments (Voous 1960). The species has longer legs and tends to probe in deeper water than does the bar-tailed godwit, with which it often associates on wintering areas, and its foraging is done in a more leisurely manner. It also evidently relies more on nonvisual methods of foraging than does that species. Both eat rather large numbers of moderate to large polychaete worms and bivalves (Burton 1974). However, this species' diet is quite diversified, and it also eats earthworms, crustaceans, insects, and seeds. Besides probing in soft ground, the birds also pick prey from stems of grasses or other plants. On the breeding grounds grasshoppers and other orthopterans are sometimes prevalent in the diet, and land snails and slugs have also been reported at times.

Social behavior. Although the birds arrive on the breeding grounds in groups of 5–30 individuals, these soon disperse, and song flights or "ceremonial may be much less and the nest may be sheltered by scrubby vegetation. There are normally 4 eggs in the clutch; of a sample of 145 clutches, 125 had 4 eggs and the rest had 3 (Haverschmidt 1963). Clutches of 5 are very seldom found, and larger clutches are almost certainly the result of two females laying in one nest. Replacement clutches after the loss of the initial one are fairly frequent and occur 5–16 days after the loss. The second nesting may occur 80–640 meters from the initial nest location. Eggs are laid on successive days. Incubation begins with the third or fourth egg and is performed by both sexes, with the male probably taking the greater share. Estimates of the incubation period range from 22 to 27 days, but the usual period is probably 23–24 days. The brood is tended by both parents, and under natural conditions the fledging period has been estimated to range from 30 to 35 days, while in captivity it may occur as early as 25–30 days after hatching (Glutz et al. 1977).

Status and relationships. Although the range of this species in Europe has greatly diminished in historical times (it was even absent as a breeding species from Great Britain for about a century), since the early 1950s it has bred in the Ouse Washes, and in the last few years there have been some minor range extensions. The population of the Icelandic subspecies has also increased since about 1920 (Voous 1960). As I note in the account of *haemastica*, it has sometimes been suggested that the North American form should be considered conspecific with *limosa*. Yet, as Voous (1960) points out, the two species occupy very different breeding habitats and also differ in wing plumage patterns, and thus there seems to be little reason at present for merging them. However, they clearly constitute a superspecies. Collectively, the godwits are probably not very distantly removed from the curlews (Jehl 1968*a*), though Ahlquist (1974) questioned this and instead suggested that *Limosa* may be closer to *Limnodromus* and *Calidris* than to either *Numenius* and *Pseudoscolopax*, with the latter leading to the typical dowitchers.

Suggested reading. Lind 1961; Haverschmidt 1963.

Hudsonian Godwit

Limosa haemastica (L.) 1758

Other vernacular names. None in general English use; barge Hudsonienne (French); Amerikanische Uferschnepfe (German); becasa de mar, aguja de mar (Spanish).

Subspecies and range. No subspecies recognized; sometimes considered a subspecies of *L. limosa.* Breeds locally from southern Alaska (Cook Inlet) and northwestern Mackenzie to northeastern Manitoba; possibly also on Southampton Island and Akimiski Island, Hudson Bay. Winters in southern South America and on the Falkland Islands. See map 118.

Measurements and weights. Wing (chord): males 196–203 mm; females 195–225 mm. Culmen (to feathering): males 69–92 mm; females 67–99 mm. Weights: 6 males 196.5–266 g, average 221.8 g; 6 females 246–358 g, average 288.8 g (Jehl and Smith 1970). Eggs ca. 55 x 37 mm, estimated weight 37.5 g (Schönwetter 1963).

DESCRIPTION

Breeding adults of both sexes have the head and neck dull cinnamon buff to pale buffy, with dusky streaking, which is heaviest on the crown and absent or nearly so on the chin, on the upper throat, and above the lores and eyes. The mantle feathers are black to fuscous, with buffy or cinnamon spotting or barring, the wing coverts are brown, sometimes spotted with buffy and cinnamon, and the greater coverts are tipped with white. The inner primaries and secondaries are dusky grayish brown, and the outer primaries and their coverts are darker grayish brown or fuscous. The lower back and rump are plain

grayish brown to fuscous, and the upper tail coverts are white anteriorly and black posteriorly. The tail is black, margined terminally and basally with white, with the white increasing outwardly. The underparts of the body are light russet to chestnut red with irregular dusky barring, sometimes with narrow white tips on the feathers. The under tail coverts are mixed white and light russet, with dusky barring, and the axillaries and under wing coverts are mostly sooty brown to fuscous, some of the wing coverts being margined with white. The iris is brown, the bill is blackish at the tip and pinkish toward the base, and the legs and toes are light bluish gray. *Females* have paler, more blotchy underparts, longer bills (usually at least 85 mm), and longer tarsal measurements (usually at least 67 mm), and they also tend to be paler on the mantle and to have more white on the undersides. They also have a purplish flesh bill color in spring, compared with a clear, bright orange in males (*Wilson Bulletin* 80:251). *Adults in winter* are uniformly grayish brown on the upperparts, and the wing coverts are grayish brown with white fringes on the inner coverts. The underparts are whitish, with a grayish brown wash on the upper breast and neck. *Juveniles* are brownish black above, with buffy to whitish fringes on the coverts and upperparts. The tertials and scapulars are broadly barred with buff and brown. *First-winter* birds resemble winter adults but retain buff-fringed inner median coverts until spring (Prater et al. 1977; Ridgway 1919).

In the hand, this form may be separated from the other godwits by the combination of its black tail, which is white basally and at the tip, and its more recurved bill, which is slightly flattened at the tip. The tail pattern is much like that of *limosa,* but that species has white rather than blackish underwings and axillaries.

In the field (15 inches), this godwit in spring plumage is distinctively chestnut-colored on the neck and underparts, while in fall or winter the white tail coverts and base of the tail provide for certain identification. In flight, the blackish under wing lining is conspicuous. The call given in flight is a low, double *ta-it* or *toe-wit'.*

NATURAL HISTORY

Habitats and foods. Judging from studies in the Churchill area of Manitoba, the breeding habitat of this species consists of ecotone areas where woods and scrub tundra intermix, and where wet sedge meadows, lakes, or ponds are in close proximity. Ac-

cording to one observer, a combination of extensive sedge marshes and meadows lying at the northern edge of tree line and not far from a tidal coast are the essentials of breeding habitat (Hagar 1966). Another description mentions a preference for fairly dry sedge meadows dotted with small birches and situated near the tree line (Jehl and Smith 1970). In winter the birds are found in aquatic habitats ranging from saline to fresh water, and from coastal to interior locations. In coastal Argentina the birds are usually found in swamps, streams, tidal flats, and shallow estuaries, foraging in flocks of up to about 30 birds. A few birds defend winter foraging territories, but these may be defended for only a few days or as long as several weeks (Myers and Myers 1979). The birds often forage by wading in water 4–6 inches deep, feeding much like dowitchers. They readily swim and thus can easily cross from one shallow area to another. Their foods are essentially unstudied but probably include worms, insects, mollusks, crustaceans, and other small marine life (Bent 1927).

Social behavior. Godwits arrive in the Churchill area about the first of June and typically spend a day or two sleeping and apparently resting from their long migration before actively beginning display behavior. Territories are extremely hard to define because of the birds' tendency to range widely over the tundra, but, even in the absence of frequent obvious territorial conflict between males, it is likely that localized social dominance does occur. The two most likely reflections of territoriality are song flights and pursuit flights. The former are performed by males and typically begin by the bird's performing a strong ascent. He utters low calls for about the first 100 feet of elevation but is silent as he continues to climb in a series of spirals or figure eights to up to 800 feet. Then he levels off and, while performing normal wingbeats, begins a series of *toe-whit'* calls gradually increasing in intensity. He may cover a half mile before turning back, and as he approaches the starting area he begins to glide while holding the wings upward at a 45° angle, making wide and irregular circles and calling in a series of high-intensity *toe-wit'* notes as he loses altitude. Suddenly the calling stops and the bird plummets downward with closed wings until he is within 40 feet of the ground. Then, spreading his wings, he glides to a graceful stop on a tussock. As he lands he raises his wings and then folds them deliberately. The next flight may occur after 20 minutes or more has elapsed. Pursuit flights involving a female and one or more chasing males are

erratic and spectacular chases over the tundra, usually started by two birds, but often joined in by several extra males, the males all shrieking their *toe-wit'* calls. In one instance a pursuit flight involving two males and a female terminated when one male broke away to begin a song flight. The second male soon veered away and landed in the tundra, while the female turned and landed at the top of a spruce near the first male, which by then was now calling and waving his wings at the climax of his display. Instead of plummeting to the ground, he resumed normal flight toward the female and landed on her back, and copulation followed. Although this occurred very late in the season and was probably not a typical copulation, it suggests the importance of the song flight as a stimulant to sexual behavior (Hagar 1966).

Reproductive biology. A typical nest of this species is a depression about 5 inches across, in or under the edge of a dwarf birch on the dry top of a hummock in a sedge marsh. Some nests are placed in grass or sedge tussocks, and sometimes they are under low willow bushes or, rarely, beside a fallen spruce. In any case the cover is usually at least 5 inches high and the nest is well hidden except from directly above (Hagar 1966). The clutch is almost invariably 4 eggs; 14 of 15 clutches in the Churchill area were of this number (Jehl and Smith 1970). Both sexes incubate, with the female typically on the nest during the day and the male at night. Incubation probably begins with the laying of the third egg and lasts 22–25 days, with the period for one clutch determined to be with-

in 5 hours of 23.5 days. The fledging period is approximately 30 days, and through the male usually is directly involved in tending the chicks the female usually guards the family by keeping a lookout from the top of a nearby tree. When the adults no longer guard them, the young drop out of sight for about 10 days, then reappear on the shorelines, fully feathered and flying (Hagar 1966).

Status and relationships. Although until as recently as about 1960 this species was regarded as threatened and perhaps on the way to extinction, this assessment was based on incomplete knowledge, and it is now known that the species is fairly common over a specialized, but not particularly restricted, range in central and northwestern Canada (Hagar 1966). The species is clearly a close relative of the Eurasian *limosa* and has at times been considered conspecific with it. However, the more reasonable approach on the basis of plumage and ecological differences is to consider the two as distinct species (Voous 1960), though they almost certainly constitute a superspecies (Mayr and Short 1970).

Suggested reading. Bannerman 1961; Hagar 1966.

Bar-tailed Godwit

Limosa lapponica (L.) 1758

Other vernacular names. Barred-rumped godwit; barge rousse (French); Pfuhlschnepfe (German); aguja colipinta (Spanish).

Subspecies and ranges. See map 119.

L. l. lapponica: Western bar-tailed godwit. Breeds in Lapland from Norway west to Sweden and northern Finland, west through the Kola Peninsula and southern Kanin Peninsula to the Taimyr Peninsula, where it intergrades with *baueri*. Winters from the North Sea and Iberian Peninsula east to the Persian Gulf and India, and south to South Africa; sometimes strays to the eastern coast of North America.

L. l. baueri: Eastern bar-tailed godwit. Breeds in northeastern Asia from about the Taimyr Peninsula east to the Chukotski Peninsula, and in North America from Point Barrow east to the Colville River and south to the Kuskowim delta. Includes *menzbieri,* which is not recognized by Vaurie (1965). Winters from China, Taiwan, and the Philippines south to the Sundas, Australia, New Zealand, and adjoining islands.

Measurements and weights. Wing: males 200–42 mm; females 214–56 mm. Culmen: males 69–94; females 75–108 mm. Weights (of *lapponica,* breeding birds): males 240–317 g; females 300–400 g (Glutz et al. 1977). Eggs ca. 53 x 38 mm, estimated weight 37–40 g (Schönwetter 1963).

DESCRIPTION

Breeding adults of both sexes have the head, neck, and underparts cinnamon to light cinnamon, with dusky streaking on the crown, hindneck, and sides of the neck and with a paler chin, throat, and supraloral stripe. The color of the underparts deepens to orange cinnamon on the breast and abdomen, and some of the feathers of the sides and underparts are streaked with sepia, while the under tail coverts and sometimes the flanks are irregularly marked with sepia. The axillaries and under wing coverts are white with ashy brown markings. The feathers of the mantle are notched and edged with pinkish cinnamon and have

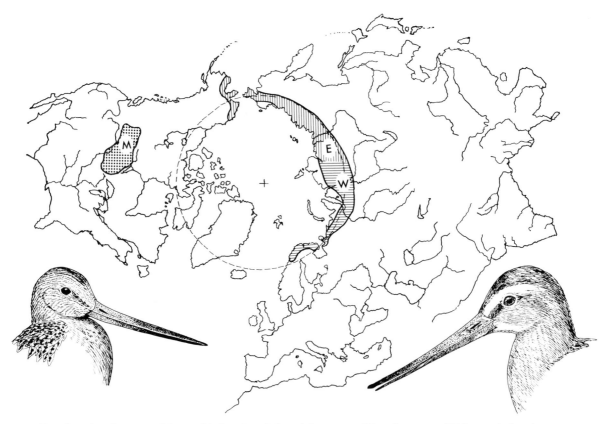

119.   Breeding distributions of the marbled godwit (M) and the eastern (E) and western (W) bar-tailed godwits.

blackish centers, while the upper tail coverts are white with variable sepia barring (heavier in *lapponica*, white predominating in *baueri*). The tail feathers are ashy brown and mottled or barred with white. The wing coverts are mostly grayish brown, with paler edges, and the greater and median coverts are broadly edged with white. The primary coverts and primaries are blackish, the inner primaries being edged with white. The secondaries are sepia with white edging, and the inner feathers are mottled with white basally. The iris is brown, the bill is dark brownish, becoming pinkish on the basal half, and the legs and toes are greenish gray. *Females* average larger in size and tend to be paler and more barred on the breast. *Adults in winter* have grayish brown upperparts with darker shaft streaks. The head, neck, and underparts are dull whitish, with slight grayish brown barring on the chest and sides. The coverts are brownish gray, with paler tips and whitish edges or fringes on the inner median coverts. *Juveniles* have brown upperparts with broad buff borders and brown coverts with buffy fringes. The underparts are whitish, with a strong buffy wash on the breast. *First-winter* birds resemble adults but retain some juvenile buff-edged inner median coverts and tertials, and they have central tail feathers that are strongly barred with brown and white (Prater et al. 1977; Witherby et al. 1941).

In the hand, the long (69–108 mm) and pinkish-based bill, which is slightly recurved, and the brown barring on the rump, upper tail coverts, and tail serve to separate this species from other godwits.

In the field (15 inches), this species differs from the generally similar black-tailed godwit and the Hudsonian godwit by the lack of a black band across the tail and by the absence of a white wing stripe. It is also more extensively white on the underparts in the spring plumage than is the black-tailed godwit, and it has less evident barring on the sides and flanks. In flight it utters *kirruc, kirruc* notes and sometimes also repeated *wik* notes.

NATURAL HISTORY

Habitats and foods. The breeding-season habitats of this species consist of marshy places in moss and shrub tundra, swampy heathlands in the willow and birch zones near the tree line, and occasionally open bogs in the extreme northern portions of the coniferous forest belt (Voous 1960). In the Kaolak River area of northern Alaska the birds occur in various tundra types, but at least at times they are associated

with jaeger territories. They also often occur near willow-edged river habitats, and they often forage along alluvial bars, apparently for land snails (Maher 1959). In a study area of about 10 square miles, about 6–8 pairs bred in each of two years, or nearly one pair per square mile. Breeding populations of 0.1–1.0 pairs per square kilometer have been estimated in Asia (Glutz et al. 1977). Outside the breeding season the birds occupy muddy coastlines and estuaries having mud flats or sandbars within the tidal zone (Voous 1960). When foraging, the species often wades in water as deep as it possibly can, often submerging the entire head and neck, walking forward and picking up crustaceans, mollusks, fishes, and insects. It also probes deeply in mud or sand. This species apparently relies more on vision for its foraging than does the black-tailed godwit, and it is generally more active as well as more frequently foraging at tideline or well up on the mud. Probing, "stitching," and "mowing" techniques are all used by this species (Burton 1974). It has considerable sexual dimorphism in its bill-length, and the longer bill of the female may be of foraging advantage. Females tend to forage in deeper water than do males, and they are equally successful in obtaining prey in shallow or deep water, while males are most successful in shallow water (Smith and Evans 1973).

Social behavior. The territorial and advertising behavior of this species are only poorly described, but the male's song flight is apparently much like that of the other godwits, alternating gliding and flying with strong wingbeats. Additionally, communal aerial display occurs on the wintering grounds and during spring migration, when individual birds sometimes dive out of flocks and plummet several hundred feet while calling loudly. The usual song flight call is a loud, ringing *ku-wew* or *vee-oo*, with strong emphasis on the second syllable. A harsh, gutteral croaking call, *kirruc*, is the usual flight note and apparently also sometimes accompanies flight displays. As in the other godwits, there are probably several types of aerial displays, including song flights and pursuit flights, but this still remains uncertain. Likewise, there seems to be no description available of copulatory behavior.

Reproductive biology. Nests of this species are usually on dry and elevated sites, such as the ridges of rolling tundra, often between clumps of grass. They are usually lined with a few lichens or birch leaves, but sometimes a substantial grass lining is added. There are normally 4 eggs; among 20 sets obtained in the Hooper Bar area, all were of this number except for one clutch each of 3 and 5 eggs (Brandt 1943). Brandt observed only females incubating, with the male usually standing guard nearby, but as is typical of godwits the female probably does the daytime incubation and the male takes over at night. The incubation period has been estimated at 20–21 days, and both sexes are known to share in rearing the brood. There is no information on the fledging period (Glutz et al. 1977).

Status and relationships. The size of the North American population is completely unknown, but there are some counts of the European segment. The heart of the wintering grounds in western Europe is in Holland, which supports about three-fourths of the approximately 26,000 birds in the Continental wintering population between Denmark and Spain. About half of the European wintering population is concentrated in Britain, with Morecambe Bay and the Wash both being very important wintering areas (Gooders 1969). The evolutionary relationships of this species are by no means self-evident (Mayr and Short 1970). It is probably more closely related to the *limosa-haematica* superspecies than to *fedoa*, but beyond that the pattern of speciation is doubtful.

Suggested reading. Bannerman 1961; Bent 1927.

Marbled Godwit

Limosa fedoa (L.) 1758

Other vernacular names. None in general English use; barge marbrée (French); Amerikanische Pfuhl-schnepfe (German); aguja moteada (Spanish).

Subspecies and range. No subspecies recognized. Breeds from the central Prairie Provinces of Canada southward through Montana, the Dakotas, and western Minnesota. Previously bred to Nebraska, Iowa, and Wisconsin. Winters from California, Texas, and Florida south to Central America, rarely to South America. See map 119.

Measurements and weights. Wing (chord): males 221–28 mm; females 212–35 mm. Culmen (to feathering): males 92–119 mm; females 88–118 mm (Blake 1977). Weights: 10 males 281–362, average 319.7 g; 9 females 240–510 g, average 420.9 g (various sources). Eggs ca. 57 x 40 mm, estimated weight 44.5 g (Schönwetter 1963).

DESCRIPTION

Breeding adults of both sexes are generally pale cinnamon on the head, with the crown heavily streaked with dusky, a dark streak from the lore through the eye, the side of the head and neck streaked with dusky, and the chest, breast, sides, and under tail coverts all narrowly and irregularly barred with dusky. The axillaries and under wing coverts are light cinnamon, with some darker barring on the axillaries. The scapulars and back are spotted with sooty black and have barring and spotting of reddish buff. The wing coverts are mostly spotted and barred with blackish and edged with warm buff, except the primary coverts, which are mostly blackish. The primaries are blackish outwardly and progressively more cinnamon inwardly, with dusky and buffy markings. The secondaries and tertials are also cinnamon, with dusky mottling. The lower back and rump are buffy with dusky crescent markings, and the upper tail coverts are similar but barred with dusky. The middle tail feathers are pale cinnamon to buffy, with dusky barring, and the outer feathers are a deeper cinnamon, with dusky bars and stripes. The iris is brown, the bill is blackish toward the tip and reddish pink basally, and the legs and toes are gray to bluish gray. *Adults in winter* are dull cinnamon on the underparts, with very little barring, and the upperparts are like the summer plumage. The median wing coverts are edged with pale buffy brown to chestnut. *Juveniles* have buffy brown median coverts, broadly edged with pale buff. The underparts are a uniform cinnamon buff, with a few bars on the flanks. *First-winter* birds have very faded median coverts, which may be replaced by spring (Prater et al. 1977; Ridgway 1919).

In the hand, this is the largest of the godwits (wing at least 217 mm), and the most cinnamon-tinted, especially on the under wing surface. Only the four outer primaries are blackish, while the inner ones are cinnamon-tinted.

In the field (16 inches), this large shorebird is noticeably mottled with brown and buff and has a long bill that is blackish toward the tip and pinkish at the base. In flight, the cinnamon under wing lining is conspicuous and resembles that of the long-billed curlew, which has a strongly decurved bill. The usual call is a loud *ger-whit,* sometimes with one or three notes, but typically accented on the second syllable.

NATURAL HISTORY

Habitats and foods. Breeding-season habitats of this species in North Dakota are prairie wetlands such as intermittent streams and various kinds of ponds and lakes ranging from fresh to strongly saline. Of 125 breeding pairs classified as to habitat, half were associated with semipermanent ponds and lakes; 36 percent were on seasonal ponds and lakes; alkali ponds and lakes supported 10 percent; and 4 percent were found on miscellaneous wetlands (Stewart 1975). Grasslands of low to medium stature seem to be preferred over tall cover, and native prairie is apparently preferred over cropland cover. An average population density for the entire state of North Dakota has been calculated at 0.52 pairs per square mile, and a maximum density of 8 per square mile, based on surveys of 130 randomly selected quarter-sections (Stewart and Kantrud 1972). In the winter the species is largely coastal in distribution and seems to prefer muddy tidal flats with adjoining wet savannas or grassy borders. In South America it is mostly associated with beaches, mud flats, marshes, and flooded fields. Foods of the species are rather poorly known, but on the breeding grounds they are believed to include both aquatic and terrestrial insects, especially grasshoppers, and also mollusks. Wintering birds have been reported to eat small mollusks, crustaceans, adult and larval insects, worms, and leeches (Bent 1927). Foraging is done by both pecking and probing, the latter typical of rather moist soil. During the breeding season the birds forage on dry uplands, in wet meadows, in roadside ditches, and in open water (Nowicki 1973).

Social behavior. Birds arrive in flocks on their North Dakota nesting grounds in May, and by the middle of that month they are already dispersing. Males initiate dispersal by flying from flocks to territories in adjacent fields, periodically performing "ceremonial flights" above a particular area chosen as a territory. Distinct territorial boundaries are apparently lacking, and frequently a displaying bird will return to the flock to chase females. These chases include both ground chasing and pursuit flights, both of which may help to break the female's flocking tendency and entice her into selecting a territory. Ground chasing is normally between males and females, with the

male the aggressor, and it may lead to his pecking the female or forcing her into flight. This results in a pursuit flight, with the pair flying almost side by side, with spread tails and shallow wingbeats. The male's ceremonial flight, performed above his territory, is much like that of other godwits. He typically ascends to 22–90 meters, then levels off and begins circling with slow wingbeats. While he circles, he utters loud *ger-whit* calls at a rate of about 5 per 7 seconds. After calling, the male may glide slowly back to earth, or he may glide for a short time and then resume the calling. Often gliding precedes a terminal "nose dive" phase, in which the wings are folded on the back and the bird plummets almost vertically toward the ground, spreading the wings just before reaching the ground and landing with the wings raised in a wing-up display posture. Unlike the black-tailed godwit, there is no "tumbling flight" phase to the ceremonial flight, but both species perform a "sexual pursuit" flight during which males will chase females or pairs that encroach on the territory. It apparently serves both territorial advertisement and sexual functions. Another flight of rather uncertain function is the "figure-eight ceremony," which includes a vertical flight phase by two birds followed by a more extended low flight. Joint flights between paired birds also occur, with the male closely following the female and both birds flying in an undulating pattern not far above the ground. Copulation is preceded by behavior resembling ground chasing, with the male attempting to peck the female's back feathers and also apparently trying to mount her. Nest-scraping is a common display between paired birds and takes a form much like that in other scolopacids. Nest-scraping by the male may attract a female to the site, and she will sometimes crouch in the male's scrape. Before leaving the scrape the male typically performs a wing-up display. Scraping activities encompass a rather large area within which the nest is eventually placed (Nowicki 1973).

Reproductive biology. Nests are usually on native prairie, sometimes at considerable distance from water. They are often in the drier portion of wet meadows or in short upland grasses, usually in sparse cover not much higher than 30 centimeters. Nesting often occurs in a semicolonial pattern, with nests as close together as 60 meters, and as many as 4 nests present within a 0.15 square kilometer area. The egg-laying period has not been established with accuracy but is probably between 24 and 72 hours. The clutch size in all of 15 North Dakota nests was 4 eggs, though there is one reported instance of a clutch of 5 eggs. Incubation apparently starts before the laying of the last egg, and typically the female incubates from midmorning until late afternoon or early evening, while the male incubates during the night. The incubation period is still unknown but is probably close to the 21–23 days of other godwit species. Both sexes tend the young, with the female usually the first to respond to disturbance. At the time the young are almost 3 weeks old, and presumably fledged or nearly fledged, the male leaves the family and probably joins the flocks of nonbreeders or unsuccessful breeders (Nowicki 1973).

Status and relationships. The range of this large, prairie-dependent species has declined considerably in recent decades, but it is still moderately common in some parts of North Dakota and southern Alberta. The estimated North Dakota population in 1967 was 37,000 breeding pairs (Stewart and Kantrud 1972). This is the largest of the godwits and the most distinctive in plumage of the four species of *Limosa*. Its relationships to the other godwits are unclear (Mayr and Short 1970). Its plumage pattern is interestingly similar to that of another large grassland-nesting shorebird, the long-billed curlew, even to the evolution of very similar cinnamon-colored under wing coverts and the exhibition of this pattern in a wing-raising display.

Suggested reading. Nowicki 1973.

Tribe Limnodromini (Dowitchers)

Short-billed Dowitcher

Dowitchers are small to medium-sized temperate to arctic-nesting shorebirds with a bill that is longer than the tarsus and more than one-third the length of the wing. The bill is distinctly expanded and pitted for the outer third of its length, the tarsus is scutellated in front and reticulated behind, there is a hind toe, and at least the outer and middle toes are connected by webbing. The downy young resemble those of snipes, with white-tipped powder-puff feathers forming paired lines on the midback, flanked by other lines on the sides of the body. The sexes are alike as adults, but seasonal variations occur in plumage. Two genera and 3 species are recognized here.

KEY TO THE SPECIES OF *Limnodromini*

A Web present between bases of inner and middle toes, rear part of back mottled like rump, wing at least 170 mm . . . . *Pseudoscolopax semipalmatus*

A′ No web present at bases of inner and middle toes, rear part of back white or nearly white, wing length no more than 160 mm . . . . (*Limnodromus*)

　B Exposed culmen shorter (usually under 58 mm in males, 60 mm in females), summer adults light pinkish cinnamon to white on underparts, the sides slightly spotted or barred with dusky . . . . *griseus*

　B′ Exposed culmen longer (usually at least 60 mm in males, 67 mm in females), summer adults deep cinnamon on underparts, the sides distinctly barred with dusky . . . . *scolopaceus*

Genus *Pseudoscolopax* Blyth 1859 (Snipe-billed Dowitcher)

The snipe-billed dowitcher is a medium-sized Asian shorebird with a bill that is nearly half as long as the wing and is strongly widened laterally toward the tip, which is ridged and pitted. The hind toe is well developed, and all the front toes are connected by webbing. The downy young resemble those of *Limnodromus*. The sexes are alike as adults. One species.

Snipe-billed Dowitcher

Pseudoscolopax semipalmatus (Blyth) 1848
(*Limnodromus semipalmatus* of Peters, 1934)

Other vernacular names. Asian dowitcher; snipe-billed godwit; limnodrome semipalmé (French); Asiatscher Schlammläufer (German).

Subspecies and range. No subspecies recognized. The breeding range is not well known but probably exists as scattered areas. These include the regions of Tara and Tyukalinsk, the eastern Kulunda steppe near Kamen, and Lake Rakity, all in western Siberia; northern Orok Nor in Mongolia, southeastern Transbaikalia in the Argun River valley, central Manchuria in the region of Lungkiang, and Lake Khanka, on the Manchuria–USSR border. Winters in the Indo-Chinese countries, Malaya, and eastern India. See map 120.

Measurements and weights. Wing: males 160–72 mm; females 163–77 mm. Culmen (from feathering): both sexes 77–87 mm. Weights: 3 males 168–94 g, average 181 g; 5 females 165–245 g, average 190 g (Shaw 1936). The weights of 6 birds of both sexes banded on the wintering grounds were 127–47 g (Ali and Ripley 1969). Eggs ca. 48 x 33 mm, average weight 26.8 g (Leonovitch 1973).

DESCRIPTION

Breeding adults of both sexes are predominantly rusty rufous in tone, the crown being streaked with black and with a blackish streak from the bill through the eyes. The underparts are rather uniformly chestnut with some whitish on the abdomen. The mantle and scapulars are brownish black with chestnut feather edging. The rump and upper tail coverts are mottled with brown and white, and the tail feathers are barred white and blackish brown. The under tail coverts are white basally and rufous with narrow brownish bars toward the tips. The axillaries and under wing coverts are white, with brown-

120. Known and presumptive breeding distributions of the snipe-billed dowitcher.

ish markings on the axillaries and marginal coverts. The primaries and secondaries are dark brown, with the primaries whitish on their inner webs, and the inner primaries are margined with white. The secondaries are banded with brown and white. The iris is brown, the bill is black, becoming brownish at the base, and the legs and toes are grayish black. *Females* average larger in size and possibly are slightly duller in color. *Adults in winter* are dark grayish brown on the upperparts, and the underparts are whitish with mottling and barring of pale brown on the breast and flanks. The wing coverts are grayish brown with white fringes. *Juveniles* are uniformly brownish on the upperparts, the feathers narrowly edged with buff. The wing coverts are light brown with pale buff edges, and the underparts are white, with a warm buff wash on the breast, upper flanks, and neck. *First-winter* birds resemble adults but retain some buff-edged inner median coverts to late winter (Prater et al. 1977).

In the hand, this species more closely resembles *Limosa* than the other *Limnodromus* species and has

a similar bill length (culmen 77–87 mm), but it differs from the other dowitchers and the godwits by its strongly swollen bill tip. It has a larger wing measurement (at least 160 mm) than the other dowitchers, but smaller than the smallest of the godwits (maximum 190 mm, vs. at least 200 in *limosa*).

In the field (14 inches), this species resembles the bar-tailed godwit but has an all-blackish bill, a more uniform brown mantle, a shorter neck and legs, and a white under wing lining except for a dusky area near the wrist. Foraging is done by strong probing behavior rather than by the guzzling activity of godwits. Calls are poorly described but apparently include a quiet, nondescript *chewsk* and a plaintive, mewing *miau* or *kiow,* as well as a more extended *kewik-kewik-kewik, ku.* It is also said to have a monotonous croaking voice that distinguishes it from all other shorebirds (Leonovitch 1973). The other two dowitcher species are both smaller and have more distinct rump and upper tail covert patches. In flight this species exhibits a pale white wing lining rather than a barred under wing surface, and the upper side of the wing has flight feathers distinctly paler than the coverts (*British Birds* 71:561).

NATURAL HISTORY

Habitats and foods. In the wintering areas, where this species is best known, it is found primarily on seacoasts and swampy plains adjacent to the coasts, especially where mud flats are available. The nesting areas on the Kulunda steppes consist of boggy shores of alkaline pools, ponds with bare mud and grassy vegetation, wet depressions overgrown with weeds (*Salsola*) or sedges, and wet meadows with rather tall grassy cover. Near Tara in western Siberia the birds have been found breeding in a grassy bog, above which there was a small ridge. When foraging, the birds probe in soft, muddy ground, and the few foods reported for them include mollusks, insect larvae, and worms (Kozlova 1962).

Social behavior. On migration, these birds apparently occur in small flocks and keep in close groups when on the ground. They are known in at least some cases to nest in small colonies; Leonovitch (1973) reported a colony of 6 nests, and he noted that

colonial nesting had earlier been reported by Velizhanin near Kamer and Barnaul. The birds are apparently rather unpredictable in their breeding distribution, easily abandoning former nesting sites and moving into areas where pond water levels fluctuate greatly from year to year (Leonovitch 1973). Thus, although Velizhanin reported nesting in the Kamen and Barnaul regions of the USSR Altai Territory, the birds reportedly no longer nest there. The displays of this species are essentially undescribed. In one early account, the male drove away other males from its mate by lowering the body and pointing its bill toward the opponent while uttering a gutteral groaning sound. No specific distraction displays have been reported, but the birds are said to hover over intruders on the nesting ground while uttering uninterrupted croaking sounds (Kozlova 1962).

Reproductive biology. Relatively few accounts are available, but the birds evidently nest in colonies of about 10–20 pairs, laying their eggs in western Siberia in late May and early June. In northeastern China eggs have been found as late as July. A colony described by Leonovitch had the nests either on mounds in shallow water among reeds, or in the

open, in hollows almost devoid of cover. Nests are often built over water from 6 to 25 centimeters deep, and such nests are constructed of grass stalks and dead leaves. All of 6 nests found by Leonovitch had 2 eggs, and other descriptions indicate that 2 eggs are the usual clutch, with 3 being exceptional. Incubation is by both members of the pair, but there seems to be a rather low level of parental defense against human intruders. The incubation period and other aspects of breeding behavior have recently been established by Tolchin and Mel'nikov (1977). These workers located a breeding population on the delta of the Selenga River in southeastern Siberia, where two colonies were nesting near a colony of white-winged black terns. Six nests were found in swampy fields overgrown with *Equisetum,* and 7 were found in sedge and *Agropyron* fields. Nests were from 30 to 100 meters apart, either on flat ground or on hummocks, and in relatively wet situations, so much so that 8 of the nests were eventually flooded. Both sexes apparently guard the clutch, with one member of the pair watching for intruders from a distance. The incubation period was found to be 22 days. Several of the nests contained only a single egg, which the authors attributed to the poor environmental conditions there. They estimated that the total population in the Selenga delta in 1974 was only about 300 pairs, and that the gradual drying of previously wet breeding habitats is contributing to the species' gradual disappearance.

Status and relationships. Leonovitch (1973) supported the idea that this is a relict and endangered species that may well be becoming extinct. The scattered nesting records and the seeming impermanence of nesting colonies support this view, and certainly there are no indications of its being abundant anywhere. Ali and Ripley (1969) report it "not uncommon" in winter on Chilka lake, Orissa, where its status should perhaps be monitored in the future. This species' relationships are as uncertain and enigmatic as its breeding distribution. It has at various times been considered a monotypic genus. Pitelka (1948) supported this view, since he doubted that it is a close relative of the dowitchers or even that it should be included in the Scolopacinae (sensu stricto) since it also exhibits some godwitlike traits. Sutton (1949) supported the idea of a monotypic genus and emphasized some anatomical distinctions from the dowitchers. Rand (1950) took exception to this view and urged the species' retention in *Limnodromus.* It has also been retained in *Limnodromus* by Jehl (1968a) and Strauch (1976), but whereas Jehl believes *Limnodromus* to be related in turn to the snipes, Strauch (1976) thinks that the genus has closer affinities with *Limosa* and *Calidris.* Burton (1974) believes that *Limnodromus* and *Limosa* are well separated from most calidridine sandpipers and probably are closely related. On the basis of all this, I have concluded that *semipalmatus* does perhaps constitute an evolutionary link between the godwits and the dowitchers and should probably be placed in a separate genus to designate this.

Suggested reading. Dementiev and Gladkov 1969; Tolchin and Mel'nikov 1977.

Genus *Limnodromus* Wien 1833 (Typical Dowitchers)

Typical dowitchers are small to medium-sized arctic-nesting shorebirds with a bill that is more than one-third the length of the wing and is slightly widened laterally toward the ridged and pitted tip. The hind toe is only weakly developed, and only the outer and middle toes are connected by webbing. The downy young have powder-puff feathers that form a pattern of paired lines on the midback, flanked by shorter lines of variable prominence on the sides of the body. The sexes are alike as adults but vary seasonally in plumage. Two species are recognized here.

Long-billed Dowitcher

Limnodromus scolopaceus (Say) 1823
(Limnodromus griseus scolopaceus of Peters 1934)

Other vernacular names. None in general English use; limnodrome à long bec (French); Grosser Schlammläufer (German).

Subspecies and range. No subspecies recognized. Breeds in Siberia on the Chukotski Peninsula and in the Anadyr Basin, and in North America along the coasts of western and northern Alaska, and in northern Mackenzie (perhaps also the northern Yukon). Winters from California and Florida southward through Mexico to Guatemala. See map 121.

Measurements and weights. Wing (chord): males 133–43 mm; females 138–51 mm. Culmen (to feathering): males 57–69 mm; females 64–76 mm. Weights (on migration): males 90–114 g, average of 7 was 99.9 g; females 93–131 g, average of 11 was 114.7 g (Pitelka 1950). The average weights of 28 males and 11 females in summer were 100 g and

121. Breeding distributions of the long-billed dowitcher (L) and the Alaskan (A), Interior (I), and Ungava (U) short-billed dowitchers.

109 g, respectively (Irving 1960). Eggs ca. 42 x 29 mm, estimated weight 17.5 g (Schönwetter 1963).

DESCRIPTION

Breeding adults of both sexes closely resemble the short-billed dowitcher, but the underparts are a deeper and more uniform cinnamon red, covering the entire abdomen, and the sides are distinctly barred with dusky. *Adults in winter* probably cannot be distinguished from *griseus* by plumage alone, though the tail feathers of this species tend to be more heavily barred, with more brown pigment than white, rather than the reverse. *Juveniles* are much grayer on the head, neck, and upper breast, and the upperparts tend to be fringed with chestnut brown rather than bright buffy brown.

In the hand, the differences mentioned above help to separate these two species, as do the measurement differences mentioned in the *griseus* account.

In the field (12 inches), this species of dowitcher shows a bright rusty cinnamon color in the spring and summer, with distinct blackish barring on the sides. In the fall and winter the two dowitcher species cannot be distinguished by plumage, but call differences (a mellow three-note *tu-tu-tu* whistle in *griseus* vs. one or more thin, piping *keek* notes in *scolopaceus*) are useful criteria. In flight both species show extensive white on the rump and tail coverts and exhibit whitish trailing edges on the secondaries.

NATURAL HISTORY

Habitats and foods. Rather little has been written on this species' breeding habitat requirements in Alaska, but in Siberia the birds breed in low, moss-sedge tundra having an abundance of lakes and temporary bodies of water during spring. In such locations on the Chukotski Peninsula densities of 5–6 pairs per square kilometer may occur. Somewhat lower densities of 3–4 pairs per square kilometer occur in very swampy, polygonal moss-sedge tundra around lakes of the lower Indigirka River valley. In forest-tundra swamps on the lower parts of the Yana River they occur in densities up to 2 pairs per square kilometer, while in swampy tundra north of the tree line the birds reach densities of up to 7 pairs per square kilometer. At the time of spring arrival the birds feed on thawing lakeshores, eating a variety of insect larvae, seeds, and even mosses or plant fibers. Later they move to more typical locations such as sedgy swamps where cranefly larvae are abundant and

46. Breeding-plumage heads and outer tail feathers of short-billed (*above*) and long-billed (*below*) dowitchers.

thereafter constitute the basic food during the late summer, the birds begin to forage in other habitats and eat a larger array of foods, including midge larvae, small gastropod mollusks, and some seeds (Kistchinski 1973). On migration in California, the long-billed dowitcher apparently prefers freshwater habitats, while the short-billed species is more common on saltwater or brackish water habitats. Apparently the longer bill of the former species allows it to feed in deeper water and thus increase the available foraging areas on a pond. There is also a greater degree of sexual dimorphism of bill length in this species, which may be ecologically advantageous in reducing competition between the two sexes (Pitelka 1950). Both species feed while wading, often partially or entirely immersing the head while probing the muddy bottoms of ponds. The long-billed species feeds without the benefit of the tidal fluctuations that aid in increasing the foraging area of the short-billed form, but both species tend to feed on much the same food items (Sperry 1940).

Social behavior. Immediately after their arrival on the nesting areas, pairs disperse, with each pair remaining in an area about 100–300 meters in diameter.

The areas are intensively defended, though occasionally neighboring pairs will forage together. However, courting flights above the area are performed from the time of arrival almost until the time the first broods begin to appear (Kistchinski 1973). The song flight is performed about 15–20 yards above the tundra, where the male hovers on quivering wings while uttering a prolonged song that has been described as *peet-peet; pee-ter-wee-too; wee-too; per-ter-wee-too; pee-ter-wee-too, wee-too, wee-too.* Frequently only fragments of this song are uttered, especially when the male is chasing a female (Bent 1927).

Reproductive biology. These birds place their nests in damp locations, often at the edges of polygonized ground, where the ridge of the polygon passes into the flooded portion. The nest is usually in a small clump of sedges, often in cover only about 6 inches high, and typically is damp at the bottom. Four eggs is the invariable clutch size (Brandt 1943). Although both sexes have brood patches, it is uncertain which sex plays the larger role. According to Kistchinski (1973), the male remains near the nest early in the incubation period to warn off intruders, while the female does the actual incubation. Later the male takes over the incubation and does the hatching. Incubation requires 20–21 days, and after hatching usually only a single parent tends the young. When brooding birds have been collected, they have proved to be males, and apparently only a few females remain with the nest until hatching or a few days thereafter (Kistchinski 1973). The fledging period is unreported.

Status and relationships. The great similarity of the two species of dowitchers makes the status of this form impossible to judge, but neither species is currently believed to be very rare. The dowitchers' relationships and tangled taxonomic history was reviewed at length by Pitelka (1950), who considered the two forms distinct species on the basis of their body measurements, the degree of sexual dimorphism in the two forms, and the absence of undoubted hybrids. Pitelka suggested that the ancestor of *scolopaceus* was isolated in the Bering Sea Yukon refugium during the maximum of Pleistocene glaciation, when it became progressively more specialized for freshwater habitat and gradually evolved a longer bill and tarsus but a relatively shorter wing. I suggest that the two be regarded as constituting a superspecies, as Mayr and Short (1970) have already done.

Suggested reading. Bent 1927; Pitelka 1950.

Short-billed Dowitcher

Limnodromus griseus (Gmelin) 1789

Other vernacular names. Common dowitcher; eastern dowitcher; red-breasted snipe; limnodrome à bec court (French); kleiner Schlammläufer (German); agujeta gris (Spanish).

Subspecies and ranges. See map 121.

L. g. hendersoni: Interior short-billed dowitcher. Breeds in British Columbia, southern Mackenzie, and northeastern Manitoba south to northern Saskatchewan and central Alberta. Winters south on both coasts of Central America to northern South America.

L. g. griseus: Ungava short-billed dowitcher. Breeds in northern Quebec. Winters on the Atlantic coast of South America south to Brazil.

L. g. caurinus: Alaskan short-billed dowitcher. Breeds in southern Alaska from Nushagak Bay to Yakutat Bay. Winters along the Pacific coast from Central America south to central Peru.

Measurements and weights. Wing (chord): males 133–51 mm; females 136–55 mm. Culmen (to feathering): males 51–62 mm; females 56–68 mm. Weights (on migration): males 91–145 g, average of 19 was 107.3 g; females 95–140 g, average of 19 was 113.1 g (Pitelka 1950). Eggs ca. 41 x 29 mm, estimated weight 17.5 g (Schönwetter 1963).

DESCRIPTION

Breeding adults of both sexes have a crown that is pinkish cinnamon with black streaking, a white superciliary stripe, a broad blackish stripe from the bill to the eye, and blackish streaking behind the eye on the ear coverts. The sides of the head are otherwise dull pinkish cinnamon, which extends to the neck and underparts, becoming variably intermixed with white, especially on the abdominal region. The axillaries and under wing coverts are white with dusky chevrons. The upperparts are mostly cinnamon with blackish streaking, especially on the back, while the scapulars are spotted with black. The wing coverts are mostly grayish brown, with whitish or buffy edging, and the secondaries are similar but have white tips and white streaks on the outer webs near the shafts. The primaries are dusky outwardly and light brownish gray inwardly, the latter margined with white. The rump and upper tail coverts are white with blackish spotting or barring. The tail is regularly barred with white and blackish, and the middle tail feathers are sometimes tinged with cinnamon. The iris is brown, the bill is blackish at the tip and greenish basally, and the legs and toes are green to greenish yellow. *Adults in winter* have the upperparts nearly plain gray, with the chest and sides also gray but intermixed with white, while the rest of the underparts are white. The wing coverts are fringed with white, especially the inner medians (Ridgway 1919). *Juveniles* have brown coverts that are tipped with pale buff or chestnut, the abdomen is pale orange buff, and the neck and breast are gray suffused with buff and speckled with brown. *First-winter* birds resemble adults but retain buff-tipped inner median coverts until spring (Prater et al. 1977).

In the hand, the long, snipelike bill (culmen 51–65 mm) and rather short tarsus (32–37 mm) separate this species from most species other than *scolopaceus,* which has a somewhat longer bill (usually over 65 mm vs. under 60 mm) and a longer tarsus (usually over 39 mm vs. under 36 mm). Furthermore, the tail feathers of *scolopaceus* are more evenly barred with brown than white, rather than having more white than brown, as is usually the case with *griseus.* Juveniles can be separated by the fact that *griseus* has the upperparts dark brown, with bright buffy feather patterning and edging, while in *scolopaceus* the upperparts have a narrow fringe of chestnut brown that tends to be scalloped (Prater et al. 1977). Further,

griseus lacks a series of ridges angling forward on the culmen above the rear end of the nostril that are usually present on *scolopaceus* (*C.F.O. Journal* 13:98).

In the field (12 inches), dowitchers are readily identified by their long, snipelike bills and the whitish rump area. In spring plumage the spotting and less rich rusty tones on the sides and flanks of this species help to separate it from the long-billed form. This species usually utters a low and mellow three-noted whistle, whereas the long-billed dowitcher usually produces a single, thin peeping note or a series of these. In young birds, the long-billed dowitcher is darker above and grayer below than this form; but differences are very small and probably the presence or absence of the *keek* note, either singly or in series, is the best field mark for nonbreeding birds (*British Birds* 61:366–72).

NATURAL HISTORY

Habitats and foods. This subarctic species is mainly associated with muskegs or similar boggy and marshy areas having low vegetation. In the Churchill area the birds breed in hummocky sedge marshes close to trees and do not extend out onto the flat marshy tundra near the coast. Apparently much the same habitats are used throughout the interior of Canada, where boggy muskegs often occur as ecological "islands" of varying size between areas of coniferous forest (Pitelka 1950). There is no obvious difference in the breeding habitats of the two dowitcher species, other than that perhaps the long-billed form more readily extends out into true tundra vegetation than does the short-billed species. Likewise, they are probably very similar in the foods they eat on the breeding grounds, with the larvae and pupae of dipterans playing an important role. A large series of stomachs from birds taken between April and December indicated that the short-billed dowitcher consumes a larger proportion of marine polychaete worms and mollusks, while the long-billed species eats more insects, a difference that is to be expected considering the more coastal wintering habitats of the short-billed dowitcher (Sperry 1940). Both species forage by probing while standing in rather shallow water, and frequently the short-billed dowitcher feeds in tidal flats, moving in and out with the tide. In their foraging behavior the dowitchers resemble snipes, but probing rates are very high, and there is rarely any pivoting around the bill or reorientation of probes (Burton 1974).

Social behavior. Surprisingly little has been written on the social behavior of this species during the breeding season. It is known that a hovering flight song is present, and that the accompanying call is a liquid, musical, contralto gurgle. A woodcocklike strutting by the male has also been observed, but remarkably little in the way of detailed descriptions is yet available (Bannerman 1961). Likewise, copulatory behavior is still undescribed. Fairly strong pair bonds are certainly formed, for nesting pairs remain close together, and pairs invariably defend their nesting territory against intruders in concert. Like tattlers, they typically leave the nest well before a human gets very near, and together they attempt to lure one away by short flights, calls, and other conspicuous behavior. Frequently the dowitchers from adjoining territories are attracted and join in the general confusion, leaving the observer with no indication of where the nest might be.

Reproductive biology. Nests are extremely hard to locate and usually are quite close to water if not actually surrounded by it. The nests are often in bogs or muskegs with soggy grass or sedge cover, and usually are as well hidden as those of snipes. In one case a total of 4 nests were found during each of two years in an area of marsh extending some 120 yards from a small lake. In one year the nests were widely separated, at least 50 yards apart, while in the following year 3 nests were found in a clearing no bigger than 60 by 25 feet. On rare occasions the nests may be as far as a quarter of a mile from a lake (Bannerman 1961). Almost invariably there are 4 eggs present; Jehl and Smith (1970) report that 8 of 9 clutches found near Churchill had 4 eggs, while the exceptional one had 5 eggs. A few other cases of 5-egg clutches are known. The incubation period is approximately 21 days, with both sexes incubating but the female taking little part in the care of the brood (Jehl and Smith 1970). The fledging period is still unreported.

Status and relationships. This species has a rather widespread breeding range in Canada, but its total population is completely unknown. Its taxonomic relationships to *scolopaceus* have been discussed in that species' account.

Suggested reading. Bannerman 1961.

Subfamily Scolopacinae
(Snipes and Woodcocks)

North American Snipe

This subfamily includes small to medium-sized shorebirds with a flexible and soft-tipped bill that is longer than the tarsus and often is more than half the length of the wing. The ear opening is anterior to the middle of the orbit, which is usually completely ringed with bone, and the posterior part of the skull is somewhat conical, being much narrower above than below. The tarsus is relatively short and is scutellated in front, and the hind toe is well developed. The downy young are usually covered with white-tipped powder-puff feathers that form irregular dark bands down the back (snipes), or they have alternating stripes of buff, dark brown, and tawny brown without white tips (woodcocks). The sexes are alike as adults. Two tribes, containing 20 species, are recognized here.

Tribe Gallinagini (Snipes and Semi-woodcocks)

Snipes are small to medium-sized shorebirds with a bill that is usually quite soft and flexible near its tip, and a short tarsus that is scutellated both in front and behind. The back is often strongly striped, and there is usually a median stripe of buff on the crown. The downy young usually have white-tipped powder-puff feathers. Three genera and 16 species are recognized here.

KEY TO GENERA OF *Gallinagini*

A Tail wedge-shaped, of 12 feathers . . . . *Limnocryptes* (1 sp.)
A' Tail rounded, usually of at least 14 feathers
  B Tarsus reticulated behind, wing under 115 mm . . . . *Coenocorypha* (1 sp.)
  B' Tarsus scutellated behind, wing over 115 mm . . . . *Gallinago* (14 spp.)

Genus *Coenocorypha* Gray 1855 (Sub-antarctic Snipe)

The sub-antarctic snipe is a small island-dwelling shorebird with a bill that is much longer than the tarsus, and with the tip of the upper mandible drooping downward over the tip of the lower mandible. The hind toe is well developed and elevated, the front toes lack webbing, and the tarsus is scutellated in front and reticulated behind. The wing is short and rounded, and the tail is also short and rounded. The back is unstriped, the crown is only indistinctly striped, and the bony orbital ring is incomplete. The downy young resemble those of *Gallinago* but are only obscurely patterned and have inconspicuous white "powder-puff" feather tips. A single species is recognized here.

Sub-antarctic Snipe

Coenocorypha aucklandica (Gray) 1845

Other vernacular names. Auckland snipe; bécassine d'Auckland (French); Auckland-Schnepfe (German).

Subspecies and ranges. See map 122.
  C. a. aucklandica: Auckland sub-antarctic snipe. Resident on the Auckland Islands (Ewing and Adams islands).
  C. a. pusilla: Chatham sub-antarctic snipe. Confined to South East Island of the Chatham Islands. Sometimes regarded as a separate species.
  C. a. huegeli: Snares sub-antarctic snipe. Resident on Snares Islands.
  C. a. iredalei: Stewart sub-antarctic snipe. Previous resident on the islands off Stewart Island, including Big South Cape and Jacky Lee; now apparently extinct.
  C. a. meinertzhagenae: Antipodes sub-antarctic snipe. Resident on Antipode Island.
  C. a. barrierensis: Barrier sub-antarctic snipe. Known only from one specimen taken on Little Barrier Island in 1870; now apparently extinct.

Measurements and weights. Wings (both sexes): 94–114 mm. Culmen (both sexes): 42–64 mm. Weights (*huegeli*): 90–125 g. Eggs ca. 39 x 28 mm (*pusilla*) to 51 x 35 mm (*aucklandica*), estimated weight 16.5 g (*pusilla*) to 30.0 g (*aucklandica*) (Schönwetter 1963). Measurements of *pusilla* are appreciably smaller (maximum culmen 47 mm, maximum wing 102 mm) than those of the other forms (minimum culmen 51 mm, minimum wing 102 mm).

DESCRIPTION

Adults of both sexes have the top of the head striped with black and reddish brown, or mostly dark brown, with a buffy stripe in the midline. There is a pale superciliary stripe and a dark brown line from the gape through the eye and beyond. The sides of the head and chin are buffy, with variable amounts of dark spotting. The foreneck, upper breast, and sides of the body are buff, spotted or blotched with

122. Breeding distributions of the Auckland (Au), Antipodes (An), Chatham Island (Ch), and Snares Island (S) sub-antarctic snipes.

dark brown. The lower breast and abdomen vary from unspotted buffy white (*pusilla*) to mostly barred or blotched, the barring sometimes covering the entire underparts or at times being limited to the flanks. The tail feathers (fourteen to sixteen) are brownish black, with cinnamon buff centrally and dusky brown laterally. The upperparts are generally brownish black, the feathers edged with tawny and cinnamon buff, the wing coverts are sandy brown with blackish centers and buffy edge markings, and the primaries are brown, sometimes with buff edges. The iris is brown, the bill grayish brown to slate colored, and the legs and feet have been variously described as grayish brown, brownish yellow, and grayish yellow. *Immatures* resemble adults, but the black markings are less distinct and the streaking on the upper breast and throat is less apparent.

In the hand, the short wing (under 115 mm), the tarsus scaling (which is scutellated in front and reticulated behind), and the snipelike bill serve to identify this species.

In the field (8–9 inches), this species is rarely seen, as it feeds in the evening and flies rarely, and then only for short distances. The call of one of the forms (*pusilla*) has been described as starting with a series of soft, low, sibilant *chirrups*, quickly working up to a series of sharp whistles repeated about 8 or 10 times in about 5 seconds.

NATURAL HISTORY

Habitats and foods. This is a somewhat crepuscular or nocturnal species, spending most of the daylight hours in holes of rocks, hollow trees, and thick undergrowth. The habitat of *huegeli* has been described as a mixture of clumped *Olearia* from 20 to 30 feet high, interspersed with much lower *Senecio* cover. Probably in all the forms the food is primarily earthworms, for which the birds probe in soft earth with their long, woodcocklike bills, but the droppings of some specimens of *huegeli* indicate that a variety of insects and other arthropods are also eaten. These birds were observed feeding mostly during morning and evening, mainly in areas of forest edges and areas of scattered scrub and tussock grasses. There they probe by pushing the bill deep into the soil or the base of tussocks, evidently feeling for insect movements (Anderson 1968).

Social behavior. These birds are known to be highly territorial during the breeding season, and they defend their territories fiercely by actual fighting (Anderson 1968). Presumably their loud calls are associated with territorial advertisement, but so far no aerial advertisement displays have been seen. Considering the short wings and limited flying abilities of these birds, it is quite possible that aerial displays have been gradually lost during evolution.

Reproductive biology. The nesting of this species has been reported only for two of the subspecies. The Stewart Island form has been found laying during late October and early November. A nest was found on scattered *Dracophyllum* needles in one instance, and another was placed on granite grit and sand with a litter of *Dracophyllum* needles and a pile of mosses, lichens, and twiglets as a base for the eggs. In the Snares Island race the nests have been found in December, with laying probably starting in November. Nests there have been found in the middle of *Poa* tussocks, about a foot above ground level, or on the ground under "solid" vegetation or at the hollow of a tree trunk. Evidently 2 brown, raillike eggs make up the clutch, and both members of the pair incubate. However, it is possible that each parent takes on the care of a single chick after they hatch, since apparently there are no records of a pair with 2 youngsters. The young evidently follow their parents for a prolonged period, long after they are fully fledged. Typical "broken-wing" injury-feigning has been reported for adults tending chicks (Anderson 1968; Oliver 1955).

Status and relationships. Apart from the extinct subspecies *barrierensis*, the status of this species is precarious. The population of *iredalei* was likewise eliminated on Jacky Lee Island by introduced wekas (*Gallirallus australis*), and on Big South Cape Island more recently by introduced rats. Populations on Adams and Ewing islands of the Auckland group, and on Antipodes and Snares of the Chatham group,

are considered reasonably secure, but on Snares Island the population is so small that it is believed that extinction resulting from the accidental introduction of rats is a real possibility (Gooders 1969). The status of the Antipodes Island population is unknown. The relationships of these birds are most intriguing. Seebohm (1888) considered them "semi-woodcocks," and together with several South American snipes (*undulata*, *imperialis*, and *stricklandii*) as primitive snipe stock, and he believed *Coenocorypha* and *stricklandii* may have a common ancestry or perhaps may be the result of parallel evolution. I certainly agree that this seems to be the most primitive of living snipes and also provides a possible evolutionary connection between the snipes and the woodcocks.

Suggested reading. Oliver 1955; Anderson 1968.

Genus *Gallinago* Brisson 1760 (Typical Snipes)

Typical snipes are small to medium-sized shorebirds with a long, flexible bill that is much longer than the tarsus and has a tip that is variably expanded and pitted, with upper mandible drooping over the tip of the lower mandible. The hind toe is small, the front toes are unwebbed, and the tarsus is scutellated in front and behind. The wing is pointed or rounded, and the tail is short and rounded, of twelve to twenty-eight feathers. The back and crown are both striped, the ears are below the middle of the eye, and the eyes are large and positioned high in the skull. The downy young have varying amounts of white-tipped powder-puff down and are marked with dark brown or blackish on a chocolate ground color, often forming a pattern similar to that of dowitchers. Fourteen species are recognized here.

KEY TO SPECIES OF *Gallinago*

A With 18–28 tail feathers, the outermost usually under 5 mm wide
  B At least 20 tail feathers, the outer 12–16 distinctly narrowed
    C Tail feathers 24–28, of which 16 are extremely narrow . . . . *stenura*
    C' Tail feathers 20–22, of which 12 are relatively narrow
      D Overall plumage grayish to gingery in tone, with white spotting on upperparts and breast . *solitaria*
      D' Overall plumage dark brownish in tone, the back unspotted and with a scaly or vermiculated pattern . . . . *megala*
  B' Tail feathers 18 (sometimes 16–20), the outer two distinctly narrowed
    C Belly distinctly barred, upper parts dark brown, bill relatively thick toward base . . . . *nemoricola*
    C' Belly white, upperparts not unusually dark, bill relatively slender toward base
      D Outermost tail feather distinctly barred, wing at least 150 mm . . . . *hardwickii*
      D' Outermost tail feather not distinctly barred, wing no more than 140 mm . . . . *nigripennis*
A' With 14–16 tail feathers (rarely 12–18), the outermost usually at least 5 mm in width
  B Lower tibia bare of feathers, abdomen white or only slightly barred
    C Exposed culmen over 90 mm, wing over 140 mm
      D Maxilla distinctly heavier toward base, wing usually over 150 mm . . . . *undulata*
      D' Maxilla relatively slender throughout, wing usually under 150 mm
        E Upper parts with little tawny, tarsus reticulated halfway to upper joint . . . . *macrodactyla*
        E' Upperparts heavily marked with tawny, tarsus reticulated two-thirds the distance to joint . *nobilis*
    C' Exposed culmen usually less than 90 mm, wing usually under 140 mm
      D Distal half of outermost tail feathers pure white or nearly white
        E Belly white, exposed culmen over 70 mm long . . . . *nigripennis*
        E' Belly barred, exposed culmen under 70 mm long . . . . *media*
      D' Distal half of outermost tail feathers distinctly barred
        E Legs bright yellow . . . . *puna*
        E' Legs greenish gray . . . . *gallinago*
  B' Tibia feathered nearly to joint, abdomen buff, variably barred with dark brown, wing usually over 150 mm
    C Upper parts heavily marked with blackish and rufous, tail feathers 12, unbarred sepia . . . . . *imperialis*
    C' Upperparts with several buffy longitudinal stripes, tail feathers 14, barred . . . . *stricklandii*

Summary of Field Marks of *Gallinago* and *Limnocryptes*
(arranged from small to large)

Jack Snipe. 8 inches, small body, very short bill, flight slow and lacks zigzags, silent, or alarm call low and weak.

Puna Snipe. 9 inches, small size, Andean puna habitat, yellow legs, *itch* alarm call (Olrog).

Snipe. 10 inches, relatively long bill, greenish legs, white edge on secondaries, explosive zigzag flight, *scape* alarm call.

Pintailed Snipe. 10½ inches, upper wing coverts more buffy than *gallinago*, underside more heavily barred, no trailing white edge on secondaries, little or no zigzag in flight, less loud *charp* or *scaap* alarm call.

African Snipe. 11 inches, bill relatively longer and dorsal surface darker than *gallinago*, outer tail feathers almost entirely white, lacks zigzag flight, alarm call *skaap* or *tchek*.

Great Snipe. 11 inches, bill shorter than *gallinago*, underparts more heavily marked, wings with distinct white lines on upper wing coverts enclosing black panels, secondaries with pale tips, flight flow and sometimes wavering, alarm call a croaking *brad*, infrequently uttered.

Forest Snipe. 11 inches, much like *gallinago*, but no white trailing edge on secondaries, more white on tail; smaller and paler above than *hardwickii*; zigzag flight, usually silent when flushed, or sometimes a gutteral grunt.

Japanese Snipe. 11 inches, much like forest snipe, but larger, darker, and utters sharp *kreck* in flight, which is swift and weaving.

Himalayan Snipe. 12 inches, broad, rounded wings, flight slower and more woodcocklike than other snipe; darker below than other Asian snipes except solitary; told from solitary by darker buff, brown and black back, lacking chestnut, heavy wooded and hilly habitat, call a gutteral croak.

Solitary Snipe. 12 inches, similar to wood snipe but paler and with longer bill, chestnut on back, buffy or white freckling on breast and upper wing coverts, wooded mountain habitat, flight erratic, but slower than *gallinago*, and call deeper and harsher.

Madagascan Snipe. 12 inches, only snipe on Madagascar.

Imperial Snipe. 12 inches, extremely rare in damp timberline habitat of Andes, very dark and rufous-colored above, sides and underparts barred with black.

Noble Snipe. 12 inches, grassy bog and savanna habitats at middle to high altitudes of Andes, bill relatively longer than that of *andina* or *gallinago*, and bird generally larger, lighter in color than *imperialis*. Call "clear and melodious."

Cordillerian Snipe. 12 inches, grassy bogs, meadows, and swampy woods, temperate and paramo zones, upperparts heavily spotted with buff and cinnamon, underparts narrowly barred, lighter and less rufous in color than *imperialis*. Call notes include a repeated *chip*.

Giant Snipe. 16–17 inches. Separated from all other snipes by extreme size, with heavy barring and striping on wings, neck, and chest. Limited to savannas and grasslands of tropical zone, calls *kek-kek-kek*, similar to but softer than clapper rail.

Giant Snipe

Gallinago undulata (Boddaert) 1783
(*Capella undulata* of Peters, 1934)

Other vernacular names. Guianan giant snipe; bécassine géante (French); Riesenbakassine (German); becassina gigante (Spanish).

Subspecies and ranges. See map 123.
  G. u. undulata: Lesser giant snipe. Resident in Colombia, Venezuela, Guyana, Surinam, French Guiana, and northern Brazil.

G. u. gigantea: Greater giant snipe. Resident in southeastern Brazil, Paraguay, and probably Uruguay.

Measurements and weights. Wing (flattened): males 170–81 mm; females 136–83 mm. Culmen (from base): males 121–36 mm; females 99–135 mm. Weights: 5 males 270–320 g, average 294 g; 3 females 282–363 g, average 332 g (Haverschmidt 1974). Eggs ca. 54 x 38 mm, estimated weight 40.5 g (Schönwetter 1963).

DESCRIPTION

Adults of both sexes are similar to other snipes in general appearance but are much more heavily patterned with brownish black and cinnamon above and with blackish below, with heavy striping on the foreneck and breast and buffy barring on the primaries and secondaries. The wing coverts, secondaries, lower back, and rump are all black, with white to buffy barring, the tail coverts are buff to cinnamon, and the under coverts have slight brown barring. The fourteen tail feathers are black with cinnamon tips and bars. The underparts are white, with the foreneck striped with black and the breast, sides, flanks, and upper legs barred with black. The underpart barring is variable, being heavier in *gigantea*, which also has broader cinnamon edging on the up-

47. Adult heads and central tail feathers of cordillerian (*above*) and giant (*below*) snipes.

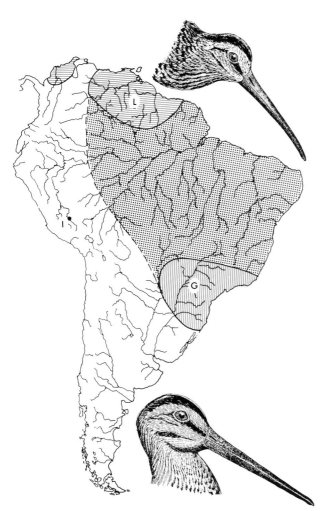

123. Known breeding distributions of the imperial snipe (I) and the lesser (L) and greater (G) giant snipes. The shaded area indicates presumptive range of uncertain subspecies.

perparts. The iris is dark brown, the bill is brownish horn, becoming paler below and black toward the tip, and the legs and toes are lead gray (also reported as dark brown). *Juvenile* plumages are apparently undescribed.

In the hand, the very large size (over 250 g) and long bill (at least 100 mm), as well as the barring of buff on the flight feathers, should identify this species.

In the field (16–17 inches), this species is restricted to the tropical zone and is the only extremely large snipe of that climatic zone. Its size alone should serve to identify it if seen; its vocalizations include *kek-kek-kek*, similar to but softer than those of a clapper rail. The repeated *kek* call is uttered on takeoff, while the display call is said to be a three-syllable *or-a-paz* or *buen-esta*.

NATURAL HISTORY

Habitats and foods. Disappointingly little is known of the ecology of this gigantic snipe, which is limited

to tropical habitats such as marshy pastures and savannas and probably rarely if ever is found at elevations higher than 1,000 meters (Blake 1977). It has been generally reported to favor swampy habitats, and Haverschmidt (1974) collected several in an area of mixed aquatic grasses and shrubbery about 1.5 meters high. However, he also found it in an area of sandy soil, where short grasses and small dense bushes and shrubs provided a dry savanna environment.

Social behavior. Unfortunately, little can be said of the species' social behavior. There is no description of the aerial display, but Helmut Sick (in litt.) reports that the species makes an enormous noise ("instrumental music") during courtship flights. It also utters a strong call independently. The display is performed in total darkness.

Reproductive biology. About the only information that has been published on this species' breeding is that its nests are built on small hillocks among the swamps (Meinertzhagen 1926) and that 2–3 eggs are the usual clutch (Maclean 1972). Haverschmidt (1974) collected a female in Surinam that was nearly ready to lay in early January.

Status and relationships. This species does not seem to be common anywhere, though Brazil is presumably the center of its abundance. It is one of a group of snipes called "semi-woodcocks" by Seebohm (1888), and has a distinctly woodcocklike bill shape. Its eggs are also said to be woodcocklike (Schön-

wetter 1963). Meinertzhagen (1926) suggested that this species and "Chubbia" (stricklandii) probably represent ancestral types of the snipe and woodcock groups, and I believe that indeed undulata approaches the Scolopax body shape and bill type perhaps more closely than any other snipe. Almost certainly its nearest relative is stricklandii, and the two might be conveniently regarded as generalized snipes that are the closest to Coenocorphya among the surviving Gallinago forms.

Suggested reading. Haverschmidt 1974.

Cordillerian Snipe

Gallinago stricklandii (Gray) 1845
(*Chubbia jamesoni and C. stricklandii* of Peters, 1934)

Other vernacular names. Andean snipe; Strickland's snipe (*stricklandii*); bécassine des páramos, bécassine de Strickland (*stricklandii*) (French); Paramoschnepfe (German); becasina Andina, agachadiza grande (Spanish).

Subspecies and ranges. See map 124.

    G. s. jamesoni: Northern cordillerian snipe. Resident in the Andes of Venezuela, Colombia, Ecuador, Peru, and Bolivia.

G. s. stricklandii: Southern cordillerian snipe. Breeds in the Andes of southern Chile and southern Argentina to Tierra del Fuego. Probably migratory, at least in the southern part of its range.

Measurements and weights. Wing (flattened): males 149–70 mm; females 144–65 mm. Culmen (from base): males 79–94 mm; females 78–96 mm. Weights: 2 males of *jamesoni* 221 and 224 g (LSU specimens); 6 specimens of *jamesoni* ranged 140–53 g and averaged 146.8 g (Tuck 1972). Eggs ca. 51 x 34 mm, estimated weight 30.6–31.7 g (Schönwetter 1963).

DESCRIPTION

Adults of both sexes have the upperparts boldly variegated with black, brown, buff, and cinnamon, but with the light dorsal markings more rufous than other snipes and the pale crown stripe poorly defined or absent. The wing coverts and inner secondaries are heavily spotted and barred with cinnamon rufous, and the primaries and outer secondaries are barred with cinnamon, buff, or whitish. The 14 tail feathers are black basally, have very little barring toward the tips, and lack rufous coloration. The undersurface is whitish to light buff (*stricklandii*) or more grayish (*jamesoni*), narrowly barred with sepia, and the foreneck and breast are boldly streaked with brown to blackish. The iris is brown, the legs and feet are grayish (also reported as yellowish and reddish brown), and the bill is brownish horn, becoming paler below and grading to black toward the tip. *Juveniles* closely resemble adults, but the feathers of the sides of the mantle and scapulars are narrowly edged with light buff (Meinertzhagen 1926).

In the hand, the presence of fourteen tail feathers, a heavy woodcocklike bill, the tibia feathered almost to the joint, and a weak or absent crown stripe should identify this distinctive species.

In the field (11½–12 inches), this large snipe is found in grassy bogs, wet meadows, and swampy woods. In Bolivia the species has been reported most common in patchwork areas of grassland and forest, at elevations of 3,300 to 3,400 meters. If seen, it is most likely to be confused with the imperial snipe, but it is lighter and less rufous in color and is less heavily barred on the sides. Call notes include a repeated *chip.* Display in this species sometimes occurs well into the night. The bird flies in circles, uttering a loud *whee-tchwu* with each syllable equally accented, at the rate of about two per second. After calling thus for 30 to 60 seconds, the bird descends, while producing a whirring sound.

NATURAL HISTORY

Habitats and foods. The northern form of this species is a highland bird, occurring at altitudes at least to 3,500 meters in Venezuela, typically in páramo vegetation, including moist grassy areas and boggy meadows. In Bolivia the species has been observed displaying at an elevation of 3,000–3,400 meters, in an ecotone area of montane forest and grassland (Vuilleumier 1969). The southern race *stricklandii* occurs in boggy areas of forest or in more open areas where marsh grasses and mosses impinge on dwarf scrub vegetation (Johnson 1965). Its foods are not known well, but they at least include beetles (Reynolds 1935).

Social behavior. There are two good descriptions of vocalizations and aerial display in this species, in-

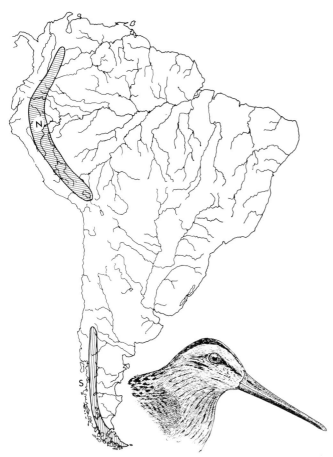

124. Breeding distributions of the northern (N) and southern (S) cordillerian snipes.

48. Adult heads and outermost tail feathers of cordillerian (*above*) and imperial (*below*) snipes.

cluding one of the southern form by Reynolds (1935) and one of the northern form by Vuilleumier (1969). Both clearly indicate that a series of long calls precedes or alternates with the drumming, including repeated *chip* notes and louder double-syllable notes. Evidently, after a rather long period of calling the bird enters a descent that is slow at first but gradually gains in velocity and produces a muffled and low-pitched sound that Vuilleumier compared to a cow's bellow and Reynolds called a whir or drone of exceedingly low pitch, almost reaching the lower limit of human hearing. It is of considerable interest that the outermost tail feathers of these birds are relatively wide and unspecialized (ranging from 5 to 7 mm). Vuilleumier was uncertain whether it is the tail or wing feathers that are vibrated, since neither show obvious specialization, but it seems likely that it is the tail feathers as in the smaller and more specialized snipe forms. Nonetheless, it should be remembered that in the American woodcock it is the three outer pairs of wing feathers that are vibrated, after a prolonged period of calling in a manner comparable to that in this snipe.

Reproductive biology. Evidently the normal clutch size in both races is 2 eggs (Maclean 1972*a*). The only

nest description seems to be that of Reynolds (1935), who found a nest among short and sparse rushes and grasses on high ground near the edge of a bog. Two nests were found, both in December.

Status and relationships. Evidently the southern race of this species is most common on the islands around Cape Horn, with decreasing numbers toward the north. Like many Andean birds, the status of both races is extremely uncertain. These two forms, often considered full species, have frequently been placed in the genus *Chubbia*. Perhaps the best justification for that is to emphasize their woodcocklike features, but to varying degrees these also occur in others of the "semi-woodcocks." Their nearest relative almost certainly is *imperialis,* and there have been some suggestions that *imperialis* may have been described on the basis of an aberrant or immature specimen of *stricklandii* or *jamesoni.* The similarities of plumage and what little is known of the aerial displays and vocalizations of these forms certainly suggest a close relationship among them.

Suggested reading. Humphrey et al. 1970; Johnson 1965.

Imperial Snipe

Gallinago imperialis Sclater and Salvin 1869
(*Chubbia imperialis* of Peters, 1934)

Other vernacular names. Banded snipe; Bogotá snipe; bécassine imperiale (French); Kaiserbekassine (German).

Subspecies and range. The total range is unknown. Two old specimens are known from Colombia, and recently the species has been observed and collected near Cuzco, Peru. See map 123.

Measurements and weights. Wing (flat): male 161 mm. Culmen (to feathering): male 94 mm. Weights: no information. Eggs: undescribed.

DESCRIPTION

Adults of both sexes differ from all other snipes in the very dark, black-barred rufous back coloration and the lack of buffy longitudinal striping along the scapulars. The sides of the head, throat, foreneck, and chest are rufescent chestnut, the first three of these heavily striped and spotted with black and the

last marked with somewhat wavy black bands. The lower breast, sides, and rear underparts are strongly barred with black and white. The tail feathers are all sepia and unbarred, the central ones being slightly paler toward their tips. The iris is dark brown, the bill is dark brownish gray, and the legs and toes are medium gray. *Immature* plumages are undescribed.

In the hand, this species is most like the cordillerian snipe in its measurements but is appreciably darker and more rufous and has a prominent rufous crown stripe. Additionally, this species has very long secondaries that completely cover the primaries when the wing is folded, while in the cordillerian snipe the tips of the primaries extend beyond the ends of the secondaries.

In the field (11½–12½ inches), this species has been found only above 3,100 meters, in damp timberline vegetation. The birds display well after sundown, and they call while flying in nearly level circles, as does the cordillerian snipe. However, in that species calling is more continuous, while in the imperial snipe calling is divided into discrete episodes and consists of a series of staccato notes that first increase in volume, then reach a peak, and gradually diminish, with each series lasting about 10 seconds. In this species a whirring sound is also produced at

the end of each song bout, while in the cordillerian snipe it is done at the end of a 30- to 60-second period (Terborgh and Weske 1972).

NATURAL HISTORY

Habitat and foods. Judging from the only available account, this species occurs entirely above 3,100 meters in the Andean timberline zone, where the weather is damp and chill at all times. Here, a low elfin forest gradually changes to larger areas of grassland, and a thick mat of sphagnum and peat is present. Evidently the birds probe in this layer of thick mosses, presumably for insects, worms, or various arthropods, but there is no obvious abundance of potential snipe foods in this layer (Terborgh and Weske 1972).

Social behavior. Terborgh and Weske (1972) made observations and recordings of the aerial display of this species during July. Most display was at dawn and dusk, with cloudiness reducing the duration and intensity of the display. The birds typically performed at 50–150 meters above the ridgetop and began the display with a series of song bouts that started with a series of rough, staccato notes that generally increased in volume. In the middle of the song there was a series of double and triple notes, followed by a group of evenly spaced single notes of declining volume. Individual song bouts lasted an average of nearly 10 seconds, with silent intervals of about 6 seconds separating them. The terminal sequence of single notes occurs during a gently sloping dive, and at the end of the dive the bird abruptly pulls out, producing a sound made by the rush of air through feathers that these authors believed to be the wing feathers. They estimated that, on the basis of a transect they made, each displaying bird (or pair) occupied a territory of about 130–60 acres along timberline ridges.

Reproductive biology. Nothing is known of this species' reproduction.

Status and relationships. The species is known from only a few specimens, and even in its recently discovered habitat the population is apparently very low. Terborgh and Weske found a total of 4 or 5 displaying birds along a mile stretch of suitable ridgetop habitat. It is presumably the rarest of the South American snipes, but it may well occur in other similar but unknown localities. The species' relationships are almost certainly with *stricklandii* (including *jamesoni*). The displays of *imperialis* obviously have close similarities to those described for *jamesoni*, and the emphasis on vocalizations during this display, together with the lack of obvious structural specialization for winnowing, suggest that this is a much more primitive type of aerial display than that typical of *gallinago* and its near relatives.

Suggested reading. Terborgh and Weske 1972.

Himalayan Snipe

Gallinago nemoricola Hodgson 1836
(*Capella nemoricola* of Peters, 1934)

Other vernacular names. Wood snipe; bécassine des bois (French); Nepalbekassine (German).

Subspecies and range. No subspecies recognized. Breeds in the Himalayas from about 76° E longitude eastward to eastern Assam, southern Tibet, and western Sikang; possibly also breeds in the mountains of northern Burma. Winters south to India, Burma, northern Thailand, northern Yunnan, and northern Tonkin. See map 125.

Measurements and weights. Wing (both sexes): 133–48 mm. Culmen (from feathering): both sexes 61–71 mm. Weights: ca. 142–98 g (Ali and Ripley 1969). Meinertzhagen (1926) reported somewhat greater culmen lengths (72–84 mm) for this species than I have been able to verify. Eggs ca. 44 x 31 mm, estimated weight 21.7 g (Schönwetter 1963).

DESCRIPTION

Adults of both sexes differ from nearly all the other Eurasian snipes in that the barring of the underparts is heavy and extends across the abdomen. Additionally, the wings are relatively broad and rounded, with narrow grayish tips on the secondaries. The upperparts are dark brown to blackish, with rufous and grayish buff markings, while the breast is fulvous with brown barring and the rest of the underparts of white with close brown barring. The tail consists of eighteen feathers, the four outer pairs narrow and the outermost pair only 3–4 mm wide and similar to those of *hardwickii*, but with a grayish ground color. The iris is dark brown, the bill is greenish brown to reddish brown, becoming darker toward the tip and yellowish at the base of the lower mandible, and the legs and toes are dark grayish green. *Juveniles* have most of the upper wing coverts medium brown, with warm chestnut buff bars and a pale buff tip, while the inner median coverts are dark brown with a narrow buffy fringe. By comparison, adults have most of the coverts dark brown with large grayish buff spots, and the inner medians have fairly wide grayish buff tips. Primaries of juveniles usually have clear white tips that become slightly worn by winter and moderately worn by spring (Prater et al. 1977).

In the hand, this species is identified by its eighteen tail feathers, with the outer three pairs progressively more narrow, and the outermost pair only 3–4 mm wide. It differs from other Eurasian snipes, except for the solitary, by its relatively large size and its barred underparts, and differs from the solitary by the absence of chestnut to ginger coloration on the upperparts.

In the field (12 inches), this species is found in thick wooded cover and relatively hilly habitats. When flushed it flies slowly and with a wavering, woodcocklike flight, with the bill pointed downward, and

125. Breeding distributions of the Himalayan (*shaded*) and forest (*hatched*) snipes.

it usually drops back into cover again after 100 yards or less. It sometimes utters a low, croaking *tok-tok* note on flushing, but it is frequently silent. Its wings are broad and relatively rounded, further increasing its similarity to a woodcock. It is known to have a "beating" display flight, but the details are unreported.

NATURAL HISTORY

Habitats and foods. In wintering range this species is usually seen in heavy cover in marshes and along streams, but on its breeding ground it is apparently always associated with wooded areas. In the Himalayas it evidently occurs from 4,200 to 14,000 feet in thick cover. There is an early report of nesting in a pine forest between 5,500 and 6,000 feet, and more recently specimens have been collected just above tree line on marshy ground overgrown with long grass or low rhododendron scrub. The little information available on foods suggests that the birds eat worms, small aquatic insects and their larvae, as well as a few seeds (Ali and Ripley 1969; Vaurie 1965).

Social behavior. There is no real information on the species' social behavior, other than that it is known to have a bleating display flight (Tuck 1972).

Reproductive biology. Almost all that is known of this species' nesting comes from the report of a single egg and female snared in 1908 on a nest in a pine forest. However, they were reported nesting at tree line in eastern Nepal in 1973 (Fleming et al. 1976), and observers have heard drumming by large snipes that very possibly were of this species, at a height of 13,000 feet in central Bhutan (Vaurie 1965). Further, since these snipes have been collected in late summer at elevations of 14,000 feet (*Ibis* 86:387), it seems quite likely that they may indeed be timberline nesters rather than forest-nesting forms.

Status and relationships. This is one of the least-known of all snipes and is generally believed to be

49. Adult heads and outermost tail feathers of Himalayan (*above*) and solitary (*below*) snipes.

very rare. It has relatively rounded wings and a rather slow and heavy flight, which together with its dark coloration give it a woodcocklike appearance. Part of this similarity is doubtless the result of convergence to a forest habitat, and I suspect that its nearest relatives are the semi-woodcocks of South America rather than the other more specialized Asian snipes.

Suggested reading. Ali and Ripley 1969.

Great Snipe

Gallinago media (Lathan) 1787
(*Capella media* of Peters, 1934)

Other vernacular names. Double snipe; bécassine double (French); Doppelschnepfe (German); agachadiza real (Spanish).

Subspecies and range. No subspecies recognized. Breeds in Scandinavia, and from Finland and Russia south to the Baltic countries and east across Siberia to the Yenisei River. Winters in tropical Africa south to South-West Africa and Natal. See map 126.

Measurements and weights. Wing: males 139–53 mm; females 141–51 mm. Culmen: males 57–69 mm; females 59–70 mm. Weights (fall migrants): males 153–225 g; both sexes average 199.0 g (Kozlova 1962). Eggs ca. 45 x 32 mm, estimated weight 23.2 g (Schönwetter 1963).

DESCRIPTION

Adults of both sexes closely resemble *gallinago*, but the mantle and scapulars are darker, with fewer buff markings and with the buff edges narrower and paler, barring on the breast is more pronounced, and barred, extending across the abdomen anteriorly. Additionally, there is more white spotting on the wing coverts, and a white wing bar is formed by the tips of the greater coverts. The tail contains sixteen (rarely fourteen or eighteen) feathers, with the outer feathers relatively broad and barred with dark brown, but the outer three or four pairs are white, without pink or russet suffusion and with little or no barring beyond the outer two-thirds of the feather. The iris is dark brown, the bill is pale yellowish at the base and dark brown toward the tip, and the legs and feet are grayish green or very pale yellowish. *Juveniles* have the outer tail-feathers with brown barring extending to within 10 mm (vs. no more than 15 mm in adults) of the tip on both webs, and the median wing coverts have relatively small white tips that are partly obscured by buff, rather than broad white tips as in adults. They also are considerably more heavily barred on the sides and underparts than most adults. *First-winter* birds might be aged by these same characteristics until about midwinter (Prater et al. 1977).

In the hand, this species may be separated from *gallinago* by the traits mentioned above, and from *nigripennis* by the barring on the anterior abdomen.

In the field (11 inches), this species is found in somewhat drier habitats than is the Eurasian snipe, and when flushed it appears larger and darker and exhibits much white on the outer tail feathers but little or no white on the trailing edge of the wing. However, the white tips of the median and greater coverts form two conspicuous parallel white lines enclosing a blackish "panel" of greater coverts, resulting in a distinctive speculum (Wallace 1977). It also flies directly, without the twisting typical of Eurasian snipes, and is usually quite silent, though it sometimes utters a guttural croaking note when flushing. Its "song" is uttered on the ground and is a twittering or warbling vocalization with a whispering quality. In contrast to those of other snipes, this

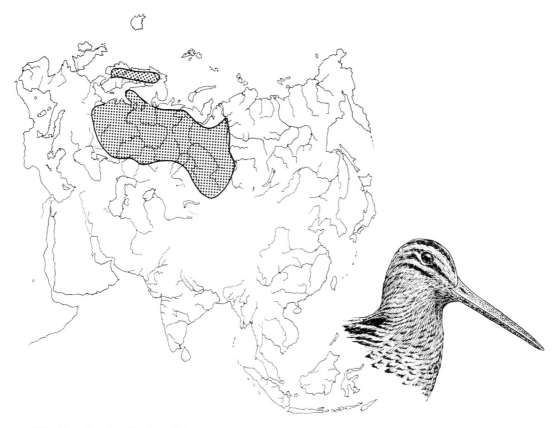

126. Breeding distribution of the great snipe.

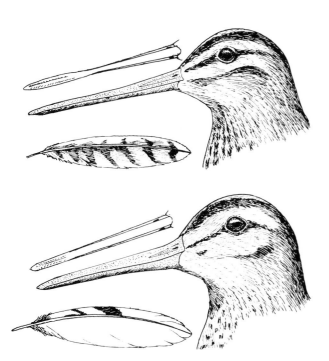

50. Adult heads and outermost tail feathers of Eurasian (*above*) and great (*below*) snipes, including dorsal view of bill.

display is typically performed by several males in an open arena, and the white outer tail feathers may be used as a visual display rather than for producing noise by vibration (Tuck 1972).

NATURAL HISTORY

Habitats and foods. During the breeding season this species is almost exclusively boreal in its climatic distribution, and it breeds on marshy and grassy grounds, in generally more wooded and less swampy habitats than does the Eurasian snipe. It especially favors woods of birches and willows, where the ground is well covered with mosses, lichens, and dead and decaying leaves. Bogs and wetlands in subalpine areas of Scandinavia are used during the breeding season as communal display sites, or leks. In one Swedish study these were areas about 800 meters above sea level, in boggy openings of coniferous forest dominated by sphagnum and sedge cover and a sparse growth of dwarf birch and willows (Lemnell 1978). Outside the breeding season the birds inhabit marshy or swampy areas that are dominated by grassy cover. In Africa the species

usually is seen in fairly large, scattered groups of up to 30 or more in a mile of swamp, and it occurs not only in swamps but also on quite dry ground and high plateaus. Such habitats include rough pastures, moorlands, sand dunes, wood borders, and sometimes also croplands or stubble fields. Foods of the species are rather little studied but include earthworms, snails, slugs, insects, and also some vegetation, including seeds and cultivated grain. Earthworms are probably the major dietary component during much of the year, though insects apparently are important in spring and summer (Glutz et al. 1977; Bannerman 1960). Besides probing for foods, the birds thus must also do a good deal of surface foraging.

Social behavior. In contrast to all other species of snipes, the great snipe performs on a terrestrial display arena, or lek. Unlike the ruff or lek-forming species of grouse, there is no sexual dimorphism of plumage or size associated with this behavior. The dominant activity of males on the lek is the "drumming display," which includes metallic "comb sounds" made as the bird stands on tiptoe, finally making some wing and tail movements. Leks hold from 5 to 30 males through the display season, which in western Sweden lasts from mid-May until July, with the onset probably determined by the time of thawing and the maximum activity occurring during the first few weeks. Early in the season there are morning and evening peaks of activity, but normally display begins at dusk, reaches a peak before midnight, and drops sharply with increasing light intensity. Individual territories are held that average about 120 square meters in area. Each male advertises its presence and proclaims its territory mainly by drumming, though some "flutter leap" displays also occur during peak display activity. Drumming is performed by a standing bird, which initially begins a songlike twittering that may carry up to 100 meters. Then the male stretches his neck as the major sound elements become shorter and stronger. In the third phase the bird opens his bill wide, raises and spreads the tail, and performs a single quick wingflap as the song reaches a peak of pitch and speed. This phase ends with a second quick opening of the bill and the uttering of a high-frequency tone that ends with a click. The final phase is a series of "whizzing" sounds of varied frequency as the tail is spread a second time. Drumming by one male frequently induces this behavior in a nearby one. Although drumming is certainly the major form of territorial advertisement, intimidation, fighting, and chases also occur during territorial defense. The presence of a female on a male's territory results in increased drumming, flutter leaping, wing-trembling, tail-flicking, and crouching while facing the female. The only precopulatory display of females is squatting. Copulation takes only a few seconds, with the male approaching from behind and seizing the female's nape while remaining balanced with wing movements. Apparently only older and experienced males are successful in copulating. A substantial number of the males are nonterritorial and occupy peripheral positions where they are often chased by resident males. Such nonterritorial birds are probably young males or those from other areas (Lemnell 1978).

Reproductive biology. Nests are typically on grassy tussocks in a swamp. The nest may be partly concealed by shrubbery or relatively exposed overhead. Sometimes the birds also nest on dry ground, and in such areas they often resemble woodcocks in their choice of nesting cover. Additionally they sometimes form nest scrapes that are apparently much like those of other more typical shorebirds. There are normally 4 eggs; 9 clutches from Norway included 8 of 4 eggs and 1 of 3. Never have more than 5 been reported. It is believed that the female alone incubates, a suspicion reinforced by the lek behavior of the males, and the incubation period has been variously estimated at from 17 to 24 days. The fledging period is probably 3–4 weeks (Glutz et al. 1977; Bannerman 1960), and, although it seems likely that all brooding care is by the female, there is at least one report of apparent male participation (Tuck 1972).

Status and relationships. This species has diminished greatly in Europe in historical times, the decline probably beginning early in the 1800s. Nesting ceased in the Netherlands by about 1830, in Denmark by 1902, and in Germany by 1926. Now most breeding outside the USSR is in Scandinavia, especially Norway, although the species is still locally common in northeastern Poland. It also breeds locally in Sweden and very rarely in Finland (Hudson 1974). The species' relationships are not at all clear, but I believe it is a rather isolated form, in some ways possibly intermediate between the more primitive forms and the typical snipes.

Suggested reading. Bannerman 1960; Lemnell 1978.

Madagascan Snipe

Gallinago macrodactyla Bonaparte 1839
(*Capella macrodactyla* of Peters, 1934)

Other vernacular names. Malagsy snipe; bécassine Malgache (French); Madagascar-Schnepfe (German).

Subspecies and range. No subspecies recognized. Resident in Madagascar and Mauritius. See map 129.

Measurements and weights. Wing (both sexes): 141–48 mm. Culmen (both sexes): 101–22 mm. Weights: 1 female 216 g (Benson et al. 1976). Eggs ca. 46 x 32 mm, estimated weight 24.3 g (Schönwetter 1963). Benson et al. (1976) report the egg size as slightly larger than these figures.

DESCRIPTION

Adults of both sexes generally resemble *gallinago*, but the axillaries and under wing coverts have sepia bars that are broader than the white bars. The tail, with sixteen feathers, is very similar in structure and color to that of *gallinago*. The species is also closely similar to *nobilis* in size and plumage but is less heavily barred with tawny, and the buffy edging of the mantle and scapular feathers is wider. The iris is "black" (presumably very dark brown), the bill is black, and the legs and toes are slate-colored. *Juveniles* are said to resemble adults but their scapulars and mantle feathers have narrower and paler buffy edges (Meinertzhagen 1926).

In the hand, this species is hard to separate from *nobilis* by its plumage (see above), but its tarsus is reticulated only halfway up, while in *nobilis* the reticulation extends two-thirds of the way. The outer tail feathers may perhaps be somewhat more suffused with brownish in this form, and the legs are said to be slate-colored rather than greenish.

In the field (12 inches), this is the only large species of snipe occurring in Madagascar and thus could hardly be confused with any other. Its calls and display behavior are undescribed except that they are apparently rather similar to those of *gallinago* (Rand 1936).

NATURAL HISTORY

Habitats and foods. Evidently this species occurs in grassy and sedge-covered marshes and swamps, between sea level and 1,800 meters in the humid eastern areas of Madagascar (Rand 1936).

Social behavior. All that is known of the species' social behavior is that Rand (1936) states that the nuptial flight was heard in September and that it is similar to that of *G. g. delicata*.

Breeding biology. Nesting evidently occurs in July and August, since Rand (1936) reported a nest with one egg and several downy young during August. The nest was found on a dry hummock in a grassy swamp. There was a faint runway through the grass between the swamp and the nest, some 3 feet away. The nest was simply a slight hollow with a scanty grass lining. Benson et al. (1976) reported a nest in late November, with a clutch of 2 eggs. It seems likely that 3 or 4 eggs represent the normal clutch.

Status and relationships. Rand (1936) regarded this as a common bird in its limited range. It is almost certainly a derivative of an ancestral form of *G. gallinago*, having evolved a larger body size and sedentary distribution. In many ways, it comes close to *G. nobilis* in appearance, and indeed Seebohm (1888) regarded these two forms as close relatives. However, it seems much more likely that they are a product of parallel evolution from very similar ancestral forms.

Suggested reading. None.

Noble Snipe

Gallinago nobilis Sclater 1856
(*Capella nobilis* of Peters, 1934)

Other vernacular names. None in general English use; bécassine noble (French); Adel-Schnepfe (German); becasina paramera (Spanish).

Subspecies and range. No subspecies recognized. Resident in the Andes of South America from western Venezuela southwest through Colombia to Ecuador. See map 127.

Measurements and weights. Wing (flattened): males 133–48 mm; females 131–45 mm. Culmen (to feathering): males 84–93 mm; females 80–102 mm. One male weighed 188 g, and a nesting female

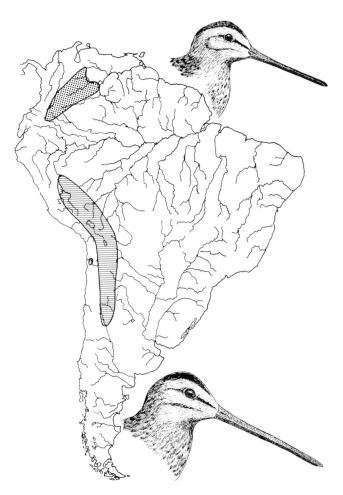

127. Breeding distributions of the noble (*shaded*) and the Peruvian (*horizontal hatching*) and Chilean (*vertical hatching*) puna snipes.

weighed 197 g (Tuck 1972). Eggs ca. 46 x 33 mm, average weight 25.5 g (Tuck 1972).

DESCRIPTION

Adults of both sexes closely resemble *gallinago* and *andina*, but the tawny coloration of the foreneck and chest is stronger, and the birds are more heavily streaked with dark brown. Additionally, the white of the under wing coverts and axillaries is regularly barred with sooty. It also closely resembles *macrodactyla*, but is more extensively barred with tawny, and the upper breast is usually suffused with deeper buff. The iris is brown, the bill is blackish at the tip and paler basally, and the legs and toes are greenish gray. In *juveniles* the buff edges of the scapulars and sides of the mantle are apparently narrower.

In the hand, the large size (exposed culmen 80–100 mm) separates this species from most snipes, and it is distinguished from the very similar *macrodactyla* by the fact that the tarsus is reticulated two-thirds rather than only halfway. It differs from the larger *imperialis* and *stricklandii* snipes in that its underparts are not completely barred.

In the field (12 inches), this species is associated with high-altitude grassy bogs and savannas in the Andes, and only the imperial snipe is of comparable size and found at these elevations. The imperial snipe is much darker dorsally than is this species, and its underparts are also heavily barred. Its vocalizations are poorly known but have been called "clean and melodious," and it is known to produce a very low, deep sound during its "bleating" display flight (Tuck 1972).

NATURAL HISTORY

Habitats and foods. This species is associated with the temperate and páramo zones of the Andes, where it occupies grassy bogs and wet savannas. Vuilleumier and Ewert (1978) found this species in a sedge bog surrounded by montane forest well below the páramo zone, but they noted that it has been collected at the same altitude and location as *jamesoni*, at over 3,000 meters elevation, suggesting that both species may breed in the same páramo vegetation in some areas. The foods and foraging behavior of this species remain undescribed.

Social behavior. There is evidently no description of the aerial displays of this species except for a comment that it is "very low and deep" in tone. The

51. Adult heads and outermost tail feathers of noble (*above*) and cordillerian (*below*) snipes.

outermost tail feathers are distinctly narrowed and average about 6.5 mm wide (Tuck 1972), clearly suggesting that such a display is well developed.

Reproductive biology. Almost all that has been known of the reproduction of this species is the existence of 2 eggs in the British Museum (Schönwetter 1963). Tuck (1972) described a nest with 2 eggs found in late July on the Páramo de Tama, Tachira, Venezuela. It was in a depression on a dry hummock of grasses and mosses in a marshy area.

Status and relationships. Nothing can be said of this species' status, since its habitat is so remote and little studied. Externally it seems to be very close to *gallinago*, especially the South American races. It seems likely to have been an offshoot of *gallinago* stock that became adapted to temperate and alpine habitats in northern South America, perhaps about the same time that a smaller derivative (*andina*) was similarly adapting to drier punalike habitats in southern parts of that continent.

Suggested reading. None.

Snipe

Gallinago gallinago L. 1758
(*Capella gallinago, C. delicata,* and *C. paraguaiae* of Peters, 1934).

Other vernacular names. Common snipe; fantail snipe; Wilson snipe; bécassine des marais (French); Sumpfschnepfe (German); becasina chillona, agachadiza común (Spanish).

Subspecies and ranges. See map 128.

G. g. gallinago: Eurasian snipe. Breeds in the British Isles and Eurasia from Scandinavia and France eastward through Russia, Siberia, Mongolia, and Manchuria to the Sea of Okhotsk, Sakhalin, and Kamchatka, and locally in northern India, northeastern Afghanistan, and probably in the Azores. May also breed in the western Aleutian Islands (*Condor* 80:311). Winters south to the equator of Africa, India, the Indo-Chinese countries, and the Greater Sundas.

G. g. faeroeensis: Faeroe snipe. Breeds in Iceland and in the Faeroes, Orkneys, and Shetlands. Winters south to the British Isles.

G. g. delicata: North American snipe. Breeds in North America from Alaska and California eastward across the northern United States and Canada to Hudson Bay, northern Quebec, Labrador, Newfoundland, New Brunswick, and southward to Utah, Colorado, Iowa, Ohio, and West Virginia. Breeds locally south to northern Mexico. Winters from British Columbia and Georgia south to Central and South America.

G. g. paraguaiae: Paraguaian snipe. Resident in South America from Trinidad south on the eastern side of the Andes to eastern Peru, Bolivia, Paraguay, Uruguay, and northern Argentina.

G. g. magellanica: Magellanic snipe. Resident in South America from northern Chile and southern Argentina south to Cape Horn. Winters to Uruguay and northern Argentina. Also breeds on the Falkland Islands. Sometimes regarded (with *paraguaiae*) as a distinct species.

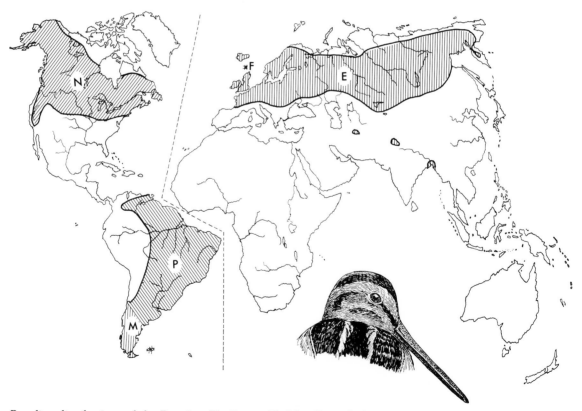

128. Breeding distributions of the Eurasian (E), Faeroe (F), Magellanic (M), North American (N), and Paraguaian (P) snipes.

Measurements and weights. Average (flat) wing measurements for most races range from 122 mm (*paraguaiae*) to 133 mm (*magellanica*), with maximum ranges for these two forms 118–37 mm. Extreme culmen lengths range from 58 mm (*delicata*) to 78 mm (*paraguaiae*). Average weights (*gallinago*) range seasonally from 97 to 125 g (Glutz et al. 1977). Weight data on other races are less complete, but adults of *delicata* average about 100 g during summer months, with females about 16 percent heavier than males. Winter weights are similar, but sex differences are not so great. The average of 98 *paraguaiae* was 112.0 g (Tuck 1972). Eggs average about 39 x 28 mm, estimated weight 16.0–20.7 g (Schönwetter 1963).

DESCRIPTION

Adults of both sexes have the crown black, with a buffy central stripe and a buff superciliary stripe, bordered below by a dark stripe from the base of the bill to the eye and beyond. The general upperpart coloration is a mixture of black, brown, and sooty ashy, forming mottling, spots, and barring, being darkest on the hindneck, back, and scapulars, and there are several buff to cinnamon stripes extending down the scapulars. The primaries and inner secondaries are uniformly dark or are tipped with white, and the rump and upper tail coverts are grayish or buff with fuscous spots or bars. The inner tail feathers are black basally and cinnamon to russet posteriorly, with narrow black bars and white tips, while the outer tail feathers are white with russet tinges and black barring. The underparts are mostly white, with the flanks, under tail coverts, under wing coverts, and axillaries barred with black. The foreneck and breast are buffy but are heavily spotted, streaked, or mottled with darker brown. The iris is brown, the bill is greenish gray (also reported as brownish and pinkish) at the base and black toward the tip, and the legs and toes are greenish gray. The races differ somewhat in the extent of barring on the sides and underparts, the amount of russet on the tail feathers, and the amount of buffy on the breast and dorsally. The number of tail feathers is typically fourteen in *gallinago* (extremes twelve to eighteen) but normally is sixteen in the North American and South American forms, with the outermost pair unusually narrow in the South American races. *Juveniles* have distinctive pale buff fringes on the median and lesser wing coverts and usually lack a dark shaft streak extending onto the vane. Additionally, young

birds show more heavily worn primaries than adults through their first winter (Prater et al. 1977). A marginal black tip on the upper wing feathers is typical of juveniles (Tuck 1972).

In the hand, this species is best recognized by the fourteen to sixteen tail feathers, the outermost of which are usually not distinctly narrowed and show some black barring as well as russet tinges. The white abdomen and nonyellow legs and feet distinguish it from *media* and *audina*, respectively, while a bill length of no more than 78 mm and russet-tinted outer tail feathers separate it from *nigripennis*.

In the field (10 inches), identification is a problem only where other similar-sized snipes occur. When it flushes, this species utters a distinctive *scape* note, and it flies both rapidly and erratically. The narrow white trailing edge on the tips of the secondaries helps to separate this species from nearly all the other Eurasian snipes, but it cannot be separated from the African snipe in flight, and the call differences must be noted to separate it from the puna snipe, as indicated in that species' account. The "winnowing" display is accompanied by a series of quickly repeated *who* sounds, which are mechanically produced but apparently are rather different between North American and European birds (Thönen 1969).

NATURAL HISTORY

Habitats and foods. The breeding habitat of the North American race of this species has been described as being restricted to organic soils, primarily peatlands, mostly within the northern forest zone. The primary peatland habitats occupied are bogs, fens, and swamps, but wet areas along ponds, rivers, brooks, or other marshy habitats are also utilized

North American Snipe winnowing

(Tuck 1972). The European race likewise occupies marshy bogs and moors, grassy or marshy shores of rivers and lakes, swampy meadows, wet hayfields, and marshy patches of shrub tundra (Voous 1960). In South America the breeding habitats range from tropical and subtropical savannas of the llanos to the pampas and steppes of Patagonia and Tierra del Fuego, where the major cover is provided by tussock grasses (Tuck 1972). In Europe, breeding densities in large areas of suitable habitat typically range from about 0.5 to 10.0 pairs per 100 hectares (Glutz et al. 1977), but much higher densities have been locally reported, such as on the St. Kilda Islands, at Kirgiz, USSR, and in other very favorable local habitats (Tuck 1972). In Canada, breeding densities in sedge bogs, fens, and alder or willow swamps generally range from about 5 to 17 pairs per 100 hectares. Wintering habitats are mostly freshwater environments, but the birds sometimes occur on coastal meadows. The wintering habitats are typically

marshy areas having mucky soils that are rich in decomposed organic matter. Additionally, cattle pastures, coastal meadows, rice fields, sugarcane fields, drainage ditches, and other areas providing a combination of grassy cover and moist, organic-rich soils are utilized (Tuck 1972). While feeding, the birds typically probe in water-saturated soils, obtaining most of their food without removing the bill from the soil. Foraging is done in small areas, with little walking, and with the bird often probing in a small semicircle by pivoting with the legs. Many probes are brief, while others may last about half a minute (Burton 1974). Animal materials compose the majority of foods, with insects, especially larval forms, ranging from about 10 percent to as much as 80 percent of total foods, on both breeding and wintering areas. Fly and beetle larvae are especially important insect foods for snipes. Additionally, earthworms, small crustaceans (especially isopods and amphipods), small snails, and occasional other invertebrate foods

416 ❖ ❖ ❖

are also eaten by the North American race. Plant fibers, seeds, and grit are also ingested in small quantities (Tuck 1972). Foods of the Eurasian race are probably very similar to this, but there is less information available on that form (Glutz et al. 1977).

Social behavior. Males are strongly territorial and select areas of wetland over which they display and attempt to attract females. Males typically arrive on the breeding grounds well in advance of the females and immediately establish territories. Territories may initially be rather large, perhaps as much as 10 hectares, but they gradually shrink to include a small area immediately around the nest itself (Tuck 1972). The most common territorial display is bleating, which may occur at other times of the year but is most intense on the breeding grounds. It is mostly a male display, although females bleat occasionally during early stage of breeding. The bird takes off, ascends to 100–200 feet, and begins a generally circular course above its territory. It then begins a 45° angle dive, its tail fanned horizontally and its wings continuing to beat. The noise generated by the vibrating outer tail feathers, as modulated by the periodic interruption of wind flow past them by the moving wings, produces the distinctive "winnowing" sound (Reddig 1978). Differences in the configuration of the outermost tail feathers are evidently responsible for acoustical differences in the sounds produced by North American and European forms (Thönen 1969). After an extended series of such dives, the bird raises his wings diagonally upward and drops rather rapidly back into the marsh. There are several other less frequent displays, of which the most important is perhaps the arched-wing display. This may occur high in the air or close to the ground, and it often occurs after a series of excited *jick-jack* or yakking calls. One or both wings are raised and the tail is fanned, causing the bird to drop quickly downward. At times the bird will perform a sideways somersault, or even fly upside-down for a brief period before resuming its normal position. This display is apparently part of pair-formation and may help entice a female into the male's territory. During pair-formation the birds often perform aerial chases, alternately calling and bleating. Sometimes the male apparently tries to force the female to the ground; he then follows close behind, and copulation may occur after a short ground chase. While chasing her, the male will strut with drooping wings and erect his fanned tail, while the female will squat with her wings drooped or outspread. The male flutters up onto her back and flutters his wings while copulating. There is no apparent postcopulatory display. Early copulations do not always occur in the territory, but gradually the male seems to become progressively concerned with keeping the female within the territorial boundaries (Tuck 1972).

Reproductive biology. The nest is usually placed in a fairly dry location, even if the surrounding area is very wet. Of 563 nests studied in Great Britain, about half were in wet or dry pastureland, 30 percent were in marshes, bogs, or fens, and most of the rest were on moorlands. Grasses, rushes, or sedges were the usual cover for these nests (Mason and McDonald 1976), while in Newfoundland sphagnum is the most frequent cover. In nearly all cases the cover is tall or thick enough to hide the nest from above, and frequently a nest canopy is formed by the birds' interweaving the tips of grasses or sedges. The clutch is normally 4 eggs in North America and Europe, while the race *faroeensis* averages slightly less, being 3.3 on St. Kilda and 3.7 on the Faeroe Islands. Apparently the usual clutch of *paraguaiae* is 4 eggs, but that of *magellanica* is said to normally contain only 2 eggs (Tuck 1972). Although males have at times been reported to help with incubation, current evidence favors the view that only the female incubates. Incubation probably begins with the laying of the third egg and requires 17–20 days, probably most often 19 days. Most of the eggs hatch within an interval of 4–8 hours. The fledging period is only 19–20 days, though some fluttering is possible by birds that are only 2 weeks old, and the flight feathers are not fully grown until 30–35 days of age. There is a protracted breeding season in Europe, which has led to the assertion that the species may be double brooded there. However, there is still no direct evidence that this is the case, and the long breeding season may reflect only renesting or late nesting by young birds (Tuck 1972).

Status and relationships. A review of North American populations of this species does not show any obvious recent trends, and the total North American population probably numbers in the low millions (Tuck 1972). In Europe the species has apparently not suffered the range retraction typical of the great snipe and jack snipe, and management for hunting purposes has a long tradition in that area. As to relationships, this species is in many ways a central component of the genus *Gallinago*. The forms *andina* and *nigripennis* are scarcely separable from *gallinago* morphologically, and, when more is known of their behavior and reproductive biology, it may prove necessary to consider them only subspecifically distinct. Only slightly more distantly removed morphologically are *macrodactyla* and *nobilis,* and while these probably are "good" species I think that they belong in the same species group with the superspecies made up of *gallinago, andina,* and *nigripennis.*

Suggested reading. Tuck 1972; Bannerman 1960.

Puna Snipe

Gallinago andina Taczanowski 1874 (1875)
(*Capella paraguaiae* in part, of Peters, 1934)

Other vernacular names. Andean snipe; bécassine du puna (French); Punaschnepfe (German); becasina andina (Spanish).

Subspecies and ranges. See map 127.

G. a. andina: Peruvian puna snipe. Breeds in the puna zone of the Andes of South America from central Peru south to northern Chile and northwestern Argentina, wintering at lower elevations.

52. Adult heads and outermost tail feathers of Paraguaian (*above*) and puna (*below*) snipes.

G. a. innotata: Chilean puna snipe. Known only from the puna zone of Antofagasta, Chile, on the lower Loa River.

Measurements and weights. Wing (flattened): males 113–25 mm; females 114–21 mm. Weights: males 65–90 g, average of 3 was 77.6 g; females 90–105 g, average of 4 was 97 g (various sources). Eggs ca. 38 x 27 mm, estimated weight 15.3 g.

DESCRIPTION

Adults of both sexes are very similar to those of *gallinago,* but the axillaries have narrower blackish bars. The under wing coverts may be heavily marked (*andina*) as in *gallinago,* or immaculate white (*innotata*). In addition, there is white edging and tipping on the two outermost (*andina*) or all (*innotata*) of the primaries. The iris is brown, the bill is yellowish with a black tip, and the legs and toes are bright yellow. *Immature* plumages are undescribed but probably resemble those of *gallinago.*

In the hand, this species is readily distinguished by the combination of its bright yellow legs and its small size (flattened wing no more than 125 mm).

In the field (9 inches), this species is associated with the high puna zone of the Andes, occurring between 13,000 and 14,000 feet (3,800–4,300 m) in Peru and at lower altitudes in northen Chile. Its call is quite different from that of *gallinago* and is said to resemble the melodious, almost cooing calls of the graybreasted seedsnipe (*Thinocorus orbignianus*). It is the only small snipe of the high Andes, but in winter it also occurs on the plains of northern Argentina where Magellanic snipes might also be found. Where both species occur, *andina* can be recognized by its higher-pitched *itch* alarm call, as compared with *atch* in *gallinago* (C. Olrog, pers. comm.).

NATURAL HISTORY

Habitat and foods. This species is found in the puna zone of the Andes, mainly along rivers flowing through valleys, rather than in wet savannas and marshes. The race *andina* is commonly found at altitudes as high as 13,000 to 14,000 feet, while *innotata* evidently occurs at considerably lower altitudes in its very limited range. Presumably its foods and foraging behavior are very much like those of *gallinago.*

Social behavior. Almost nothing has been noted on this. François Vuilleumier (in litt.) informs me that he heard the display calls of this form only once, in the Cordilla Blaca of Peru, at 4,050 meters at Lagunas Queshque, in Huascarán National Park. Snipes were not rare up to 4,200 meters, and at dusk and dawn of late October and early November they were regularly heard during their display flights. The associated calls are a repeated *kek-kek-kek . . .*, but Dr. Vuilleumier did not remember any more of the display details.

Reproductive biology. Apparently the only record of breeding is of an egg and a young bird from the Parinacota marshes of Arica, Chile, found in late September. Johnson (1965) reported that this represents an extremely early nesting period, a full two months earlier than most birds of that region.

Status and relationships. Evidently the race *innotata* is extremely restricted and must therefore have a very limited population, but Johnson (1965) believed it might occur at several localities along the Loa River. The other race is much more widespread and presumably fairly common in the puna zone. This form is of questionable specific status and is usually considered a subspecies of *gallinago* or, if *paraguaiae* is regarded as a distinct species, as a race of the latter. About the only evidence for judging this is the condition of the outermost tail feathers, which are slightly less barred than in *paraguaiae*, and Olrog's (in litt.) observation that *andina* and *gallinago magellanica* have distinctly different alarm calls. According to data of Tuck (1972), *andina* has a slightly shorter and wider outer tail feather than does *magellanica* or *paraguaiae* and tends to be smaller in measurements of wing length, bill length, and tarsus length as well. Although I personally suspect it could well be considered a race of *gallinago*, I have followed Blake (1977), giving it species status for the present.

Suggested reading. None.

African Snipe

Gallinago nigripennis (Bonaparte) 1839

Other vernacular names. Ethiopian snipe; bécassine Africaine (French); Afrikanische Bekassine (German).

Subspecies and ranges. See map 129.
  G. n. nigripennis: East African snipe. Resident in Africa, from South Africa to Ethiopia.

G. n. angolensis: West African snipe. Resident in Africa, from Angola and South-West Africa (Namibia) to Zambia and Rhodesia, where it intergrades with *nigripennis*.

Measurements and weights. Wing (both sexes): 120–41 mm. Culmen (both sexes): 73–88 mm. Weights: 22 adults of both sexes 93–164 g, average 114 g (Britton 1970). Skead (1977) indicates slightly lighter weights for South Africa. Eggs ca. 42 x 30 mm, estimated weight 19.0 (Schönwetter 1963).

DESCRIPTION

Adults of both sexes closely resemble *gallinago*, but the upperparts including the bands on the head are much darker, and the ground color of the mantle and scapulars is almost black. The tail is also distinctive,

129. Breeding distributions of the Madagascan (*shaded*) and the east African (E) and west African (W) snipes.

53. Adult heads and outermost tail feathers of African (*above*) and Japanese (*below*) snipes.

with sixteen to eighteen feathers, and at least the three outer pairs of feathers are predominantly white rather than tinged with brownish. The outermost pair is also distinctly narrower than in *gallinago* and has only a few dusky spots or broken bars on the outer web, rather than having broad and continuous bars of dark brown across both webs. The iris is brown, the bill is greenish black to black, and the legs are greenish brown or yellow green. *Juveniles* resemble adults, but the feathers of the mantle and scapulars have narrower edges.

In the hand, the tail characteristics mentioned above will separate this species from the closely similar *gallinago*, and the relatively longer bill (culmen at least 73 mm) helps distinguish it from *media*, which also has a barred belly and lacks white edging on the outermost primary. It is extremely similar in measurements and appearance to *hardwickii*, which also has sixteen to eighteen tail feathers, but the outermost feathers of that species are distinctly barred, while *nigripennis* has only a few dusky spots or broken bars.

In the field (11 inches), this species might be easily confused with either the Eurasian snipe or the great snipe, but it is darker dorsally than either of these. Its alarm note has been reported both as a quiet *tchek*

and a *gallinago*-like *skaap*. It is also said to produce a dull but far-carrying piping call or whistle, usually repeated several times, as well as aerial drumming noises, during the breeding season. For additional field identification information, see *Scopus* 4:1–5.

NATURAL HISTORY

Habitats and foods. This species is also associated with the moist edges of lakes, vleis, and marshes, and with riverbanks. However, it nests mainly in moist upland areas. Areas of flooded shortgrass are favored feeding ground, where the birds eat larvae of beetles, dragonflies, and flies, as well as small crustaceans and mollusks. The birds also probe for worms, and they sometimes eat seeds as well (Clancey 1967; McLachlan and Liversidge 1957).

Social behavior. The aerial displays of this form have not been critically compared with those of *gallinago*, but they are known to consist of stooping while producing a tremulous bleating sound. A close comparison of this display and the bird's vocalization would be extremely useful.

Reproductive biology. Nesting in South Africa occurs from April to October, during the relatively cool winter months. In Zambia breeding has been reported for May and June, and in Ethiopia the species has been reported breeding in July and August. Nesting in Kenya is evidently prolonged as it has been noted for January, April, June, and September. The nest is very similar to that of *gallinago*, and the is 2 to 3 eggs in central Africa, but perhaps only 2 in South Africa. Almost certainly in other aspects its breeding biology is comparable to that of *gallinago*, but there are no details (Mackworth-Praed and Grant 1952; Clancey 1967).

Status and relationships. This species is certainly locally abundant, such as in the marshes of Kenya and Tanzania between October and March, but like most snipes its actual status is impossible to judge. It is often considered a subspecies of *gallinago* and, except for its darker color and longer bill, and its additional outer tail feathers and their more specialized condition than in the Old World form of *gallinago*, this might well be the most obvious conclusion. Yet, until more information on the form becomes available, it seems prudent to follow the tradition of African ornithologists and consider these distinct species.

Suggested reading. Clancey 1967.

Japanese Snipe

Gallinago hardwickii (Gray) 1831
(*Capella hardwickii* of Peters, 1934)

Other vernacular names. Latham's snipe; Australian snipe; bécassine du Japon (French); Kreischbekassine (German).

Subspecies and range. No subspecies recognized. Breeds on southern Sakhalin, in the southern Kurils, and in Japan from Hokkaido south to central Honshu. Winters in Australia. See map 130.

Measurements and weights. Wing: both sexes 154–69 mm. Culmen: both sexes 74–84 mm (Rand and Gilliard 1968). Weights: both sexes 95–189 g, average of 122 was 154 g (D. Purchase, pers. comm.). The average of 249 males was 151.1 g, and that of 250 females was 161.8 g (Frith 1976). Eggs ca. 44 x 30 mm, estimated weight 20.2 g (Schönwetter 1963).

DESCRIPTION

Adults of both sexes are generally similar to those of *gallinago*, but with relatively longer wings and tail, and closer in size and color to *megala*. However, this species has sixteen to eighteen rather than twenty tail feathers, with the outermost pair 4–6 mm wide and the second pair 6–8 mm wide. Both of these feather pairs are barred with brown and white. Additionally, there is a broader trailing white edge on the wing feathers (at least 2 mm, vs. 1 mm or less). Generally, the plumage is heavily marked with brown, black, and buff on the upperparts, including the rump. Dark brown stripes, with buff centers and edging, extend from the forehead over the crown. The underparts are of gray and softer browns, with a pale throat. The iris is brown, the bill is black, and the legs and toes are grayish olive. *Females* average slightly heavier than males, and the outermost pair of tail feathers is slightly shorter and broader. *Juveniles* have a distinct pale buff fringe on their median upper wing coverts, and most of the primaries have definite white tips. By comparison, adults have median coverts that are brown and buffish brown, with distinct dark shaft streaks forming distal spots, and the primaries are indistinctly tipped with paler coloration (Prater et al. 1977).

In the hand, the tail characteristics mentioned

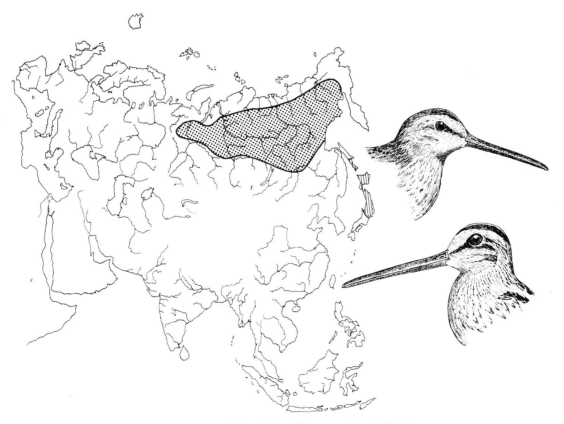

130. Breeding distributions of the pintailed (*shaded*) and Japanese (*hatched*) snipes.

above help separate this species from *megala*, the absence of breast and abdominal barring distinguishes it from *nemoricola*, and the absence of gingery tones on the breast and upperparts separates it from *solitaria*. It has two more tail feathers than are normally present on *gallinago*, and the outermost pair is narrower.

In the field (11 inches), this species is found in swampy areas and grassy fields. When flushed, it flies in a zigzag manner, usually uttering a loud *arrk*, *shek*, or *cresk* note. The display behavior of this species has been described as a circular flight, performed fairly close to the ground, accompanied by harsh, repeated *rack* notes interrupted by tremulous sounds apparently produced by feather vibration (Tuck 1972). The notes given before "drumming" have also been described as a harsh, repeated *khha*, and a sucking sound resembling *kee-oow, wee-oow*.

NATURAL HISTORY

Habitats and foods. On its Australian wintering grounds, this species is usually found in freshwater areas, but it sometimes occurs on saline or brackish places such as mangrove creeks. Fresh meadows, seasonal fresh swamps, semipermanent fresh swamps, and open fresh water are the major wintering habitats; soft mud, shallow water, and some cover seem to be the essentials (Frith et al. 1977). In the Kurils the birds breed in wet glades of open pine forests, while in Japan breeding usually occurs in dry areas on well-drained soils, often far from water. On

southern Sakhalin they frequent river valleys with dry meadows, and the birds prefer to nest in hayfields and pastures (Nechaev 1973). Foods there include various insects and their larvae, plus grass leaves and roots, while a more extensive sample from Australia indicates that about 55–60 percent of the food is of animal origin and the rest is vegetable material. About half the plant food was seeds and the remainder plant fibers, while the animal foods were nearly all insects, earthworms, and spiders (Frith et al. 1977).

Social behavior. Several observers have commented on this species' aerial displays, usually noting the dives or plunges made from considerable height. Fennell's (1953) description indicates that it is strongest at dusk and consists of a circular flight 25–30 feet above ground, accompanied by harsh *zrack* notes that are regularly uttered and interrupted by the sounds made during the "power dives," which he referred to as "calls" but which certainly are instrumental sounds as in other species. Nechaev's (1973) description indicated that the birds display singly or in groups of 4–6, and also that the circling occurs at a greater height (70–100 meters), from which the birds dive down to 30–40 meters and utter quick chirping sounds at the bottom of the plunge. According to him, this species sometimes displays in groups, with one bird after another swooping downward, then catching up to the rest of the flock. During a 20–30 minute interval the flight might make up to 10 plunges, after which the birds land on the ground, in trees, or on telegraph poles. They sometimes also call from these locations, in the manner of the forest snipe.

Reproductive biology. Nests with eggs of this species are sometimes found as early as late April and as late as July in Japan, while on Sakhalin they have been found in May and June. The nest is usually built in dense grass, or sometimes under overhanging shrubs. It has been found nesting as high as 1,380 meters in the mountains of Japan, and in one case newly hatched chicks were observed on a dry mountain slope some 2.5 kilometers from the nearest water. The normal clutch is 4 eggs, and only females have been found incubating (Kozlova 1962; Nechaev 1973). There seems to be no information on brooding behavior, other than that females perform strong distraction displays when flushed from the nest. The young birds apparently remain on the nesting grounds until fall departure, with groups of up to 4 individuals usually being found in meadows with marshes and varied herbage present. The remains of small insects were noted in the stomachs of 2 nestling birds (Nechaev 1973).

Status and relationships. Although there is no certainty, it is believed that this species is declining in Japan and also has become less common on its Australian wintering grounds (Frith et al. 1977). Breeding-ground changes associated with industrialization have probably adversely affected the birds, and they are also hunted extensively in Australia, so the species should be watched closely. The evolutionary relationships of this species seem clearly to be with *megala*; Larson (1957) considered *hardwickii* to be clearly derived from *megala*, and both the behavior and the ecology of the two forms have a good deal in common. It does have considerably longer wings than does *megala*, but this is obviously associated with its strong migratory pattern.

Suggested reading. Frith et al. 1977.

Forest Snipe

Gallinago megala Swinhoe 1861
(*Capella megala* of Peters, 1934)

Other vernacular names. Chinese snipe (Australia); marsh snipe; pin-tailed snipe (Australia); Swinhoe's snipe; wood snipe; bécassine de Swinhoe (French); Waldbekassine (German).

Subspecies and range. No subspecies recognized. Breeds in central Siberia from about 82° E longitude eastward to Lake Baikal and southwestern Transbaikalia. It apparently also breeds in southern Ussuriland and perhaps on Sakhalin. Winters from the Phillipines and western Micronesia to India, the Indo-Chinese countries, the Sundas, New Guinea, and Australia. See map 125.

Measurements and weights. Wing: males 131–38 mm; females 131–39 mm. Culmen: males 57–68 mm; females 60–70 mm. Weights: both sexes 112–64 g, 6 males averaged 138 g (Shaw 1936). Eggs ca. 42 x 31 mm, average weight 21.7 g (Naumov 1962).

DESCRIPTION

Adults of both sexes are generally similar to those of *gallinago*, and especially to those of *stenura*. The tail is of twenty to twenty-two feathers (rarely eighteen to twenty-six), with the outermost pair 2–6 mm wide, the second pair 4–6 mm wide, and the outer six pairs distinctly narrowed, though not so much as in *stenura*. There is a narrow (ca. 1 mm wide) whitish

54. Adult heads and outermost tail feathers of forest (*above*) and pintailed (*below*) snipes.

trailing edge to the secondaries, and the wing coverts are often more clearly barred than in other species. The iris is dark brown, the bill is yellowish brown, becoming more yellow at the base of the upper mandible, and the legs and toes are yellowish gray to leaden gray. *Juveniles* have brown median coverts with whitish buff fringes, and the tertials have large, pale buff mottled areas, with a whitish buff fringe around the distal third. By comparison, adults have median coverts that are barred and distally spotted with dull buff, and with a definite central shaft streak. Adult tertials have clear brown and buff barring, without a buffy fringe (Prater et al. 1977).

In the hand, this species is most likely to be confused with *stenura*, from which it can be separated by the number and relative narrowness of the outer tail feathers and its longer tarsus (32–35 mm, vs. 29–31 mm). It might also be confused with *hardwickii*, but is slightly smaller, has twenty rather than eighteen tail feathers; also, in *hardwickii* the inner secondaries are shorter than the folded wing.

In the field (11 inches), this species is found in grassy localities near marshes and streams. It flies in a zigzag manner, sometimes uttering a sharp rasping note. On the breeding grounds it has a display flight in which it soars to a great height, utters a *tchiki-tchiki-tchiki* call, then descends while making a whistling noise. It sometimes also makes the same call from trees.

NATURAL HISTORY

Habitat and foods. Although in its wintering areas this species is sometimes found in marshy areas and paddy fields, this is not the case during the breeding season, when it is limited to well-forested plains, river valleys, and clearings or margins of well-wooded regions. It has also been reported in alpine meadows at timberline. It apparently avoids wet and boggy sites and breeds in both coniferous and deciduous forests. Its food consists of earthworms, adult and larval insects, and terrestrial mollusks. The birds often forage among hummocks in grass and on mud flats around seepage areas (Kozlova 1962).

Social behavior. Courtship display begins with spring arrival and persists well past the appearance of young birds. The birds display solitarily, often with considerable distances between them, suggesting strong territoriality. With its chirring *chiki-chiki-chiki*, the bird first ascends and then closes the wings and plunges sharply downward, producing an in-

creasingly stronger sound resembling a twirling metal object. It breaks its plunges only a few meters from the ground, quickly ascending again. These activities are strongest in the evening until about midnight, and again from about 2:00 a.m. until daylight. Later the bird sometimes sits on a stump, fallen tree, or dead tree, still chirring strenuously. Apparently the birds are monogamous, and they display directly above the nesting locality, but reportedly males do not participate in incubation (Kozlova 1962).

Reproductive biology. Nests are among shrubbery, in meadows, or in swamps or bogs on slight elevations or other situations that allow for a dry substrate. The nest is a simple structure, and usually consists of 4 eggs, but an incubated set of 2 has also been reported. Like other snipes, the female sits very close and performs wing-drooping distraction displays if flushed from the nest. Eggs have been found from May to mid-August, suggesting a rather long nesting season, and young have been reported as early as late May (Kozlova 1962). It has been suggested that the male shares in the care of the chicks and that there is a fledging period of 18–20 days (Tuck 1972).

Status and relationships. As with other snipes, it is difficult to judge the rarity of this species, as it is inconspicuous and easily confused with similar forms. Tuck (1972) indicates that *megala* is most similar in appearance and behavior to *hardwickii*; Vaurie has listed it between *hardwickii* and *stenura*. Larson (1957) believes that *hardwickii* is "undoubtedly" a derivative of *megala*, and their allopatric ranges suggest that they constitute a superspecies. I believe that these two species not only are closely related but also are fairly close to *stenura*.

Suggested reading. Dementiev and Gladkov 1969.

Pintailed Snipe

Gallinago stenura (Bonaparte) 1830
(*Capella stenura* of Peters, 1934)

Other vernacular names. Pin-tailed snipe; bécassine à queue pointue (French); Spiessbekassine (German).

Subspecies and range. No subspecies recognized. Breeds from extreme northwestern Russia (western foothills of Urals) eastward through Siberia to western Anadyrland and the coast of the Sea of Okhotsk. Winters from southeastern China and Taiwan south to the Indo-Chinese countries and Malaya, India, and the Sundas. See map 130.

Measurements and weights. Wing: males 120–30 mm; females 123–32 mm. Culmen: males 57–64 mm; females 58–68 mm. Weights: 3 males 106–37 g, average 123 g; 3 females 124–55 g, average 136 g (Shaw 1936). A group of 472 winter specimens of both sexes averaged ca. 113 g (Ali and Ripley 1969). Eggs ca. 41 x 29 mm, average weight 14.4 g (Berman and Kuz'min 1965).

DESCRIPTION

Adults of both sexes are very similar to *gallinago,* but the white tips on the secondaries are only about 1 mm wide, and the tail has twenty-six feathers (range twenty-four to twenty-eight), of which the outer eight pairs (extremes six to nine) are pinlike and only 1–2 mm wide. The under wing coverts and axillaries are more heavily barred than in *gallinago,* but apart from the tail condition these two species are extremely similar in plumage. The iris is brown, the bill is horn brown, becoming blackish toward the tip and pale greenish toward the base. The legs and toes are greenish gray to bluish gray. *Juveniles* have pale whitish buff fringes on the lesser and median wing coverts, and the inner eight primaries are tipped with pale buff; by comparison, adult lesser and median coverts have a dark central shaft streak, producing a spotted appearance, and the primaries have narrow and indistinct whitish buff tips. The primaries of first-year birds are slightly worn in winter, and moderately worn in spring, while in adults at least some of the primaries are freshly grown and unworn in winter (Prater et al. 1977).

In the hand, the large number of tail feathers and the pinlike condition of the outer ones readily separate this species from all others.

In the field (10½ inches), pintailed snipes are associated with marshy areas and wet fields, and are found in generally damp areas rather than the moist substrates typical of the Eurasian snipe. In flight this species flies more slowly and heavily than does that species, but with some zigzags, and its call is a short, raspy *squik* or *squok* similar to that of the Eurasian snipe. Unlike the Eurasian snipe, the under wing surface is very heavily barred, and there is no prominent white trailing edge on the secondary feathers; additionally the upper wing surface is marked with pale

buffy (Madge 1977). As in the great snipe, there is a communal display. but it takes place in the air rather than on the ground. It consists of vertical plunges and associated short metallic calls, *tcheka-tcheta-tcheka,* which merge with the sounds made by the vibrating tail feathers (Tuck 1972).

NATURAL HISTORY

Habitats and foods. In wintering areas, this species is often found in marshy areas and flooded paddy fields, but it reportedly tends to occupy somewhat drier sites than does *gallinago,* using damp rather than wet substrates for its foraging. Its breeding grounds include almost the entire forested area of Siberia, and its nesting habitats are said to include grassy swamps in taiga, damp meadows along river valleys, sphagnum bogs, timberline, and areas of tundra having patches of dwarf birch at altitudes of about 2,300 meters. However, it is not nearly so much a montane species as is *solitaria,* and it is obviously closer to *megala* in its ecology. Its foods are not well known, but they include mollusks, insects and their larvae, and earthworms. Muddy stream banks and muddy shorelines of swamps are often used for foraging (Kozlova 1962).

Social behavior. These birds are usually seen in only small groups during migration and on the wintering grounds. However, it has what is perhaps the most interesting of all snipe aerial displays, judging from recent observations by Berman and Kuz'min (1965). In mid-May they observed a mass mating flight that started as the day cooled and continued until complete darkness. Groups of 10–15 males flew together, maneuvering in synchrony and plunging more and more vertically while uttering short metallic *tcheka-tcheka-tcheka* calls. These calls soon merged with the noises made by their tail feathers, the sounds becoming stronger, higher, and longer as each bird descended toward the ground, pulled out, and soared back upward. This leklike display was presumably for mating purposes; courtship flights have also been seen on the breeding grounds as late as July and August, where they presumably serve as territorial advertisement signals. Kozlova (1962) states that the display usually lasts until late July and occurs above the area where the birds mate or where the nest is situated. Calling at times also occurs while a bird is standing on some elevated site.

Reproductive biology. Several nests of this species have been described, and have been located in such sites as a larch forest among dead grass and a *Vaccinium* mat, in a tuft of sedges and grasses near a dry hummocky bog, among alders and dwarf birches on a burned-over forest slope, and on a dry moss hummock in tundra. Two Siberian nests were among dwarf birches on dry slopes or ridges. The usual clutch size is 4 eggs, but at least one 3-egg clutch has also been found. Apparently no males have been collected on eggs or while attending young (Kozlova 1962), but the chicks are reportedly cared for by both parents (Berman and Kuz'min 1965).

Status and relationships. This is probably one of the more abundant Asian snipes, but as usual there is no good basis for judging its actual population. Its tail structure is the most specialized among all the snipes, and indeed among all the shorebirds, but in all other respects it is hardly separable from *gallinago* or *megala*. I believe it is a close relative of both these forms, with the trend of specialization being in the sequence *gallinago*: *megala*: *stenura*, and with side branches leading to *hardwickii* and *solitaria*. There seems to be considerable overlap in the ranges of *stenura* and *gallinago*, and the two forms must interact considerably on their breeding grounds. This need for species distinction may account for *gallinago's* usually wide outer rectrices in Eurasia, whereas elsewhere in its range the outer rectrices are relatively narrow.

Suggested reading. Dementiev and Gladkov 1969; Tuck 1972.

Solitary Snipe

Gallinago solitaria Hodgson 1831
(*Capella solitaria* of Peters, 1934)

Other vernacular names. Tibet snipe; bécassine solitaire (French); Tibetbekassine (German).

Subspecies and ranges. See map 131.

G. s. solitaria: Western solitary snipe. Breeds in the mountains from southern Lake Baikal to the Soviet Altai and in northwestern Mongolia. May breed in the Tian Shan Mountains and other ranges of Russian Turkestan. Also breeds in the Himalayas from western Kashmir east to Sikkim and probably southern Tibet. In eastern Tibet or Sikang it possibly intergrades with *japonica*, though Kozlova (1962) assigns all these birds to *solitaria*. Winters in the foothills of the Himalayas and sometimes south to Burma.

G. s. japonica: Eastern solitary snipe. Breeds in the mountains of Sakhalin and western China, probably from Sikang and Tisinhai east to the Tsinling Range. Probably also breeds in the Kamchatka, Dzhugdzhur, Sikhote Alin, Kolymski, and Stanovoi ranges (Kozlova 1962). Winters in Amurland, in Kamchatka and on Sakhalin, and less frequently in Korea, Japan, and eastern China. Possibly not a valid subspecies (Meinertzhagen 1926).

Measurements and weights. Wing (both sexes): 147–69 mm. Culmen (both sexes): 67–77 mm (Kozlova 1962). Weights: males 130–48 g; females 126–59 g (Shaw 1936). Kozlova (1962) gives a similar range, but Ali and Ripley (1969) report heavier weights (142–227 g). Eggs ca. 41 x 31 mm, average weight 18.7 g (Zubarovsii 1976).

DESCRIPTION

Adults of both sexes have a distinctive white to pale cream crown stripe and conspicuous longitudinal stripes down the mantle and scapulars. The upperparts are barred with ginger and lack solid black feather centers. The upper breast is ginger to umber brown, with the feathers tipped and edged with white, the lower breast, sides, and flanks are barred with umber brown, and the upper tail coverts are uniform umber brown. There is only a small clear white abdomen, and there are very narrow pale tips on the primaries and secondaries. The tail usually contains twenty feathers (extremes sixteen to twenty-eight), with the outermost pair only 2–3 mm shorter than the central pair. The race *japonica* has narrower white dorsal stripes and is more barred below than is the nominate race. The iris is brown, the bill is olive brown to yellowish brown, darker toward the tip and yellowish at the base of the lower mandible, and the legs and toes are dull olive to pale yellowish green. *Juveniles* are very difficult to distinguish but possibly have more barring on the underparts, and the mantle and scapulars may be more heavily vermiculated and barred with tawny, with narrower white edges (Meinertzhagen 1926; Prater et al. 1977).

In the hand, the distinctive tail, usually of twenty feathers (of which twelve are relatively narrow) the relatively long wings (usually over 150 mm), and the gingery brown plumage, with white spotting and an absence of blackish or creamy tones, serve to identify this distinctive species.

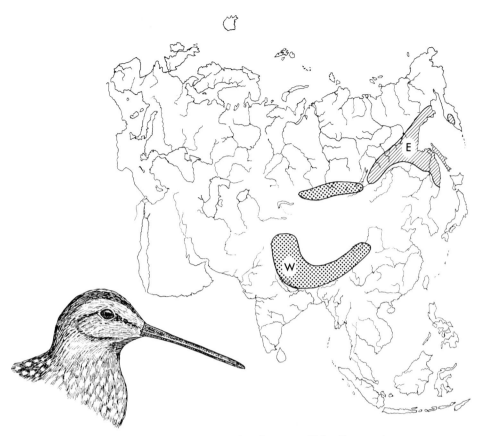

131.   Breeding distributions of the eastern (E) and western (W) solitary snipes.

In the field (12 inches), this species is usually found in grassy cover around swamps at rather high elevations. When flushed, the birds fly with zigzag twists as does the Eurasian snipe, but the flight is somewhat slower and heavier. The alarm call is a *scape* or *pench*, similar to that of the Eurasian snipe and the Japanese snipe. The display flight is evidently much like that of *gallinago*, being performed at considerable height, producing a harsh buzzing sound, and having an associated call sounding like *chock-a, chock-a, chock-a* (Tuck 1972).

NATURAL HISTORY

Habitats and foods. During the winter this species can be found in heavy cover along mountain streams, or among reed beds surrounding ponds at lower elevations of the mountains, and it probably only rarely occurs near sea level. It is usually found near running water, with dry gravelly or sandy banks. During summer the birds are found in the Himalayas between 2,800 and 4,600 meters in boggy mountain streams, where there are grass hummocks, rhododendrons, and similar scrub growth (Ali and Ripley 1969). Farther north the birds are found at lower altitudes during the breeding season, but they are nonetheless associated with the alpine zone, such as nesting at 2,400 meters in the Altai range and at timberline (2,500 meters) in the Khangai Mountains of northern Mongolia. They have been found nesting in Sikkim at about 2,500 meters elevation, but there are also many late summer or fall records of birds at timberline or alpine zones at comparable heights. Their foods are not well studied but are known to include snails, insect larvae, and flies (Kozlova 1962; Dementiev and Gladkov 1969; Ali and Ripley 1969).

Social behavior. This rather sedentary species gradually moves in spring from its wintering areas to its breeding grounds, which may be from a few dozen to a hundred or more kilometers apart. The migration is normally a solitary one, but in favored areas hundreds of these birds sometimes aggregate. The birds apparently seek out mountain streams with alpine meadows, patches of alpine vegetation, and gravelly stream beds, avoiding both muddy and boggy areas (Kozlova 1962). In such areas the birds perform aerial swooping displays, which are similar to those of *megala,* but the bird does not call while on the ground, and additionally it sometimes terminates its droning descent and temporarily suspends itself in the air with a series of gutteral notes that gradually change to a repeated *chock-a* sound. Evidently the sound produced by the rectrices is a harsh buzzing similar to but louder than that of *gallinago,* based on our limited knowledge to date (Tuck 1972).

Reproductive biology. There are few authentic nesting records. One nest found in the Munko Sardik range of Mongolia was on a mountain slope amid a willow thicket and contained 4 eggs. However, Kozlova (1962) questioned the identity of this nest, suggesting that it was actually that of *G. megala.* Recently Zubarovskii (1976) has described the nest, eggs, and nestlings, based on observations in the Altai Mountains of the USSR, where the species nests at 2,000–2,500 meters. The normal clutch there is evidently 4 eggs. The entire nesting range of the questionable eastern subspecies (*japonica*) is somewhat debatable, since there seem to be no actual records of nests or young available (Kozlova 1962).

Status and relationships. The solitary nature and montane habitats of this species make its status impossible to judge. In spite of its seemingly wide breeding range, it remains very little known and apparently rather rare everywhere. Its relationships are also problematic. It is probably closest to *megala* in its tail structure, but it might also be thought of as a larger, more southerly and montane-adapted version of *stenura.* For the present at least, it seems reasonable to consider *stenura* and *solitaria* a superspecies.

Suggested reading. Zubarovskii 1976; Kozlova 1962.

Genus *Limnocryptes* Boie 1826 (Jack Snipe)

The jack snipe is a small Eurasian shorebird with a long, laterally compressed and flexible bill that is much longer than the tarsus, with a tip that is expanded and pitted and that droops downward over the tip of the lower mandible. The hind toe is small, the front toes lack webbing, and the tarsus is scutellated in front and behind. The wing is pointed, and the tail is wedge-shaped and composed of twelve pointed feathers. The back is striped with buff and glossy green, and the crown is strongly striped. The downy young resemble those of *Gallinago.* One species.

Jack Snipe

Limnocryptes minimus (Brunnich) 1764

Other vernacular names. Half-snipe; bécassine sourde (French); Zwergschnepfe (German); agachadiza chica (Spanish).

Subspecies and range. No subspecies recognized. Breeds from northern Norway east across Scandinavia through Russia and Siberia at least to the Taimyr Peninsula. Also breeds in the Kolyma delta and probably in Anadyrland, but there is a gap in breeding records east of the Taimyr Peninsula. Winters from the British Isles, Europe, the Caspian, Turkistan, and southeastern Asia south to tropical Africa, India, the Indo-Chinese countries, and the Andamans. See map 132.

Measurements and weights. Wing: males 109–19 mm; females 106–17 mm. Culmen (to feathering): males 45–53 mm; females 46–52 mm. Weights (fall migrants): males 60–91 g; females 53–71 g (Glutz et al. 1977). The average of 328 wintering birds of both sexes was 63.4 g (Bird 1909). Eggs ca. 38 x 27 mm, estimated weight 14.0 g (Schönwetter 1963).

DESCRIPTION

Adults of both sexes have a median black line from the base of the upper mandible to the crown, where it expands to form a broad band reaching the nape. There is a light buffy band below this, also extending from the bill to the nape. The feathers above the eye are edged with blackish, forming a superciliary stripe, a second black streak runs from the lores to

132. Breeding distribution of the jack snipe.

the eye, and there is a third from the lower mandible through the ear coverts to the nape. Otherwise the sides of the head and chin are white to buffy, as are the lower breast and abdomen. The upper breast and sides of the neck are streaked with brownish black, the scapulars and back feathers are black glossed with purple and green, the anterior upperparts are barred or spotted with cinnamon, buff, and white, and the rump is black, glossed with purple and edged with white. The tail is blackish centrally and dusky brown outwardly, with irregular paler markings. The primaries and their coverts are dark brown, the secondaries are similar but are distinctly barred and broadly tipped with white, and most of the coverts are edged with white or buffy. The iris is brown, the bill is yellowish with a black tip and grayish at the sides and base of the lower mandible, and the legs and feet are light greenish blue gray. *Juveniles* have white under tail coverts, with indistinct yellowish brown to grayish brown subterminal shaft streaks (which in adults are more distinct and blackish brown), and their second and third pairs of tail feathers counting outward) are narrow and finely

pointed, rather than relatively broad and rounded (Prater et al. 1977).

In the hand, the wedge-shaped tail, with only twelve feathers, is unique to this genus. The absence of a pale central crown streak is also characteristic.

In the field (8 inches), these birds are found in swamps and bogs and infrequently are visible. In flight they appear smaller than other snipes, have a shorter bill, have a more direct flight that is usually very brief, and are silent when flushed. The typical call is a muffled drumming note, uttered during display flights and while on the ground and reportedly sounding something like the bubbling of a spring or the boiling of a kettle. There is also a precipitous display flight, during which a sound is produced reminiscent of horses hooves on hard dry ground.

NATURAL HISTORY

Habitats and foods. This species is distinctly boreal in its breeding distribution, mainly nesting in grassy marshes and bogs having swampy ground, especially in coniferous forests, but also in willow marshes and

55.   Adult heads and central tail feathers of jack snipe (*above*) and Eurasian snipe (*below*), including dorsal view of bill.

wet alder woods. Large swamps are apparently preferred, but sometimes a pair will occur in a bog as small as an acre in extent. Population densities seem to be quite low on the breeding grounds, with one estimate of an average density of only 0.2 pairs per square kilometer in northern Finnish Lapland (Glutz et al. 1977). It reaches as far north as the arctic tree line extends, and it favors the northern birch zone. Outside the breeding season it is usually found along flooded muddy and densely overgrown wide riverbanks, grassy marshlands, and rice paddies (Voous 1960). It does not inhabit seashores, and it never forages on open mud but favors areas that have dense vegetation and subject to shallow flooding. Like other snipes, it forages by probing, although compared with the more typical snipes its bill is relatively short and stout. Its foods are rather poorly known, but they include annelids, small freshwater and land mollusks, adult and larval insects, and the seeds or vegetative parts of various plants, especially aquatic or shoreline forms. Feeding is apparently done much as in the European snipe, but in one case the species was seen to raise and lower the body while walking forward, halting and then making a series of three probes from right to middle to left. In another instance a jabbing or tapping movement of

the bill was evidently used to cause springtails to come to the surface. It has been suggested that this species takes more items from the surface than does the European snipe (Witherby et al. 1941; Glutz et al. 1977).

Social behavior. Although the species is extremely quiet and unobtrusive outside the breeding season, it becomes conspicuous when the males establish territories after spring arrival. The birds produce a "cantering" call both while on the ground and while in the air, which has been described as sounding like the cantering of a distant horse, *clockety-clockety, clockety-clock,* or *lock-toggi, lock-toggi.* Although typically done at dawn or dusk, flight displays occur in bright sunshine as well as on overcast days and are generally done at greater heights under relatively bright conditions. Typically the male ascends diagonally upward to 50–60 meters and begins to fly in rather wide circles. Suddenly he enters a rather steep dive that brings him to about 20 meters above the substrate. He utters the cantering call as he starts the descent and continues it beyond the shorter ascending glide phase that brings him back to about 30–40 meters high. He then begins a short period of silent flight as he moves slowly forward on stiffened wings for about 10–20 yards. Finally he performs a series of silent and abrupt undulating movements, as if he were being jerked up and down on an invisible string. The flights are performed 12–15 times at intervals of 2–10 minutes. Occasionally two birds will perform the display in concert, suggesting that the behavior is not limited to males. Male territories are about 20 hectares in area (Nilsson and Nilsson 1978). Unlike the *Gallinago* snipes, there is no real evidence that any of the noise produced during the display flight originates from tail vibration. One rather mechanical ticking sound has been reported during the display flight, but this has been attributed to bill-snapping (Bannerman 1961; Glutz et al. 1977).

Reproductive biology. This species has a surprisingly long breeding season, especially considering its northerly distribution. In both Finland and Norway a two-month spread in egg dates has been reported. It has been suggested that the species may normally be double brooded, although there is so far no direct evidence on this point. Nests are usually in extremely damp situations, often on hummocks of sphagnum or grass tussocks only a few inches above the surrounding water. Others have been found in drier locations, such as where dwarf birches and heather were growing on a peat base. There are usually 4

eggs; all but one of 19 clutches in Finland had this complement, while the exceptional clutch contained 3 eggs. It is believed that only the female incubates, and the male does not participate in brood care. Two minimal estimates of the incubation period were 21 and 24 days. There is no detailed information on brooding behavior, nor is the fledging period known (Bannerman 1961; Glutz et al. 1977).

Status and relationships. Virtually nothing can be said of the status of this extremely elusive species. Morphologically it is much like *Gallinago*, but in addition to its several plumage differences (iridescent dorsal feathers, wedge-shaped tail feathers) it also differs in its bill shape and in having two pairs of sternal notches, rather than a single pair as in *Gallinago*. Additionally, the syrinx is distinctly different from that of *Gallinago* and indeed apparently is unique among shorebirds (Pycraft 1912) in having a supplementary tracheal semiring. I thus believe that generic separation is warranted, though Strauch (1976) has recently merged *Limnocryptes* with *Gallinago*.

Suggested reading. Bannerman 1961; Nilsson and Nilsson 1978.

Tribe Scolopacini (Woodcocks)

American Woodcock

Woodcocks are small to medium-sized terrestrially adapted shorebirds with a long and relatively stout but flexible bill whose tip is pitted and droops downward over the tip of the lower mandible. The crown is transversely barred with dark brown and buff, and the back is usually striped with ashy gray. The eyes are very large and placed high on the skull, the openings of the ears are under the anterior margins of the eyes, and the posterior part of the skull is much reduced. The short tarsus is scutellated in front and reticulated behind, a small hind toe is present, and the front toes lack webbing. The wing is broad and rounded, and the tail is short, rounded, and composed of twelve feathers. The downy young are buffy with dark brown markings that form dorsal stripes and lack white powder-puff tips. One genus and 4 species are recognized here.

Genus *Scolopax* Linné 1758 (Woodcocks)

This genus of 4 species has the characteristics of the tribe Scolopacini.

KEY TO SPECIES OF *Scolopax*

A Wing under 145 mm, outer three primaries with vanes distinctly narrowed .... *minor*
A' Wing at least 145 mm, outer three primaries with vanes of normal width
  B Breast barred with dusky and white, wing under 170 mm .... *saturata*
  B' Breast not barred with dusky and white, wing over 170 mm
    C Breast buffy with narrow brown barring, wing usually over 190 mm .... *rusticola*
    C' Breast clear buffy, wing usually under 190 mm .... *rochussenii*
      D Back finely barred with black and buff .... *r. celebensis**
      D' Back boldly patterned with black and buffy .... *r. rochussenii*

Dusky Woodcock

Scolopax saturata Horsfield 1821

Other vernacular names. East Indian woodcock; Horsfield woodcock; bécasse de Java (French); Malaienschnepfe (German).

Subspecies and ranges. See map 135.
  S. s. saturata: Java dusky woodcock. Resident in the mountains of Java and Sumatra.
  S. s. rosenbergii: New Guinea dusky woodcock. Resident in the mountains of New Guinea.

Measurements and weights. Wing (both sexes): 154–88 mm. Culmen (both sexes): 74–85 mm. Weights: a female of *rosenbergii* weighed 220 g (Ripley 1964). Eggs ca. 43 x 34 mm, estimated weight 26.5 g (Schönwetter 1963).

DESCRIPTION

Adults of both sexes have the crown rufous, with several narrow black bars on the forehead and three broader black bars on the crown and hindneck; the lores and lower cheeks are also blackish. The rest of the upperparts including the wing coverts are rufous, with a heavy black mottling in the form of blotches and bars. The flight feathers are uniform blackish, with a few small rufous spots on the outer vanes of the primaries and larger spots on the secondaries. The tail feathers are black, with ashy gray tips above and silvery gray below, and with a few subterminal rufous markings. The underparts are mostly barred with black and rufous, except for a whitish patch on the lower lores and chin, and there are whitish tips on the abdominal and flank feathers. The axillaries and under wing coverts are blackish, with a few small bars or spots of rufous. The iris is brown, the bill is brownish black, tinged with fleshy at the tip and at the base, and the legs and toes are dark gray. *Juvenile* plumages are undescribed, but Ripley (1964) noted that an apparently young bird had heavier and more rufous barring on the upperparts, less well defined black areas dorsally, and the underparts uniformly barred and lacking the pale patch.

In the hand, the broad black bars on the crown identify this as a woodcock, and this species is the

* Often considered a full species

darkest and most heavily barred below of any of the species of *Scolopax*.

In the field (12 inches), the limited range of this species and its unusually blackish coloration should allow for ready identification. The calls include various nasal notes uttered by males in flight.

NATURAL HISTORY

Habitat and foods. These birds are associated with montane forests; however, they probably do not occur in dense forest, but rather are found where forests adjoin grasslands or other rather open areas. Rand and Gilliard (1968) reported that most of the birds they observed were flushed from the edges of little patches of forest in alpine grassland, and that at least in New Guinea they occur between 5,000 and 12,000 feet of elevation. Judging from the contents of 2 stomachs, their foods consist at least in part of caterpillars and moth pupae.

Social behavior. Rand and Gilliard (1968) reported seeing birds performing aerial courtship displays during August and September. At dawn and dusk the males covered a regular beat over habitats that included open country, forest clearings, and the forest itself. While in such horizontal flight the male would occasionally utter a loud, nasal *queet*, followed by a series of *quo* notes and finally punctuated and ended with a louder and nasal *queeet*.

Reproductive biology. The only available information on nesting in this species comes from Java, where Hellebrekers and Hoogerwerf (1967) reported that the clutch size is only 2 eggs. Two clutches were reported for March, and 1 for April, from Mount Pangerango, West Java.

Status and relationships. Virtually nothing is known of the status of this nocturnal and elusive species. Judging from specimens, it seems in its plumage to be the most snipelike of the woodcocks, and it bears a more than passing resemblance to such generalized Andean snipes as *G. imperialis*. It is smaller than the other woodcocks but is somewhat intermediate between *minor* and *rusticola* in wing shape and in the pigmentation patterning on the primaries, as noted by Seebohm (1888). It is impossible to do more than hazard a guess on its evolutionary history, but my own judgment is that it might be considered the most generalized of the woodcocks, and perhaps not

distantly removed from the "semi-woodcocks" of the genus *Gallinago*.

Suggested reading. Mayr and Rand 1937.

133.   Breeding distribution of the American woodcock.

American Woodcock

Scolopax minor Gmelin 1789
(*Philohela minor* of Peters, 1934)

Other vernacular names. Timberdoodle; bécasse Américaine (French); Amerikanischer Waldschnepfe (German).

Subspecies and range. No subspecies recognized. Breeds in North America from Manitoba, Ontario, Quebec, Newfoundland, and New Brunswick south through the United States from the Mississippi Valley eastward, and south to Louisiana and Florida. Winters from the southern parts of its breeding range south to the Gulf coast. See map 133.

Measurements and weights. Wing: males 118–28 mm; females 129–43 mm. Culmen (to feathering): males 60–66 mm; females 64–74 mm. Weights (in summer): adult males 127–65 g, average of 31 was 147 g; females 162–216 g, average of 48 was 186 g (Sheldon 1967). The average of 1,481 males was 164.5 g, and that of 1,371 females was 210.7 g (various sources). Eggs ca. 38 x 29 mm, estimated weight 17.0 g (Schönwetter 1963).

DESCRIPTION

Adults of both sexes have three wide blackish bands extending across the rear of the crown, with intervening narrow brownish bands. The face is mostly gray, with a dark line from the bill through the eye and another on the side of the cheek. The upper parts are generally rusty brown, with lighter tranverse spots and dusky lines. The scapulars and back have irregular black spots or blotches, with this area largely enclosed in a well-defined light grayish V-shaped mark. The rump is rusty to cinnamon with dusky barring, as are the upper tail coverts, which are tipped with light gray. The underparts are mostly pale buffy grayish to white medially, with barring of blackish brown. The primaries are dusky, their outer webs having pale cinnamon spots, and the tail feathers are black with rusty edge markings and pale gray tips. The bill is pinkish basally, grading to olive brown at the tip. The iris is dark brown, and the legs and toes are reddish pink. *Females* average somewhat larger in measurements and weights than do males, and the vanes of their outer three primaries are slightly wider (the combined width of the three primaries at least 12.6 mm, vs. no more than 12.4 mm in males). *Juveniles* have the undersides of the middle four secondaries with clear white tips and a contrasting dark brown subterminal zone, usually with a clear dark point in the center, while adults have only slight contrast between the distal light brown area and the buffy subterminal zone (Prater et al. 1977).

In the hand, this woodcock is easily distinguished from other *Scolopax* species by the extreme narrowness of the outer three primaries.

In the field (11 inches), woodcocks are unlikely to be mistaken for any other shorebirds, and in North

America no other similar species occurs. Calls are essentially limited to the breeding season and include *peent* notes, a muffled *took-oo* call, and a twittering flight "song" caused by wing vibration.

NATURAL HISTORY

Habitats and foods. Throughout their range and during the entire year woodcocks are generally confined to young forests with scattered openings on poorly drained land. The openings serve both as singing grounds and as nocturnal roost sites. The male's breeding area must have a clearing in which the bird can fly, with the size of the clearing directly related to the height of the surrounding trees. Plant life in this clearing is usually of the early woody or low shrubby stage, with some herbaceous cover for alighting. Scattered trees up to 15 feet tall are sometimes present. Shrubby habitats having a herbaceous ground cover support large numbers of earthworms and, especially where soil moisture is adequate but not extreme and where soil texture favors moisture retention, also support the largest numbers of woodcocks. Earthworms apparently are the staple diet of woodcocks throughout the entire year, and in one study of 261 stomachs they composed 68 percent of all the food taken, with two other studies of 63 and 70 stomachs each reporting that earthworms composed 86 percent of the total foods. Beetles and fly larvae are taken when available, but plant foods are unimportant. It thus seems that woodcock habitats essentially must support high earthworm populations, though some observers have reported that cover characteristics rather than earthworm numbers dictate woodcock distribution on feeding fields. When woody vegetation grows 6–10 feet high over more than 60 percent of an area, it is abandoned as breeding habitat. Likewise, when ground vegetation grows too dense and much more than 6 feet tall it too is abandoned as winter cover and the birds move to fallow fields, especially those that are burned or grazed. Pine uplands, especially those with openings formed by abandoned home sites, are also favored by wintering birds, especially where there are moist slopes or creek bottoms nearby for foraging. The birds not only need moist soil for earthworms, but apparently are also very dependent on water for drinking, which they apparently do by sucking rather than by lifting their bills in the usual manner of birds (Sheldon 1967).

Social behavior. Woodcocks arrive on their northern breeding grounds in March and April, with males tending to precede the females slightly. The males soon establish territories that center on singing grounds, which range in size from 0.25 to 100 acres (0.1–40.5 hectares). The size of the area that is actually defended is probably rather small and in one instance was estimated to be from 4.38 to 7.85 acres (Weeks 1969). Territorial advertisement consists of dawn and dusk display, which is initiated by a series of ground calls. These are buzzing *peent* notes, preceded by barely audible *tuko* sounds, uttered at rates varying from 7 to 28 times per minute before takeoff. Then the bird ascends quickly, and as it rises in wide circles it begins to produce mechanical wing-twittering noises. Finally, at nearly 300 feet in the air, the bird begins to hover and utters his aerial "song," a series of liquid chirping notes that are continued as he descends on a zigzag course back to the ground. The flight lasts 30–60 seconds, with the bird typically repeating the performance 10–20 minutes and apparently serves to attract females into the display area for copulation. Judging from observations of males with decoys, the birds raise their wings vertically and stiffly walk or dash toward the female, immediately attempting to copulate. Sometimes the male will raise and spread his tail, and immediately before copulation the wings are lowered and dropped turkeylike toward the ground. The male flutters his wings during copulation. Immediately afterward the female may fly away, or she may remain in the area, and another copulation can occur in the same display session. Males are apparently promiscuous and evidently play no further role in breeding (Sheldon 1967; Weeks 1969).

Reproductive biology. The nest constructed by the female is a fairly rudimentary structure, placed with no effort for concealment. It is often at the base of a small tree or shrub, and often within a few yards of a brushy field edge. Commonly it is placed in a mixed growth of birches, aspens, or other hardwoods and conifers, but sometimes it is in pure hardwood cover, and less often it is in brushland, old fields, or blueberry cover. In any case, the surrounding cover is of only light to medium density. Nests are often within 150 yards of an occupied singing ground, but at times females may nest as far away as three-fourths of a mile from the singing ground where copulation probably occurred earlier. Almost invariably the clutch is 4 eggs; of 115 nests found in Maine only 4 late nests containing 3 eggs each were exceptions to this pattern. There is apparently a rather high incidence of renesting by unsuccessful females, but there is no indication that two broods are ever produced in a single season. Eggs are laid at the rate of one per day, with incubation probably beginning with the final egg. Females are "tight" sitters and typically leave their nests at dusk, probably to feed, returning after dark. The incubation period is about 21 days, and unlike most other birds the young are said to emerge from the shell by splitting it longitudinally through muscular action by arching the thoracic and cervical vertebrae against the inside of the shell. The young are able to fly short distances when only 14–15 days old, and they are nearly full grown at the end of 4 weeks. Chicks have been observed probing for earthworms as early as 3 days after hatching. Although there are a large number of published accounts of woodcock females carrying their young, at least some of these probably are the result of the female's "decoy flight," in which she depresses her tail and flies in a labored manner, as if carrying something. On the other hand, a number of reliable naturalists have reported seeing actual carrying (Sheldon 1967).

Status and relationships. Population estimates based on singing ground surveys made between 1963 and 1973 have not shown any significant trends, while data from wing collections have shown considerable fluctuations in apparent harvests but also no obvious trends (Sanderson 1977);. The affinities of this species are probably closest to *rusticola*; Mayr and Short (1970) believe that the two are biologically "good" species but that they are clearly geographical replacement forms and members of the same species group. I concur with this idea and suspect that *rusticola* is closer to the ancestral woodcock type, as well as to the East Indian forms.

Suggested reading. Sheldon 1967.

436 ❖ ❖ ❖

Woodcock

Scolopax rusticola L. 1758

Other vernacular names. European woodcock; bécasse des bois (French); Waldschnepfe (German); chocha perdiz (Spanish).

Subspecies and ranges. See map 133.

S. r. mira: Amami woodcock. Resident in the northn Ryukyus, on Amami Oshima and probably Tokuno Shima. Regarded by Vaurie (1965) as a distinct species.

S. r. rusticola: Eurasian woodcock. Breeds in the Azores, Madeira, the Canaries, the British Isles, and on the mainland of Eurasia from Scandinavia south to Spain and eastward across Russia and Siberia to Sakhalin and Japan. Also breeds in China, including Szechwan, Kansu, and eastern Sikang, probably in northern Burma, in northern India, Russian Turkistan, the Caucasus, Georgia, Transcaucasia, and perhaps Armenia. Winters south to the Mediterranean Basin, Iraq, Iran, Afghanistan, India, the Indo-Chinese countries, and the Ryukyus.

Measurements and weights. (of *rusticola*). Wing: males 190–214 mm; females 186–218 mm. Culmen (from feathering): males 63–77 mm; females 63–84 mm. Weights: males 185–400 g; females 230–405 g (Glutz et al. 1977). Eggs ca. 44 x 34 mm, estimated weight 26.5 g (Schönwetter 1963).

DESCRIPTION

Adults of both sexes (of rusticola) have a buffy pink forehead, a narrow blackish median streak from the base of the bill to the crown, and the rest of the crown black with two or three transverse bars of buffy or bright cinnamon. There is a broad dark stripe from the base of the bill to the eye, and the rest of the face is mostly pinkish buff with sepia spotting. The lower throat, breast, and underparts are buff to pinkish cinnamon, with sepia barring that is broadest on the breast, sides, and flanks. The scapulars, back, and wing coverts are orange cinnamon, with black and buff to grayish spots and patches, and the lower back to upper tail coverts are also orange cinnamon with black mottling. The tail feathers are black, with tips of smoky gray above and whitish below and

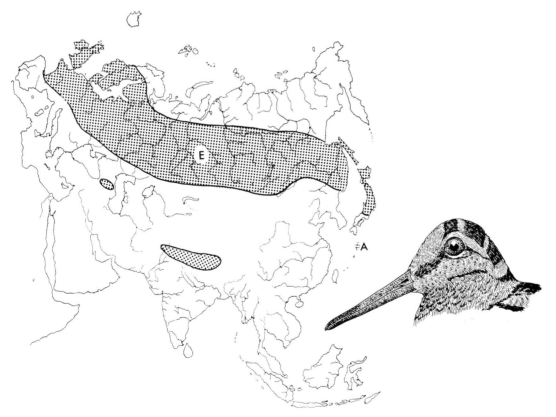

134. Breeding distributions of the Eurasian (E) and Amami (A) woodcocks.

56. Adult heads and outer (*seventh*) primaries of Eurasian (*above*), and North American (*below*) woodcocks.

orange cinnamon notches on the outer webs. The primaries, primary coverts, and secondaries are brownish black, tipped with orange cinnamon or buff, and most feathers are notched with pinkish buff on both webs. The iris is dark brown, the bill is dull pinkish, becoming dark toward the tip, and the legs and toes are dull grayish pink. The patterning of *mira* is different from this description, being considerably more reddish and lacking gray tones, and *mira* is sometimes considered a separate species. *Juveniles* (at least of *rusticola*) have primary coverts with a broad distal pale spot that is about the same color as other spots on that covert, while adults have a distal fringe that is paler than the spots of the covert. Additionally, juveniles have shorter tail feathers than adults, and each feather has a dull grayish white distal spot. The adult feathers are generally darker brown, with clear small paler patches rather than indistinct mottling (Prater et al. 1977).

In the hand, the large size (wings over 185 mm) of flight with the bill carried in a drooping manner, and forms.

In the field (14 inches), the rounded wings, twisting flight with the bill carried in a drooping manner, and generally russet coloration identify this as a woodcock immediately, and no other species of woodcocks occur in its range. Its calls include a low croaking sound, a thin *tsiwick*, and some twittering notes. The birds are usually silent when flushed, but they sometimes utter a short *pier* note.

NATURAL HISTORY

Habitats and foods. Breeding in this species mainly occurs in broad-leaved or mixed forests, but it also occurs in coniferous forests having an undergrowth of broad-leaved bushes and ferns, together with a moist ground cover of decaying leaves. Swampy forests with mossy ground and brooks or other watercourses are sometimes also used (Voous 1960). In Great Britain the most typical breeding habitat is mature deciduous woodland or mixed woodland (Morgan and Shorten 1974). The best habitats provide both dry, warm resting places and nearby moist areas for foraging, together with open areas providing flight paths. The distribution of earthworms is probably an important habitat consideration throughout the year, but in addition woodcocks eat other invertebrates that live in soft, moist soil. These include numerous families of insects and their larvae, millipedes, spiders, and freshwater mollusks. Earthworms probably are of major importance during spring, whereas insects and other invertebrates are eaten in greater numbers in summer and fall and are likewise more important to young chicks than are earthworms. Additionally, small bivalve mollusks and crustaceans are sometimes eaten by migrating birds, and a very limited amount of vegetative material is also eaten. Woodcocks have a refined method of trial probing, which is geared to leg movement, so that as the weight is transferred to the front foot with each step the bill is inserted for about a third of its length (Burton 1974). As in other woodcocks, the distal portion of the upper mandible is highly flexible and can be lifted to grasp a prey item without opening the entire bill.

Social behavior. In spite of the species' abundance over much of Europe, several aspects of its social behavior and reproduction remain uncertain. For example, some authorities regard it as monogamous, but it has also been reported as polygamous, polygynous, and even polyandrous. Judging from the fact that the male does not participate in incubation or brood care, and from similarities with the North American species, it seems likely that polygyny or promiscuity is the typical reproductive strategy. Males are territorial, and their territories are apparently advertised as well as essentially defined by their daily "roding" flights, normally performed near dusk or dawn. If the territory size can be estimated by the area traversed by such flights, it probably averages about 9 hectares (Tester and Watson 1973), with the size varying somewhat according to amount of open land present. During the flight the male flies fairly rapidly but with slow and owllike wingbeats at a height of 30–200 feet over open ground, for periods of from 15 minutes to nearly an hour. The usual call given during roding is a three-element vocalization, *quorr-quorr'k-psswitt*, with the first portion low and croaking, and the last part shrill (Nemetschek 1977). Roding behavior is apparently limited to male birds and occurs primarily at dusk and dawn, though it rarely also occurs during daylight hours. The purpose of the flights is probably a dual one, including both territorial proclamation or defense and also searching for females. Nemetschek (1977) has suggested that it is primarily for locating females, but that they are sometimes found visually, when the females fly up from roosting sites to meet them. Additionally, flying males are apparently called to earth by females, which utter shrill notes that cause the male to "drop like a stone." Thereafter a ground display by the male follows, which has been described as a strutting with dropped wings, the tail raised and spread, and some of the head or body feathers puffed. In one observed case the male followed the female with a low, crouching gait and swaying motions until the female crouched and copulation followed. It has also been suggested that the fanned tail, with its reflective white feather tips, may be the stimulus for copulation. In one case copulation was observed near a nest with an incomplete clutch, but the male remains with a single female only until her clutch is complete, and then resumes his roding, indicating a mating system approaching successive polygyny (Hirons 1980).

Reproductive biology. Nests are usually placed in wooded but rather open situations, often at the base of a tree or near a dead branch or log. The ground cover immediately around the nest is frequently composed of bracken fern, dead leaves, or brambles (Morgan and Shorten 1974). Of 33 nests studied in Sweden, more than half were near the foot of a tree, fallen trunk, bushes, or the like. Only one was in fully open terrain. Young stands of deciduous woods are preferred for nesting, while older and denser stands are less used, and dense monocultures of conifers are also used only very slightly (Marcström and Sundgren 1977). In the vast majority of cases there are 4 eggs; among 33 Swedish clutches the only exceptions were 1 nest each with 2 and 5 eggs. A British sample of 132 clutches had 111 of 4 eggs, and the range was 2–5 eggs. The egg-laying interval is usually 24 to 48 hours, and a clutch of 4 eggs is often completed in 4½ to 5 days. The egg-laying season is greatly extended over much of the species' range, which has given rise to speculation that double brooding may be regularly performed. Yet there is no direct evidence to support this view, and there seems to be no indication of double peaks of either egg-laying or hatching periods. All the incubation is performed by the female, and various estimates of it have ranged from 20 to 24 days, with an average of several clutches being slightly more than 22 days. As in the American woodcock, hatching is achieved by a longitudinal splitting of the eggshell by pressure exerted by the chick's back. Fledging occurs remarkably early, with short flights sometimes made by birds as young as 10 days, and the birds are able to fly considerable distances when only 2–3 weeks old (Shorten 1974; Glutz et al. 1977). There are a surprising number of observations of young being carried by adult birds, usually between their legs, and though such accounts have often been disregarded by naturalists, there is some recent anatomical evidence supporting the idea that this behavior may indeed be possible (Ingram 1978).

Status and relationships. There do not appear to be any clear trends in the status of this species. In recent decades numbers are believed to have increased in Denmark, northern West Germany, and Holland, decreased in southern West Germany, Czechoslovakia, Poland, East Germany, and Finland, and remained about the same in Norway, Sweden, and Austria (Shorten 1974). The relationships of the woodcocks are certainly very obscure, but I suspect

that *saturata* may be the most generalized form and is perhaps nearest the ancestral type from which *rusticola* evolved. The North American species *minor* would represent a later separation, and still later *mira* (sometimes considered a separate species) presumably separated.

Suggested reading. Shorten 1974; Alexander 1945–47.

Indonesian Woodcock

Scolopax rochussenii Schlegel 1866
(*S. celebensis* and *S. rochussenii* of Peters, 1934)

Other vernacular names. Obi woodcock; bécasse des Célèbes; becasse des Moluques (French); Celebes-schnepfe, Molukkenschnepfe (German).

Subspecies and ranges. See map 135.

    S. r. rochussenii: Obi Indonesian woodcock. Resident on Obi Island, and probably also Batjan, in the Moluccas.

    S. r. celebensis: Celebes Indonesian woodcock. Resident in the mountains of Celebes. Includes *heinrichi;* these forms are often considered a distinct species.

Measurements and weights. Wing (both sexes): 187–99 mm. Culmen (both sexes): 76–92 (*heinrichi* is said to have a maximum culmen of 80 mm, and *celebensis* a minimum of 86 mm). Eggs: undescribed.

DESCRIPTION

Adults of both sexes have a sandy buff forehead, terminated by a broad black band above the eyes, behind which there is another on the nape and two more on the hindneck, the latter extending forward to the ear coverts. There is also a heavy black stripe from the bill to the eyes. The rest of the upperparts are mostly blackish brown, ochre, and rufous in various combinations, the rufous being reduced in *rochussenii*, with the ochre expanding to form large spots and bars, while in *celebensis* the size of the barring and spotting is greatly reduced and rufous tones

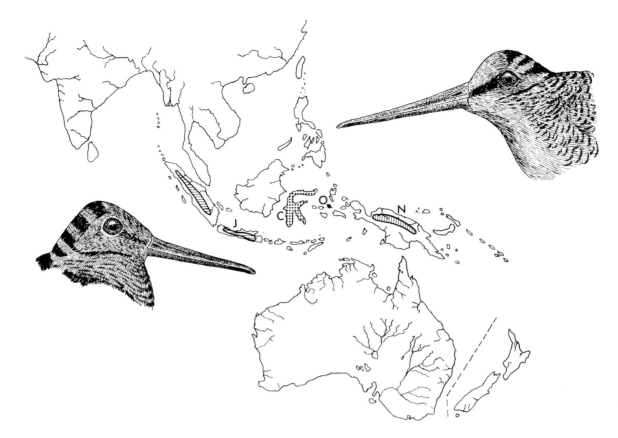

135. Breeding distributions of the Celebes (C) and Obi (O) Indonesian woodcocks, and the Java (J) and New Guinea (N) dusky woodcocks.

predominate over the ochre ones. The tail is blackish with a lighter tip, which is grayish in *celebensis* and tinged with rufous ochre in *rochussenii*. The latter also has extensive barring on the outer webs of the tail. The sides of the face and underparts are ochre buff, with some black barring on the cheeks, breast, and flanks, especially in *rochussenii*, where the breast markings are strongly developed. The primaries and secondaries are blackish, notched with small (*celebensis*) or large (*rochussenii*) ochre spots that become barring on the inner secondaries of the latter form. The under wing coverts are rufous ochre, with black barring. The iris is brown, the bill is bluish gray, and the legs and toes are gray to bluish gray. *Juvenile* plumages are undescribed.

In the hand, the combination of a very long bill (at least 76 mm), a relatively large size (wing 185–200 mm), and fully developed outer primaries separates this species from *rusticola* and *minor*, and the absence of heavy black barring on the underparts distinguishes it from *saturata*.

In the field (13 inches), the very limited range and the woodcocklike appearance should provide for easy recognition.

NATURAL HISTORY

Habitats and foods. Stresemann (1941) states that the habitat of the Celebes race of this species consists of high mature forests at about 2,000 meters elevation. The birds hide in the thick undergrowth there, apparently favoring ridges on which there are areas with little undergrowth and rich soils. He flushed birds at about 2,200 meters elevation in the Latimodjong Mountains, where wallows made by wild pigs caused temporary water puddles to form, in which the birds presumably foraged. He believed that the birds are almost exclusively terrestrial, taking flight only to escape from danger and then flying only short distances.

Social behavior. Nothing has been written on this species' social behavior.

◈ ◈ ◈ 441

link between the typical woodcocks and the snipes. Seebohm (1888) stated that in its unbarred breast *rochussenii* shows affinities with *minor*, while its barred primaries indicate a relationship with *rusticola*.

Suggested reading. None.

57. Adult heads of dusky (*top*), Obi (*middle*), and Celebes (*bottom*) woodcocks.

Reproductive biology. Nothing is known of this species' reproduction.

Status and relationships. Stresemann (1941) states that this is one of the least known of all the indigenous species of Celebes birds, and its status can only be guessed. The race limited to Obi Island is obviously extremely limited in numbers. The relationships of these woodcocks are also somewhat dubious. They have often been placed in the genus *Neoscolopax*, which has been recognized on the basis of relative thigh feathering, but Stresemann (1932) has rejected this. He noted that these tropical forms have much stronger feet, relatively longer bills, and shorter wings than does *rusticola*. Certainly, in their elusive and highly terrestrial behavior the birds seem very snipelike, but it is rather questionable whether the forms can be considered in any way a transitional

Head Profile Identification Guide

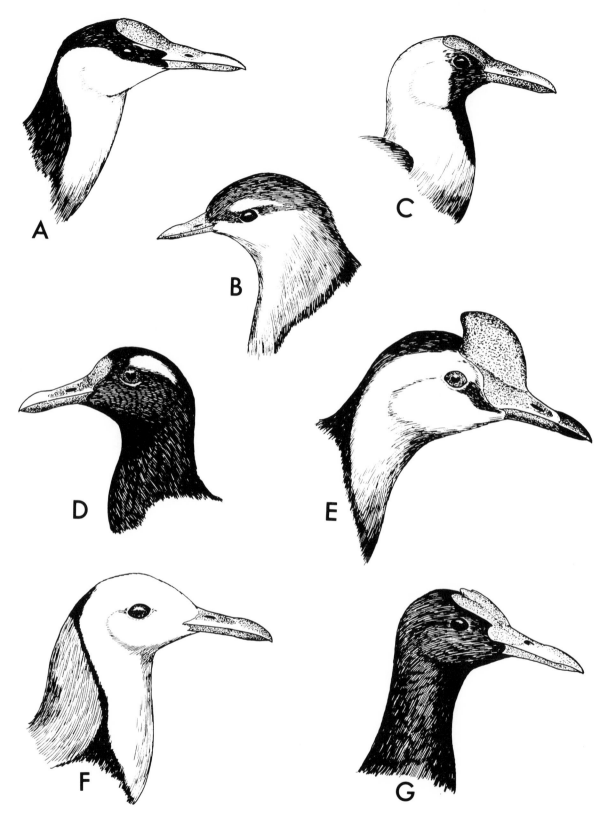

58. Heads of jacanas, including *A*, African; *B*, lesser; *C*, Madagascan; *D*, Bronze-winged; *E*, comb-crested; *F*, pheasant-tailed; and *G*, American.

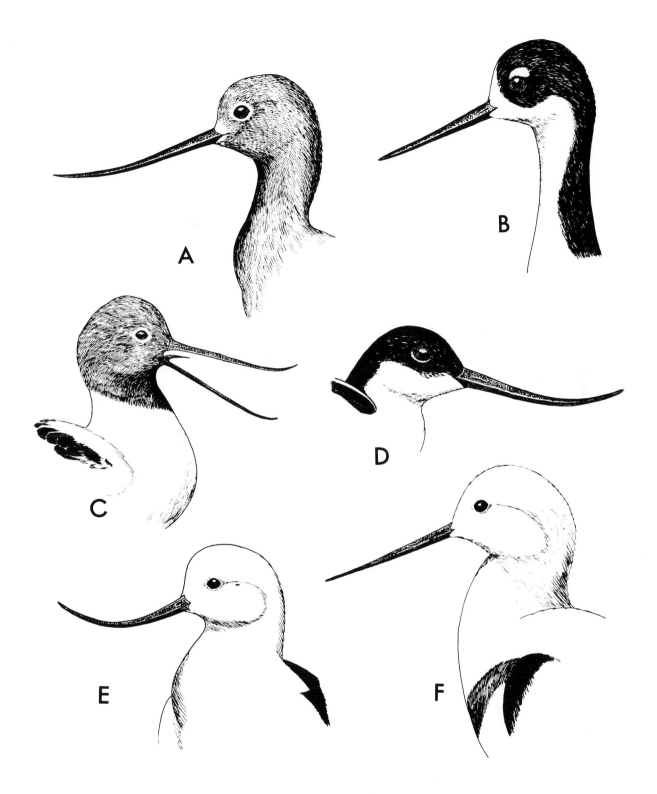

59. Heads of avocets and stilts, including *A*, American avocet; *B*, northern black-necked stilt; *C*, red-necked avocet; *D*, Eurasian avocet; *E*, Andean avocet; *F*, banded stilt.

60. Heads of African lapwings, including *A*, Senegal; *B*, black-headed; *C*, spot-breasted; *D*, white-crowned; *E*, long-toed; *F*, crowned; *G*, spur-winged; and *H*, blacksmith plover.

61. Heads of non-African lapwings, including *A*, red-wattled; *B*, banded; *C*, Eurasian; *D*, Javanese; *E*, masked; *F*, southern; *G*, pied.

62. Heads of miscellaneous lapwings, including *A*, brown-chested; *B*, lesser black-winged; *C*, yellow-wattled; *D*, gray-headed; *E*, white-tailed; *F*, sociable; *G*, Andean; and *H*, black-winged.

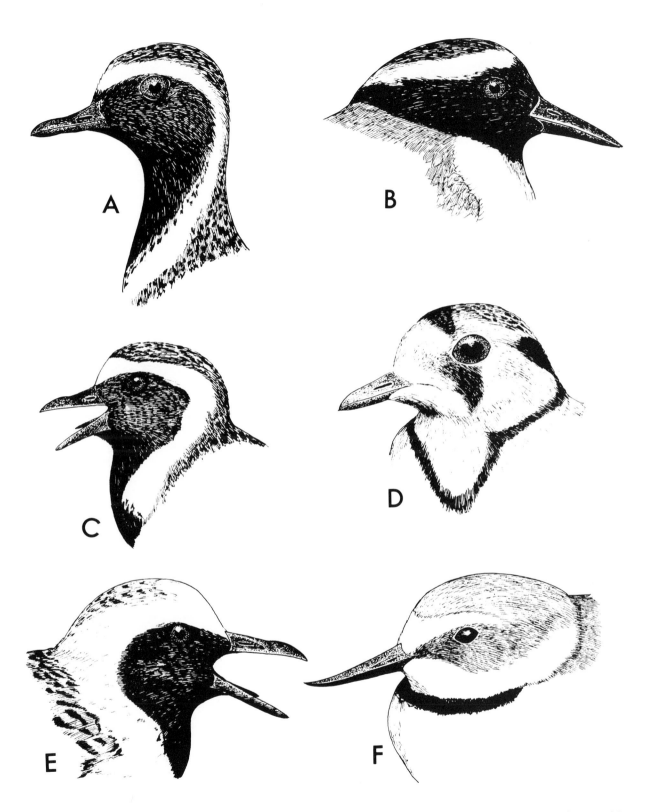

63. Heads of aberrant plovers, including *A*, greater golden plover; *B*, New Zealand shore plover; *C*, lesser golden plover; *D*, inland dotterel; *E*, gray plover; and *F*, wrybill.

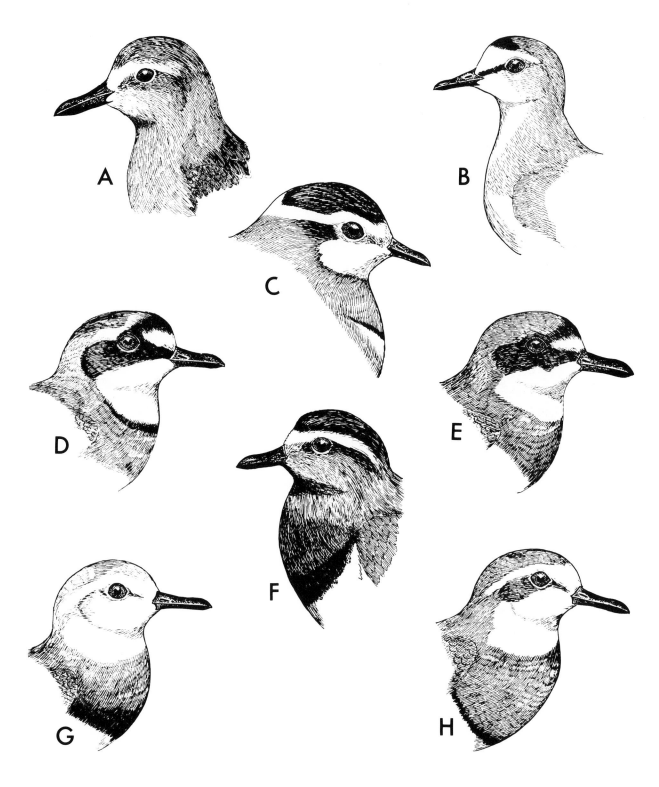

64. Heads of nonringed plovers, including *A*, red-breasted dotterel; *B*, dotterel; *C*, mountain plover; *D*, greater sand plover; *E*, Mongolian plover; *F*, rufous-chested dotterel; *G*, greater oriental plover; and *H*, lesser oriental plover.

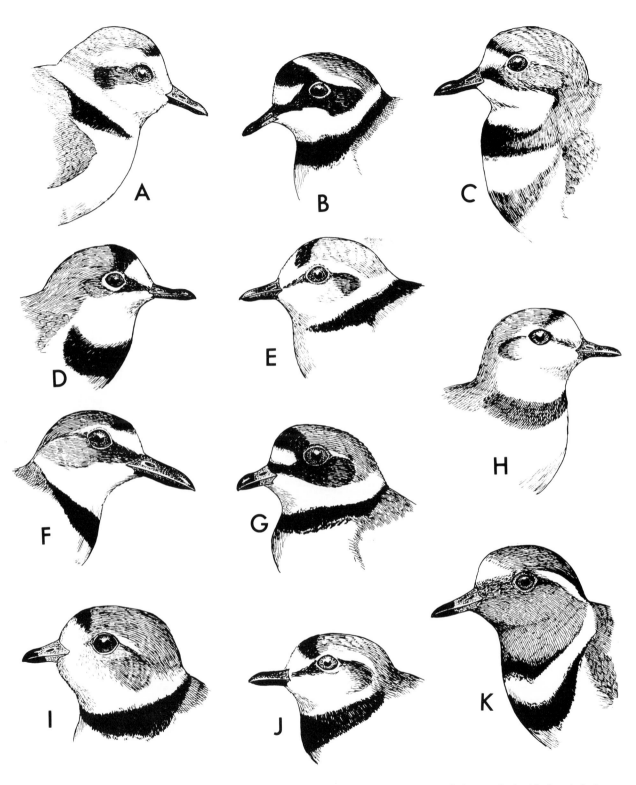

65. Heads of ringed plovers and sandplover, including *A*, sandplover; *B*, little ringed plover; *C*, double-banded plover; *D*, collared plover; *E*, Malayasian sandplover; *F*, thick-billed plover; *G*, ringed plover; *H*, chestnut-breasted sand-plover; *I*, piping plover; *J*, Madagascan sandplover; and *K*, three-banded plover.

66. Heads of miscellaneous plovers, including *A*, killdeer; *B*, Kittlitz plover; *C*, two-banded plover; *D*, black-fronted dotterel; *E*, red-kneed dotterel; *F*, hooded dotterel; *G*, diademed sandpiper-plover; and *H*, tawny-throated dotterel.

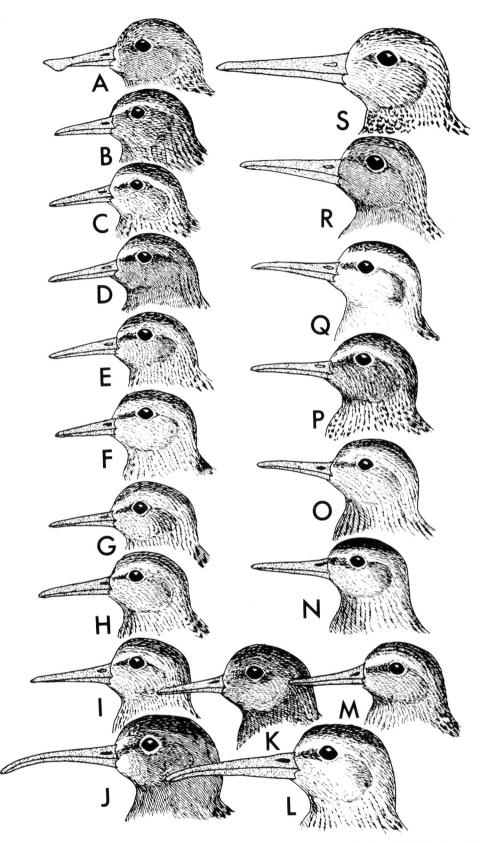

67. Heads of *Calidris* species, including *A*, spoon-billed sandpiper; *B*, Temminck stint; *C*, little stint; *D*, rufous-necked sandpiper; *E*, semipalmated sandpiper; *F*, least sandpiper; *G*, long-toed stint; *H*, western sandpiper; *I*, Baird sandpiper; *J*, curlew sandpiper; *K*, sanderling; *L*, dunlin; *M*, white-rumped sandpiper; *N*, sharp-tailed sandpiper; *O*, pectoral sandpiper; *P*, purple sandpiper; *Q*, rock sandpiper; *R*, red knot; *S*, great knot. Drawn to scale.

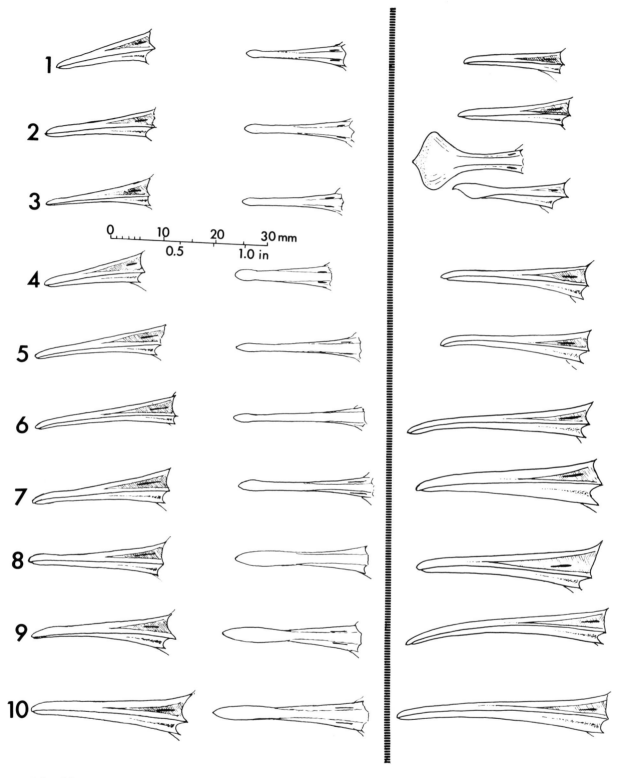

68. Bills of *(left row)* *1,* Temminck stint; *2,* little stint; *3,* least sandpiper; *4,* semipalmated sandpiper; *5,* Baird sandpiper; *6,* white-rumped sandpiper; *7,* western sandpiper; *8,* sanderling; *9,* sharp-tailed sandpiper; *10,* pectoral sandpiper; *(right row)* *1,* buff-breasted sandpiper; *2,* rufous-necked stint; *3,* spoon-billed sandpiper; *4,* purple sandpiper; *5,* rock sandpiper; *6,* dunlin; *7,* ruff; *8,* red knot; *9,* curlew sandpiper; *10,* stilt sandpiper. Drawn to scale.

69. Heads of *A*, rufous-necked stint; *B*, Baird sandpiper; *C*, western sandpiper; *D*, least sandpiper; *E*, semipalmated sandpiper; *F*, spoon-billed sandpiper; *G*, Temminck stint; *H*, little stint.

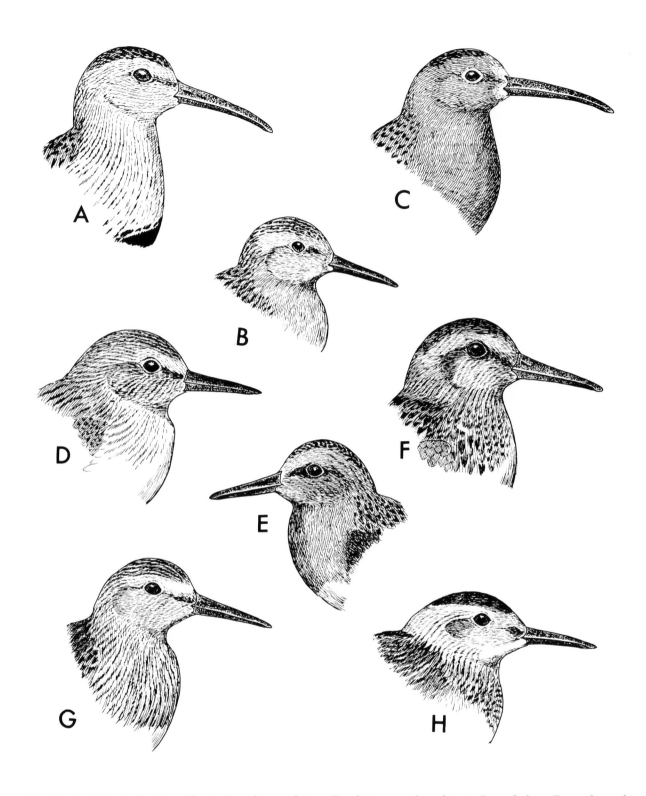

70. Heads of *A*, dunlin; *B*, red knot; *C*, curlew sandpiper; *D*, white-rumped sandpiper; *E*, sanderling; *F*, purple sandpiper; *G*, pectoral sandpiper; *H*, sharp-tailed sandpiper.

71. Heads of aberrant sandpipers, including *A*, black turnstone; *B*, ruff; *C*, ruddy turnstone; *D*, stilt sandpiper; *E*, broad-billed sandpiper; *F*, surfbird; *G*, buff-breasted sandpiper.

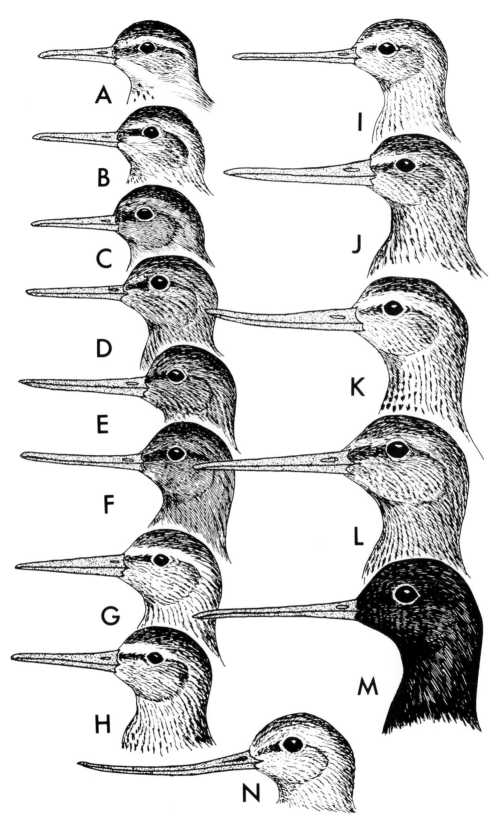

72. Heads of tattlers, including *A*, spotted sandpiper; *B*, wood sandpiper; *C*, solitary sandpiper; *D*, green sandpiper; *E*, redshank; *F*, lesser yellowlegs; *G*, Siberian tattler; *H*, wandering tattler; *I*, marsh sandpiper; *J*, greenshank; *K*, spotted greenshank; *L*, greater yellowlegs; *M*, spotted redshank; *N*, Terek sandpiper. Drawn to scale.

73. Heads of *A*, Eurasian sandpiper; *B*, solitary sandpiper; *C*, spotted redshank; *D*, spotted sandpiper; *E*, wandering tattler; *F*, marsh sandpiper; *G*, willet; *H*, Terek sandpiper.

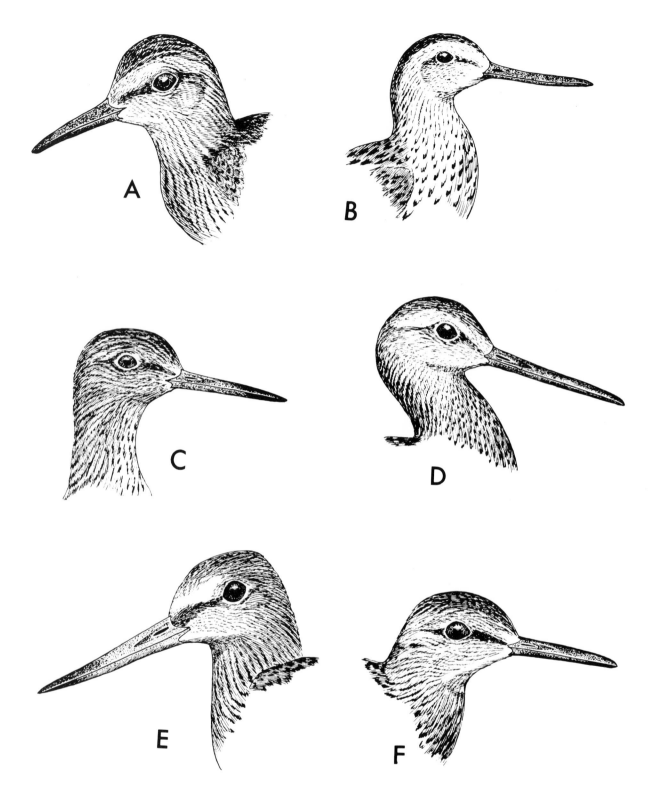

74. Heads of *A*, wood sandpiper; *B*, spotted greenshank; *C*, redshank; *D*, greater yellowlegs; *E*, greenshank; *F*, lesser yellowlegs.

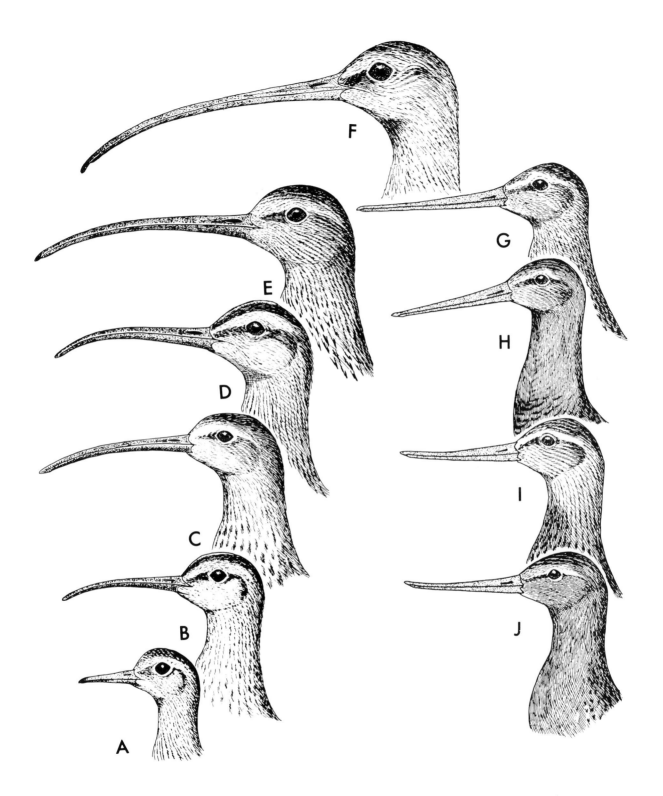

75. Heads of *A*, upland sandpiper; *B*, Eskimo curlew; *C*, slender-billed curlew; *D*, whimbrel; *E*, curlew; *F*, long-billed curlew; *G*, marbled godwit; *H*, black-tailed godwit; *I*, Hudsonian godwit; *J*, bar-tailed godwit.

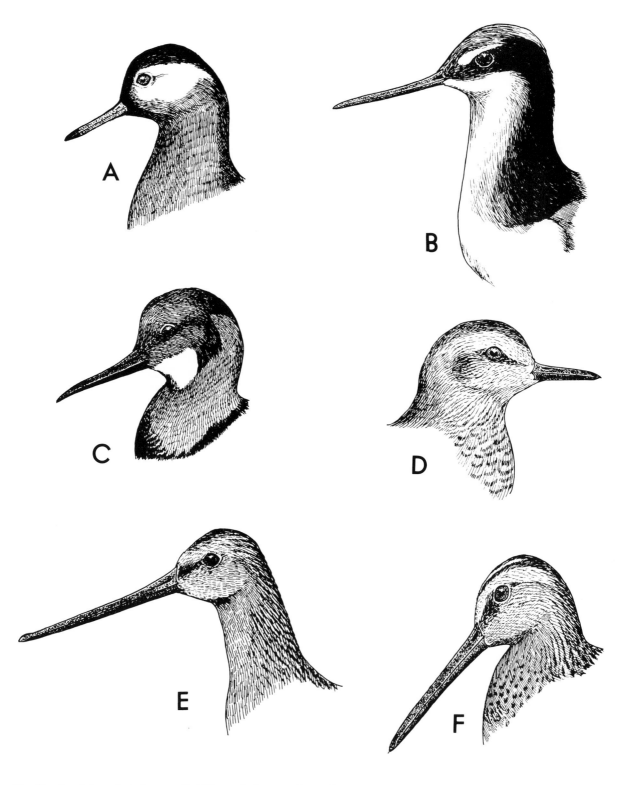

76. Heads of *A*, red phalarope; *B*, Wilson phalarope; *C*, northern phalarope; *D*, Tuamotu sandpiper; *E*, snipe-billed dowitcher; *F*, long-billed dowitcher.

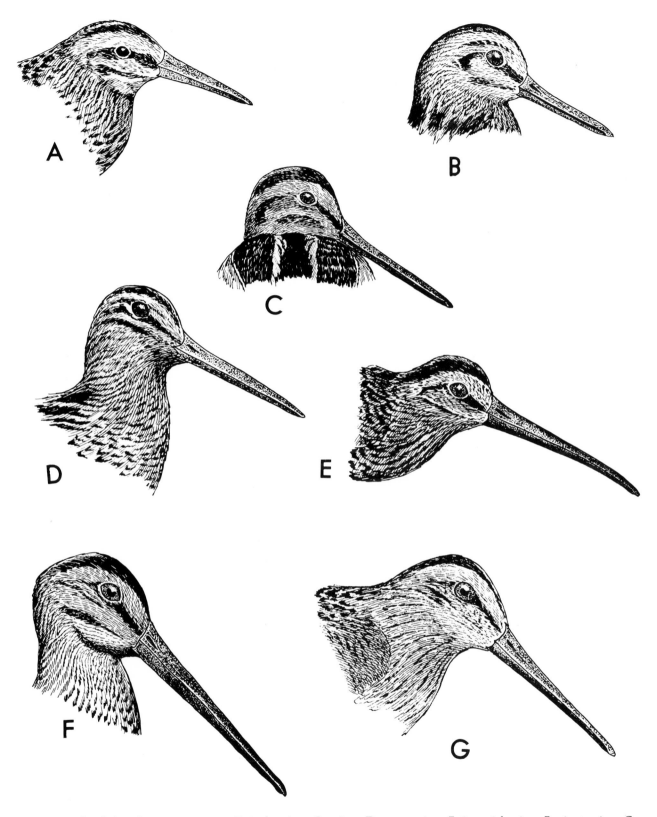

77. Heads of *A*, sub-antarctic snipe; *B*, jack snipe; *C*, snipe; *D*, great snipe; *E*, imperial snipe; *F*, giant snipe; *G*, cordillerian snipe.

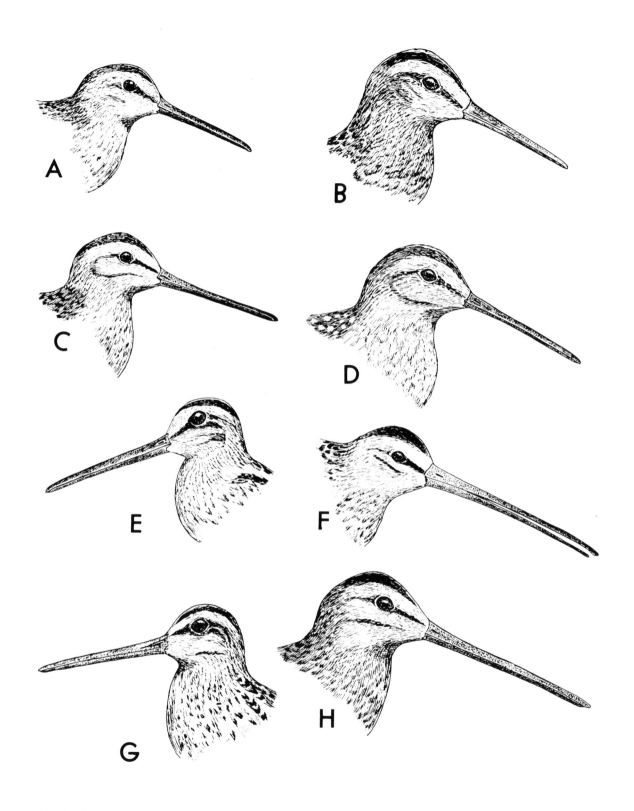

78. Heads of *A*, pintailed snipe; *B*, Himalayan snipe; *C*, forest snipe; *D*, solitary snipe; *E*, Japanese snipe; *F*, Madagascan snipe; *G*, African snipe; *H*, noble snipe.

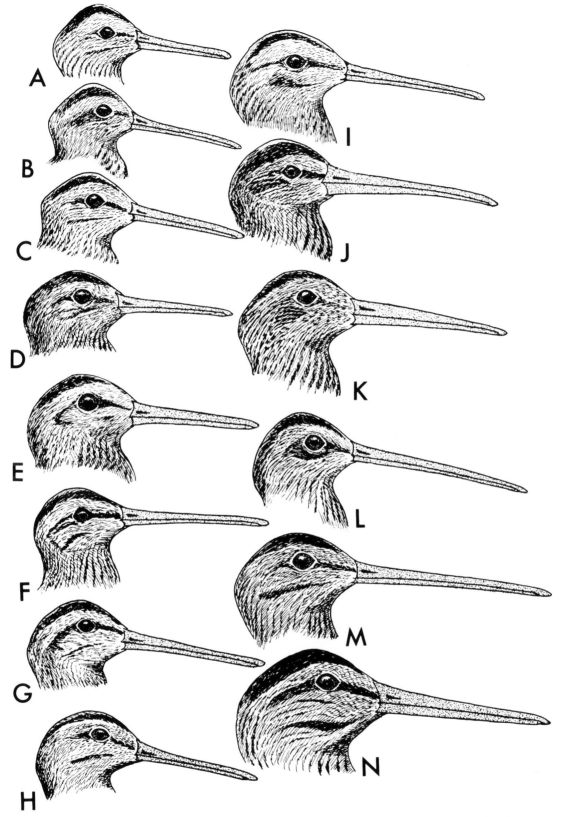

79. Heads of *A*, pintailed snipe; *B*, puna snipe; *C*, forest snipe; *D*, Himalayan snipe; *E*, great snipe; *F*, African snipe; *G*, Japanese snipe; *H*, snipe; *I*, solitary snipe; *J*, imperial snipe; *K*, cordillerian snipe; *L*, Madagascan snipe; *M*, noble snipe; and *N*, giant snipe. Drawn to scale.

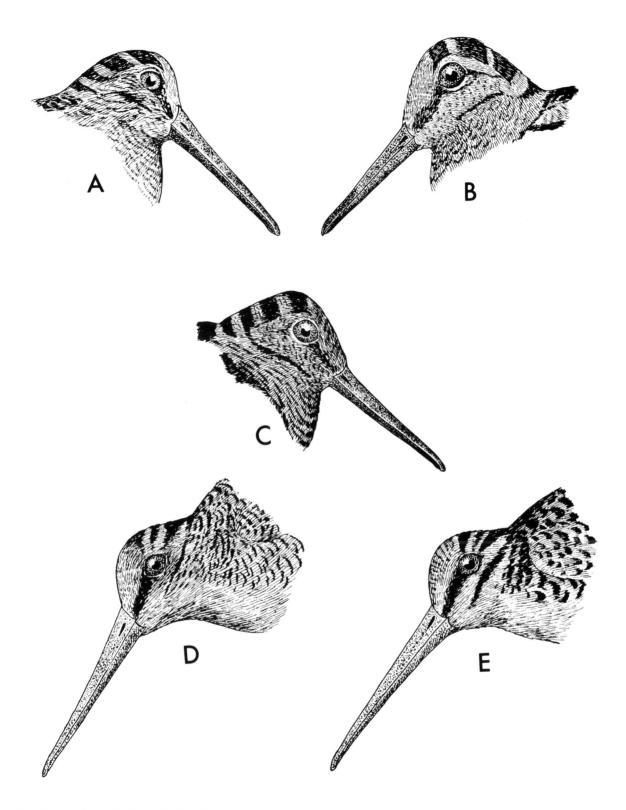

80. Heads of woodcocks, including *A*, American; *B*, Eurasian; *D*, dusky; *D*, Celebes Indonesian; *E*, Obi Indonesian.

Derivations of Generic and Specific Names

(Including those subspecies that are often considered full species)

Anarhynchus—from the Greek prefix *ana-*, up or upon, and *rhynchus*, bill or beak. The vernacular name wrybill means having a twisted bill.

> *frontalis*—from Latin, the front, and probably referring to the breastband.

Aphriza—a coined Greek construction from *aphros*, seafoam, and *zao*, to live.

> *virgata*—from Latin, striped, and referring to the striped breast.

Arenaria—from the Latin feminine form of *arenarius*, related to sand.

> *interpres*—from Latin, a go-between or agent, referring to the sentrylike behavior of this species.
>
> *melanocephala*—from the Greek *melanos*, black, and *kephale*, head

Bartramia—after William Bartram (1739–1823), American botanist and ornithologist.

> *longicauda*—from Latin *longus*, long, and *caudum*, tail.

Calidris—An Aristotelian name (*skalidris*) for a gray, speckled shorebird. The English vernacular name stint, used for several small *Calidris* species, probably relates to their small stature. The American vernacular term "peep" refers to their calls.

> *acuminatus*—from Latin, pointed.
>
> *alba*—from the feminine of Latin *albus*, white. This species is sometimes placed in the genus *Crocethia*, a word of uncertain origin. Sanderling is from the Icelandic *sand* and *erla*, a "sand-wagtail."
>
> *alpina*—from Latin, alpine. Dunlin is a variant of "dunling," meaning a small dun-colored bird. This species and several others are often placed in the genus *Erolia*, a word of obscure origin.
>
> *bairdii*—after S. F. Baird (1823–87), American ornithologist and secretary of the Smithsonian Institution.
>
> *canutus*—Latinized form of the Danish king Canute (Cnut), after whom the bird was named by Linnaeus. The common name knot has the same origin.
>
> *ferruginea*—from Latin, rusty red. The alternate name *testacea* is from the Latin *testaceus*, brick or tile, and presumably also refers to the reddish color.
>
> *fuscicollis*—from Latin *fuscus*, dark, and *collum*, neck.
>
> *maritima*—from Latin, maritime.
>
> *mauri*—after E. Mauri (1791–1836), Italian botanist and director of the Rome Botanical Gardens.
>
> *melanotos*—from Greek *melanos*, black, and *noton*, back.
>
> *minuta*—from Latin *minutus*, small.
>
> *minutilla*—corrupted from Latin *minutulus*, very small.
>
> *ptilocnemis*—from Greek *ptilon*, feather, and *knemis*, shin, referring to the feathered tibia.
>
> *pusilla*—from Latin, puerile or petty, meaning small.
>
> *pygmeus*—Latin for dwarf. This species is often placed in the genus *Eurynorhynchus*, from Greek *eurono*, to widen, and *rhynchos*, beak.
>
> *ruficollis*—from Latin *rufer*, red, and *collum*, neck.
>
> *subminuta*—from Latin, very small.
>
> *temminckii*—after G. J. Temminck (1775–1858), Dutch ornithologist.
>
> *tenuirostris*—from Latin *tenuis*, slender or thin, and *rostrum*, bill or beak.

Catoptrophorus—a latinized form of the Greek *katophophoros*, mirror-bearing, referring to the white wing patches. The common name willet is onamatopoetic.

> *semipalmatus*—from Latin, referring to the semipalmated feet.

Charadrius—from Greek *kharadrios*, an Aristotelian name for a waterbird nesting in ravines. The vernacular name plover comes from Latin *pluvia*, rain.

> *alexandrinus*—of Alexandria.

alticola—from Latin *altus*, high, and *colo*, to inhabit.

asiaticus—of Asia.

bicinctus—from the Latin prefix *bi-*, two, and *cinctus*, banded or girdled.

collaris—from Latin, collared.

cucullatus—from Latin, hooded.

dubius—from Latin, uncertain or questionable.

falklandicus—of the Falkland Islands.

forbesi—after W. A. Forbes (1855–83), English naturalist.

hiaticula—possibly a mistake for Latin *hiaticola*, a cliff-dweller, but perhaps a translation of Greek *kharadrios*, from which *Charadrius* was derived.

leschenaultii—after J. B. Leschenault (1773–1826), French naturalist.

marginatus—from Latin, edged or bordered.

melanops—from Greek *melas*, black, and *ops*, face.

melodus—latinized form of the Greek *melodos*, melodius.

modestus—from Latin, calm or unassuming. This species is often placed in the genus *Zonibyx*, from Greek *zone*, girdle, and *ibyx*, a variant of ibis.

mongolus—of Mongolia.

montanus—from Latin, of the mountains. This species is often placed in the genus *Eupoda*, from Greek *eu*, good, and *pous*, foot.

morinellus—a latinized diminutive of the Greek *moros*, foolish. The vernacular name dotterel is from the English "dote," to be foolish, with a diminutive ending.

novaeseelandiae—of New Zealand. The generic name *Thinornis* is from Greek *thinos*, beach, and *ornis*, bird.

obscurus—from Latin, obscure.

occidentalis—from Latin, western.

pallidus—from Latin, pallid. The alternate name for this species, *venustus*, is Latin for charming or elegant.

pecuarius—from Latin, a grazer. The vernacular name Kittlitz plover is after F. H. von Kittlitz (1779–1874), German naturalist.

peronii—after François Peron (1775–1810), French naturalist.

placidus—from Latin, placid.

ruficapillus—from Latin *rufus*, reddish, and *capillis*, hair.

sanctaehelenae—of St. Helena Island.

semipalmatus—from Latin, semipalmated.

thoracicus—from Greek *thorakos*, breastplate.

tricollaris—from Latin, of three colors.

veredus—from Latin, fleet-footed.

vociferus—from Latin, vocal. The vernacular name killdeer is onamatopoetic.

wilsonius—after Alexander Wilson (1776–1813), father of American ornithology.

Cladorhynchus—from Greek *leucos*, curving branch, and *rhyncos*, beak.

leucocephalus—from Greek *leucos*, white, and *cephala*, head.

Coenocorypha—from Latin *coenum*, dirt, and Greek *koryphe*, head or summit.

aucklandica—of the Auckland Islands.

Erythrogonys—from Greek *erythros*, reddish, and *gonys*, knee.

cinctus—from Latin, banded or girdled.

Gallinago—from Latin *gallina*, hen. The alternative generic name *Capella* is the Latin diminutive for a female goat and refers to the "bleating" behavior. The English name snipe comes from "snipper" and refers to the long beak.

andina—of the Andes.

gallinago—as above.

hardwickii—after General Hardwick (?–1835) of Tasmania, who discovered the species. The vernacular name Latham snipe is after its describer, John Latham (1740–1837), English naturalist.

imperialis—from Latin, of the empire or emperor.

jamesoni—after William Jameson (1796–1873) of the University of Quito.

macrodactyla—from Greek *makros*, long, and *daktylos*, fingers or toes.

media—from Latin *medius*, the middle.

megala—from Greek *megas*, great.

nemoricola—from Latin *nemoralis*, wood or grove, and *colere*, to inhabit.

nigripennis—from Latin *niger*, dark or black, and *penn*, feather or wing.

nobilis—from Latin, famous or well known.

paraguaiae—of Paraguay.

solitaria—from Latin *solitarius*, in solitude.

stenura—from Greek *stenos*, narrow or straight, and *oura*, tail.

stricklandii—after H. E. Strickland (1811–53), English naturalist and author of the Strickland code of taxonomy. This species is often placed in the genus *Chubbia*, after Charles Chubb

(1851–1924) of the British Museum (Natural History).

undulata—from Latin undulatus, wavelike.

Haematopus—probably from Greek haima, blood, and pous, foot, but possibly referring to the eye. The vernacular name oystercatcher is derived from ostralegus and is inappropriate, since the birds rarely eat oysters.

ater—from Latin, black.

bachmani—after Rev. John Bachman (1790–1874), friend and collaborator of J. J. Audubon.

chathamensis—of the Chatham Islands.

finschi—after Dr. Otto Finsch, nineteenth-century German ornithologist.

fuliginosus—from Latin, sooty.

leucopodus—from Greek leukon, white, and pous, foot.

meadewaldoi—after E. G. B. Meade-Waldo, nineteenth-century English naturalist, who first collected this form.

moquini—after C. H. B. A. Moquin-Tandon, nineteenth-century French professor of natural history in Paris and student of Canary Island birds.

ostralegus—from Greek ostreon, oyster, and lego, to gather.

palliatus—from Latin, cloaked.

reischeki—after A. Reischek, who first collected this form in 1885.

unicolor—from Latin, of one color.

Hemiparra—from the Greek prefix hemi-, half, and Latin parra, bird of evil omen.

crassirostris—from Latin crassus, thick or heavy, and rostrum, beak.

Himantopus—a latinized form of Greek himantopous, meaning thong-footed. The vernacular name stilt comes from the Middle English stilte, which in turn comes from various Scandinavian words such as stilta and sylta, meaning wooden poles on which to stand.

ceylonensis—of Ceylon.

knudseni—after Vlademar Knudsen, who collected the first specimens.

leucocephalus—from Greek leukos, white, and kephale, head.

melanurus—from Greek melas, black, and oura, tail.

meridionalis—from Latin, southern.

mexicanus—of Mexico.

novaeseelandiae—of New Zealand.

Hoploxypterus—from Greek hoplon, tool or weapon, oxys, sharp, and pteron, wing.

cayanus—of Cayenne.

Hydrophasianus—from the Greek prefix hydro-, water, and phasianus, pheasant.

chirurgus, from Latin chirurgicus, surgical, probably in reference to the lancelike wing spur.

Ibidorhynchus—from Greek ibis, ibis, and rhynchos, beak.

struthersii—after Dr. John Struthers (?–after 1855), Scottish physician.

Jacana—a Portuguese word derived from jassana, the Tupi-Guarani Indian name for the bird.

spinosa—from Latin, spiny.

Limicola—from Latin limus, mud, and colere, to inhabit.

falcinellus—from Latin, a small scythe, in reference to the bill shape.

Limnocryptes—from Greek limne, marsh, and krypto, hider. "Jack" is a common English diminutive term and the origin of the vernacular name jack snipe.

minimus—from Latin, smallest.

Limnodromus—from Greek limne, marsh, and dromos, runner. The vernacular name dowitcher is derived from the Iroquoian word for the bird.

griseus—from Latin, gray or grizzled.

scolopaceus—a latinized form of the Greek skolopax, woodcock or snipe.

Limosa—from Latin limus, mud, and referring to the habitat. The vernacular name godwit is of uncertain origin but probably is from the Old English "god whit," meaning a good creature, and referring to its table qualities.

fedoa—probably a latinized version of some English word for godwit.

haemastica—from Greek haimastikos, blood red.

lapponica—of Lapland, where the species was first collected.

limosa—see above.

Metopidius—probably from Greek metopion, having a high or broad forehead.

africana—of Africa. This and the following species are often placed in the genus Actophilornis, from Greek aket, a seashore or promontory, philos, fond of, and ornis, bird.

albinucha-from the feminine of Latin albus, white, and the Latin nucha, nape.

capensis—of the Cape of Good Hope. This

species is often placed in the genus *Micro-parra*, which is from the Greek *mikros*, small, and *parra*, bird of evil omen.

gallinacea—from Latin *gallinaceous*, fowllike. This species is often placed in the genus *Irediparra*, after the Australian naturalist Tom Iredale (1880–1972), and *parra*, bird of evil omen.

indicus—of India.

Micropalama—from Greek *mikros*, small, and *palame*, web, meaning slightly webbed.

himantopus—the latinized form of Greek *himantopous*, meaning thong-footed.

Numenius—from Greek *noumenios*, new moon, referring to the shape of the bill. The vernacular name curlew refers to the bird's call and derives from the Old French *courlieus*.

americana—of America.

arquata—from Latin *arquatus*, bent.

borealis—from Latin, northern.

madagascariensis—referring to Madagascar, an erroneous judgment of the species' range.

minutus—from Latin, little.

phaeopus—from Greek *phaios*, gray, and *pous*, foot. The vernacular name whimbrel is reportedly from its "whimpering" call.

tahitiensis—of Tahiti, part of the species' winter range.

tenuirostris—from Latin *tenuis*, thin, and *rostrum*, bill.

Oreopholus—from Greek *oreos*, mountain, and (probably) *pholeos*, hiding place.

ruficollis—from Latin *rufer*, red, and *collum*, neck.

Peltohyas—from Greek *pelte*, shield, and *hyas*, plover.

australis—from the south (Australia).

Phalaropus—from Greek *phalaris*, coot, and *pous*, foot, referring to the cootlike lobed feet.

fulicarius—from Latin, cootlike.

lobatus—from Latin lobed.

tricolor—from Latin, three-colored. The vernacular name Wilson phalarope refers to Alexander Wilson (1776–1813), father of American ornithology.

Phegornis—from Greek *phegos*, a kind of tree, and *ornis*, bird.

mitchellii—after D. W. Mitchell (?–1859), onetime secretary of the Zoological Society of London. The vernacular name diademed sandpiper-plover refers to the distinctive crown pattern.

Philomachus—from Greek *philomakhos*, warlike, and referring to the males' behavior. The name reeve refers to the female and derives from the Anglo-Saxon gerifa, one in authority (from which sheriff also derives). The name ruff refers to the male and is in reference to the male's ruffled cape.

pugnax—from Latin, pugnacious.

Pluvialis—from Latin *pluviarius*, rain bird.

apricaria—from Latin *apricus*, exposed to the sun or sun-tinged.

dominica—of Santo Domingo.

squatarola—of uncertain origin, but perhaps a local Venetian name for a plover.

Pluvianellus—from Latin *pluvia*, rain, and the diminutive suffix -*ellus*.

socialis—from Latin, sociable. The vernacular name Magellanic plover refers to the Strait of Magellan.

Prosobonia—probably from Greek *prosopon*, face or front.

cancellatus—from Latin, latticed. This species is often placed in the genus *Aechmorhynchus*, from Greek *aichme*, spear, and *rhynchos*, beak.

leucoptera—from Greek *leukos*, white, and *pteron*, wing.

Pseudoscolopax—from Latin *pseudo*, false, and *scolopax*, pointed (see below).

semipalmatus—from Latin, semipalmated, referring to the feet.

Recurvirostra—from Latin *recurvus*, turned up, and *rostrum*, beak or bill. The vernacular name avocet comes from Italian *avocetta*.

americana—of America.

andina—of the Andes.

avocetta—from Italian *avocetta* or *avosetta*, but of unknown earlier origin.

novaehollandiae—of New Holland (Australia).

Rostratula—diminutive of Latin *rostratus*, beaked.

benghalensis—of Bengal.

semicollaris—from Latin, meaning half-collared. This species is often placed in the genus *Nycticryphes*, from Greek *nyktero*, noctural, and *kryphos*, a hiding place.

Scolopax—from the Aristotelian *scolopax*, meaning pointed, presumably referring to the long beak. The vernacular name woodcock means "cock of the woods." "Cock" is from the Middle English "cok" and suggests the sound of the bird.

celebensis—of the Celebes Islands.

minor—from Latin, smaller or lesser. This species is often placed in the genus *Philohela*, from Greek *philos*, loving, and *helos*, swamp or marsh.

mira—feminine of Latin *mirus*, wonderful.

rochussenii—after J. J. Rochussen (1797–1872), governor general of the Dutch East Indies.

rusticola—an incorrect version of Latin *rusticula*, referring to a bird that runs on the ground.

saturata—from Latin, saturated, in reference to the coloration.

Tringa—from the Greek *tryngas*, used by Aristotle for a "waterbird with a white rump." The vernacular name tattler refers to the loud and frequent calling of the birds when alarmed.

brevipes—from Latin *brevis*, short, and *pes*, foot. This species is sometimes placed in the genus *Heteroscelus*, which is derived from Greek

heteroskeles, meaning "different legs," referring to the scutellation pattern.

erythropus—from Greek *erythros*, red and *pous*, foot.

flavipes—from Latin *flavus*, yellow, and *pes*, foot.

glareola—a Latinism from *glarea*, gravel, apparently meaning a gravel bird.

guttifer—from Latin, spotted. The vernacular name Nordmann (1803–66), German naturalist.

hypoleucos—from Greek *hypo*, less than usual, and *leukos*, white. This species is sometimes placed in the genus *Actitis*, which derived from Greek *aktites*, "dweller on the coast."

incana—from Latin, gray.

macularia—a Latinism from *macula*, spot, referring to the spotted breast.

melanoleuca—from *melas* or *melanos*, black, and *leukos*, white, meaning spotted with black and white.

nebularia—from Latin, dark or clouded.

ochropus—from Greek *ochra*, yellow ochre, and *pous*, foot.

solitaria—from Latin *solitarius*, in solitude.

stagnatalis—from Latin *stagnatus*, referring to a pool of still water.

totanus—a Latinism derived from Italian *totano*, moorhen.

Tryngites—a Greek word coined from *trynga* and meaning "like a *Tringa*."

subruficollis—from the Latin prefix *sub-*, somewhat, *rufer*, red, and *collum*, neck.

Vanellus—a misspelled diminutive form of Latin *vannus*, or fan, thus a little fan. The vernacular name lapwing is from the Old English *hleapewince*, meaning to waver or wince while leaping or running, or to turn while in flight.

albiceps—from Latin *albus*, white, and *ceps*, head. This species is often placed in the genus *Xiphidiopterus*, from Greek *xiphidion*, sword, and *pterus*, wing.

armatus—from Latin, armed. This and other species are often placed in the genus *Hoplopterus*, from Greek *hoplon*, weapon, and *pteryx*, wing.

chilensis—of Chile. This species is often placed in the genus *Belonopterus*, from Greek *belone*, needle, and *pterus*, wing.

cinereus—from Latin, ash-colored. This species is often placed in the genus *Microsarcops*, from Greek *mikros*, small, and *sarkos*, flesh.

coronatus—from Latin, crowned.

duvaucelii—after Alfred Duvaucel (1796–1824) of Sumatra.

gregaris—from Latin, belonging to a flock or herd.

indicus—of India.

leucurus—from Greek *leukon*, white, and *oura*, tail.

lugubris—from Latin, sorrowful or gloomy, in reference to the color.

macropterus—from the Latin prefix *macro-*, large, and Greek *pteron*, wing.

malabaricus—from the Malabar coast. This species is often placed in the genus *Lobipluvia*, from Greek *lobos*, a lobe, and Latin *pluvia*, rain.

melanocephalus—from Greek *melas*, black, and *kephale*, head. This species is often placed in the genus *Tylibyx*, from Greek *tylos*, callus, and *ibyx*, a variant of ibis.

melanopterus—from Greek *melas*, black, and *pteron*, wing.

miles—from Latin, soldier. This species is often placed in the genus *Lobibyx*, from Greek *lobos*, lobe, and *ibyx*, a variant of ibis.

novaehollandiae—of New Holland (Australia).

resplendens—from Latin, glittering or shining. This species is often placed in the genus *Ptilosceles*, from Greek *ptilon*, feather, and *skelos*, leg.

senegallus—of Senegal. This species is often placed in the genus *Afribyx*, from *Afer*, African, and *ibyx*, a variant of ibis.

spinosus—from Latin, spiny.

superciliosus—from the Latin *super,* above, and *cilium,* eyelid. This species is often placed in the genus *Anomalophrys,* from Greek *anomalos,* uneven, and *ophryx,* brown or eyebrow.

tectus—from the Latin *tectum,* roof or covering. This species is often placed in the genus *Sarciophorus,* from Greek *sarkos,* flesh, and *phoros,* bearing.

tricolor—from Latin, of three colors. This species is often placed in the genus *Zonifer,* from Latin *zona,* band, and *fero,* I bear.

vanellus—as above.

Xenus—from Greek *xenos,* stranger or foreigner.

cinereus—from Latin, ashy or ash-colored. The vernacular name Terek sandpiper refers to the Terek Basin of the northern Caucasus.

Sources Cited

Abdusalyamov, I. A. 1971. [Fauna of the Tadjik Soviet Socialist Republic. Vol. 19, pt. 1 Birds]. Dushanbe, USSR (In Russian).

Ade, B. 1979. Some observations on the breeding of crowned plovers *Bokmakierie* 31:9–16.

Ahlquist, J. E. 1974. The relationships of the shorebirds (Charadriiformes). Ph.D. diss., Yale University.

Ailes, M. 1976. Ecology of the upland sandpiper in central Wisconsin. M.S. thesis, University of Wisconsin, Stevens Point.

Alexander, W. B. 1945–47. The woodcock in the British Isles. *Ibis* 87:512–50; 88:1–24, 159–79, 271–86, 427–44; 89:1–28.

Ali, S. 1962. *The birds of Sikkim.* Madras: Oxford University Press.

Ali, S., and Ripley, D. 1969. *Handbook of the birds of India and Pakistan.* Vol. 2. London: Oxford University Press.

Allen, A. A., and Kyllingstad, H. 1949. The eggs and young of the bristle-thighed curlew. *Auk* 66: 343–50.

Anderson, R. A. 1968. Notes on the Snares Island snipe. *Notornis* 15:223–27.

Appert, O. 1971. Die Limikolen des Mangokygebietes in Sudwest-Madagaskar. *Ornithologische Beobachter* 68:53–77.

Ashkenazie, S., and Safriel, U. N. 1979. Breeding cycle and behavior of the semipalmated sandpiper at Barrow, Alaska. *Auk* 96:56–67.

Ashmole, M. J. 1970. Feeding of western and semipalmated sandpipers in Peruvian winter quarters. *Auk* 87:131–35.

Baker, A. J. 1973. Distribution and numbers of New Zealand oystercatchers. *Notornis* 20:128–44.

———. 1974*a*. Prey-specific feeding methods of New Zealand oystercatchers. *Notornis* 21:219–33.

———. 1974*b*. *Ecological and behavioral evidence for the systematic status of New Zealand oystercatchers (Charadriiformes: Haematopodidae).* Life Sciences Contribution 96. Toronto: Royal Ontario Museum. 34 pp.

———. 1974*c*. Criteria for aging and sexing New Zealand oystercatchers. *New Zealand Journal of Marine and Freshwater Research* 8:211–21.

———. 1975. Morphological variation, hybridization and systematics of New Zealand oystercatchers. *Journal of Zoology* 175:357–90.

———. 1977. Multivariate assessment of the phenetic affinities of Australian oystercatchers (Aves: Charadriiformes). *Bijdragen Tot de Dierkunde* 47: 156–64.

Baker, M. C., and Baker, A. E. M. 1973. Niche relationships among six species of shorebirds on the wintering and breeding ranges. *Ecological Monographs* 43:192–212.

Bannerman, D. A. 1951. *The birds of tropical West Africa.* Vol. 8. Edinburgh: Oliver and Boyd.

———. 1960, 1961. *The birds of the British Isles.* Vols. 9 and 10. Edinburgh: Oliver and Boyd.

———. 1963. *Birds of the Atlantic Islands.* Vol. 1. Edinburgh: Oliver and Boyd.

Barlow, M. 1972. The establishment, dispersal and distribution of the spur-winged plover in New Zealand. *Notornis* 19:201–11.

Barlow, M.; Mutter, P. M.; and Sutton, R. R. 1972. Breeding data on the spur-winged plover in Southland, New Zealand. *Notornis* 19:212–49.

Bates, R. S. P., and Lowther, E. H. N. 1952. *Breeding birds of Kashmir.* London: Oxford University Press.

Bengtson, S.-A. 1968. The breeding behavior of the gray phalarope in West Spitsbergen. *Vår Fågelvårld* 27:1–13.

———. 1970. Breeding behavior of the purple sandpiper *Calidris maritima* in West Spitsbergen. *Ornis Scandinavica* 1:17–25.

———. 1975. [Observations on the breeding biology of the purple sandpiper *Calidris maritima*, on Svalbard]. *Fauna* 28:81–86. (In Norwegian, English summary.)

Bengtson, S.-A., and Fjellberg, A. 1975. Summer food of the purple sandpiper (*Calidris maritima*) in Spitzbergen. *Astarte* 8:1–6.

Bengtson, S.-A., and Svensson, B. 1968. Feeding habits of *Calidris alpina* L. and *C. minuta* Leisel. in relation to the distribution of marine shore invertebrates. *Oikos* 19:152–57.

Benson, C. W.; Brooke, R. K.; Dowsett, R. J.; and Irwin, M. P. S. 1971. *The birds of Zambia.* London: Collins.

Benson, C. W.; Colebrook-Robjent, J. F. R.; and Williams, A. 1976. Contribution de l'ornithologie de Madagascar. *L'Oiseau* 46:103–34, 209–42.

Bent, A. C. 1927, 1929. *Life histories of North American shore birds.* Two parts. Bulletins 142 and 146. Washington, D.C.: United States National Museum.

Bergman, G. 1946. Der Steinwälzer in seiner Beziehung zur Umwelt. *Acta Zoologica Fennica* 47: 1–151.

Bergman, R. D.; Howard, R. L.; Abraham, H. F.; and Weller, M. W. 1977. *Water birds and the wetland resources in relation to oil development at Storkersen Point, Alaska.* Fish and Wildlife Service Resource Publication 129, Washington, D.C.: United States Department of Interior.

Berman, D. I., and Kuz'min, I. F. 1965. [The pintail snipe in the Tuva, USSR] *Ornitologiya* 7:209–16. (In Russian.)

Betts, B. J. 1973. A possible hybrid wattled jacana x northern jacana in Costa Rica. *Auk* 90:687–89.

Beven, G. and England, M. D. 1977. Studies of less familiar birds. 181. Turnstone. *British Birds* 70: 23–32.

Beven, J. O. 1913. Notes and observations on the painted snipe (*Rostratula capensis*) in Ceylon. *Ibis,* ser. 10, 1:527–34.

Bianki, V. V. 1977. *Gulls, shorebirds and alcids of Kandalasksha Bay.* Translated from Russian by the Israel Program for Scientific Translations, Jerusalem. 250 pp.

Bicak, T. K. 1977. Some eco-ethological aspects of a breeding population of long-billed curlews in Nebraska. M.S. thesis, University of Nebraska, Omaha.

Bird, M. C. H. 1909. The average weight of snipe. *British Birds* 2:312.

Blake, E. R. 1977. *Manual of neotropical birds.* Vol. l. Chicago: University of Chicago Press.

Blaker, D. 1966. Notes on the sandplovers *Charadrius* in southern Africa. *Ostrich* 37:95–102.

Bock, W. 1958. A generic review of the plovers (Charadriinae, Aves). *Bulletin of the Museum of Comparative Zoology* 118:25–97.

———. 1959. The status of the semipalmated plover. *Auk* 76:98–100.

———. 1964. The systematic position of the Australian dotterel, *Peltohyas australis. Emu* 63:383–404.

Boetticher, H. V. 1954. Note sur la classification des Vanneaux. *L'Oiseau* 24:175–79.

Boyd, H. 1962. Mortality and fertility of European Charadrii. *Ibis* 104:368–87.

Boyd, R. L. 1972. Breeding biology of the snowy plover at Cheyenne Bottoms Wildlife Management Area, Barton Country, Kansas. M.S. thesis, Kansas State Teachers College, Emporia.

Brandt, H. 1943. *Alaska bird trails.* Cleveland: Bird Research Foundation.

Britton, P. L. 1970. Some non-passerine bird weights from East Africa. *Bulletin of the British Ornithologists' Club* 90:140–44, 152–93.

Britton, P. L., and Dowsett, R. J. 1969. More bird weights from Zambia. *Ostrich* 40:55–60.

Browning, M. R. 1977. Geographic variation in dunlins, *Calidris alpina,* of North America. *Canadian Field-Naturalist* 71:391–93.

Bryant, C. E. 1947. Notes on avocets breeding near Melbourne. *Emu* 46:241–45.

———. 1948. More observations on nesting avocets. *Emu* 48:89–92.

Bub, H. 1958. Unterschungen an einer Population des Flussregenpfeifers (*Charadrius dubius coronicus* Gr.). *Beiträge zur Vogelkunde* 5:268–83.

Bunni, M. K. 1959. The killdeer, *Charadrius v. vociferus* Linnaeus, in the breeding season: Ecology, behavior and the development of homoiothermism. Ph.D. diss., University of Michigan.

Burger, J., and Shisler, J. 1978. Nest-site selection of willets in a New Jersey salt marsh. *Wilson Bulletin* 90:599–607.

Burton, P. J. K. 1971. Comparative anatomy of head and neck in the spoon-billed sandpiper, *Eurynorhynchus pygmeus* and its allies. *Journal of Zoology* 163:145–63.

———. 1974. *Feeding and the feeding apparatus in waders: A study of anatomy and adaptations in the Charadrii.* London: British Museum (Natural History).

Buss, I. O. 1951. The upland plover in southwestern Yukon Territory. *Arctic* 4:204–13.

Buss, I. O., and Hawkins, A. S. 1939. The upland plover at Faville Grove, Wisconsin. *Wilson Bulletin* 51:202–20.

Cadbury, C. J., and Olney, P. J. S. 1978. Avocet population dynamics in England. *British Birds* 70: 102–21.

Cairns, W. E. 1977. Breeding biology and behavior of the piping plover (*Charadrius melodus*) in southern Nova Scotia. M.S. thesis, Dalhousie University.

Campbell, B. 1974. *The dictionary of birds in colour.*

London: Michael Joseph.

Chapin, J. P. 1939. The birds of the Belgian Congo. Part II. *Bulletin of the American Museum of Natural History* 75:1–632.

Chen Tso-hsin. 1973. *A distributional list of Chinese birds.* Vol. I. *Non-passeriformes.* Arlington, Va.: Joint Publications Research Service.

Clancey, P. A. 1960. A new race of crowned plover *Vanellus (Stephanibyx) coronatus* (Boddaert) from South-West Africa. *Bulletin of the British Ornithologists' Club* 80:13–16.

———. 1967. *Gamebirds of southern Africa.* New York: Elsevier.

———. 1971. Miscellaneous taxonomic notes on African birds, 33. *Durban Museum Novitates* 9:109–29.

———. 1975. Miscellaneous taxonomic notes on African birds, 44. *Durban Museum Novitates* 11: 1–24.

———. 1979*a*. Miscellaneous taxonomic notes on African birds, 53. *Durban Museum Novitates* 12: 1–17.

———. 1979*b*. An overlooked race of wattled plover. *Durban Museum Novitates* 12:5–6.

Condon, H. T. 1975. *Checklist of the birds of Australia.* Part 1. *Non-passerines.* Melbourne: Royal Australian Ornithologists' Union.

Conway, W. G., and Bell, J. 1969. Observations on the behavior of Kittlitz's sandplovers at the New York Zoological Park. *Living Bird* 7:57–70.

Cunningham, J. M. 1973. The banded dotterel, *Charadrius bicinctus:* Pohowera or tuturiwhatu? Call notes and behavior. *Notornis* 20:21–27.

Cunningham–van Someren, G. R., and Robinson, C. 1962. Notes on the African lily-trotter, *Actophilornis africanus* (Gmelin). *Bulletin of the British Ornithologists' Club* 82:67–72.

Dabelsteen, T. 1978. An analysis of the song-flight of the lapwing (*Vanellus vanellus* L.) with respect to causation, evolution and adaptations to signal functions. *Behavior* 66:136–78.

Davis, W. A., and Reid, A. J. 1964. Mating display of red-capped dotterel. *Emu* 63:332.

Dean, A. R.; Forteg, J. E.; and Phillips, E. G. 1977. White-tailed plover: New to Britain and Ireland. *British Birds* 70:465–71.

Dementiev, G. P., and Gladkov, N. A. 1969. *The birds of the Soviet Union.* Vol. 5. Jerusalem: Israel Program for Scientific Translations.

Dhondt, A. A. 1975. Note sur les échassier (Charadrii) de Madagascar. *L'Oiseau* 45:73–83.

Dixon, J. 1918. The nesting grounds and nesting habits of the spoon-billed sandpiper. *Auk* 35: 387–404.

———. 1927. The surf bird's secret. *Condor* 29: 3–16.

———. 1933. Nesting of the wandering tattler. *Condor* 35:173–79.

Drury, W. H., Jr. 1961. The breeding biology of shorebirds on Bylot Island, Northwest Territories, Canada. *Auk* 78:176–219.

Eck, S. 1976. Randbermerkungen zur taxonomic der Sandregenpfeifer. *Beitrage zur Vogelkunde* 22:38–48.

Edgar, A. T. 1969. Estimated population of the red-breasted dotterel. *Notornis* 16:85–100.

Ehlert, W. 1964. Zur Ökologie und Biologie der Ernahrung einiger Limikolen. *Journal für Ornithologie* 105:1–53.

Elliott, H. F. J. 1956. Some field-notes on the Caspian plover. *British Birds* 49:282–83.

Emlen, S. T., and Oring, L. W. 1977. Ecology, sexual selection, and the evolution of mating systems. *Science* 197 (4300):215–23.

Ennion, E. A. R. 1949. *The lapwing.* London: Methuen.

Etchécopar, R. D., and Hüe, F. 1967. *The birds of North Africa from the Canary Islands to the Red Sea,* trans. P. A. D. Hollom. Edinburgh: Oliver and Boyd.

Evans, W. 1920. The nest of a Mesopotamian plover. *Ibis,* ser. 11, 2:731–32.

Falla, R. A. 1978. Banded dotterel at the Auckland Islands: Description of a new subspecies. *Notornis* 25:101–08.

Farrand, J., Jr. 1977. What to look for: Eskimo and little curlew compared. *American Birds* 31:137–38.

Fay, F. H., and Cade, T. J. 1959. An ecological analysis of the avifauna of St. Lawrence Island, Alaska. *University of California Publications in Zoology* 63:73–150.

Feare, C. J. 1966. The winter feeding of the purple sandpiper. *British Birds* 59:165–79.

Fennell, C. M. 1953. Notes on the birds of Daikokumima, Hokaido, Japan. *Condor* 55:38–42.

Fitzner, J. N. 1978. The ecology and behavior of the long-billed curlew (*Numenius americanus*) in southeastern Washington. Ph.D. diss., Washington State University, Pullman.

Fjeldså, J. 1977. *Guide to the young of European precocial birds.* Tisvildelje, Denmark: Scarv Nature Publications.

Fleetwood, R. J. 1973. Jacana breeding in Brazoria County, Texas. *Auk* 90:422–23.

Fleming, R. L., Sr.; Fleming, R. L., Jr.; and Bengdel, L. S. 1976. *Birds of Nepal, with references to Kashmir and Sikkim.* Katmandu: Published by the authors.

Flint, V. E., ed. 1973*a*. [Fauna and ecology of waders] 2 vols. Moscow: Moscow University Press. (In Russian.)

―――. 1973*b*. [On the biology of the broad-billed sandpiper, *Limicola falcinellus sibiricus*]. (In Russian.) In Flint 1973*a*, 1: 98–99.

Flint, V. S., and Kistchinski, A. A. 1973. [On the biology of the sharp-tailed sandpiper, *Calidris acuminata*]. (In Russian.) in Flint 1973*a*, 1:100–104.

Friedmann, H., and Smith, F. D. 1975. A further contribution to the ornithology of northeastern Venezuela. *Proceedings of the United States National Museum* 104:463–524.

Frisch, R. 1978. Surfbirds in Ogilvie and Richardson Mountains, Yukon Territory. *Canadian Field-Naturalist* 92:401–03.

Frith, H. J., ed. 1969. *Birds in the Australian high country.* Sydney: A. H. and A. W. Reed.

―――. 1976. *Reader's Digest complete book of Australian birds.* Sydney: Reader's Digest Services.

Frith, H. J.; Crome, F. H. J.; and Brown, B. K. 1977. Aspects of the biology of the Japanese snipe *Gallinago hardwickii. Australian Journal of Ecology* 2: 341–68.

Gabrielson, I. N., and Lincoln, F. C. 1959. *Birds of Alaska.* Washington and Harrisburg: Wildlife Management Institute and Stackpole Company.

Gerrits, H. A. 1956. Breeding of the black-breasted plover. *Avicultural Magazine* 62:219–23.

Gibson, F. 1971. The breeding biology of the American avocet (*Recurvirostra americana*) in central Oregon. *Condor* 73:444–54.

Gladkov, N. A. 1957. Der Rotkehlige Strandläufer ist eine selbstandige Art. *Journal für Ornithologie* 98:195–203.

Glutz von Blotzheim, U. N.; Bauer, K. M.; and Bezzel, E. 1975, 1977. Handbuch der Vögel Mitteleuropas. Parts 6 and 7. Wiesbaden: Akademische Verlagsgesellschaft.

Gooders, J., ed. 1969–71. *Birds of the world.* 9 vols. London: IPC (published in serial form).

Goss-Custard, J. D. 1969. The winter feeding ecology of the redshank. *Ibis* 111:338–56.

Goss-Custard, J. D., and Jones, R. E. 1976. The diets of redshank and curlew. *Bird Study* 23:233–43.

Graul, W. D. 1973. Adaptive aspects of the mountain plover social system. *Living Bird* 12:69–94.

―――. 1975. Breeding biology of the mountain plover. *Wilson Bulletin* 87:6–31.

Graul, W. D., and Webster, L. E. 1976. Breeding status of the mountain plover. *Condor* 78:265–67.

Greenway, J. C. 1958. *Extinct and vanishing birds of the world.* New York: American Committee for International Wild Life Protection.

Greer, J. K., and Greer, M. 1967. Notes on hatching and growth of the southern lapwing in Chile. *Auk* 84:121–22.

Grosskopf, G. 1958, 1959. Zur Biologie des Rotschenkels (*Tringa t. totanus*). *Journal für Ornithologie* 99:1–17; 100:210–36.

Groves, S. 1978. Age-related differences in ruddy turnstone foraging and aggressive behavior. *Auk* 95:95–103.

Haedo Rossi, J. A. 1969. Notas ornitologicas. 7. Observaciones sobre del tero en semicautividad. *Acta Zoologica Lilloane* 25:109–19. (English summary.)

Hagar, J. A. 1966. Nesting of the Hudsonian godwit at Churchill, Manitoba. *Living Bird* 5:5–43.

Hale, W. G. 1971. A revision of the taxonomy of the redshank *Tringa totanus. Zoological Journal of the Linnaean Society* 50:199–268.

Hall, B. P. 1974. *Birds of the Harold Hall Australian expeditions, 1962–1970.* London: British Museum (Natural History).

Hall, K. R. L. 1959*a*. Nest records and additional behavior for Kittlitz's sand plover *Charadrius pecuarius* in the S.W. Cape Province. *Ostrich* 30: 33–38.

―――. 1959*b*, 1964. A study of the blacksmith plover (*Hoplopterus armatus* (Burchell)) in the Cape Town area. *Ostrich* 30:117–25; 35:3–16.

Hamilton, R. C. 1975. *Comparative behavior of the American avocet and the black-necked stilt (Recurvirostridae).* A.O.U. Monographs, no. 17.

Hanna, G. D. 1921. The Pribilov sandpiper. *Condor* 23:50–57.

Harrell, B., ed. 1978. *The birds of South Dakota: An annotated check list.* Vermillion: South Dakota Ornithologists' Union and W. H. Over Museum.

Harrington, B. A., and Morrison, R. I. G. 1979. Semipalmated sandpiper migration in North America. *Studies in Avian Biology* 2:83–99.

Harris, M. P. 1967. The biology of oystercatchers *Haematopus ostralegus* on Skokholm Island, S. Wales. *Ibis* 109:180–93.

Hartwick, E. B. 1974. Breeding ecology of the black oystercatcher (*Haematopus bachmani* Audubon). *Syesis* 7:83–92.

Haverschmidt, F. 1963. *The black-tailed godwit.* Leiden: E. J. Brill.

_____. 1968. *Birds of Surinam.* Edinburgh: Oliver and Boyd.

_____. 1974. The occurrence of the giant snipe *Gallinago undulata* in Surinam. *Bulletin of the British Ornithologists' Club* 94:132–34.

Hays, H. 1972. Polyandry in the spotted sandpiper. *Living Bird* 11:43–57.

Heather, B. D. 1973. The black-fronted dotterel (*Charadrius melanops*) in the Wairarapu. *Notornis* 20:251–61.

Hellebrekers, W. P. J., and Hoogerwerf. A. 1967. A further contribution to our zoological knowledge of the island of Java (Indonesia). *Zoologische Verhandlungen* 88:1–164.

Helversen, O. V. 1963. Beobachtungen zum Verhalten und zer Brutbiologie des Spornkeibitzes (*Hoplopterus spinosus*). *Journal für Ornithologie* 104:89–96.

Heppleston, P. B. 1970. The function of oystercatcher piping behavior. *British Birds* 63:113–15.

_____. 1971. Nest site selection by oystercatchers (*Haematopus ostralegus*) in the Netherlands and Scotland. *Netherlands Journal of Zoology* 21:208–11.

_____. 1972. The comparative breeding ecology of oystercatchers (*Haematopus ostralegus* L.) in inland and coastal habitats. *Journal of Animal Ecology* 41:23–51.

_____. 1973. The distribution and taxonomy of oystercatchers. *Notornis* 30:102–12.

Higgins, K. F., and Kirsch, K. M. 1975. Some aspects of the breeding biology of the upland sandpiper in North Dakota. *Wilson Bulletin* 87:96–102.

Hilden, O. 1975. Breeding system of Temminck's stint, *Calidris temminckii. Ornis Fennica* 52:117–46.

_____. 1978*a*. Population dynamics in Temminck's stint *Calidris temminckii. Oikos* 30:17–28.

_____. 1978*b*. [Occurrence and breeding biology of the little stint *Calidris minuta* in Norway.] *Anser,* suppl. 3, pp. 96–100. (In Swedish, English summary.)

Hilden, O., and Vuolanto, S. 1972. Breeding biology of the red-necked phalarope *Phalaropus lobatus* in Finland. *Ornis Fennica* 49:57–85.

Hindwood, H. A. 1940. Notes on the distribution and habits of the jacana or lotusbird. *Emu* 39:261–67.

Hirons, G. 1980. The significance of roding by woodcock *Scolopax rusticola:* an alternative explanation based on observations of marked birds. *Ibis* 122:350–54.

Hobbs, J. N. 1972. Breeding of red-capped dotterel at Fletcher's Lake, Dareton, N.S.W. *Emu* 72:121–25.

Hobson, W. 1972. The breeding biology of the knot. *Proceedings of the Western Foundation of Vertebrate Zoology* 2:1–29.

Hoffmann, A. 1950. Über die brutpflege des polyandrischen Wasserfasan, *Hydrophasianus chirurgus* (Scop.). *Zoologische Jahrbuch (Systematik, Ökologie und Geographie)* 78:367–403.

Hogan-Warburg, A. J. 1968. Social behaviour of the ruff *Philomachus pugnax* (L.). *Ardea* 54:109–229.

Höhn, E. O. 1967. Observations on the breeding biology of Wilson's phalarope (*Steganopus tricolor*) in central Alberta. *Auk* 84:220–44.

_____. 1968. Some observations on the breeding of northern phalaropes at Scammon Bay, Alaska. *Auk* 85:316–17.

_____. 1971. Observations on the breeding behaviour of gray and red-necked phalaropes. *Ibis* 113:335–48.

_____. 1975. Notes on black-headed ducks, painted snipe and spotted tinamous. *Auk* 92:566–75.

Höhn, E. O., and Barron, J. R. 1963. The food on Wilson's phalarope (*Steganopus tricolor*) during the breeding season. *Canadian Journal of Zoology* 41:1171–73.

Holmes, R. H., and Pitelka, F. A. 1962. Behavior and taxonomic position of the white-rumped sandpiper. In *Proceedings 12th Alaskan Science Conference* (Science in Alaska, 1961), College, Alaska: AAAS, pp. 19–20.

_____. 1964. Breeding behavior and taxonomic relationships of the curlew sandpiper. *Auk* 81:362–79.

_____. 1968. Food overlap among coexisting sandpipers on northern Alaskan tundra. *Systematic Zoology* 17:305–18.

Holmes, R. T. 1966*a*. Breeding ecology and annual cycle adaptations of the red-backed sandpiper (*Caladris alpina*) in northern Alaska. *Condor* 68:3–46.

_____. 1966*b*. Feeding ecology of the red-backed sandpiper (*Calidris alpina*) in arctic Alaska. *Ecology* 47:32–45.

_____. 1971. Density, habitat, and the mating systems of the western sandpiper (*Calidris mauri*). *Oecologia* 7:191–208.

_____. 1972. Ecological factors influencing the breeding season schedule of western sandpipers (*Calidris mauri*) in subarctic Alaska. *American*

Midland Naturalist 87:472–90.

_____. 1973. Social behavior of breeding western sandpipers *Calidris mauri. Ibis* 115:107–23.

Hoogerwerf, A. 1966. On the validity of *Charadrius alexandrinus javanicus* Chasen and the occurrence of *Charadrius alexandrinus ruficapillus* Temm. and of *Charadrius peronii* Schl. on Java and New Guinea. *Philippine Journal of Science* 95:209–14.

Hopcraft, J. B. D. 1968. Some notes on the chick-carrying behavior in the African jacana (*Actophilornis africanus*). *Living Bird* 7:85–88.

Hosking, E., and Ferguson-Lees, I. J. 1959. Photographic studies of some less familiar birds: Terek sandpiper. *British Birds* 52:85–90.

Howe, F. E., and Ross, J. A. 1931. Eggs of the banded stilt. *Emu* 31:63–65.

Howe, M. A. 1975a. Social interactions in flocks of courting Wilson's phalaropes. *Condor* 77:24–33.

_____. 1975b. Behavioral aspects of the pair bond in Wilson's phalarope. *Wilson Bulletin* 87:248–70.

Hoy, G. 1967. The eggs and nesting ground of the puna plover. *Auk* 84:130–31.

Hudson, R., ed. 1974. *Threatened birds of Europe.* London: Macmillan.

Hudson, W. H. 1920. *Birds of La Plata.* 2 vols. New York: E. P. Dutton.

Humphrey, P. S.; Bridge, D.; Reynolds, P. W.; and Peterson, R. T. 1970. *Birds of Isla Grande (Tierra del Fuego).* Washington: Smithsonian Institution.

Hussell, D. J. T., and Page, G. W. 1976. Observations on the breeding biology of black-bellied plovers on Devon Island, N.W.T., Canada. *Wilson Bulletin* 88:632–53.

Huxley, J. S. 1912. A first account of the courtship of the redshank (*Totanus calidris* L.). *Proceedings of the Zoological Society* 191:647–55.

Huxley, J. S., and Montague, F. A. 1926. Studies on the courtship and sexual life of birds. 6. The black-tailed godwit (*Limosa limosa* L.). *Ibis,* ser. 12, 2:1–25.

Ingram, C. 1978. Carriage of the young and related adaptations in the anatomy of the woodcock *Scolopax rusticola. Ibis* 120:67.

Irving, L. 1960. *Birds of Anaktuvuk Pass, Kobuk, and Old Crow: A study in arctic adaptation.* Bulletin 217. Washington, D.C.: United States National Museum.

Jayakar, S. D., and Spurway, H. 1965a. The yellow-wattled lapwing, a tropical dry-season nester (*Vanellus malabaricus* Boddaert, Charadriidae). I. The locality, and the incubatory adaptations. *Zoologische Jahrbuch (Systematik, Okologie und Geographie)* 92:53–72.

_____. 1965b. The yellow-wattled lapwing, *Vanellus malabaricus* (Boddaert), a tropical dry-season nester. II. Additional data on breeding biology. *Journal of the Bombay Natural History Society* 62:1–14.

_____. 1968. The yellow-wattled lapwing, *Vanellus malabaricus* (Boddaert), a tropical dry-season nester. III. Two further seasons' breeding. *Journal of the Bombay Natural History Society* 65:369–83.

Jeffery, R. G., and Liversidge, R. 1951. Notes on the chestnut-banded sandplover. *Ostrich* 22:68–76.

Jehl, J. R., Jr. 1968a. Relationships in the Charadrii (shorebirds): A taxonomic study based on color patterns of the downy young. *Memoirs of the San Diego Society of Natural History* 3:1–54.

_____. 1968b. The systematic position of the surfbird, *Aphriza virgata. Condor* 70:206–10.

_____. 1970. Sexual selection for size differences in two species of sandpipers. *Evolution* 24:311–19.

_____. 1973. Breeding biology and systematic relationships of the stilt sandpiper. *Wilson Bulletin* 85:115–47.

_____. 1975. *Pluvianellus socialis:* Biology, ecology, and relationships of an enigmatic Patagonian shorebird. *Transactions of the San Diego Society of Natural History* 18:25–74.

_____. 1979. The autumn migration of Baird's sandpiper. *Studies in Avian Biology* 2:55–68.

Jehl, J. R., Jr., and Smith, B. A. 1970. *Birds of the Churchill region, Manitoba.* Special Publication no. 1. Manitoba: Manitoba Museum of Man and Nature.

Jenkins, C. F. H. 1975. Nesting of banded stilts at Lake Ballard. *Western Australian Naturalist* 13:94–95.

Jenni, D.A. 1974. Evolution of polyandry in birds. *American Zoologist* 14:129–44.

Jenni, D. A., and Betts, B. J. 1978. Sex differences in nest construction, incubation and parental behavior in the polyandrous American jacana (*Jacana spinosa*). *Animal Behaviour* 26:207–18.

Jenni, D. A., and Collier, G. 1972. Polyandry in the American jacana (*Jacana spinosa*). *Auk* 89:743–65.

Jenni, D. A.; Gamb, R. D.; and Betts, B. J. 1974. Acoustic behavior of the northern jacana. *Living Bird* 13:193–210.

Johns, J. E. 1969. Field studies of Wilson's phalarope. *Auk* 86:660–70.

Johnsgard, P. A. 1979. *Birds of the Great Plains: Breeding species and their distribution.* Lincoln: University of Nebraska Press.

_____. 1980. Where have all the curlews gone? *Natural History* 89(8):30–33.

Johnson, A. W. 1965. *The birds of Chile and adjacent regions of Argentina, Bolivia and Peru.* Vol. 1. Buenos Aires: Platt Establecimientos Gráficos.

Johnson, O. W., and Morton, M. L. 1976. Fat content and flight range in shorebirds summering on Enewetak Atoll. *Condor* 78:144–45.

Jones, J. 1945. The banded stilt. *Emu* 45:1–36, 110–18.

Junge, G. C. A. 1939. Description of a new bird from Simalur. *Zoologische Medelingen* 22:120.

Jungfer, W. 1954. Über Paartreue, Nistplatztreue und Alter des Austernfischers (*Haematopus o. ostralegus*) auf Mellum. *Vogelwarte* 17:6–15.

Kagarise, C. M. 1979. Breeding biology of the Wilson's phalarope in North Dakota. *Bird-Banding* 50:12–22.

Kenyon, K. W. 1961. Birds of Amchitka Island, Alaska. *Auk* 78:305–26.

Kessel, B., and Gibson, D. D. 1979. Status and distribution of Alaska birds. *Studies in Avian Biology* 1:1–100.

Kessel, B., and Schaller, G. B. 1960. *Birds of the upper Sheenjek Valley, northeastern Alaska.* Biological Papers, University of Alaska no. 4. College: University of Alaska.

King, B. 1972. Food of buff-breasted sandpiper in the Isle of Scilly. *British Birds* 65:444.

Kirchner, K. 1969. *Die Uferschnepfe.* Neue Brehm-Bucherei 413. Wittenberg: Ziemsen Verlag.

_____. 1978. *Bruchwasserläufer and Waldwasserläufer.* Neue Brehm Bucherei 309. Wittenberg: Ziemsen Verlag.

Kistchinski, A. A. 1973. [Foods of the Siberian sharp-tailed sandpiper, *Calidris acuminata*, in the tundras of northeast Yukutia.] (In Russian.) In Flint 1973*a*, 1:46–48.

_____. 1974. [The biology and behavior of the pectoral sandpiper in the tundra of eastern Siberia.] *Byulleten Moskovskogo Obshchestva Ispytatelei Prirody Otdelenie Biologicheskii* 79:73–88. (In Russian, English summary.)

_____. 1975. Breeding biology and behaviour of the gray phalarope *Phalaropus fulicarius* in East Siberia. *Ibis* 117:285–301.

Kistchinski, A. A., and V. E. Flint. 1973. [A case of double-nesting involving the little stint]. (In Russian.) In Flint 1973*a*, 1:56–57.

Kitson, A. R. 1978. Identification of long-toed stint, pintail snipe and Asiatic dowitcher. *British Birds* 71:558–62.

Klomp, H. 1954. [Habitat selection in the lapwing]. *Ardea* 42:1–139. (In Dutch, English summary.)

Kokhanov, V. D. 1973. [On the ecology of Temminck's stint in the Kandalaksha Bay]. (In Russian.) In Flint 1973*a*, 1:66–70.

Kozlova, E. V. 1961, 1962. [The fauna of the USSR. Birds, vol. 2, part 1, sections 2 and 3]. *Instituta Zoologii Akademii Nauk SSSR.* N.S. 80:1–500; 81:1–432. (In Russian.)

Kuenzel, W. J., and Wiegert, R. G. 1973. Energetics of a spotted sandpiper feeding on brine fly larvae (*Paracoenia,* Diptera; Ephydridae) in a thermal spring community. *Wilson Bulletin* 85:473–76.

Kuroda, N. 1936. [On a new breeding ground for *Pseudototanus guttifer*]. *Tori* 9(43):232–36. (In Japanese, with English summary.)

Kuzyakin, A. P. 1959. [The semipalmated sandpiper in the east part of the Chukotski Peninsula]. *Ornitologiya* 2:130–34. (In Russian.)

Kyllingstad, H. C. 1948. The secret of the bristle-thighed curlew. *Arctic* 1:113–18.

Labutin, Y. V. 1959. [The little curlew of the Verkoyansk region]. *Ornitologiya* 2:111–14. (In Russian.)

Lacan, F., and Mougin, J. -L. 1974. Les oiseaux des Iles Gambier et de quelques atolls orientaux de l'Archipel des Tuamotu (Océan Pacifique). *L'Oiseau* 44:191–280.

Larson, S. 1957. The suborder Charadrii in arctic and boreal areas during the Tertiary and Pleistocene. *Acta Vertebratica* 1:1–84.

Laven, B. 1941. Beobachtungen über Balz und Brut beim Kiebitz (*Vanellus vanellus* L.). *Journal für Ornithologie* 89:1–64.

Lehmann, H. 1969. The greater sand plover in Asia Minor. *Oological Record* 43:30–54.

Lemnell, P. E. 1978. Social behavior of the great snipe *Capella media* at the arena display. *Ornis Scandinavica* 9:146–63.

Leonovitch, V. V. 1973*a*. [On the distribution and biology of the long-toed stint, *Calidris subminuta*]. (In Russian.) In Flint 1973*a*, 1:78–80.

_____. 1973*b*. [New breeding area of the snipe-billed godwit, *Limnodromus semipalmatus*]. (In Russian.) In Flint 1973*a*, 1:81–82.

Leonovitch, V. V., and Kretzschmar, A. W. 1966. Zur Biologie des Graubürzelwasserläufers. *Falke* 13:154–56.

Le'vêque, R. 1964. Notes sur la reproduction des oiseaux aux Iles Galapagos. *Alauda* 32:5–44.

Lind, H. 1961. *Studies on the behaviour of the black-tailed godwit.* (Limosa limosa [L.]). Copenhagen: Munksgaard.

Little, J. DeV. 1968. Some aspects of the behaviour of the wattled plover *Afribyx senegallus* (Linnaeus). *Ostrich* 38:259–80.

Liversidge, R. 1965. Egg-covering in *Charadrius marginatus Ostrich* 36:59–61.

Lowe, P. R. 1915. Studies on the Charadriiformes. 1. On the systematic position of the ruff (*Machetes pugnax*) and the semipalmated plover (*Ereunetes pusillus*), together with a review of some osteological characters which differentiate the Eroliinae (dunlin group) from the Tringinae (redshank group). *Ibis*, ser. 10, 3:609–16.

————. 1925. A preliminary note on the classification of the Charadriiformes (Limicolae and Laro-Limicolae) based on this character, viz., the morphology of the quadrato-tympanic articulation. *Ibis*, ser. 12, 1:144–47.

————. 1927. On the anatomy and systematic position of *Aechmorhynchus cancellatus* (Gmelin), together with some notes on the genera *Bartramia* and *Mesoscolopax*; the subfamily Limosinae; and the pterylosis of *Scolopax*. *Ibis*, ser. 12, 3:114–32.

————. 1931a. On the relations of the Gruimorphae to the Charadriimorphae and Rallimorphae, with special reference to the taxonomic postion of Rostratulidae, Jacanidae, and Burhinidae (Oedicnemidae *olim*); with a suggested new order (Telmatomorphae). *Ibis*, ser. 13, 1:491–534.

————. 1931b. An anatomical review of the "waders" (Telmatomorphae) with special reference to the families, subfamilies and genera within the suborders Limicolae, Grui-Limicolae, and Lari-Limicolae. *Ibis*, ser. 13, 1:712–71.

Lowe, V. T. 1963. Observations on the painted snipe. *Emu* 62:221–37.

McClure, H. E. 1974. *Migration and survival of the birds of Asia.* Bangkok: SEATO Medical Research Laboratory.

Macdonald, J. D. 1973. *Birds of Australia: A summary of information.* London: Witherby.

McFarlane, R. W. 1963. The taxonomic significance of avian sperm. *Proceedings XIII International Ornithological Congress*, pp. 91–102.

McGilp, J. N., and Morgan, A. M. 1931. The nesting of the banded stilt *(Cladorhynchus leucocephala). South Australian Ornithologist* 11:37–52.

McKenzie, H. R. 1953. Nesting of New Zealand dotterel, 1951. *Notornis* 5:121–22.

————. 1978. New Zealand dotterel banding report number one. *Notornis* 25:186–94.

Mackworth-Praed, C. W., and Grant, C. H. B. 1952.

Birds of eastern and northeastern Africa. London: Longmans.

————. 1970. *Bird of west central and western Africa.* London: Longmans.

McLachlan, G. P., and Jeffery, R. G. 1949. Nesting of the chestnut-banded sandplover. *Ostrich* 20:36–37.

McLachlan, G. P., and Liversidge, R. 1957. *Roberts' Birds of South Africa.* Cape Town: Central News Agency.

Maclean, G. L. 1972a. Clutch size and evolution in the Charadrii. *Auk* 89:299–324.

————. 1972b. Problems of display postures in the Charadrii (Aves: Charadriiforms). *Zoologica Africana* 7:57–74.

————. 1972c. Waders of waterside vegetation: The African jacana and Ethiopian snipe. *African Wild Life* 26:163–67.

————. 1973. A review of the biology of the Australian desert waders, *Stiltia* and *Peltohyas*. *Emu* 73:61–70.

————. 1976. A field study of the Australian dotterel. *Emu* 76:207–15.

————. 1977. Comparative notes on black-fronted and red-kneed dotterels. *Emu* 77:199–207.

Maclean, G. L., and Moran, V. C. 1965. The choice of nest site in the white-fronted sandplover *Charadrius marginatus*. *Ostrich* 36:63–72.

MacLean, S. F., Jr., and Holmes, R. T. 1971. Bill-length, wintering areas, and taxonomy of North American dunlins, *Calidris alpina*. *Auk* 88:893–901.

Madge, S. C. 1977. Field identification of pintail snipe. *British Birds* 70:146–52.

Maher, W. J. 1959. Habitat distribution of birds breeding along the upper Kaolak River, northern Alaska. *Condor* 61:351–68.

Makkink, G. F. 1936. An attempt at an ethogram of the European avocet. *Ardea* 25:1–74.

————. 1942. Contribution to the knowledge of the behavior of the oystercatcher (*Haematopus ostralegus*). *Ardea* 31:23–74.

Mambetjumayev, A. W., and Ametov, M. 1973. [On the nesting biology of the stilt in the valley of Lower Amu-Darya River]. (In Russian.) In Flint 1973a, 1:83.

Marcström, V., and Sundgren, F. 1977. On the reproduction of the European woodcock. *Viltrevy* 10:27–40.

Martin, J. 1972. Nesting habits of three resident sandplovers. *Bokmakierie* 24:40–41.

Mason, A. G. 1947. Territory in the ringed plover.

British Birds 40:66–70.

Mason, C. F., and McDonald, S. M. 1976. Aspects of the breeding biology of the snipe. *Bird Study* 23:33–38.

Mathew, D. N. 1964. Observations on the breeding habits of the bronze-winged jacana (*Metopidius indicus* [Latham]. *Journal of the Bombay Natural History Society* 61:295–301.

Mayfield, H. F. 1979. Red phalaropes breeding on Bathurst Island. *Living Bird* 17:7–40.

Mayr, E., and Rand, A. L. 1937. Results of the Archibold expeditions. 14. Birds of the 1933–1934 Papuan expedition. *Bulletin of the American Museum of Natural History* 73:1–248.

Mayr, E., and Short, L. L., Jr. 1970. Species taxa of North American birds: A contribution to comparative systematics. *Nuttall Ornithological Club Publications*, no. 9.

Medway, Lord, and Wells, D. R. 1976. *The birds of the Malay Peninsula*. Vol. 5. London: Witherby.

Meinertzhagen, A. C. 1926. A review of the subfamily Scolopacinae. *Ibis*, ser. 12, 2:477–521.

Miller, E. H. 1977. Breeding biology of the least sandpiper, *Calidris minutilla* (Vieill.), on Sable Island, Nova Scotia. Ph.D. diss., Dalhousie University, Halifax.

———. 1979*a*. Functions of display flights by males of the least sandpiper, *Calidris minutilla* (Vieill.). on Sable Island, Nova Scotia. *Canadian Journal of Zoology* 57:879–93.

———. 1979*b*. Egg size in the least sandpiper, *Calidris minutilla*, on Sable Island, Nova Scotia. *Ornis Scandinavica* 10:10–16.

Miller, E. H., and Baker, A. J. 1980. Displays of the Magellanic oystercatcher (*Haematopus leucopodus*). *Wilson Bulletin* 92:149–68.

Miller, J. R., and Miller, J. T. 1948. Nesting of the spotted sandpiper at Detroit, Michigan. *Auk* 65:558–67.

Miller, W. T. 1951. The bird that walks on water. *African Wild Life* 5:283–89.

Milon, P.; Petter, J. -J.; and Randrianosola, G. 1973. Oiseaux. *Faune Madagascar* 35:1–263.

Monroe, B. L. 1968. *A distributional survey of the birds of Honduras*. A.O.U. Monographs, no. 7.

Moon, G. J. H. 1967. *Refocus on New Zealand birds*. Wellington: A. H. and A. W. Reed.

Morgan, R., and Shorten, M. 1974. Breeding of the woodcock in Britain. *Bird Study* 21:193–99.

Morony, J. J., Jr.; Bock, W. J.; and Farrand, J., Jr. 1975. *Reference list of the birds of the world*. New York: American Museum of Natural History.

Morrison, R. I. G. 1975. Migration and morphometrics of European knot and turnstone on Ellesmere Island, Canada. *Bird-Banding:* 46:290–301.

Muller, K. 1975. Threat display of the Australian painted snipe. *Emu* 75:28–30.

Murie, A. 1946. Observations on birds of Mount McKinley National Park, Alaska. *Condor* 48:253–61.

Murie, O. J. 1924. Nesting records of the wandering tattler and surf bird in Alaska. *Auk* 41:235–37.

———. 1959. *Fauna of the Aleutian Islands and Alaska Peninsula*. North American Fauna 61. Washington, D.C.: U.S. Fish and Wildlife Service.

Murphy, R. C. 1936. *Oceanic birds of South America*. 2 vols. New York: American Museum of Natural History.

Murray, B. G. Jr., and Jehl, J. R., Jr. 1964. Weights of autumn migrants from coastal New Jersey. *Bird-Banding* 35:253–63.

Myers, J. P. 1979. Leks, sex, and buff-breasted sandpipers. *American Birds* 33:823–85.

Myers, J. P.; Conners, P. G.; and Pitelka, F. A. 1979. Territoriality in non-breeding shorebirds. *Studies in Avian Biology* 2:231–46.

Myers, J. P., and Myers, L. P. 1979. Shorebirds of coastal Buenos Aires Province, Argentina, *Ibis* 121:186–200.

Naik, R. M.; George, P. V.; and Dixit, D. B. 1961. Some observations on the behavior of the incubating red-wattled lapwing, *Vanellus indicus indicus* (Bodd.). *Journal of the Bombay Natural History Society* 58:223–30.

Naumov. R. L. 1962. [The biology of Latham's snipe (*Capella megala*) in middle Siberia]. *Ornitologiya* 4:160–68. (In Russian.)

Nechaev, V. A. 1973. [The Latham's snipe (*Gallinago hardwickii*) on the Sakhalin]. (In Russian.) In Flint 1973*a* 1: 87–89.

———. 1978. [A contribution to the biology and behaviour of *Tringa guttifer* on the Sakhalin Island]. *Zoologigheskii Zhurnal* 57:727–37. (In Russian, English summary.)

Nelson, T. 1939. The biology of the spotted sandpiper (*Actitis macularia* [Linn.]). Ph. D. diss., University of Michigan.

Nemetschek, G. 1977. Beobachtungen zur Flugbalz der Waldschnepfe (*Scolopax rusticola*). *Journal für Ornithologie* 118:68–86.

Nethersole-Thompson, D. 1951. *The greenshank*. London: Collins; 2d ed., 1979.

———. 1973. *The dotterel*. London: Collins.

Nettleship, D. N. 1973. Breeding ecology of turnstones *Arenaria interpres* at Hazen Camp, Ellesmere Island, N.W.T. *Ibis* 115:202–17.

_____. 1974. The breeding of the knot *Calidris canutus* at Hazen Camp, Ellesmere Island, N.W.T. *Polarforschung* 44:8–26.

Neufeldt, I.; Krechmae, A. V.; and Ivanov, A. I. 1961. Studies of less familar birds. 110. Grey-rumped sandpiper. *British Birds* 54:30–33.

New Zealand Checklist Committee. 1970. *Annotated checklist of the birds of New Zealand, including the birds of the Ross Dependency.* Wellington: A. H. and A. W. Reed.

Nichols, J. T. 1923. Yellow-legs skeletons. *Auk* 40:593–95.

Nielsen, B. P. 1971. Migration and relationships of four Asiatic plovers Charadriinae. *Ornis Scandinavica* 2:137–42.

Niethammer, G. 1953. Zur Vogelwelt Boliviens. *Bonner Zoologische Beiträge:* 4:195–303.

_____. 1966. Sexualdimorphismus am Ösophagus von *Rostratula. Journal für Ornithologie* 107:201–04.

Nilsson, S. G., and Nilsson,, I. N. 1978. [Population, habitat, and display activity of the jacksnipe, *Lymnocryptes minimus,* in southern Sweden]. *Vår Fågelvårld* 37:1–8. (In Swedish, English summary.)

Nisbet, I. C. T. 1961. Studies of less familiar birds. 113. Broad-billed sandpiper. *British Birds* 54:320–23.

North, M. E. W. 1937. Breeding habitats of the crested wattled plover. *Journal of the East Africa Natural History Society and National Museum* 13:132–45.

Norton, D. W. 1972. Incubation schedules of four species of calidrine sandpipers at Barrow, Alaska. *Condor* 74:164–76.

Nowicki, T. 1973. A behavioral study of the marbled godwit in North Dakota. M.S. thesis, Central Michigan University.

Okugawa, K. T.; Ishii, T.; Mitsuno, M.; Hasegawa, S.; Tsukamoto, K.; Aoki, M.; Yamashita, S.; and Yamamoto, S. 1973. [An ecological study of *Macrosarcops cinereus* (Blyth), gray-headed lapwing (Charadriidae), of the Ogura Farm area, Kyota]. *Bulletin of the Kyoto University Education,* ser. B, no. 37 (1970), pp. 3–87. (In Japanese, English summary.)

Oliver, W. R. B. 1955. *New Zealand birds.* Wellington: A. H. and A. W. Reed.

Olney, P. J. S. 1970. Studies of avocet behaviour. *British Birds* 63:206–9.

Oring, L. W. 1964. Displays of the buff-breasted sandpiper at Norman, Oklahoma. *Auk* 81:83–86.

_____. 1968. Vocalizations of the green and solitary sandpipers. *Wilson Bulletin* 80:395–420.

_____. 1973. Solitary sandpiper early reproductive behavior. *Auk* 90:652–63.

Oring, L. W., and Knudson, M. L. 1972. Monogamy and polyandry in the spotted sandpiper. *Living Bird* 11:59–73.

Oring, L. W., and Maxon, S. J. 1978. Instances of simultaneous polyandry by a spotted sandpiper (*Actitis macularia*). *Ibis* 120:349–53.

Orr, R. T. 1942. A study of the birds of the Big Basin region of California. *American Midland Naturalist* 27:273–337.

Osborne, D. R., and Bourne, G. R. 1977. Breeding behavior and food habits of the wattled jacana. *Condor* 79:98–105.

Panov, E. 1963. [Taxonomic position of the Ussuri plover, *Charadrius hiaticula placidus* Gray and Gray on the basis of ethological data.] *Zoologischeseii Zhurnal* 42:1546–53. (In Russian).

Parmelee, D. F. 1970. Breeding behavior of the sanderling in the Canadian high arctic. *Living Bird* 9:97–146.

Parmelee, D. F., and MacDonald, S. D. 1960. The birds of west-central Ellesmere Island and adjacent areas. *Bulletin of the National Museums of Canada* 169:1–103.

Parmelee, D. F., and Payne, R. B. 1973. On multiple broods and the breeding strategy of arctic sanderlings. *Ibis* 115:218–26.

Parmelee, D. F.; Stephens, H.A.; and Schmidt, R. H. 1967. The birds of southeastern Victoria Island and adjacent small islands. *Bulletin of the National Museums of Canada* 222:1–229.

Peters, J. L. 1934. *Check-list of birds of the world.* Vol. 2. Cambridge: Harvard University Press.

Phillips, A. R. 1975. Semipalmated sandpiper: Identification, migrations, summer and winter ranges. *American Birds* 29:799–806.

Phillips, B. T. 1945. Photographing the ibis-bill (*Ibidorhynchus struthersii* Gould). *Journal of the Bombay Natural History Society* 45:347–52.

Phillips, R. E. 1972. Sexual and agonistic behavior in the killdeer. *Animal Behavior* 20:1–9.

_____. 1977. Notes on the behaviour of the New Zealand shore plover. *Emu* 77:23–27.

_____. 1980. Behavior and systematics of New Zealand plovers. *Emu* 80:177–97.

Piechocki, R. 1967. Der Zwergbrachvogel. *Falke* 3:82–87.

Pienkowski, M. W., and Green, G. H. 1976. Breeding biology of sanderlings in north-east Greenland.

British Birds 69:165–77.

Pitelka, F. A. 1948. The problematical relationships of the Asiatic shorebird *Limnodromus semi-palmatus*. *Condor* 50:259–69.

———. 1950. Geographic variation and the species problem in the shore-bird genus *Limnodromus*. *University of California Publications in Zoology* 50:1–108.

———. 1959. Numbers, breeding schedules, and territoriality in pectoral sandpipers of northern Alaska. *Condor* 61:233–64.

Pitelka, F. A.; Holmes, R. T.; and MacLean, S. F., Jr. 1974. Ecology and evolution of social organization in arctic sandpipers. *American Zoologist* 14:185–204.

Pitman, C. R. S. 1965. The eggs and nesting habits of the St. Helena sand-plover or wirebird, *Charadrius pecuarius sanctae-helenae* (Harting). *Bulletin of the British Ornithologists' Club* 85:121–29.

Portenko, L. A. 1957. Studien an einigen selten Limicolen aus dem nördlichen und ostlichen Sibirien. I. Die Löffelschnepfe-*Eurynorhynchus pygmaeus* (L.). *Journal für Ornithologie* 98:454–66.

———. 1959. Studien an einigen seltenen Limicolen aus dem nördlichen und ostlichen Sibirien. II. Der Sichelstrandlaufer-*Erolia ferruginea* (Pontopp.). *Journal für Ornithologie* 100:141–72.

———. 1968. Studien an einigen selten Limicolen aus dem nördlichen und östlichen Sibirien. III. Der Graubruststrandläufer-*Heteropygia melanotos* (Vieill.). *Journal für Ornithologie* 109:96–115.

———. 1972. [The birds of Chukotski and Wrangell Island]. Leningrad: Edition "Science." (In Russian.)

Potter, J. 1934. Lotus-birds found breeding on Hawkesbury River, N.S.W. *Emu* 33:298–305.

Prater, A. J. 1974a. Breeding biology of the ringed plover. In *Proceedings IWRB Wader Symposium, Warsaw, 1974*, pp. 241–51.

———. 1974b. The distribution of coastal waders in Europe and North Africa. Paper presented to the International Conference on Conservation of Wetlands and Waterfowl, Heiligenhafen, 1974.

Prater. A.J.: Marchant, J.H.; and Vuorinen, J. 1977. *Guide to the identification and aging of Holarctic waders*. Field Guide 17. Tring: British Trust for Ornithology.

Preble, E. A., and McAtee, W.L. 1923. A biological survey of the Pribilov Islands, Alaska. *North American Fauna*, no. 46.

Prevett, J. P., and Barr, J. F. 1976. Lek behavior of the buff-breasted sandpiper. *Wilson Bulletin* 88:500–03.

Princepe, W. L., Jr. 1977. A hybrid American avocet x black-necked stilt. *Condor* 79:128–29.

Pycraft, W. P. 1912. Remarks on the syrinx of the Scolopacidae. *Ibis*, ser. 9, 9:334–41.

Quellet, H.; McNeil, R.; and Burton, J. 1973. The western sandpiper in Quebec and the Maritime Provinces, Canada. *Canadian Field-Naturalist* 87:291–300.

Rand, A. L. 1936. The distribution and habits of Madagascar birds. *Bulletin of the American Museum of Natural History*, 72:143–499.

———. 1950. Critical notes on *Limnodromus semi-palmatus*. *Condor* 52:228–31.

Rand, A. L., and Gilliard, E. T. 1968. *Handbook of New Guinea birds*. Garden City: Natural History Press.

Raner, L. 1972. [Polyandry in the red-necked phalarope, *Phalaropus lobatus*, and the spotted red-shank, *Tringa erythropus*] *Fauna och Flora* 67:135–38. (In Swedish.)

Ratcliffe, D. A. 1977. Observations on the breeding of the golden plover in Great Britain. *Bird Study* 23:63–116.

Recher, H. F. 1966. Some aspects of the ecology of migrant shorebirds. *Ecology* 47:393–407.

Reddig, E. 1978. Der Ausdrucksflug der Bekassine (*Capella gallinago gallinago*). *Journal für Ornithologie* 19:357–87.

Reynolds, J. F. 1968. Observations on the white-headed plover. *East African Wildlife Journal* 6:142–44.

Reynolds, P. W. 1935. Notes on the birds of Cape Horn. *Ibis*, ser. 13, 5:65–101.

Ridgway, R. 1919. *The birds of North and Middle America. Part VIII*. Bulletin 50. Washington, D.C.: United States National Museum.

Ridley, M. W. 1980. The breeding behaviour and feeding ecology of grey phalarope *Phalaropus fulicarius* in Svalbard. *Ibis* 122:210–26.

Riley, J. W., and Rookse, K. B. 1962. Sociable plover in Dorset. *British Birds* 55:233–51.

Ripley, S. D. 1964. A systematic and ecological study of birds in New Guinea. *Peabody Museum of Natural History Bulletin* 19:1–85.

Rittinghaus, H. 1961. *Der Seeregenpfeifer*. Neue Brehm-Bucherei 282. Wittenberg: A. Ziemsen Verlag.

Roberts, T. S. 1932. *The birds of Minnesota*. Vol. 1. Minneapolis: University of Minnesota Press; rev. ed., 1936.

Ross, H. A. 1979. Multiple clutches and shorebird egg and body weight. *American Naturalist* 113:618–21.

Rowan, W. 1929. Notes on Alberta waders included on the British list, 7. *British Birds* 23:2–17.

Sadler, D. A., and Maher, W. J. 1976. Notes on the long-billed curlew in Saskatchewan. *Auk* 93:382–44.

Sakane, M. 1957–58. Notes on the gray-headed lapwing in Kinki, W. Honshu. *Tori* 14:25–37; 15:13–17.

Salt, W. R., and Salt, J. R. 1976. *The birds of Alberta.* Edmonton: Hurtig Publishers.

Sanderson, G. C., ed. 1977. *Management of migratory shore and upland game birds in North America.* Washington, D.C.: International Association of Fish and Wildlife Agencies.

Sanft, K. 1970. Gewichte sudamerikanischer Vögel-Nonpasseres. *Beiträge zur Angewandten Vögelkunde* 16:344–54.

Sauer, E. G. F. 1962. Ethology and ecology of golden plovers on St. Lawrence Island, Bering Sea. *Psychologische Forschung* 26:399–470.

Saunders, C. R. 1970. Observations on breeding of the long-toed or white-winged plover *Hemiparra crassirostris leucoptera* (Reichenow). *Honeyguide,* no. 62, pp. 27–29.

Schamel, D., and Tracy, D. 1977. Polyandry, replacement clutches, and site tenacity in the red phalarope (*Phalaropus fulicarius*) at Barrow, Alaska. *Bird-Banding* 48:314–24.

Schönwetter, M. 1962–63. *Handbuch der Oologie.* Lieferung 6 and 7, pp. 370–418. Berlin: Akademie Verlag.

Schwartz, C. W., and Schwartz, E. R. 1951. The Hawaiian stilt. *Auk* 68:505–06.

Seebohm, H. 1888. *The geographical distribution of the family Charadriidae, or the plovers, sandpipers, snipes, and their allies.* London: Henry Sotheran.

Serle, W. 1939. Field observations on some nothern Nigerian birds. *Ibis,* ser. 14, 3:654–99.

———. 1956. Notes on *Anomalophrys superciliosus* (Reichenow) in West Africa with special reference to its nidification. *Bulletin of the British Ornithologists' Club* 76:101–04.

Serventy, D. L., and Whittel, H. M. 1962. *Birds of western Australia.* Perth: Lambert Publications.

Sharpe, R. B. 1896. *Catalogue of the Limicolae in the collection of the British Museum.* Vol. 24. London: British Museum (Natural History).

Shaw, Tsen-hwang. 1936. The birds of Hopei Province. *Zoologica Sinica,* ser. B, 15:1–974.

Sheldon, W. G. 1967. *The book of the American woodcock.* Amherst: University of Massachusetts Press.

Shepard, J. M. 1976. Factors influencing female choice in the lek mating system of the ruff. *Living Bird* 14:87–111.

Shorten, M. 1974. *The European woodcock (Scolopax rusticola): A search of the literature since 1940.* Fordingbridge: Game Conservancy.

Sibley, C. G., and Alquist, J. E. 1972. *A comparative study of the egg white proteins of non-passerine birds.* Bulletin 39. New Haven: Yale University, Peabody Museum of Natural History.

Sibley, C. G.; Corbin, K. W.; and Alquist, J. E. 1968. The relationships of the seed-snipe (Thinocoridae) as indicated by their egg-white proteins and hemoglobins. *Bonner Zoologische Beiträge* 19:235–48.

Sibson, R. B. 1943. Observations on the distribution of the wrybill in the North Island, New Zealand. *Emu* 43:49–62.

Sick, H. 1962. Die Buntschnepfe, *Nycticryphes semicollaris,* in Brasilien. *Journal für Ornithologie* 103:102–07.

Simmons, K. E. L. 1953a. The aggressive behavior of three closely related plovers (*Charadrius*). *Ibis* 95:115–27.

———. 1953b. Some studies of the little ringed plover. *Avicultural Magazine* 59:191–207.

———. 1956. Territory in the little ringed plover *Charadrius dubius. Ibis* 93:390–97.

Skead, C. J. 1955. A study of the crowned plover. *Ostrich* 26:88–98.

Skead, D. M. 1977. Weights of birds handled at Barberspan. *Ostrich,* suppl. 12, pp. 117–31.

Skeel, M. A. 1976. Nesting strategies and other aspects of the breeding biology of the whimbrel (*Numenius phaeopus*) at Churchill, Manitoba. M.Sc. thesis, University of Toronto.

———. 1978. Vocalizations of the whimbrel on its breeding grounds. *Condor* 80:194–202.

Slud, P. 1964. The birds of Costa Rica. *Bulletin of the American Museum of Natural History* 128:1–430.

Smith, H. G. 1969. Polymorphism in ringed plovers. *Ibis* 111:177–88.

Smith, P. C., and Evans, P. R. 1973. Studies of shorebirds at Lindisfarne, Northumberland. 1. Feeding ecology and behaviour of the bar-tailed godwit. *Wildfowl* 24:135–39

Smithies, B. E. 1968. *The birds of Borneo.* Edinburgh: Oliver and Boyd.

Snow, D. W. 1978. *An atlas of speciation in African*

non-passerine birds. London: British Museum (Natural History).

Snyder, L. L. 1957. *Arctic birds of Canada.* Toronto: University of Toronto Press.

Soikkeli, M. 1967. Breeding cycle and population dynamics in the dunlin (*Calidris alpina*). *Annales Zoologici Fennici* 4:158–98.

Sordahl, T. A. 1979. Vocalizations and behavior of the willet. *Wilson Bulletin* 91:551–74.

Southern, H. N., and Lewis, W. A. S. 1938. The breeding behaviour of Temminck's stint. *British Birds* 31:314–21.

Spencer, K. G. 1953. *The lapwing in Britain.* London: A. Brown and Sons.

Sperry, C. C. 1940. Food habits of a group of shorebirds: Woodcock, snipe, knot and dowitcher. Wildlife Research Bulletin 1. Washington, D. C.: United States Biological Survey.

Stegmann, B. C. 1978. Relationships of the superorders Alectoromorphae and Charadrimorphae (Aves): A comparative study of the avian hand. *Publications of the Nuttall Ornithological Club,* no. 17.

Stenzel, L. E.; Huber, H. R.; and Page, G. W. 1976. Feeding behavior and diet of the long-billed curlew and willet. *Wilson Bulletin* 88:314–31.

Stepanjan, L. E., and Flint, V. E. 1973. [On the systematic status of the rock sandpiper *Calidris ptilocnemis*]. (In Russian.) In Flint 1973*a*, 1:20–22.

Stevenson, H. M. 1975. Identification of difficult birds. Part 3, Semipalmated and western sandpipers. *Florida Field Naturalist* 3:39–44.

Stewart, R. E. 1975. *Breeding birds of North Dakota.* Fargo: Tri-college Center for Environmental Studies.

Stewart, R. E., and Kantrud, H. A. 1972. Population estimates of breeding birds in North Dakota. *Auk* 89:766–88.

Stout, G. A., ed. 1967. *The shorebirds of North America.* New York: Viking Press.

Strauch, J. G., Jr. 1976. The cladistic relationships of the Charadriiformes. Ph.D. diss., University of Michigan.

_____. 1978. The phylogeny of the Charadriiformes (Aves): A new estimate using the method of character compatability analysis. *Transactions of the Zoological Society of London* 34:203–45.

Stresemann, E. 1932. Vorläufiges über die ornithologischer Ergebnisse der Expedition Heinrich 1930–32. 7. Zur Ornithologie von Südost Celebes. *Ornithologische Monatsbericht* 40:104–115.

_____. 1941. Die Vögel von Celebes, Part 3. *Journal für Ornithologie* 89:1–112.

Summers, R. W., and Cooper, J. 1977. The population, ecology and conservation of the black oystercatcher, *Haematopus moquini. Ostrich* 48:28–40.

Summers, R. W., and Waltner, M. 1979. Seasonal variation in the mass of waders in southern Africa, with special reference to migration. *Ostrich* 30:21–37.

Sutton, G. M. 1932. The birds of Southampton Island. *Memoirs of the Carnegie Museum* 12 (pt. 2, sec. 2):1–275.

_____. 1949. Validity of the shorebird genus *Pseudoscolopax. Condor* 51:259–61.

_____ 1967. Behaviour of the buff-breasted sandpiper at the nest. *Arctic* 20:3–7.

Sutton, G. M., and Parmelee, D. F. 1955. Breeding of the semipalmated plover on Baffin Island. *Birdbanding* 26:137–47.

Taka-Tsukasa, N. 1967. *The birds of Nippon.* Tokyo: Matuzen.

Temple, S., ed. 1979. *Red data book.* Vol. 2. *Aves.* Morges, Switzerland: IUCN.

Terborgh, J., and Weske, J. S. 1972. Rediscovery of the imperial snipe in Peru. *Auk* 89:497–505.

Tester, J. R., and Watson, A. 1973. Spacing and territoriality of woodcock *Scolopax rusticola* based on roding behavior. *Ibis* 115:135–38.

Thomas, D. G. 1969. Breeding biology of the Australian spur-winged plover. *Emu* 69:81–102.

Thomas, D. G., and Dartnall, A. J. 1971*a*. Ecological aspects of the feeding behavior of two calidridine sandpipers wintering in south-eastern Tasmania. *Emu* 71:20–26.

_____. 1971*b*. Moult of the red-necked stint. *Emu* 71:49–53.

Thönen, W. 1969. Auffallenden unterschied zwischen den instrumentalen Balzlauten der europäischer und nordamerikanischer Bekassine *Gallinago gallinago. Ornithologische Beobachter* 66:6–13.

Tinbergen, N. 1935. Field observations of East Greenland birds. 1. The behaviour of the red-necked phalarope (*Phalaropus lobatus* L.) in spring. *Ardea* 24:1–42.

Todd, W. E. C. 1953. A taxonomic study of the American dunlin (*Erolia alpina* subspp.). *Journal of the Washington Academy of Science* 43:85–88.

Tolchin, V. 1976. [Distribution and ecology of the marsh sandpiper in central Siberia]. *Biologicheskie Nauki* 1976(5):42–48. (In Russian.)

Tolchin, V., and Mel'nikov, V. I. 1977. [Nesting

habits of the snipe-billed godwit, *Limnodromus semipalmatus*, in eastern Siberia]. *Vestnik Zoologii* 43(3):16–19. (In Russian.)

Tomkins, I. R. 1944. Wilson's plover in its summer home. *Auk* 61:259–69.

_____. 1965. The willets of Georgia and South Carolina. *Wilson Bulletin* 77:151–67.

Tuck, L. M. 1972. *The snipes: A study of the genus* Capella. Ottawa: Monograph Series no. 5. Canadian Wildlife Service.

Tyler, S. 1978. Observations on the nesting of the three-banded plover *Charadrius tricollaris*. *Scopus* 2:39–41.

Urban, E. K. 1978. *Ethiopia's endemic birds.* Addis Ababa: Ethiopian Tourist Organization.

Urban, E. K.; Brown, L. H.; Buer, C. E.; and Plage, G. D. 1972. Four descriptions of nesting, previously undescribed, for Ethiopia. *Bulletin of the British Ornithologists' Club* 90:162–64.

Uspenski, S. M. 1969. *Die Strandlaufer Eurasiens* (*Gattung* Calidris). Neue Brehm Bucherei 420. Wittenberg: A. Ziemsen Verlag.

van Rhijn, J. G. 1973. Behaviorial dimorphism in male ruffs, *Philomachus pugnax* (L.). *Behaviour* 47:153–229.

van Tets, G. F.; D'Andria, A. H.; and Slater, E. 1967. Nesting distribution and nomenclature of Australasian vanelline plovers. *Emu* 67:85–93.

Vaurie, C. 1964. Systematic notes on Palaearctic birds. No. 53. Charadriidae: The genera *Charadrius* and *Pluvialis*. *American Museum Novitates* 2177.

_____. 1965. *The birds of the Palaearctic fauna: Non-passeriformes.* London: Witherby.

Verheyen, R. 1953. Exploration du Parc National de l'Upemba. Brussels: Institut des Parcs Nationaux du Congo Belge.

_____. 1957. Contribution au démembrement de l'ordo artificiel des Gruiformes (Peters 1934). 3. Les Jacaniformes. *Institut Royal des Sciences Naturel de Belgique Bulletin* 33(48):1–19.

Vernon, C. J. 1973. Polyandrous *Actophilornis africanus*. *Ostrich* 44:85.

Vogt, W. 1938. Preliminary notes on the behavior and the ecology of the eastern willet. *Proceedings of the Linnaean Society of New York* 49:8–42.

von Frisch, O. 1956. Zur Brutbiologie und Jungendenwicklung des Brachvogels (*Numenius arquata* L.). *Zeitschrift für Tierpsychologie* 13:50–81. (English summary.)

Voous, K. H. 1960. *Atlas of European birds.* Amsterdam: Thomas Nelson.

_____. 1973. List of Recent Holarctic bird species: Non-passerines. *Ibis* 115:612–38.

Vorobiev, K. A. 1963. [Birds of Yakutia]. Moscow: Akademii Nauk USSR. (In Russian).

Vuilleumier, F. 1969. Field notes on some birds from the Bolivian Andes. *Ibis* 111:599–608.

Vuilleumier, F., and Ewert, D. N. 1978. The distribution of birds in Venezuelan paramos. *Bulletin American Museum of Natural History* 162:47–90.

Vuolanto, S. 1968. On the breeding biology of the turnstone (*Arenaria interpres*) at Norrskär, Gulf of Bothnia. *Ornis Fennici* 45:19–24.

Wallace, D. I. M. 1974. Field identification of small species in the genus *Calidris*. *British Birds* 67:1–12.

_____. 1977. Further definition of great snipe characters. *British Birds* 70:283–89.

Walters, J. 1979. Interspecific aggressive behaviour by long-toed lapwings (*Vanellus crassirostris*). *Animal Behaviour* 27:969–81.

_____. 1980. Cooperative breeding in southern lapwings, *Ibis.* 122:505–9.

Webster, J. D. 1941. The breeding of the black oystercatcher. *Wilson Bulletin* 53:141–56.

Weeden, R. B. 1959. A new breeding record for the wandering tattler in Alaska. *Auk* 76:230–32.

_____. 1965. Further notes on wandering tattlers in central Alaska. *Condor* 67:87–89.

Weeks, H. P. 1969. Courtship and territorial behavior of some woodcocks. *Proceedings of the Indiana Academy of Science* 79:162–71.

Wetmore, A. 1925. *Food of American phalaropes, avocets and stilts.* Bulletin 1359. Washington, D.C.: United States Department of Agriculture.

_____. 1926. *Birds of Argentina, Paraguay, Uruguay, and Chile.* Bulletin 133. Washington, D.C.: Smithsonian Institution.

_____. 1960. A classification for the birds of the world. *Smithsonian Miscellaneous Collections* 139, no. 11:1–37.

_____. 1965. The birds of the Republic of Panama. Part 1. Tinamidae to Rhynchopidae. *Smithsonian Miscellaneous Collections* 150:1–484.

Wilcox, L. 1959. A twenty year banding study of the piping plover. *Auk* 75:129–52.

Wilson, G. 1974. Incubating behaviour of the African jacana. *Ostrich* 45:185–87.

Winterbottom, J. M. 1963. Comments on the ecology and breeding of sandplovers *Charadrius* in southern Africa. *Revue de Zoologie et de Botanique Africaines* 67:1–2.

Witherby, H. F.; Jourdain, F. C. R.; Ticehurst, N. F.; and Tucker, B. W. 1941 *The handbook of British birds.* Vol. 4. London: Witherby; rev. ed., 1943.

Wolters, H. E. 1974. Aus der ornithologischen Sammlung des Museums Alexander Koenig. III. *Bonner Zoologische Beiträge* 25:283–91.

_____. 1975. *Die Vogelarten der Erde.* Leiferung 1. Berlin: Verlag Paul Parey.

Woods, R. W. 1975. *The birds of the Falkland Islands.* Oswestery: Anthony Nelson.

Yudin, K. A. 1965. [Phylogeny and classification of Charadriiformes]. Fauna SSSR. Moscow: Akademii Nauk SSR, Zoologii Instituta, no. 91. Birds II, part 1, no. 1:1–262. (In Russian.)

Zubarovskii, Y. M. 1976. [Nesting of the solitary snipe (*Gallinago solitaria* Hodg.) in the Altai]. *Vestnik Zoologii* 43(1):28–32.

Zusi, R. L., and Jehl, J. R., Jr. 1970. The systematic position of *Aechomorhynchus, Prosobonia,* and *Phegornis* (Charadriiformes: Charadii). *Auk* 87: 760–80.

Index

The following index is limited to those shorebirds that are individually described in the text; other bird species are not indexed, nor are most subspecies indexed. However, a few vernacular names applied to certain subspecies that sometimes are considered full species are included, as are some scientific names that are not utilized in this book but that are sometimes applied to particular species or species groups. Complete indexing is limited to the entries that correspond to the vernacular names utilized in this book; in these cases the primary descriptive account is indicated in italics. The plates are not indexed. Scientific names are indexed to the section of the principal account only.

SCIENTIFIC NAMES